Formulas from Analytic Geometry

Slope of line: $m = \dfrac{y_2 - y_1}{x_2 - x_1}$

Equation of line: $y - y_1 = m(x - x_1)$

Distance formula: $d = \sqrt{(x_2 - x_1)^2 + (y_2 - y_1)^2}$

Circle: $(x - x_0)^2 + (y - y_0)^2 = r^2$

Ellipse: $\dfrac{(x - x_0)^2}{a^2} + \dfrac{(y - y_0)^2}{b^2} = 1$

Formulas and Definitions from Differential Calculus

The statement $\lim_{x \to a} f(x) = L$ means that for any $\varepsilon > 0$ there is a $\delta > 0$ such that $|f(x) - L| < \varepsilon$ whenever $0 < |x - a| < \delta$.

A function f is *continuous* at x if $\lim_{h \to 0} f(x + h) = f(x)$.

If $\lim_{h \to 0}[f(x + h) - f(x)]/h$ exists, it is denoted by $f'(x)$ or $d f(x)/dx$ and is termed the *derivative* of f at x.

$(f + g)' = f' + g'$

$(fg)' = fg' + f'g$

$(f/g)' = (gf' - fg')/g^2$

$(f \circ g)' = (f' \circ g)g'$

$\dfrac{d}{dx}x^\alpha = \alpha x^{\alpha - 1}$

$\dfrac{d}{dx}e^x = e^x$

$\dfrac{d}{dx}\sec x = \sec x \tan x$

$\sinh x = \tfrac{1}{2}(e^x - e^{-x})$

$\cosh x = \tfrac{1}{2}(e^x + e^{-x})$

$\dfrac{d}{dx}\ln x = x^{-1}$

$\dfrac{d}{dx}\sin x = \cos x$

$\dfrac{d}{dx}\cos x = -\sin x$

$\dfrac{d}{dx}\tan x = \sec^2 x$

$\dfrac{d}{dx}\arcsin x = (1 - x^2)^{-1/2}$

$\dfrac{d}{dx}\arctan x = (1 + x^2)^{-1}$

$\dfrac{d}{dx}\sinh x = \cosh x$

$\dfrac{d}{dx}\cosh x = \sinh x$

NUMERICAL MATHEMATICS AND COMPUTING

Third Edition

Ward Cheney
David Kincaid
The University of Texas at Austin

Brooks/Cole Publishing Company
Pacific Grove, California

I(T)P ™ The trademark ITP is used under license.

Image representation of three views of the surface height of a penny created in MATLAB, courtesy of The MathWorks, Inc., Natick, MA. Data courtesy of NIST.

Brooks/Cole Publishing Company
A Division of Wadsworth, Inc.

© 1980, 1985, 1994 by Wadsworth, Inc., Belmont, California 94002. All rights reserved.
No part of this book may be reproduced, stored in a retrieval system, or transcribed,
in any form or by any means—electronic, mechanical, photocopying, recording,
or otherwise—without the prior written permission of the publisher,
Brooks/Cole Publishing Company, Pacific Grove, California 93950,
a division of Wadsworth, Inc.

Printed in the United States of America
10   9   8   7   6   5   4   3

Library of Congress Cataloging in Publication Data
Cheney, E. W. (Elliott Ward), [date]
    Numerical mathematics and computing / Ward Cheney & David Kincaid.
    —3rd ed.
        p.   cm.
    Includes index.
    ISBN 0-534-20112-1
    1. Numerical analysis—Data processing.   I. Kincaid, David (David
Ronald)   II. Title.
QA297.C426   1994
519.4'0285'51—dc20   93-43850
                                                                                      CIP

Sponsoring Editor: *Jeremy Hayhurst*
Marketing Representative: *Ragu Raghavan*
Editorial Assistant: *Elizabeth Rammel*
Production Coordinator: *Marlene Thom*
Production: *Integre Technical Publishing Co., Inc.*
Manuscript Editors: *Carol Reitz*
Permissions Editor: *Carline Haga*
Interior Design: *Vernon T. Boes*
Cover Design: *Katherine Minerva*
Art Coordinator: *Lisa Torri*
Interior Illustration: *Lori Heckelman*
Printing and Binding: *Arcata Graphics/Fairfield*

PREFACE

In preparing the third edition of this book, we have adhered to the basic objectives of the previous editions—namely, to acquaint students of science and engineering with the potentialities of the modern computer for solving the numerical problems that will arise in their professions. A secondary objective is to give students an opportunity to hone their skills in programming and problem solving. A third objective is to help students arrive at an understanding of the important subject of *errors* that inevitably accompany scientific computing and to arm them with methods for detecting, predicting, and controlling these errors.

Since the book is to be accessible to students who are not necessarily very advanced in their formal study of mathematics, we have tried to achieve an elementary style of presentation, including numerous examples and fragments of computer code for illustrative purposes. Believing that most students at this level need a *survey* of the subject of numerical mathematics, we have presented a wide diversity of topics, including some rather advanced ones that play an important role in current scientific computing.

Features in the Third Edition

All sections of the book have been revised to some degree. Major new features are as follows:

- The computer algorithms have been revised to make them computer-language independent. The emphasis is on the mathematical algorithms rather than on the computer language used to implement them.
- Many examples are solved using either Maple V® or MATLAB® to illustrate just two of the powerful software tools now available for symbolic, numeric, and graphical results.[*]

---

[*]Maple is a registered trademark of Waterloo Maple Software and MATLAB is the registered trademark of The MathWorks, Inc.

- Many of the problems have been revised to be consistent with the new style of the book.

- Some chapters and sections have been reordered for logical and pedagogical reasons.

Suggestions for Use

Numerical Mathematics and Computing, Third Edition, can be used in a variety of ways, depending on the emphasis the instructor prefers. Problems have been supplied in abundance to enhance the book's versatility. They are divided into two categories: *Problems* and *Computer Problems*. In the first category, there are more than 800 exercises in analysis that require pencil, paper, and possibly a calculator. In the second category, there are approximately 450 problems that involve writing a program and testing it on a computer. Readers can often follow a model or example in the text to assist them in working out exercises, but in other cases they must proceed on their own from a mathematical description given in the text or in the problems. In most of the computer problems, there is something to be learned beyond simply writing code—a *moral*, if you like. Some computing problems are designed to give experience in using preprogrammed or *canned* library codes. Computer subprograms from general-purpose software libraries (available on many computer systems) can be used with this text. Two such products are the IMSL Library and the NAG Library.

When exploring the problems in this book, the reader may wish to utilize one or more of the useful scientific and engineering software tools available for computation, visualization, and data analysis. Noteworthly examples are the computer algebra system Maple and the MATLAB system with a variety of application toolboxes. In particular, the Symbolic Math Toolbox provides access to the entire Maple kernel using an extension of the MATLAB language.

The pseudocode displayed in this text has been coded in several programming languages and is available on the Internet by anonymous ftp from `ftp.brookscole.com` or from `math.utexas.edu`. Also available from the publisher is a Solution's Manual for instructors who adopt the book.

Our own recommendations for courses based on this text are as follows:

- A one-term course carefully covering Chapters 1 through 9 (possibly omitting Sections 4.2, 6.4, 7.3–7.4, 8.3, and 9.3, for example), followed by a selection of material from the remaining chapters as time permits.

- A one-term survey rapidly covering all chapters in the text, omitting some of the more difficult sections.

- A full-year course carefully covering all chapters.

Acknowledgments

In preparing the third edition, we have been able to profit from advice and suggestions kindly offered by a large number of colleagues, students, and users of the second edition. It is our pleasure to thank them and others who helped with the task of preparing the new edition. Kata Carbone typed most of our revisions in the manuscript with great care and attention to detail. Belinda Trevino and Katy Burrell also typed parts of the revised book. Also, we wish to thank Don DeLand for providing the LATEX macros and his help putting the book into final form.

Valuable comments and suggestions were made by our colleagues and friends. In particular, David Young has been very generous with suggestions for improving the accuracy and clarity of the exposition. Katherine Hua Guo, Kwang-il In, and Hidajaty Thajeb helped in revising and testing the computer codes as well as proofreading parts of the manuscript. Also, Bi Roubolo Vona assisted by producing some computer plots. Adarsh Beohar helped with some of the Maple examples. Thanks go to Jason Brazile for help with computer-related questions.

Several reviewers provided detailed critiques of a draft of this new edition. These were Neil Berger, Jose E. Castillo, Charles Cullen, Elias Y. Deeba, Terry Feagin, Leslie Foster, John Gregory, Bruce P. Hillam, Patrick Lang, Edward Neuman, and Roy Nicolaides, Many individuals took the trouble to write us with suggestions and criticisms. We thank Steve Batterson, Wayne Dymacek, Paul Enigenbury, Peter Fraser, Dianne O'Leary, P. W. Manual, Juan Meza, George Minty, Jeff Nunemacher, Granville Sewell, Thiab Taha, Perry Wong, and Rick Zaccone. The staff of Brooks/Cole Publishing Company has been most understanding and patient in bringing this book to fruition. In particular, we thank Jeremy Hayhurst, Marlene Thom, and Elizabeth Rammel for their efforts on behalf of this project.

We would appreciate any comments, questions, criticisms, or corrections that readers may take the trouble of communicating to us. E-mail is especially efficient. Our addresses are cheney@math.utexas.edu and kincaid@cs.utexas.edu

Ward Cheney
David Kincaid

Dedicated to David M. Young, Jr.,
on the occasion of his 70th birthday

CONTENTS

3   Locating Roots of Equations 82

4   Interpolation and Numerical Differentiation 119

5   Numerical Integration   165

8 Ordinary Differential Equations 325

Initial-Value Problem: Analytical vs. Numerical Solution *325*

9 Systems of Ordinary Differential Equations 359

12 Boundary Value Problems for Ordinary Differential Equations   **440**

13 Partial Differential Equations   **458**

1 INTRODUCTION

The Taylor series for $\ln(1 + x)$ gives us

$$\ln 2 = 1 - \frac{1}{2} + \frac{1}{3} - \frac{1}{4} + \frac{1}{5} - \frac{1}{6} + \frac{1}{7} - \frac{1}{8} + \cdots$$

Adding together the eight terms shown, we obtain 0.63452, which is a poor approximation to $\ln 2 = 0.69315\ldots$. On the other hand, the Taylor series for $\ln[(1 + x)/(1 - x)]$ gives us

$$\ln 2 = 2\left(3^{-1} + \frac{3^{-3}}{3} + \frac{3^{-5}}{5} + \frac{3^{-7}}{7} + \cdots\right)$$

Now, adding the four terms shown in the parentheses, we obtain 0.69313. This illustrates the fact that rapid convergence of a Taylor series can be expected *near* the point of expansion but not at remote points. Taylor series and Taylor's Theorem are two of the principal topics we discuss in this chapter. They are ubiquitous features in much of numerical analysis. Also, we review some commonsense suggestions and hints for good programming and for coding in a readable style.

The objective of this text is to help the reader to understand some of the many methods for solving scientific problems on a modern computer. We intentionally limit ourselves to the typical problems that arise in science, engineering, and technology. Thus, we do not touch upon problems of accounting, modeling in the social sciences, information retrieval, artificial intelligence, and so on.

Usually our treatment of problems will not begin at the source, for that would take us far afield into such areas as physics, engineering, and chemistry. Instead, we consider problems after they have been cast into certain standard mathematical forms. The reader is asked, therefore, to accept on faith the assertion that the chosen topics are indeed important ones in scientific computing.

To survey a large number of topics, some must be covered briefly and therefore compressed somewhat to make them fit comfortably into this book's format. Obviously our treatment of some topics must be superficial. But it is hoped that the reader will acquire a good bird's-eye view of the subject and therefore be better prepared for a further, deeper study of numerical analysis.

For each principal topic, we list good current sources for more information. In any realistic computing situation, considerable thought should be given to the choice of method to be employed. Although most procedures presented here are useful and important, they may not be the optimum ones for a particular problem. In choosing among available methods for solving a problem, the analyst or programmer should consult recent references.

Becoming familiar with basic numerical methods without realizing their limitations would be foolhardy. Numerical computations are almost invariably contaminated by errors, and it is important to understand the source, propagation, magnitude, and rate of growth of these errors. While we cannot help but be impressed by the speed and accuracy of the modern computer, we should temper our admiration with generous measures of skepticism. As a wise observer commented on the computer age: *Never in the history of mankind has it been possible to produce so many wrong answers so quickly.* Thus, one of our goals is to help the reader arrive at this state of skepticism, armed with methods for detecting, estimating, and controlling errors.

The reader is expected to be familiar with the rudiments of programming. Algorithms are presented as pseudocode and no particular programming language is adopted. The pseudocode used is quite similar to Fortran 90. Although Fortran is the most commonly used programming language for scientific applications, other languages are suitable for the computing problems presented here.

Preliminary Remarks

We will begin with some remarks on evaluating a polynomial efficiently and on rounding and chopping real numbers.

Nested Multiplication

To evaluate the polynomial

$$p(x) = a_0 + a_1 x + a_2 x^2 + \cdots + a_{n-1} x^{n-1} + a_n x^n \tag{1}$$

we group the terms in a **nested multiplication**:

$$p(x) = a_0 + x(a_1 + x(a_2 + \cdots + x(a_{n-1} + x(a_n)) \cdots))$$

The pseudocode that evaluates $p(x)$ starts with the innermost parentheses and works outward. It can be written as

```
p ← a_n
for i = n − 1 to 0 step −1 do
    p ← a_i + xp
end do
```

The left-pointing arrow ($\leftarrow$) means that the value on the right is stored in the location named on the left. Here we assume that the coefficients $a_0, a_1, \ldots, a_n$ are stored in a linear array. The final value of p is the value of the polynomial at x. This nested multiplication procedure is also known as *Horner's method* or *synthetic division*.

The polynomial in Equation (1) can be written in an alternative form by utilizing the mathematical symbols for sum $\sum$ and product $\prod$; namely,

$$p(x) = \sum_{i=0}^{n} a_i x^i = \sum_{i=0}^{n} \left(a_i \prod_{j=1}^{i} x \right)$$

Recall that if $n \leqq m$, we write

$$\sum_{k=n}^{m} x_k = x_n + x_{n+1} + \cdots + x_m$$

and

$$\prod_{k=n}^{m} x_k = x_n x_{n+1} \cdots x_m$$

By convention whenever $m < n$, we define

$$\sum_{k=n}^{m} x_k = 0$$

and

$$\prod_{k=n}^{m} x_k = 1$$

Rounding and Chopping

We recommend the following rule for *correct rounding to n decimal digits* in pencil-and-paper calculations: If the digits beyond the nth digit are greater than $50000\ldots$ (to the number of digits being carried), then round up the nth digit; if they are less than $50000\ldots$, then round down the nth digit; if they are equal to $50000\ldots$, then round the nth digit so that it is even (which should cause either a round up or a round down about half of the time). On the other hand, *chopping to n decimal digits* means to retain all digits up to and including the nth digit and discard all digits following it. On the computer, the user sometimes has the option of selecting to have all arithmetic operations done with either chopping or rounding. The latter is usually preferable, of course.

The maximum difference in absolute value (maximum *absolute error*) between a decimal number x and the number $\tilde{x}$ rounded to n decimal places is $\frac{1}{2} \times 10^{-n}$, whereas the difference for the number $\hat{x}$ chopped to n decimal places is twice as large, or 10^{-n}. Some of the subjects that will be discussed in more detail in Chapter 2 are absolute error, relative error, roundoff error, and floating-point arithmetic.

1.1 Programming Suggestions

The following programming suggestions and techniques for coding should be considered in context. They are not intended to be complete, and some good programming suggestions have been omitted to keep the discussion brief. Our purpose is to encourage the reader to be attentive to considerations of efficiency, economy, readability, and roundoff errors.

Programming and Coding Advice

Since the programming of numerical schemes is essential to understanding them, we offer here a few words of advice on good programming practices.

Strive to write programs carefully and correctly. Before beginning the coding, write out in complete detail the mathematical algorithm to be used in **pseudocode** such as that used in this text. The pseudocode serves as a bridge between the mathematics and the computer program. It need not be defined in a formal way as is done for a computer language, but it should contain sufficient detail so that the implementation is straightforward. When writing the pseudocode, use a style that is easy to read and understand. For maintainability, it should be easy for a person unfamiliar with the code to read and understand what it does. Check the code thoroughly for errors and omissions before beginning to edit on a computer terminal. Spend time checking the code before running it to avoid executing the program, showing the output, discovering an error, correcting the error, and repeating the process *ad nauseam*. Modern computing environments may allow the user to accomplish this process in only a few seconds, but this advice is still valid if for no other reason than that it is dangerously easy to write programs that may *work* on a simple test but not on a more complicated one; no function key or mouse can tell you what is wrong!

If the code can be written to handle a slightly more general situation, then in many cases it is worth the extra effort to do so. A program written for only a particular set of numbers must be completely rewritten for another set. For example, only a few additional statements are required to write a program with an arbitrary step size compared with a program in which the step size is fixed numerically. However, one should be careful not to introduce too much generality into the code because it can make a simple programming task overly complicated.

Build a program in steps by writing and testing a series of subprograms or procedures or functions; that is, write self-contained subtasks as separate routines. Try to keep these subprogram segments reasonably small, less than a page whenever possible, to make reading and debugging easier.

After writing the pseudocode, check and trace through it using pencil-and-paper calculations on a typical yet simple example. Checking boundary cases, such as the values of the first and second iterations in a loop and the processing of the first and last elements in a data structure, will often reveal embarrassing errors. These same sample cases can be used as the first set of test cases on the computer.

Print out intermediate results and diagnostic messages to assist in debugging and understanding the program's operation. Always echo-print the input data unless it is impractical to do so, such as with a large amount of data. Using the default read and print commands frees the programmer from errors associated with misalignment of data. Fancy output formats are not necessary, but some simple labeling of the output is recommended. A robust program always warns the user of a situation that it is not designed to handle. In general, write programs so that they are easy to debug when the inevitable bug appears.

It is often helpful to assign meaningful names to the variables because they may have greater mnemonic value than single-letter variables. There is perennial confusion between the characters O (letter "oh") and 0 (number zero) and between l (letter "ell") and 1 (number one).

All variables should be listed in type declarations at the beginning of each program or subprogram. Implicit type assignments can be ignored when one writes declaration statements that include all variables used. Historically in Fortran, variables beginning with I/i, J/j, K/k, L/l, M/m, and N/n are integer variables, and ones beginning with other letters are floating-point real variables. It may be a good idea to adhere to this scheme so that one can immediately recognize the type of a variable without looking it up in the type declarations. In this book, we present algorithms using pseudocode and therefore do not follow this advice.

Comments within a routine are helpful for recalling at some later time what the program does. Extensive comments are not necessary, but we recommend a preface to each program or subprogram explaining the purpose, the input and output variables, and the algorithm used, as well as a few comments between major segments of the code. Indent each block of code a consistent number of spaces to improve readability. Inserting blank comment lines and blank spaces can greatly improve the readability of the code as well. To save space, we have not included any comments in the pseudocode in this book.

Never put unnecessary statements within loops. Move expressions and variables outside a loop from inside a loop if they do not depend on the loop or do not change. Also, indenting loops can add to the readability of the code, particularly for nested loops. Use a nonexecutable statement as the terminator of a loop so that the code may be easily altered.

Use a parameter statement to assign the values of key constants. Parameter values correspond to constants that do not change throughout the routine. Such parameter statements are easy to change when one wants to rerun the program with different values. Also, they clarify the role key constants play in the code and make the routines more readable and easier to understand.

Use data structures that are natural to the problem at hand. If the problem adapts more easily to a three-dimensional array than to several one-dimensional arrays, then a three-dimensional array should be used.

The elements of arrays, whether one-, two-, or higher-dimensional, are usually stored in consecutive words of memory. Since the compiler maps the value of an index for two- and higher-subscripted arrays into a single subscript value that is used as a pointer to determine the location of elements in storage, the use of two- and higher-dimensional arrays can be considered a notational convenience for the user. However, any advantage in using only a one-dimensional array and performing complicated subscript calculation is slight. Such matters are best left to the compiler.

When writing a procedure or subprogram, one often wants to dimension arrays for the largest problem that the routine will need to handle and use only a portion of the arrays for smaller problems. An array can be passed as an argument to a subprogram to achieve this objective. Since the memory space has already been allocated for the array in the calling program, this needs to be communicated to the subprogram. For example, a portion of a two-dimensional array already dimensioned in a main program could be used in a subprogram if the subprogram has a pointer to the beginning of the array and the extent (the length of the columns) in the subarray. Alternatively, it may be possible to allocate space dynamically for workspace arrays in the subprogram at run time.

In scientific programming languages, many built-in mathematical functions are available for common functions such as *sin*, *log*, *exp*, *cosh*, and so on. Also, numeric functions such as *integer*, *real*, *complex*, and *imaginary* are usually available for type conversion. One should utilize these and others as much as possible. Some of these intrinsic functions accept arguments of more than one type and return a result whose type may vary depending on the type of the argument used. Such functions are called **generic functions**, for they represent an entire family of related functions. Of course, care should be taken not to use the wrong argument type.

In preference to one you might write yourself for a programming project, a *preprogrammed* routine from a program library should be used when applicable. Such routines can be expected to be state-of-the-art software, well tested, and, of course, completely debugged.

In the pseudocode used in this book, we do not include type conversions, double-precision declarations, and other coding details because these are things the reader can easily take care of when doing the coding in a computer language.

Do not sacrifice the clarity of a program in an effort to make the code run faster. Clarity of code may be preferable to "optimization of code" when the two criteria conflict.

Case Studies

When a long list of floating-point numbers is added in the computer, there will generally be less roundoff error if the numbers are added in order of increasing magnitude. (In Chapter 2, roundoff errors are discussed in detail.)

Since it is easy to mistype a long sequence of digits in mathematical constants, such as the real number π,

$pi \leftarrow 3.14159\,26535\,89793$

the use of simple calculations involving mathematical functions is recommended. For example, the real numbers π and e can be easily and safely entered with nearly full machine precision by using standard intrinsic functions such as

$pi \leftarrow 4.0 \cdot \arctan(1.0)$
$e \leftarrow \exp(1.0)$

Another reason for this advice is to avoid the problem that arises if one uses a short approximation such as $pi \leftarrow 3.14159$ on a computer with limited precision but later moves the code to another computer with more precision. If you overlook changing this assignment statement, then all results that depend on this value will be less accurate than they should be.

In coding for the computer, exercise some care in writing statements that involve exponents. The general function x^y is computed on many computers as $\exp(y \ln x)$ whenever y is not an integer. Sometimes this is unnecessarily complicated and may contribute to roundoff errors. For example, it is preferable to write code with integer exponents such as 5 rather than 5.0. Similarly, using exponents such as $1/2$ or 0.5 is not recommended because either it may *not* work in some computer languages or the built-in function *sqrt* may be used.

In general, one should avoid mixing real and integer expressions in the computer code. *Mixed expressions* are formulas in which variables and constants of different types appear together. If the floating-point form of an integer variable is needed, use a function such as *real*. Similarly, a function such as *integer* is generally available for obtaining the integer part of a real variable. In other words, use the intrinsic type conversion functions whenever converting from complex to real, real to integer, or vice versa. For example, in floating-point calculations, m/n should be coded as $\text{real}(m)/\text{real}(n)$ when m and n are integer variables so that it computes the correct real value of m/n. Similarly, $1/m$ should be coded as $1.0/\text{real}(m)$ and $1/2$ as 0.5 and so on.

In the usual mode of representing numbers in a computer, one word of storage is used for each number. This mode of representation is called **single precision**. In calculations that require greater precision (called **double precision** or **extended precision**), it is possible to allot two or more words of storage to each number. On a 32-bit computer, approximately seven decimal places of precision can be obtained in single precision and approximately 17 decimal places of precision in double precision. Double precision is usually more time consuming than single precision because it may use software rather than hardware to carry out the arithmetic. However, if more accuracy is needed than single precision can provide, then double or extended precision should be used. This is particularly true on computers with severely limited precision such as a 32-bit computer, where roundoff errors can quickly accumulate in long computations and reduce the accuracy to only three or four decimal places! (This topic is discussed in Chapter 2.)

Usually, two words of memory are used to store the real and imaginary parts of a complex number. Complex variables and arrays must be explicitly declared as being of complex type. Expressions involving variables and constants of complex type are evaluated according to the normal rules of complex arithmetic. Intrinsic functions such as *complex*, *real*, and *imaginary* should be used to convert between real and complex types.

When using loops, write the code so that fetches are made from *adjacent* words in memory. To illustrate, suppose we want to store values in a two-dimensional array (a_{ij}) in which the elements of each column are stored in consecutive memory locations. Using i and j loops with the ith loop as the innermost one would process elements down the columns. For some programs and computer languages, this detail may be of only secondary concern. However, some computers have immediate access to only a portion or a few *pages* of memory at a time. In this case, it is advantageous to process the elements of an array so they are taken from or stored in adjacent memory locations.

There is rarely any need for a calculation such as $j \leftarrow (-1)^k$ because there are better ways of obtaining the same result. For example, in a loop we can write $j \leftarrow 1$ before the loop and $j \leftarrow -j$ inside the loop.

Although the mathematical description of an algorithm may indicate that a sequence of values is computed, thus seeming to imply the need for an array, it is often possible to avoid arrays. (This is especially true if only the final value of a sequence is required.) For example, the theoretical description of Newton's method (Chapter 3) reads

$$x_{n+1} = x_n - f(x_n)/f'(x_n)$$

but the pseudocode can be written simply as

$$x \leftarrow x - f(x)/f'(x)$$

where x is a real variable and function procedures for f and f' have been written. Such an assignment statement automatically effects the replacement of the value of the *old* x with the *new* numerical value of $x - f(x)/f'(x)$.

In a repetitive algorithm, one should always limit the number of permissible steps by the use of a loop with a control variable in order to prevent endless cycling due to unforeseen problems (e.g., programming errors and roundoff errors). For example, in Newton's method above, one might write

while $|f(x)| > \frac{1}{2} \times 10^{-6}$ **do**
  $x \leftarrow x - f(x)/f'(x)$
  output x
end do

If the function involves some erratic behavior, there is a danger here in not limiting the number of repetitions. It is better to use a loop with a control variable:

for $k = 1$ **to** *nsteps* **do**
  $x \leftarrow x - f(x)/f'(x)$

if $|f(x)| \le \frac{1}{2} \times 10^{-6}$ **then** exit loop
end do

where k and *nsteps* are integer variables and the value of *nsteps* contains the number of desired repetitions.

The sequence of steps in a routine should not depend on whether two floating-point numbers are equal. Instead, reasonable tolerances should be permitted to allow for floating-point arithmetic roundoff errors. For example, a suitable branching statement for n decimal digits of accuracy might be

if $|x - y| < \varepsilon$ **then** ...

provided it is known that x and y have magnitude comparable to 1. Here x, y, and ε are real variables with $\varepsilon = \frac{1}{2} \times 10^{-n}$. This corresponds to requiring that the *absolute error* between x and y is less than ε. However, if x and y are not of the same magnitude, then the *relative error* between x and y would be needed, as in the branching statement

if $|x - y| < \varepsilon \max\{|x|, |y|\}$ **then** ...

In some situations, notably in solving differential equations (see Chapter 8), a variable t assumes a succession of values equally spaced a distance of h apart along the real line. One way of coding this is

$t \leftarrow t_0$
for $i = 1$ **to** n **do**
$\quad \vdots$
$\quad t \leftarrow t + h$
end do

Here i and n are integer variables and t_0, t, and h are real variables. An alternative way is

for $i = 1$ **to** n **do**
$\quad \vdots$
$\quad t \leftarrow t_0 + ih$
end do

In the first pseudocode, n additions occur, each with possible roundoff error. In the second, this situation is avoided but at the added cost of n multiplications. Which is better depends on the particular situation at hand. (See Computer Problem **10**.)

When values of a function at arbitrary points are needed in a program, several ways of coding this are available. For example, suppose values of the function $f(x) = 2x + \ln x - \sin x$ are needed. A simple approach is to use an assignment statement such as

$$y \leftarrow 2x + \ln(x) - \sin(x)$$

at appropriate places within the program. Here x and y are real variables. Equivalently, an *internal* function procedure corresponding to the pseudocode

$$f(x) \leftarrow 2x + \ln(x) - \sin(x)$$

could be used with

$$y \leftarrow f(2.5)$$

or whatever value of x is desired. Finally, a function subprogram can be used such as in the following pseudocode:

real function $f(x)$
real x
$\quad f \leftarrow 2x + \ln(x) - \sin(x)$
end function f

Which implementation is best? It depends on the situation at hand. The assignment statement is simple and safe. An internal or external function procedure can be used to avoid duplicating code. A separate external function subprogram is the best way to avoid difficulties that inadvertently occur when someone must insert code into another's program. In using program library routines, the user may be required to furnish an external function procedure to communicate function values to the library routine. If the external function procedure f is passed as an argument in another procedure, then a special *interface* must be used to designate it as an external function.

Programming Experiment

We conclude with a short programming experiment involving numerical computations. Here we consider, from the computational point of view, a familiar operation in calculus—namely, taking the derivative of a function. Recall that the derivative of a function f at a point x is defined by the equation

$$f'(x) = \lim_{h \to 0} \frac{f(x+h) - f(x)}{h}$$

A computer has the capacity of imitating the limit operation by using a sequence of numbers h such as $h = 4^{-1}, 4^{-2}, 4^{-3}, \ldots, 4^{-n}, \ldots$, for they certainly approach zero rapidly. Of course, many other simple sequences are possible, such as $1/n$, $1/n^2$, and $1/10^n$. The sequence $1/4^n$ consists of machine numbers in a binary computer and, for this experiment on a 32-bit computer, will be sufficiently close to zero when n is 10.

The following is a pseudocode to compute $f'(x)$ at the point $x = 0.5$, with $f(x) = \sin x$.

```
program first
integer parameter n ← 10
integer i
real error, h, x, y
x ← 0.5
h ← 1
for i = 1 to n do
    h ← 0.25h
    y ← [sin(x + h) − sin(x)]/h
    error ← |cos(x) − y|
    output i, h, y, error
end do
end program first
```

We have neither shown the output from this pseudocode nor explained the purpose of the experiment. We invite the reader to discover this by coding and running it (or one like it) on a computer. (See Computer Problems **1** and **2**.)

PROBLEMS 1.1

1. The value of π can be generated by the computer to near machine precision by the assignment statment $pi \leftarrow 4.0 \cdot \arctan(1.0)$. Determine at least four other ways to compute π using basic functions on your computer system.

2. Criticize the following pseudocode and write improved versions.

   a. real array $(a_i)_n$
   integer i, n
   real x, z
   $\vdots$
   for $i = 1$ to n do
       $x \leftarrow z^2 + 5.7$
       $a_i \leftarrow x/i$
   end do

   b. real array $(a_{ij})_{n \times n}$
   integer i, j, n
   $\vdots$
   for $i = 1$ to n do
       for $j = 1$ to n do
           $a_{ij} \leftarrow 1/(i + j − 1)$
       end do
   end do

c. real array $(a_{ij})_{n \times n}$
integer i, j, n

$\vdots$

for $j = 1$ **to** n **do**
  for $i = 1$ **to** n **do**
    $a_{ij} \leftarrow 1/(i + j - 1)$
  end do
end do

3. A doubly subscripted array (a_{ij}) can be added in any order. Write the pseudocode for each of the following parts. Which is best?

   a. $\sum_{i=1}^{n} \sum_{j=1}^{n} a_{ij}$

   b. $\sum_{j=1}^{n} \sum_{i=1}^{n} a_{ij}$

   c. $\sum_{i=1}^{n} \left(\sum_{j=1}^{i} a_{ij} + \sum_{j=1}^{i-1} a_{ji} \right)$

   d. $\sum_{k=0}^{n-1} \sum_{|i-j|=k} a_{ij}$

   e. $\sum_{k=2}^{2n} \sum_{i+j=k} a_{ij}$

4. How many multiplications occur in executing the following pseudocode?

   real array $(a_{ij})_{n \times n}, (b_{ij})_{n \times n}$
   integer i, j, n
   real x

   $\vdots$

   for $j = 1$ **to** n **do**
     for $i = 1$ **to** j **do**
       $x \leftarrow x + a_{ij} b_{ij}$
     end do
   end do

5. Count the number of operations involved in evaluating a polynomial using nested multiplication. Do not count subscript calculations.

6. For small x, show that $(1 + x)^2$ can sometimes be more accurately computed from $(x + 2)x + 1$. Explain. What other expressions can be used to compute it?

7. Show how these polynomials can be efficiently evaluated.

   a. $p(x) = x^{32}$

   b. $p(x) = 3(x - 1)^5 + 7(x - 1)^9$

   c. $p(x) = 6(x + 2)^3 + 9(x + 2)^7 + 3(x + 2)^{15} - (x + 2)^{31}$

   d. $p(x) = x^{127} - 5x^{37} - 5x^{37} + 10x^{17} - 3x^7$

8. Using the exponential function $\exp(x)$, write an efficient pseudocode for the statement $y = 5e^{3x} + 7e^{2x} + 9e^x + 11$.

9. Write pseudocode to evaluate the expression

$$z = \sum_{i=1}^{n} b_i^{-1} \prod_{j=1}^{i} a_j$$

10. Write pseudocode to evaluate the following expressions efficiently.

   a. $p(x) = \sum_{k=0}^{n-1} k x^k$

   b. $z = \sum_{i=1}^{n} \prod_{j=1}^{i} x^{n-j+1}$

   c. $z = \prod_{i=1}^{n} \sum_{j=1}^{i} x_j$

   d. $p(t) = \sum_{i=1}^{n} a_j \prod_{j=1}^{i-1} (t - x_j)$

11. Using summation and product notation, write mathematical expressions for the following pseudocode.

   a. **real array** $(a_i)_n$
      integer i, n
      real v, x

      $\vdots$

      $v \leftarrow a_0$
      for $i = 1$ **to** n **do**
        $v \leftarrow v + x a_i$
      end do

   b. **real array** $(a_i)_n$
      integer i, n
      real v, x

      $\vdots$

      $v \leftarrow a_n$
      for $i = 1$ **to** n **do**
        $v \leftarrow vx + a_{n-i}$
      end do

   c. **real array** $(a_i)_n$
      integer i, n
      real v, x

      $\vdots$

      $v \leftarrow a_0$
      for $i = 1$ **to** n **do**
        $v \leftarrow vx + a_i$
      end do

   d. **real array** $(a_i)_n$
      integer i, n
      real v, x, z

      $\vdots$

$$v \leftarrow a_0$$
$$z \leftarrow x$$
for $i = 1$ **to** n **do**
$\quad v \leftarrow v + za_i$
$\quad z \leftarrow xz$
end do

e. real array $(a_i)_n$
   integer i, n
   real v
   $\vdots$
   $v \leftarrow a_n$
   for $i = 1$ **to** n **do**
   $\quad v \leftarrow (v + a_{n-i})x$
   end do

12. Express in mathematical notation without parentheses the final value of z in the following pseudocode.

real array $(b_i)_n$
integer k, n
real z
$\vdots$
$z \leftarrow b_n + 1$
for $k = 1$ **to** $n - 2$ **do**
$\quad z \leftarrow zb_{n-k} + 1$
end do

COMPUTER PROBLEMS 1.1

1. Write a computer program that corresponds to the pseudocode program *first* described in the text and *interpret the results*.

2. (Continuation) Select a function f and a point x and carry out a computer experiment like the one given. Interpret the results. Do not select too simple a function. For example, you might consider $1/x$, $\log x$, e^x, $\tan x$, $\cosh x$, or $x^3 - 23x$.

3. The limit $e = \lim_{n \to \infty}(1 + 1/n)^n$ defines the number e in calculus. Estimate e by taking the value of this expression for $n = 8, 8^2, 8^3, \ldots, 8^{20}$. Compare with e obtained from $e \leftarrow \exp(1.0)$. *Interpret the results.*

4. It is not difficult to see that the numbers $p_n = \int_0^1 x^n e^x \, dx$ satisfy the inequalities $p_1 > p_2 > p_3 > \cdots > 0$. Establish this fact. Next, use integration by parts to show that $p_{n+1} = e - (n + 1)p_n$ and that $p_1 = 1$. In the computer, generate the first 20 values of p_n and explain why the inequalities above are violated. Do not use subscripted variables. (See Dorn and McCracken [1972], pp. 120–129.)

5. (Continuation) Let $p_{20} = 1/8$ and use the formula in Problem **4** to compute $p_{19}, p_{18}, \ldots, p_2$, and p_1. Do the numbers generated obey the inequalities

$1 = p_1 > p_2 > p_3 > \cdots > 0$? Explain the difference in the two procedures. Repeat with $p_{20} = 20$ or $p_{20} = 100$. Explain what happens.

6. Write an efficient routine that accepts as input a list of real numbers $a_1, a_2, \ldots, a_n$ and then computes the following:

$$\text{Arithmetic mean} \quad m = \frac{1}{n} \sum_{k=1}^{n} a_k$$

$$\text{Variance} \quad v = \frac{1}{n-1} \sum_{k=1}^{n} (a_k - m)^2$$

$$\text{Standard deviation} \quad \sigma = \sqrt{v}$$

Test the routine on a set of data of your choice.

7. (Continuation) Show that $v = \left[\sum_{k=1}^{n} a_k^2 - nm^2 \right] / (n - 1)$. Of the two given formulas for v, which is more accurate in the computer? Verify on the computer with a data set. *Hint*: Use a large set of real numbers that vary in magnitude from very small to very large.

8. Let a_1 be given. Write a program to compute for $1 \leq n \leq 1000$ the numbers $b_n = na_{n-1}$ and $a_n = b_n/n$. Print the numbers $a_{100}, a_{200}, \ldots, a_{1000}$. Do *not* use subscripted variables. What should a_n be? Account for the deviation of fact from theory. Determine four values for a_1 so that the computation does deviate from theory on your computer. *Hint*: Consider extremely small and large numbers and print to full machine precision.

9. In the computer, it can happen that $a + x = a$ when $x \neq 0$. Explain why. Describe the set of n for which $1 + 2^{-n} = 1$ in your computer. Write and run appropriate programs to illustrate the phenomenon.

10. Write a program to test the programming suggestion concerning the roundoff error in the computation of $t \leftarrow t + h$ versus $t \leftarrow t_0 + nh$. For example, use $h = 1/10$ and compute $t \leftarrow t + h$ in double precision for the correct single-precision value of t; print the absolute values of the differences between this calculation and the values of the two procedures. What is the result of the test when h is a machine number, such as $h = 1/128$, on a binary computer (with more than seven bits per word)?

11. Consider the following pseudocode.

```
integer i
real x, y, z
for i = 1 to 20 do
    x ← 2 + 1.0/8^i
    y ← arctan(x) − arctan(2)
    z ← 8^i y
    output x, y, z
end do
```

What is the purpose of this program? Is it achieved? Explain. Code and run it to verify your conclusions.

12. **a.** What is the difference between the following two assignment statements? Write a code that contains them and illustrate with specific examples to show that sometimes $x = y$ and sometimes $x \neq y$.

    integer m, n
    real x, y

    $\vdots$

    $x \leftarrow \text{real}(m/n)$
    $y \leftarrow \text{real}(m)/\text{real}(n)$
    output x, y

    b. What value will n receive?

    integer n
    real x, y
    $x \leftarrow 7.4$
    $y \leftarrow 3.8$
    $n \leftarrow x + y$
    output n

    What if the last statement is replaced with the following?

    $n \leftarrow \text{integer}(x) + \text{integer}(y)$

13. Write a computer code that contains the following assignment statements exactly as shown. Analyze the results.

    a. Print these values first using the default format and then with an extremely large format field.

    real p, q, u, v, w, x, y, z
    $x \leftarrow 0.1$
    $y \leftarrow 0.01$
    $z \leftarrow x - y$
    $p \leftarrow 1.0/3.0$
    $q \leftarrow 3.0p$
    $u \leftarrow 7.6$
    $v \leftarrow 2.9$
    $w \leftarrow u - v$
    output x, y, z, p, q, u, v, w

    b. What values would be computed for x, y, and z?

    integer n
    real x, y, z

```
for n = 1 to 10 do
    x ← (n − 1)/2
    y ← n²/3.0
    z ← 1.0 + 1/n
    output x, y, z
end do
```

  c. What values would the following assignment statements produce?

```
integer i, j
real c, f, x, half
x ← 10/3
i ← integer(x + 1/2)
half ← 1/2
j ← integer(half)
c ← (5/9)(f − 32)
f ← 9/5c + 32
output x, i, half, j, c, f
```

  d. Discuss what is wrong with the following code.

```
real area, circum, radius
radius ← 1
area ← (22/7)(radius)²
circum ← 2(3.1416)radius
output area, circum
```

14. The Russian mathematician P. L. Chebyshev (1821–1894) spelled his name Чебышев. Many transliterations from the Cyrillic to the Latin alphabet are possible. *Cheb* can be rendered as Ceb, Tscheb, or Tcheb. The *y* can be rendered as *i*. *Shev* can be rendered as schef, cev, cheff, or scheff. Taking all combinations of these variants, program a computer to print all possible spellings.

15. Compute $n!$ using logarithms and using (a) integer arithmetic and (b) double-precision floating-point arithmetic. For each part, print a table of values for $0 \leq n \leq 30$ and determine the largest correct value.

16. Given two arrays, a real array $v = (v_1, v_2, \ldots, v_n)$ and an integer permutation array $p = (p_1, p_2, \ldots, p_n)$ of integers $1, 2, \ldots, n$, can we form a new permuted array $v = (v_{p_1}, v_{p_2}, \ldots, v_{p_n})$ by overwriting v and not involving another array in memory? If so, write and test the code for doing it. If not, use another array and test.
Case 1. $v = (6.3, 4.2, 9.3, 6.7, 7.8, 2.4, 3.8, 9.7)$, $p = (2, 3, 8, 7, 1, 4, 6, 5)$
Case 2. $v = (0.7, 0.6, 0.1, 0.3, 0.2, 0.5, 0.4)$, $p = (3, 5, 4, 7, 6, 2, 1)$

17. Criticize the following pseudocode for evaluating $\lim_{x \to 0} \arctan(|x|)/x$. Code and run it to see what happens.

```
integer i
real x, y
```

```
x ← 1
for i = 1 to 24 do
    x ← x/2.0
    y ← arctan(|x|)/x
    output x, y
end do
```

18. Using a computer algebra system (e.g., Maple, DERIVE, Mathematica, etc.), print 200 decimal digits of $\sqrt{10}$.

19. What does the assignment $x \leftarrow sqrt(sqrt(x))$ do? Is it allowed or does it involve a recursive use of the square-root function? Run some numerical experiments to verify your conclusions.

20. In 1706 Machin used the formula

$$\pi = 16 \arctan\left(\frac{1}{5}\right) - 4 \arctan\left(\frac{1}{239}\right)$$

to compute 100 digits of π. Derive this formula. Reproduce Machin's calculations by using suitable software. *Hint*: Let $\tan\theta = 1/5$ and use standard trigonometric identities.

1.2 Review of Taylor Series

Since Taylor series and Taylor's Theorem are frequently needed in numerical analysis, we review these topics here.

Taylor Series

Familiar (and useful) examples of Taylor series are the following:

$$e^x = 1 + x + \frac{x^2}{2!} + \frac{x^3}{3!} + \cdots = \sum_{k=0}^{\infty} \frac{x^k}{k!} \qquad (|x| < \infty) \tag{1}$$

$$\sin x = x - \frac{x^3}{3!} + \frac{x^5}{5!} - \cdots = \sum_{k=0}^{\infty} (-1)^k \frac{x^{2k+1}}{(2k+1)!} \qquad (|x| < \infty) \tag{2}$$

$$\cos x = 1 - \frac{x^2}{2!} + \frac{x^4}{4!} - \cdots = \sum_{k=0}^{\infty} (-1)^k \frac{x^{2k}}{(2k)!} \qquad (|x| < \infty) \tag{3}$$

$$\frac{1}{1-x} = 1 + x + x^2 + x^3 + \cdots = \sum_{k=0}^{\infty} x^k \qquad (|x| < 1) \tag{4}$$

$$\ln x = (x - 1) - \frac{(x - 1)^2}{2} + \frac{(x - 1)^3}{3} - \cdots$$

$$= \sum_{k=1}^{\infty} (-1)^{k-1} \frac{(x - 1)^k}{k} \qquad (0 < x \leq 2) \qquad (5)$$

For each case, the series represents the given function and converges in the interval given. Series of this type are often used to compute good approximate values of complicated functions at specific points. For example, taking $x = 1.1$ in the first five terms of Equation (5) gives us

$$\ln(1.1) \approx 0.1 - \frac{0.01}{2} + \frac{0.001}{3} - \frac{0.0001}{4} + \frac{0.00001}{5} = 0.09531\,03333\ldots$$

where $\approx$ means approximate equality. This value is correct to six decimal places of accuracy.

On the other hand, suppose that we try to compute e^8 by using the Series (1). The result is $e^8 = 1 + 8 + 64/2 + 512/6 + 4096/24 + 32768/120 + \cdots$. It is apparent that many terms will be needed to compute e^8 with reasonable precision. (By repeated squaring, we find $e^2 = 7.389056$, $e^4 = 54.5981500$, and $e^8 = 2980.957987$.) These examples illustrate the general rule that a Taylor series will converge *rapidly* near the point of expansion and *slowly* (or not at all) at more remote points.

A graphical depiction of the phenomenon can be obtained by graphing a few partial sums of a Taylor series. In Figure 1.1, we show the function $y = \sin x$ and

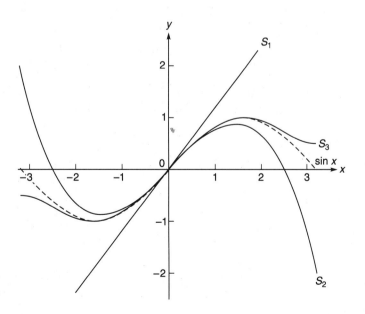

FIGURE 1.1
Approximations to $\sin x$

the partial-sum functions $S_1 = x$, $S_2 = x - x^3/6$, and $S_3 = x - x^3/6 + x^5/120$, which come from Equation (2). Software packages such as Maple, MATLAB, Mathematica, or DERIVE are particularly useful in plotting these functions. A code for producing this plot using Maple follows:

```
f(x)  := sin(x);
g(x)  := x;
h(x)  := g(x) -(x^3)/6;
k(x)  := h(x) + (x^5)/120;
interface(plotdevice=postscript);
interface(plotoutput=sin.ps);
plot({f,g,h,k},-Pi..Pi,
    title='Approximations of sin x');
```

Then the postscript file `sin.ps` is sent to a laser printer, resulting in Figure 1.1.

All of the series illustrated above are examples of the following general series:

FORMAL TAYLOR SERIES FOR f ABOUT c

$$f(x) \sim f(c) + f'(c)(x - c) + \frac{f''(c)}{2!}(x - c)^2$$

$$+ \frac{f'''(c)}{3!}(x - c)^3 + \cdots = \sum_{k=0}^{\infty} \frac{f^{(k)}(c)}{k!}(x - c)^k \qquad (6)$$

Here we have written $\sim$ to indicate that we are not allowed to assume that $f(x)$ *equals* the series on the right. All we have at the moment is a formal series that can be written down provided that the successive derivatives f', f'', f''', $\ldots$ exist at the point c. The Series (6) is called the "Taylor series of f at the point c." The reader should recall the **factorial** notation

$$n! = 1 \cdot 2 \cdot 3 \cdot 4 \cdot \cdots \cdot n$$

for $n \geq 1$ with the special definition of $0! = 1$.

In the special case $c = 0$, the Series (6) is also called a **Maclaurin series**:

$$f(x) \sim f(0) + f'(0)x + \frac{f''(0)}{2!}x^2 + \frac{f'''(0)}{3!}x^3 + \cdots$$

$$= \sum_{k=0}^{\infty} \frac{f^{(k)}(0)}{k!}x^k \qquad (7)$$

EXAMPLE 1   What is the Taylor series of the function

$$f(x) = 3x^5 - 2x^4 + 15x^3 + 13x^2 - 12x - 5$$

at the point $c = 2$?

Solution   In order to compute the coefficients in the series, we need the numerical values of $f^{(k)}(2)$ for $k \geq 0$. Here are the details of the computation:

$$f(x) \quad = 3x^5 - 2x^4 + 15x^3 + 13x^2 - 12x - 5 \qquad f(2) \quad = 207$$
$$f'(x) \quad = 15x^4 - 8x^3 + 45x^2 + 26x - 12 \qquad f'(2) \quad = 396$$
$$f''(x) \quad = 60x^3 - 24x^2 + 90x + 26 \qquad f''(2) \quad = 590$$
$$f'''(x) \quad = 180x^2 - 48x + 90 \qquad f'''(2) \quad = 714$$
$$f^{(4)}(x) = 360x - 48 \qquad f^{(4)}(2) = 672$$
$$f^{(5)}(x) = 360 \qquad f^{(5)}(2) = 360$$

and $f^{(k)}(2) = 0$ for $k \geq 6$. Therefore,

$$f(x) \sim 207 + 396(x - 2) + 295(x - 2)^2$$
$$+ 119(x - 2)^3 + 28(x - 2)^4 + 3(x - 2)^5$$

In this example it is not difficult to see that $\sim$ may be replaced by $=$. Simply expand all the binomials in the Taylor series and collect terms in order to get the original form for f. Taylor's Theorem, discussed next, will allow us to draw this conclusion without doing any work!

With the following two commands, we can use Maple to verify our results.

```
f := 3*x^5 - 2*x^4 + 15*x^3 + 13*x^2 - 12*x - 5;
taylor(", x=2);
```
□

Taylor's Theorem in Terms of $(x - c)$

In practical computations with Taylor series, it is always necessary to **truncate** the series because it is not possible to carry out an infinite number of additions. A series is said to be truncated if we ignore all terms after a certain point. Thus, if we truncate the exponential Series (1) after seven terms, the result is

$$1 + x + \frac{x^2}{2!} + \frac{x^3}{3!} + \frac{x^4}{4!} + \frac{x^5}{5!} + \frac{x^6}{6!}$$

This no longer represents e^x except when $x = 0$. But the truncated series should *approximate* e^x. Here is where we need Taylor's Theorem. With its help, we can assess the difference between a function f and its truncated Taylor series.

> ## TAYLOR'S THEOREM FOR $f(x)$
>
> If the function f possesses continuous derivatives of orders $1, 2, \ldots, (n + 1)$ in a closed interval $I = [a, b]$, then for any c in I,
>
> $$f(x) = \sum_{k=0}^{n} \frac{f^{(k)}(c)}{k!}(x - c)^k + E_{n+1} \qquad (8)$$
>
> where x is any value in I and
>
> $$E_{n+1} = \frac{f^{(n+1)}(\xi)}{(n + 1)!}(x - c)^{n+1}$$
>
> Here $\xi = \xi(x)$ is a point that lies between c and x.

(We will not prove Taylor's Theorem here.) The final term E_{n+1} in Equation (8) is the **remainder** or **error term**. The given formula for E_{n+1} is valid when we assume only that $f^{(n+1)}$ exists at each point of the open interval (a, b). The error term is similar to the terms preceding it, but notice that $f^{(n+1)}$ must be evaluated at a point other than c. This point ξ depends on x. Other forms of the remainder are possible; the one given here is **Lagrange's** form.

EXAMPLE 2   Derive the Taylor series for e^x at $c = 0$, and prove that it converges to e^x by using Taylor's Theorem.

Solution   If $f(x) = e^x$, then $f^{(k)}(x) = e^x$ for $k \geq 0$. Therefore, $f^{(k)}(c) = f^{(k)}(0) = e^0 = 1$ for all k. From Equation (8),

$$e^x = \sum_{k=0}^{n} \frac{x^k}{k!} + \frac{e^\xi}{(n + 1)!}x^{n+1} \qquad (9)$$

Now let us consider all the values of x in some symmetric interval around the origin, for example, $-s \leq x \leq s$. Then $|x| \leq s$, $|\xi| \leq s$, and $e^\xi \leq e^s$. Hence, the remainder term satisfies this inequality:

$$\lim_{n \to \infty} \left| \frac{e^\xi}{(n + 1)!}x^{n+1} \right| < \lim_{n \to \infty} \frac{e^s}{(n + 1)!}s^{n+1} = 0$$

Thus, if we take the limit as $n \to \infty$ on both sides of Equation (9), we obtain

$$e^x = \lim_{n \to \infty} \sum_{k=0}^{n} \frac{x^k}{k!} = \sum_{k=0}^{\infty} \frac{x^k}{k!} \qquad \qquad \square$$

This example illustrates how we can establish, in specific cases, that a formal Taylor Series (6) actually represents the function. Let's examine another example to see how the formal series can *fail* to represent the function.

EXAMPLE 3   Derive the formal Taylor series for $f(x) = \ln(1 + x)$ at $c = 0$, and determine the range of positive x for which the series represents the function.

Solution   We need $f^{(k)}(x)$ and $f^{(k)}(0)$ for $k \geqq 1$. Here is the work:

$$
\begin{aligned}
f(x) &= \ln(1 + x) & f(0) &= 0 \\
f'(x) &= (1 + x)^{-1} & f'(0) &= 1 \\
f''(x) &= -(1 + x)^{-2} & f''(0) &= -1 \\
f'''(x) &= 2(1 + x)^{-3} & f'''(0) &= 2 \\
f^{(4)}(x) &= -6(1 + x)^{-4} & f^{(4)}(0) &= -6 \\
&\quad\vdots & &\quad\vdots \\
f^{(k)}(x) &= (-1)^{k-1}(k - 1)!(1 + x)^{-k} & f^{(k)}(0) &= (-1)^{k-1}(k - 1)!
\end{aligned}
$$

Hence by Taylor's Theorem,

$$
\ln(1 + x) = \sum_{k=1}^{n} (-1)^{k-1} \frac{(k - 1)!}{k!} x^k + \frac{(-1)^n n!(1 + \xi)^{-n-1}}{(n + 1)!} x^{n+1}
$$

$$
= \sum_{k=1}^{n} (-1)^{k-1} \frac{x^k}{k} + \frac{(-1)^n}{n + 1}(1 + \xi)^{-n-1} x^{n+1} \tag{10}
$$

In order that the *infinite* series represent $\ln(1 + x)$, it is necessary and sufficient that the error term converge to zero as $n \rightarrow \infty$. Assume $0 \leqq x \leqq 1$. Then $0 \leqq \xi \leqq x$ because zero is the point of expansion and thus $0 \leqq x/(1 + \xi) \leqq 1$. Hence, the error term converges to zero in this case. If $x > 1$, the terms in the series do not approach zero, and the series does not converge. Hence, the series represents $\ln(1 + x)$ if $0 \leqq x \leqq 1$ but *not* if $x > 1$. [The series also represents $\ln(1 + x)$ for $-1 < x < 0$ but not if $x \leqq -1$.] $\square$

Mean-Value Theorem

The special case $n = 0$ in Taylor's Theorem is known as the **Mean-Value Theorem**. It is usually stated, however, in a somewhat more precise form.

MEAN-VALUE THEOREM

If f is a continuous function on the closed interval $[a, b]$ and possesses a derivative at each point of the open interval (a, b), then

$$
f(b) = f(a) + (b - a)f'(\xi)
$$

for some ξ in (a, b).

Hence, the ratio $[f(b) - f(a)]/(b - a)$ is equal to the derivative of f at some point ξ between a and b; that is,

$$f'(\xi) = \frac{f(b) - f(a)}{b - a}$$

The right-hand side could be used as an approximation for $f'(x)$ at any x within the interval (a, b). The approximation of derivatives is discussed more fully in Section 4.3.

Taylor's Theorem in Terms of *h*

Other forms of Taylor's Theorem are often useful. These can be obtained from the basic formula (8) by changing the variables.

TAYLOR'S THEOREM FOR *f(x + h)*

If the function f possesses continuous derivatives of order $1, 2, \ldots, (n + 1)$ in a closed interval $I = [a, b]$, then for any x in I,

$$f(x + h) = \sum_{k=0}^{n} \frac{f^{(k)}(x)}{k!} h^k + E_{n+1} \tag{11}$$

where h is any value such that $x + h$ is in I and where

$$E_{n+1} = \frac{f^{(n+1)}(\xi)}{(n + 1)!} h^{n+1}$$

with ξ between x and $x + h$.

The form (11) is obtained from Equation (8) by replacing x by $x + h$ and c by x. Notice that because h can be positive or negative, "with ξ between x and $x + h$" means $x < \xi < x + h$ if $h > 0$ or $x + h < \xi < x$ if $h < 0$.

The error term E_{n+1} depends on h in two ways: First, h^{n+1} is explicitly present, and second, the point ξ generally depends on h. As h converges to zero, E_{n+1} converges to zero with essentially the same rapidity that h^{n+1} converges to zero. For large n, this is quite rapid. To express this qualitative fact, we write

$$E_{n+1} = \mathcal{O}(h^{n+1})$$

as $h \to 0$. This is shorthand notation for the inequality

$$|E_{n+1}| \leq C|h|^{n+1}$$

where C is a constant. In the present circumstances, this constant could be any number for which $|f^{(n+1)}(t)|/(n+1)! \leq C$ with t being arbitrary in the initially given interval.

It is important to realize that this is an entire sequence of theorems, one for each value of n. For example, we can write out the cases $n = 0, 1, 2$ as follows:

$$f(x + h) = f(x) + f'(\xi_1)h$$
$$= f(x) + \mathcal{O}(h)$$

$$f(x + h) = f(x) + f'(x)h + \frac{1}{2!} f''(\xi_2)h^2$$
$$= f(x) + f'(x)h + \mathcal{O}(h^2)$$

$$f(x + h) = f(x) + f'(x)h + \frac{1}{2!} f''(x)h^2 + \frac{1}{3!} f'''(\xi_3)h^3$$

$$= f(x) + f'(x)h + \frac{1}{2!} f''(x)h^2 + \mathcal{O}(h^3)$$

For a fixed value of n, the subscript on ξ is usually omitted.

The importance of the error term in Taylor's Theorem cannot be stressed too much. In later chapters, many situations call for an estimate of errors in a numerical process by use of Taylor's Theorem. Here are some elementary examples.

EXAMPLE 4　Expand $\sqrt{1 + h}$ in powers of h. Then compute $\sqrt{1.00001}$ and $\sqrt{0.99999}$.

Solution　Let $f(x) = x^{1/2}$. Then $f'(x) = \frac{1}{2}x^{-1/2}, f''(x) = -\frac{1}{4}x^{-3/2}, f'''(x) = \frac{3}{8}x^{-5/2}$, and so on. Now use Equation (11) with $x = 1$. Taking $n = 2$ for illustration, we have

$$\sqrt{1 + h} = 1 + \frac{1}{2}h - \frac{1}{8}h^2 + \frac{1}{16}h^3\xi^{-5/2} \qquad (12)$$

where ξ is an unknown number that satisfies $1 < \xi < 1 + h$, if $h > 0$.

In Equation (12), let $h = 10^{-5}$. Then

$$\sqrt{1.00001} \approx 1 + 0.5 \times 10^{-5} - 0.125 \times 10^{-10} = 1.00000\,49999\,87500$$

By substituting $-h$ for h in the series, we obtain

$$\sqrt{1 - h} = 1 - \frac{1}{2}h - \frac{1}{8}h^2 - \frac{1}{16}h^3\xi^{-5/2}$$

Hence,

$$\sqrt{0.99999} \approx 0.99999\,49999\,87500$$

Since $1 < \xi < 1 + h$, the absolute error does not exceed

$$\frac{1}{16}h^3\xi^{-5/2} < \frac{1}{16}10^{-15} = 0.00000\,00000\,00000\,0625$$

and both numerical values are correct to all 15 decimal places shown. In this and the previous example, it is important to notice that the function $f(x) = \sqrt{x}$ possesses derivatives of all orders at any point $x > 0$.

Using the following Maple program, we are able to check our calculations.

```
y := taylor(sqrt(1+h), h=0);
p := convert(", polynom);
Digits := 40;
subs(h=0.00001, p);
evalf(");
subs(h=-0.00001, p);
evalf(");
```

We obtain the numerical results:

$$1.00000\,49999\,87500\,06249\,96093\,77734\,37500\,0000$$
$$.99999\,49999\,87499\,93749\,96093\,72265\,62500\,00000$$

Are these answers accurate to all digits shown?   □

Alternating Series

Another theorem from calculus is often useful in establishing the convergence of a series and in estimating the error involved in truncation. This theorem applies only to **alternating series**—that is, series in which the successive terms are alternately positive and negative.

ALTERNATING SERIES THEOREM

If $a_1 \geq a_2 \geq \cdots \geq a_n \geq \cdots 0$ for all n and $\lim_{n\to\infty} a_n = 0$, then the alternating series

$$a_1 - a_2 + a_3 - a_4 + \cdots$$

converges; that is,

$$\sum_{k=1}^{\infty}(-1)^{k-1}a_k = \lim_{n\to\infty}\sum_{k=1}^{n}(-1)^{k-1}a_k = \lim_{n\to\infty} S_n = S$$

where S is its sum and S_n is the nth partial sum. Moreover, for all n,

$$|S - S_n| \leq a_{n+1}$$

Thus, the error in truncation is no larger than the magnitude of the first *omitted* term.

EXAMPLE 5   If the sine series is to be used in computing sin 1 with an error less than $\frac{1}{2} \times 10^{-6}$, how many terms are needed?

Solution   From Equation (2),

$$\sin 1 = 1 - \frac{1}{3!} + \frac{1}{5!} - \frac{1}{7!} + \cdots$$

If we stop at $1/(2n - 1)!$, the error does not exceed the first neglected term, which is $1/(2n + 1)!$. Thus, we should select n so that

$$\frac{1}{(2n + 1)!} < \frac{1}{2} \times 10^{-6}$$

Using logarithms to base 10, we obtain $\log(2n + 1)! > \log 2 + 6 = 6.3$. By consulting a table of $\log n!$, we find that $\log 10! \approx 6.6$, so $n \geqq 5$.   □

EXAMPLE 6   If the logarithmic series (5) is to be used for computing ln 2 with an error of less than $\frac{1}{2} \times 10^{-6}$, how many terms will be required?

Solution   To compute ln 2, we take $x = 2$ in the series and, using $\approx$ to mean approximate equality, we have

$$S = \ln 2 \approx 1 - \frac{1}{2} + \frac{1}{3} - \frac{1}{4} + \cdots + \frac{(-1)^{n-1}}{n} = S_n$$

Then the error involved when the series is truncated with n terms is

$$|S - S_n| \leqq \frac{1}{n + 1}$$

We select n so that

$$\frac{1}{n + 1} < \frac{1}{2} \times 10^{-6}$$

Hence, more than two million terms would be needed! We conclude that this method of computing ln 2 is not practical! (See Problems **7** through **9** for several good alternatives.)   □

A word of caution is needed about this technique of calculating the number of terms to be used in a series by just making the $(n + 1)$st term less than some tolerance. This procedure is valid only for alternating series in which the terms decrease in magnitude to zero, although it is occasionally used to get rough estimates in other cases. For example, it can be used to identify a nonalternating series as one that converges slowly. When this technique cannot be used, a bound on the remaining terms of the series has to be established. This may be somewhat difficult.

EXAMPLE 7   It is known that

$$\frac{\pi^4}{90} = 1^{-4} + 2^{-4} + 3^{-4} + \cdots$$

How many terms should we take in order to compute $\pi^4/90$ with an error of at most $\frac{1}{2} \times 10^{-6}$?

Solution   A naive approach is to take

$$1^{-4} + 2^{-4} + 3^{-4} + \cdots + n^{-4}$$

where n is chosen so that the next term, $(n + 1)^{-4}$, is less that $\frac{1}{2} \times 10^{-6}$. This value of n is 37, but this is an erroneous answer because the partial sum

$$S_{37} = \sum_{k=1}^{37} k^{-4}$$

differs from $\pi^4/90$ by approximately 6×10^{-6}. What we should do, of course, is to select n so that *all* the omitted terms add up to less that $\frac{1}{2} \times 10^{-6}$; that is,

$$\sum_{k=n+1}^{\infty} k^{-4} < \frac{1}{2} \times 10^{-6}$$

By a technique familiar from calculus (see Figure 1.2), we have

$$\sum_{k=n+1}^{\infty} k^{-4} < \int_{n}^{\infty} x^{-4}\, dx = \left. \frac{x^{-3}}{-3} \right|_{n}^{\infty} = \frac{1}{3n^3}$$

Thus, it suffices to select n so that $1/(3n^3) < \frac{1}{2} \times 10^{-6}$, or $n \geqq 88$.   □

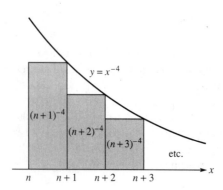

FIGURE 1.2

**PROBLEMS
1.2**

1. Determine the Taylor series for cosh x about zero. Evaluate cosh(0.7) by summing four terms. Compare with the actual value.

2. Determine the first two nonzero terms of the series expansion about zero for the following.

   a. $e^{\cos x}$

   b. $\sin(\cos x)$

   c. $(\cos x)^2(\sin x)$

3. Determine a Taylor series to represent $\cos(\pi/3 + h)$. Evaluate $\cos(60.001°)$ to eight decimal places (rounded).

4. Find the smallest nonnegative integer m such that the Taylor series about m for $(x - 1)^{1/2}$ exists. Determine the coefficients in the series.

5. Determine how many terms are needed to compute e correctly to 15 decimal places (rounded) using Series (1) for e^x.

6. (Continuation) If $x < 0$ in Problem **5**, what are the signs of the respective terms in the series? Cancellation of significant digits can be a serious problem in using the series. Will the formula $e^{-x} = 1/e^x$ be helpful in circumventing the difficulty? Explain. (See Section 2.3 for further discussion of this type of difficulty.)

7. Show how the simple equation $\ln 2 = \ln[e(2/e)]$ can be used to speed up the calculation of $\ln 2$ in Series (10).

8. What is the series for $\ln(1 - x)$? What is the series for $\ln[(1 + x)/(1 - x)]$?

9. (Continuation) In the series for $\ln[(1 + x)/(1 - x)]$, determine what value of x to use if we wish to compute $\ln 2$. Estimate the number of terms needed for ten digits (rounded) of accuracy. Is this method practical?

10. Use the Alternating Series Theorem to determine the number of terms in Equation (5) needed for computing $\ln 1.1$ with error less than $\frac{1}{2} \times 10^{-8}$.

11. L'Hôpital's rule states that under suitable conditions,

$$\lim_{x \to a} \frac{f(x)}{g(x)} = \frac{f'(a)}{g'(a)}$$

It is true, for instance, when f and g have continuous derivatives in an open interval containing a and $f(a) = g(a) = 0 \neq g'(a)$. Establish L'Hôpital's rule using the Mean-Value Theorem.

12. (Continuation) Use Problem **11** to show that

   a. $\lim_{x \to 0} \dfrac{\sin x}{x} = 1$

   Also evaluate the following.

   b. $\lim_{x \to 0} \dfrac{\arctan x}{x}$

   c. $\lim_{x \to \pi} \dfrac{\cos x + 1}{\sin x}$

13. Verify that if we take only the terms up to and including $x^{2n-1}/(2n-1)!$ in Series (2) for $\sin x$ and if $|x| < \sqrt{6}$, then the error involved does not exceed $|x|^{2n+1}/(2n+1)!$. How many terms are needed to compute $\sin(23)$ with an error of at most 10^{-8}? What problems do you foresee in using the series to compute $\sin(23)$? Show how to use periodicity to compute $\sin(23)$. Show that each term in the series can be obtained from the preceding one by a simple arithmetic operation.

14. Expand the *error function*

$$\operatorname{erf}(x) = \frac{2}{\sqrt{\pi}} \int_0^x e^{-t^2}\, dt$$

in a series by using the exponential series and integrating. Obtain the Taylor series of $\operatorname{erf}(x)$ about zero directly. Are the two series the same? Evaluate $\operatorname{erf}(1)$ by adding four terms of the series and compare with the value $\operatorname{erf}(1) \approx 0.8427$, which is correct to four decimal places. *Hint:* Recall from the *Fundamental Theorem of Calculus* that

$$\frac{d}{dx} \int_0^x f(t)\, dt = f(x)$$

15. Establish the validity of the Taylor series

$$\arctan x = \sum_{k=1}^{\infty} (-1)^{k+1} \frac{x^{2k-1}}{2k-1} \qquad (-1 \le x \le 1)$$

Is it practical to use this series directly to compute $\arctan(1)$ if ten decimal places of accuracy are required? How many terms of the series would be needed? Will loss of significance occur? *Hint:* Start with the series for $1/(1 + x^2)$ and integrate term by term. Note that this procedure is only formal; the convergence of the resulting series can be proved by appealing to certain theorems of advanced calculus.

16. It is known that

$$\pi = 4 - 8 \sum_{k=1}^{\infty} (16k^2 - 1)^{-1}$$

Discuss the numerical aspects of computing π by means of this formula. How many terms would be needed to yield ten decimal places of accuracy?

17. Determine a Taylor series to represent $\sin(\pi/4 + h)$. Evaluate $\sin(45.0005°)$ to nine decimal places (rounded).

18. Write the Taylor series for the function $f(x) = x^3 - 2x^2 + 4x - 1$, using $x = 2$ as the point of expansion; that is, write a formula for $f(2 + h)$.

19. Determine the first four nonzero terms in the series expansion about zero for $f(x) = (\sin x) + (\cos x)$ and $g(x) = (\sin x)(\cos x)$. Then find approximate

values for $f(0.001)$ and $g(0.0006)$. Compare the accuracy of these approximations to those obtained from tables or via a calculator.

20. Verify this Taylor series and prove that it converges on the interval $-e < x \leq e$.

$$\ln(e + x) = 1 + \frac{x}{e} - \frac{x^2}{2e^2} + \frac{x^3}{3e^3} - \frac{x^4}{4e^4} + \cdots = 1 + \sum_{k=1}^{\infty} \frac{(-1)^{k-1}}{k} \left(\frac{x}{e}\right)^k$$

21. How many terms are needed in Series (3) to compute $\cos x$ for $|x| < \frac{1}{2}$ accurate to 12 decimal places (rounded)?

22. A function f is defined by the series

$$f(x) = \sum_{k=1}^{\infty} (-1)^k \left(\frac{x^k}{k^4}\right)$$

Determine the minimum number of terms needed to compute $f(1)$ with error less than 10^{-8}.

23. Verify that the partial sums $s_k = \sum_{i=0}^{k} x^i/i!$ in the series for e^x, Equation (1), can be written recursively as $s_k = s_{k-1} + t_k$, where $s_0 = 1, t_1 = x$, and $t_k = (x/k)t_{k-1}$.

24. What is the fifth term in the Taylor series of $(1 - 2h)^{1/2}$?

25. Show that if $E = \mathcal{O}(h^n)$, then $E = \mathcal{O}(h^m)$ for any nonnegative integer $m \leq n$. Here $h \to 0$.

26. Why do the following functions not possess Taylor series expansions at $x = 0$?

a. $f(x) = \sqrt{x}$
b. $f(x) = |x|$
c. $f(x) = \arcsin(x - 1)$
d. $f(x) = \cot x$
e. $f(x) = \log x$
f. $f(x) = x^\pi$

27. Show how $p(x) = 6(x + 3) + 9(x + 3)^5 - 5(x + 3)^8 - (x + 3)^{11}$ can be efficiently evaluated.

28. What is the second term in the Taylor series of $\sqrt[4]{4x - 1}$ about 4.25?

29. How would you compute a table of $\log n!$ for $1 \leq n \leq 1000$?

30. The Taylor series for $(1 + x)^n$ is also known as the **binomial series**. It states that

$$(1 + x)^n = 1 + nx + \frac{n(n - 1)}{2!}x^2 + \frac{n(n - 1)(n - 2)}{3!}x^3 + \cdots \qquad (x^2 < 1)$$

Derive the series and, for $n = 2, n = 3$, and $n = \frac{1}{2}$, give its particular form. Then use the last form to compute $\sqrt{1.0001}$ correct to 15 decimal places (rounded).

31. (Continuation) Use the series in Problem **30** to obtain Series (4). How could this series be used on a computing machine to produce x/y if only addition and multiplication are built-in operations?

32. (Continuation) Use Problem **31** to obtain a series for $(1 + x^2)^{-1}$.

33. For small x the approximation $\sin x \approx x$ is often used. For what range of x is this good to a *relative* accuracy of $\frac{1}{2} \times 10^{-14}$?

34. In the Taylor series for the function $3x^2 - 7 + \cos x$ (expanded in powers of x), what is the coefficient of x^2?

35. In the Taylor series (about $\pi/4$) for the function $\sin x + \cos x$, find the third nonzero term.

36. By using Taylor's Theorem, one can be sure that for all x that satisfy $|x| < \frac{1}{2}$, $|\cos x - (1 - x^2/2)|$ is less than or equal to what numerical value?

37. Find the value of ξ that serves in Taylor's Theorem when $f(x) = \sin x$, $x = \pi/4, c = 0$, and $n = 4$.

38. Use Taylor's Theorem to find a linear function that approximates $\cos x$ best in the vicinity of $x = 5\pi/6$.

39. For the alternating series $S_n = \sum_{k=0}^{n}(-1)^k a_k$, with $a_0 > a_1 > \cdots > 0$, show by induction that $S_0 > S_2 > S_4 > \cdots$, that $S_1 < S_3 < S_5 < \cdots$, and that $0 < S_{2n} - S_{2n+1} = a_{2n+1}$.

40. What is the Maclaurin series for the function $f(x) = 3 + 7x - 1.33x^2 + 19.2x^4$? What is the Taylor series for this function about $c = 2$?

41. In the text it was asserted that $\sum_{k=0}^{6} x^k/k!$ represents e^x only at the point $x = 0$. Prove this.

42. Determine the first three terms in the Taylor series in terms of h for e^{x-h}. Using three terms, one obtains $e^{0.999} \approx Ce$, where C is a constant. Determine C.

43. What is the least number of terms required to compute π as 3.14 (rounded) using the series

$$\pi = 4 - \frac{4}{3} + \frac{4}{5} - \frac{4}{7} + \cdots$$

44. Using the Taylor series expansion in terms of h, determine the first three terms in the series for $e^{\sin(x+h)}$. Evaluate $e^{\sin 90.01°}$ accurately to ten decimal places as Ce for constant C.

45. Develop the first two terms and the error in the Taylor series in terms of h for $\ln(3 - 2h)$.

46. Establish the first three terms in the Taylor series for $\csc(\pi/6 + h)$. Compute $\csc(30.00001°)$ to the same accuracy as the given data.

47. Establish the Taylor series in terms of h for the following.

   a. e^{x+2h}

   b. $\sin(x - 3h)$

   c. $\ln[(x - h^2)/(x + h^2)]$

48. Determine the first three terms in the Taylor series in terms of h for $(x - h)^m$, where m is an integer constant.

49. Given the series

$$-1 + 2^{-4} - 3^{-4} + 4^{-4} - \cdots$$

How many terms are needed to obtain four decimal places (chopped) of accuracy?

50. How many terms are needed in the series

$$\operatorname{arccot} x = \frac{\pi}{2} - x + \frac{x^3}{3} - \frac{x^5}{5} + \frac{x^7}{7} - \cdots$$

to compute $\operatorname{arccot} x$ for $x^2 < 1$ accurate to 12 decimal places (rounded)?

51. Determine the first three terms in the Taylor series to represent $\sinh(x + h)$. Evaluate $\sinh(0.0001)$ to 20 decimal places (rounded) using this series.

52. Determine a Taylor series to represent C^{x+h} for constant C. Use the series to find an approximate value of $(10)^{1.0001}$ to five decimal places (rounded).

53. Stirling's formula states that $n!$ is greater than, and very close to, $\sqrt{2\pi n}\, n^n e^{-n}$. Use this to find an n for which $1/n! < \frac{1}{2} \times 10^{-14}$.

54. Develop the first two nonzero terms and the error term in the Taylor series in terms of h for $\ln[1 - (h/2)]$. Approximate $\ln(0.9998)$ using these two terms.

COMPUTER PROBLEMS 1.2

1. Verify that $x^y = e^{y \ln x}$. Try to find values of x and y for which these two expressions differ in your computer. *Interpret the results.*

2. (Continuation) For $\cos(x - y) = (\cos x)(\cos y) + (\sin x)(\sin y)$, repeat Computer Problem **1.**

3. Design and carry out an experiment to check the computation of x^y on your computer. *Hint*: Compare the computations of some examples, such as $32^{2.5}$ and $81^{1.25}$, to their correct values. A more elaborate test can be made by comparing single-precision results to double-precision results in various cases.

4. Everyone knows the quadratic formula $(-b \pm \sqrt{b^2 - 4ac})/(2a)$ for the roots of the quadratic equation $ax^2 + bx + c = 0$. Using this formula, by hand and by computer, solve the equation $x^2 + 10^8 x + c = 0$ with $c = 1$ and 10^8. *Interpret the results.*

5. The number of combinations of n distinct items taken m at a time is given by the **binomial coefficient**

$$\binom{n}{m} = \frac{n!}{m!\,(n - m)!}$$

for integers m and n, with $0 \le m \le n$. Recall that $\binom{n}{0} = \binom{n}{n} = 1$.

a. Write

integer function *ibin*(*n*, *m*)

which uses the definition above to compute $\binom{n}{m}$.

b. Verify the formula

$$\binom{n}{m} = \prod_{k=1}^{\min(m,n-m)} \left(\frac{n-k+1}{k}\right)$$

for computing the binomial coefficients. Write

integer function *jbin*(*n*, *m*)

based on this formula.

c. Verify the formulas (*Pascal's triangle*)

$$\begin{cases} a_{1j} = a_{j1} = 1 & (j \geq 1) \\ a_{ij} = a_{i-1,j} + a_{i,j-1} & (i,j \geq 2) \end{cases}$$

for computing an array of binomial coefficients

$$\binom{i}{j} = a_{i-j+1,j+1}$$

Write

integer function *kbin*(*n*, *m*, (*a*$_{ij}$))

that does an array lookup after computing the array (*a*$_{ij}$).

6. The length of the curved part of a unit semicircle is π. We can approximate π by using triangles and elementary mathematics. Consider the semicircle with the arc bisected as in Figure (a). The hypotenuse of the right triangle is $\sqrt{2}$. Hence, a rough approximation to π is given by $2\sqrt{2} \approx 2.8284$. In Figure (b), we consider an angle θ that is a fraction $1/k$ of the semicircle. The secant shown has length $2\sin(\theta/2)$, and so an approximation to π is $2k\sin(\theta/2)$. From trigonometry,

$$\sin^2\frac{\theta}{2} = \frac{1-\cos\theta}{2} = \frac{1-\sqrt{1-\sin^2\theta}}{2} = \frac{\sin^2\theta}{2+2\sqrt{1-\sin^2\theta}}$$

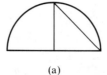

(a)

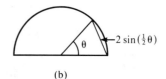

(b)

Now let θ_n be the angle that results from division of the semicircular arc into 2^{n-1} pieces. Next let $S_n = \sin^2 \theta_n$ and $P_n = 2^n \sqrt{S_{n+1}}$. Show that $S_{n+1} = S_n/(2 + 2\sqrt{1 - S_n})$ and P_n is an approximation to π. Starting with $S_2 = 1$ and $P_1 = 2$, compute S_{n+1} and P_n recursively for $2 \leqq n \leqq 20$.

7. The irrational number π can be computed by approximating the area of a unit circle as the limit of a sequence $p_1, p_2, \ldots$ described as follows. Divide the unit circle into 2^n sectors. (The figure shows the case $n = 3$.) Approximate the area of the sector by the area of the isosceles triangle. The angle θ_n is $2\pi/2^n$. The area of the triangle is $\frac{1}{2} \sin \theta_n$. (Verify.) The nth approximation to π is then $p_n = 2^{n-1} \sin \theta_n$. Prove that $\sin \theta_n = \sin \theta_{n-1}/\{2[1 + (1 - \sin^2 \theta_{n-1})^{1/2}]\}^{1/2}$ by means of well-known trigonometric identities. Use this recurrence relation to generate the sequences $\sin \theta_n$ and p_n ($3 \leqq n \leqq 20$) starting with $\sin \theta_2 = 1$. Compare with the computation $4.0 \cdot \arctan(1.0)$.

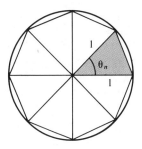

8. Calculate π by a method similar to that of Problem **7**, where the area of the unit circle is approximated by a sequence of trapezoids as illustrated by the figure.

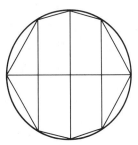

9. Write a routine in double or extended precision to implement the following algorithm for computing π.

real a, b, c, d, e
$a \leftarrow 0$
$b \leftarrow 1$
$c \leftarrow 1/\sqrt{2}$
$d \leftarrow 0.25$
$e \leftarrow 1$

for $k = 1$ **to** 5 **do**
  $a \leftarrow b$
  $b \leftarrow (b + c)/2$
  $c \leftarrow \sqrt{ca}$
  $d \leftarrow d - e(b - a)^2$
  $e \leftarrow 2e$
  $f \leftarrow b^2/d$
  $g \leftarrow (b + c)^2/(4d)$
end do

After each cycle, print $|f - \pi|$ and $|g - \pi|$. Which converges faster, f or g? How accurate are the final values? Also compare with the double- or extended-precision computation of $4.0 \cdot \arctan(1.0)$. *Hint*: The value of π correct to 36 digits is

$$3.14159\ 26535\ 89793\ 23846\ 26433\ 83279\ 50288$$

Note: A new formula for computing π was discovered in the early 1970s. This algorithm is based on that formula, which is a direct consequence of a method developed by Gauss for calculating elliptic integrals and of Legendre's elliptic integral relation, both known for over 150 years! The error analysis shows that rapid convergence results in the computation of π with the number of significant digits doubling after each step. (The interested reader should see Salamin [1976] and Brent [1976].)

10. Another quadratically convergent scheme for computing π was discovered by Borwein and Borwein [1984] and can be written as

$a \leftarrow \sqrt{2}$
$b \leftarrow 0$
$x \leftarrow 2 + \sqrt{2}$
for $k = 1$ **to** 5 **do**
  $t \leftarrow \sqrt{a}$
  $b \leftarrow t(1 + b)/(a + b)$
  $a \leftarrow \frac{1}{2}(t + 1/t)$
  $x \leftarrow xb(1 + a)/(1 + b)$
end do

Numerically verify that $|x - \pi| \leq 10^{-2k}$.

11. The Fibonacci sequence $0, 1, 1, 2, 3, 5, 8, 13, 21, \ldots$ is defined by the linear recurrence relation

$$\begin{cases} \lambda_0 = 1 \quad\quad \lambda_1 = 1 \\ \lambda_n = \lambda_{n-1} + \lambda_{n-2} \quad\quad (n \geq 2) \end{cases}$$

A formula for the nth Fibonacci number is

$$\lambda_n = \frac{1}{\sqrt{5}}\left[\left(\frac{1 + \sqrt{5}}{2}\right)^n - \left(\frac{1 - \sqrt{5}}{2}\right)^n\right]$$

Compute λ_n $(0 \leq n \leq 50)$, using both the recurrence relation and the formula. Write three programs that use integer, single-precision, and double-precision arithmetic, respectively. For each n, print the results using integer, single-precision, and double-precision formats, respectively.

12. (Continuation) Repeat the experiment of Computer Problem **11** using the sequence given by the recurrence relation

$$\begin{cases} \alpha_0 = 1 \qquad \alpha_1 = \left(\dfrac{1 + \sqrt{5}}{2} \right) \\[2ex] \alpha_n = \alpha_{n-1} + \alpha_{n-2} \qquad (n \leq 2) \end{cases}$$

A closed-form formula is

$$\alpha_n = \left(\frac{1 + \sqrt{5}}{2} \right)^n$$

13. (Continuation) Change $+\sqrt{5}$ to $-\sqrt{5}$ and repeat the computation of α_n from Computer Problem **12**. Explain the results.

14. The Bessel functions J_n are defined by

$$J_n(x) = \frac{1}{\pi} \int_0^\pi \cos(x \sin \theta - n\theta) \, d\theta$$

Establish that $|J_n(x)| \leq 1$. It is known that $J_{n+1}(x) = 2nx^{-1}J_n(x) - J_{n-1}(x)$. Use this equation to compute $J_0(1), J_1(1), \ldots, J_{20}(1)$, starting from known values $J_0(1) \approx 0.76519\,76865$ and $J_1(1) \approx 0.44005\,05857$. Account for the fact that the inequality $|J_n(x)| \leq 1$ is violated.

15. Use a computer algebra system to obtain graphs of the first five partial sums of the series

$$\arctan x = \sum_{k=1}^\infty (-1)^{k+1} \frac{x^{2k-1}}{2k - 1}$$

16. A calculus student is asked to determine $\lim_{n \to \infty} (100^n / n!)$ and writes pseudocode to evaluate $x_0, x_1, x_2, \ldots, x_n$ as follows:

```
integer parameter n ← 100
integer i
real x
x ← 1
for i = 1 to n do
    x ← 100x/i
    output i, x
end do
```

The numbers printed become ever larger, and the student concludes that $\lim_{n \to \infty} x_n = \infty$. What is the moral here?

17. Use a graphical computer package to reproduce the graphs in Figure 1.1 as well as the next two partial sums—that is, S_4 and S_5. Analyze the results.

18. Use a computer algebra system to obtain the Taylor series given in Equations (1)–(5) obtaining the final form at once, without computing all the derivatives.

19. Use a computer algebra system to carry out Example 4 to fifty decimal places. Are these answers correct to all digits obtained? Repeat using $\sqrt{x}$ expanded about $x_0 = 1$.

20. Use a computer algebra system to verify the results in Examples 5 and 7.

2 NUMBER REPRESENTATION AND ERRORS

Computers usually do not use base-10 arithmetic for storage or computation. Hence, it is necessary to know something about converting numbers into different systems of representation, such as the binary or octal systems. Numbers that have a finite expansion in one number system may have an infinite expansion in another system; for example,

$$(0.1)_{10} = (0.0\,0011\,0011\,0011\,0011\,0011\,0011\,0011\,0011\,\ldots)_2$$

In this chapter, we explain the floating-point number system and develop basic facts about roundoff errors. Loss of significance, which occurs when nearly equal numbers are subtracted, is studied and shown to be avoidable by careful programming.

2.1 Representation of Numbers in Different Bases

The familiar decimal notation for numbers uses the digits 0, 1, 2, 3, 4, 5, 6, 7, 8, and 9. When we write a whole number such as 37294, the individual digits represent coefficients of powers of 10 as follows:

$$37294 = 4 + 90 + 200 + 7000 + 30000$$
$$= 4 \times 10^0 + 9 \times 10^1 + 2 \times 10^2 + 7 \times 10^3 + 3 \times 10^4$$

Thus, in general, a string of digits represents a number according to the formula

$$a_n a_{n-1} \ldots a_2 a_1 a_0 = a_0 \times 10^0 + a_1 \times 10^1 + \cdots + a_{n-1} \times 10^{n-1} + a_n \times 10^n$$

This takes care of the positive whole numbers. A number between 0 and 1 is represented by a string of digits to the right of a decimal point. For example,

$$0.7215 = \frac{7}{10} + \frac{2}{100} + \frac{1}{1000} + \frac{5}{10000}$$

$$= 7 \times 10^{-1} + 2 \times 10^{-2} + 1 \times 10^{-3} + 5 \times 10^{-4}$$

In general, we have the formula

$$0.b_1 b_2 b_3 \ldots = b_1 \times 10^{-1} + b_2 \times 10^{-2} + b_3 \times 10^{-3} + \cdots$$

Note that there can be an infinite string of digits to the right of the decimal point, and indeed there *must* be an infinite string to represent some numbers. For example,

$$\sqrt{2} = 1.41421\,35623\,73095\,04880\,16887\,24209\,69\ldots$$

$$e = 2.71828\,18284\,59045\,23536\,02874\,71352\,66\ldots$$

$$\pi = 3.14159\,26535\,89793\,23846\,26433\,83279\,50\ldots$$

$$\ln 2 = 0.69314\,71805\,59945\,30941\,72321\,21458\,17\ldots$$

$$\tfrac{1}{3} = 0.33333\,33333\,33333\,33333\,33333\,33333\,33\ldots$$

For a real number of the form

$$a_n a_{n-1} \ldots a_1 a_0 . b_1 b_2 b_3 \ldots = \sum_{k=0}^{n} a_k 10^k + \sum_{k=1}^{\infty} b_k 10^{-k}$$

the **integer part** is the first summation in the expansion and the **fractional part** is the second.

Base β Numbers

The foregoing discussion pertains to the usual representation of numbers with base 10. Other bases are also used, especially in computers. For example, the **binary** system uses 2 as the base, the **octal** system uses 8, and the **hexadecimal** system uses 16.

In the octal representation of a number, the digits used are 0, 1, 2, 3, 4, 5, 6, and 7. If ambiguity can arise, a number represented in octal is signified by enclosing it in parentheses and adding a subscript 8. Thus,

$$(21467)_8 = 7 + 6 \times 8 + 4 \times 8^2 + 1 \times 8^3 + 2 \times 8^4$$

$$= 7 + 8(6 + 8(4 + 8(1 + 8(2))))$$

$$= 9015$$

A number between 0 and 1, expressed in octal, is represented with combinations of 8^{-1}, 8^{-2}, and so on. For example,

$$
\begin{aligned}
(0.36207)_8 &= 3 \times 8^{-1} + 6 \times 8^{-2} + 2 \times 8^{-3} + 0 \times 8^{-4} + 7 \times 8^{-5} \\
&= 8^{-5}(3 \times 8^4 + 6 \times 8^3 + 2 \times 8^2 + 7) \\
&= 8^{-5}(7 + 8^2(2 + 8(6 + 8(3)))) \\
&= \frac{15495}{32768} \\
&= 0.47286\,987\ldots
\end{aligned}
$$

We shall see presently how to convert easily to decimal form without having to find a common denominator.

If we use another base, say β, then numbers represented in the β-system look like this:

$$
(a_n a_{n-1} \ldots a_1 a_0 . b_1 b_2 b_3 \ldots)_\beta = \sum_{k=0}^{n} a_k \beta^k + \sum_{k=1}^{\infty} b_k \beta^{-k}
$$

The digits are $0, 1, \ldots, \beta - 2$, and $\beta - 1$ in this representation. If $\beta > 10$, it is necessary to introduce symbols for $10, 11, \ldots, \beta - 1$.

Conversion of Integer Part

We now formalize the problem of converting a number from one base to another. It is advisable to consider separately the integer and fractional parts of a number. Consider, then, an integer N in the number system with base α:

$$
N = (a_n a_{n-1} \ldots a_1 a_0)_\alpha = \sum_{k=0}^{n} a_k \alpha^k
$$

Suppose that we wish to convert this to the number system with base β and that the calculations are to be performed in arithmetic with base β. Write N in its nested form:

$$
N = a_0 + \alpha(a_1 + \alpha(a_2 + \cdots + \alpha(a_{n-1} + \alpha(a_n))\cdots))
$$

and then replace each of the numbers by its representation in base β. Next, carry out the calculations in β-arithmetic. The replacement of the a_k's and α by equivalent base-β numbers requires a table showing how each of the numbers $0, 1, \ldots, \alpha$ appears in the β-system. Moreover, a base-β multiplication table may be required.

To illustrate this procedure, consider the conversion of the decimal number 3781 to binary form. Using the decimal-binary equivalences and longhand multiplication in base 2, we have

$$3781 = 1 + 10(8 + 10(7 + 10(3)))$$

$$= (1)_2 + (1\,010)_2((1\,000)_2 + (1\,010)_2((111)_2 + (1\,010)_2(11)_2))$$

$$= (111\,011\,000\,101)_2$$

This arithmetic calculation in binary is easy for a computer that operates in binary but tedious for humans. Another procedure should be used for hand calculations. Write down an equation containing the coefficients $c_0, c_1, \ldots, c_m$ that we seek:

$$N = (c_m c_{m-1} \ldots c_1 c_0)_\beta = c_0 + \beta(c_1 + \beta(c_2 + \cdots + \beta(c_m) \cdots))$$

Next, observe that if N is divided by β, then the *remainder* in this division is c_0 and the *quotient* is

$$c_1 + \beta(c_2 + \cdots + \beta(c_m) \cdots)$$

If *this* number is divided by β, the remainder is c_1, and so on. Thus, we divide repeatedly by β, saving remainders $c_0, c_1, \ldots$ and quotients.

EXAMPLE 1   Convert the number 3781 to binary form using the division algorithm.

Solution   As indicated above, we divide repeatedly by 2, saving the remainders along the way. Here is the work:

$$
\begin{array}{rl}
\textit{Quotients} & \textit{Remainders} \\
2)\overline{3781} & \\
2)\overline{1890} & 1 = c_0 \qquad \downarrow \\
2)\overline{945} & 0 = c_1 \\
2)\overline{472} & 1 = c_2 \\
2)\overline{236} & 0 = c_3 \\
2)\overline{118} & 0 = c_4 \\
2)\overline{59} & 0 = c_5 \\
2)\overline{29} & 1 = c_6 \\
2)\overline{14} & 1 = c_7 \\
2)\overline{7} & 0 = c_8 \\
2)\overline{3} & 1 = c_9 \\
2)\overline{1} & 1 = c_{10} \\
0 & 1 = c_{11}
\end{array}
$$

Here the symbol $\downarrow$ is used to remind us that the digits c_i are obtained beginning with the digit next to the binary point; that is,

$$(3781.)_{10} = (111\,011\,000\,101.)_2$$

not the other way around $(101\,000\,110\,111.)_2$!   □

EXAMPLE 2   Convert the number $N = (111\,011\,000\,101)_2$ to decimal form by nested multiplication.

Solution
$$N = 1 \times 2^0 + 0 \times 2^1 + 1 \times 2^2 + 0 \times 2^3 + 0 \times 2^4 + 0 \times 2^5$$
$$+ 1 \times 2^6 + 1 \times 2^7 + 0 \times 2^8 + 1 \times 2^9 + 1 \times 2^{10} + 1 \times 2^{11}$$
$$= 1 + 2(0 + 2(1 + 2(0 + 2(0 + 2(0 + 2(1 + 2(1 + 2(0$$
$$+ 2(1 + 2(1 + 2(1)))))))))))$$
$$= 3781$$

The *nested multiplication* with repeated multiplication and addition can be carried out on a pocket calculator more easily than can the previous form with exponentation.   □

Another conversion problem exists in going from an integer in base α to an integer in base β when using calculations in base α. As before, the unknown coefficients in the equation

$$N = c_0 + c_1\beta + c_2\beta + \cdots + c_m\beta^m$$

are determined by a process of successive division, and this arithmetic is carried out in the α-system. At the end, the numbers c_k are in base α, and a table of α-β equivalents is used.

For example, we can convert a binary integer into decimal form by repeated division by $(1\,010)_2$ [which equals $(10)_{10}$], carrying out the operations in binary. The table of binary-decimal equivalents is used at the end. However, since binary division is easy only for computers, we shall develop another procedure presently.

Conversion of Fractional Part

We can convert a fractional number such as 0.372 to binary by using a direct yet naive approach as follows:

$$(0.372)_{10} = 3 \times 10^{-1} + 7 \times 10^{-2} + 2 \times 10^{-3}$$

$$= \frac{1}{10}\left(3 + \frac{1}{10}\left(7 + \frac{1}{10}(2)\right)\right)$$

$$= \frac{1}{(1\,010)_2}\left((011)_2 + \frac{1}{(1\,010)_2}\left((111)_2 + \frac{1}{(1\,010)_2}((010)_2)\right)\right)$$

Dividing in binary arithmetic is not straightforward, so we look for easier ways of doing this conversion.

Suppose that x is in the range $0 < x < 1$ and that the coefficients c_k in the representation

$$x = \sum_{k=1}^{\infty} c_k \beta^{-k} = (0.c_1 c_2 c_3 \ldots)_\beta$$

are to be determined. Observe that

$$\beta x = (c_1.c_2 c_3 c_4 \ldots)_\beta$$

because it is only necessary to shift the radix point when multiplying by base β. Thus, the unknown coefficient c_1 can be described as the **integer part** of βx. It is denoted by $I(\beta x)$. The **fractional part**, $(0.c_2 c_3 c_4 \ldots)_\beta$, is denoted by $\mathcal{F}(\beta x)$. The process is repeated in the following pattern:

$$
\begin{aligned}
d_0 &= x \\
d_1 &= \mathcal{F}(\beta d_0) \qquad c_1 = I(\beta d_0) \quad \downarrow \\
d_2 &= \mathcal{F}(\beta d_1) \qquad c_2 = I(\beta d_1)
\end{aligned}
$$

etc.

In this algorithm, the arithmetic is carried out in the decimal system.

EXAMPLE 3 Use the preceding algorithm to convert the decimal number $x = 0.372$ to binary form.

Solution The algorithm consists in repeatedly multiplying by 2 and removing the integer parts. Here is the work:

$$
\begin{array}{r}
0.372 \\
\underline{2} \\
\downarrow \qquad c_1 = \boxed{0}.744 \\
\underline{2} \\
c_2 = \boxed{1}.488 \\
\underline{2} \\
c_3 = \boxed{0}.976 \\
\underline{2} \\
c_4 = \boxed{1}.952 \\
\underline{2} \\
c_5 = \boxed{1}.904 \\
\underline{2} \\
c_6 = \boxed{1}.808
\end{array}
$$

etc.

Thus, $(0.372)_{10} = (0.010\,111\ldots)_2$. $\square$

Base Conversion 10 ↔ 8 ↔ 2

Most computers use the binary system (base 2) for their internal representation of numbers. The octal system (base 8) is particularly useful when converting from the decimal system (base 10) to the binary system, and vice versa. With base 8, the positional values of the number are $8^0 = 1, 8^1 = 8, 8^2 = 64, 8^3 = 512, 8^4 = 4096$, and so on. Thus, for example,

$$(26031)_8 = 2 \times 8^4 + 6 \times 8^3 + 0 \times 8^2 + 3 \times 8 + 1$$

$$= ((((2)8 + 6)8 + 0)8 + 3)8 + 1$$

$$= 11289$$

and

$$(7152.46)_8 = 7 \times 8^3 + 1 \times 8^2 + 5 \times 8 + 2 + 4 \times 8^{-1} + 6 \times 8^{-2}$$

$$= (((7)8 + 1)8 + 5)8 + 2 + 8^{-2}((4)8 + 6)$$

$$= 3690 + \frac{38}{64}$$

$$= 3690.59375$$

When numbers are converted between decimal and binary form by hand, it is convenient to use octal representation as an intermediate step. In the octal system, the base is 8 and, of course, the digits 8 and 9 are not used. Conversion between octal and decimal proceeds according to the principles already stated. Conversion between octal and binary is especially simple: Groups of three binary digits can be translated directly to octal according to the following table:

Binary	000	001	010	011	100	101	110	111
Octal	0	1	2	3	4	5	6	7

This grouping starts at the binary point and proceeds in both directions. Thus,

$$(101\,101\,001.110\,010\,100)_2 = (551.624)_8$$

To justify this convenient sleight of hand, consider, for instance, a fraction expressed in binary form:

$$x = (0.b_1b_2b_3b_4b_5b_6\ldots)_2$$

$$= b_1 2^{-1} + b_2 2^{-2} + b_3 2^{-3} + b_4 2^{-4} + b_5 2^{-5} + b_6 2^{-6} + \cdots$$

$$= (4b_1 + 2b_2 + b_3)8^{-1} + (4b_4 + 2b_5 + b_6)8^{-2} + \cdots$$

In the last line of this equation, the parentheses enclose numbers from the set $\{0, 1, 2, 3, 4, 5, 6, 7\}$ because the b_i's are either 0 or 1. Hence, this must be the octal representation of x.

Conversion of an octal number to binary can be done in a similar manner but in reverse order. It is easy! Just replace each octal digit with the corresponding three binary digits. Thus, for example,

$$(5362.74)_8 = (101\,011\,110\,010.111\,100)_2$$

EXAMPLE 4   What is 2576.35546875 in octal and binary forms?

Solution   We convert the original number first to octal and then to binary. For the integer part, we repeatedly divide by 8:

$$
\begin{array}{r}
8\,)2576 \\
8\,)322\ \ 0 \qquad \downarrow \\
8\,)40\ \ 2 \\
8\,)5\ \ 0 \\
0\ \ 5
\end{array}
$$

Thus,

$$2576. = (5020.)_8 = (101\,000\,010\,000.)_2$$

using the rules for grouping binary digits. For the fractional part, we repeatedly multiply by 8:

$$
\begin{array}{r}
0.35546875 \\
8 \\
\hline
\downarrow \qquad \boxed{2}.84375000 \\
8 \\
\hline
\boxed{6}.75000000 \\
8 \\
\hline
\boxed{6}.00000000
\end{array}
$$

so that

$$0.35546875 = (0.266)_8 = (0.010\,110\,110)_2$$

Again, we obtain the result:

$$2576.35546875 = (101\,000\,010\,000.010\,110\,110)_2 \qquad \square$$

Although this approach is longer for this example, we feel that it is easier, in general, and less likely to lead to error because one is working with single-digit numbers most of the time.

Base 16

Some computers whose word lengths are multiples of 4 use the hexadecimal system (base 16) in which A, B, C, D, E, and F represent 10, 11, 12, 13, 14, and 15, respectively, as given in the following table of equivalences:

Hexadecimal	0	1	2	3	4	5	6	7
Binary	0000	0001	0010	0011	0100	0101	0110	0111

Hexadecimal	8	9	A	B	C	D	E	F
Binary	1000	1001	1010	1011	1100	1101	1110	1111

Conversion between binary numbers and hexadecimal numbers is particularly easy. We need only regroup the binary digits from groups of three to groups of four. For example, we have

$$(010\,101\,110\,101\,101)_2 = (0010\,1011\,1010\,1101)_2 = (2BAD)_{16}$$

and

$$(111\,101\,011\,110\,010.110\,010\,011\,110)_2 = (1010\,1111\,0010.1100\,1001\,1110)_2$$
$$= (7AF2.C9E)_{16}$$

More Examples

Continuing with more examples, let us convert $(0.276)_8$, $(0.C8)_{16}$, and $(492)_{10}$ into different number systems. We show one way for each number and invite the reader to work out the details for other ways and to verify the answers by converting them back to the original base.

$$(0.276)_8 = 2 \times 8^{-1} + 7 \times 8^{-2} + 6 \times 8^{-3}$$
$$= 8^{-3}[((2)8 + 7)8 + 6]$$
$$= (0.37109\,375)_{10}$$

$$(0.C8)_{16} = (0.110\,010)_2$$
$$= (0.62)_8$$
$$= 6 \times 8^{-1} + 2 \times 8^{-2}$$
$$= 8^{-2}[(6)8 + 2]$$
$$= (0.78125)_{10}$$

$$(492)_{10} = (754)_8$$
$$= (111\,101\,100)_2$$
$$= (1EC)_{16}$$

because

$$
\begin{array}{r}
8\,)\overline{492} \\
8\,)\overline{61}\ \ 4 \quad \downarrow \\
8\,)\overline{7}\ \ 5 \\
0\ \ 7
\end{array}
$$

It might seem that there are four or five different procedures for converting between number systems. Actually, there are only *two* basic techniques. The first procedure for converting the number $(N)_\alpha$ to base β can be outlined as follows: (i) Express $(N)_\alpha$ in nested form using powers of α; (ii) replace each digit by the corresponding base-β numbers; and (iii) carry out the indicated arithmetic in base β. This outline holds whether N is an integer or a fraction. The second procedure is either the divide-by-β and *remainder-quotient-split* process for N an integer or the multiply-by-β and *integer-fraction-split* process for N a fraction. The first procedure is preferred when $\alpha < \beta$ and the second when $\alpha > \beta$. Of course, the $10 \leftrightarrow 8 \leftrightarrow 2 \leftrightarrow 16$ base conversion procedure should be used whenever possible because it is the easiest way to convert numbers between the decimal and hexadecimal systems.

PROBLEMS 2.1

1. Convert $(110\,111\,001.101\,011\,101)_2$ to hexadecimal, to octal, and to decimal.
2. Convert $(45653.127664)_8$ to binary and to decimal.
3. Convert the following decimal numbers to octal numbers.
   a. 23.58
   b. 75.232
   c. 57.321
4. Convert $(1\,001\,100\,101.011\,01)_2$ to hexadecimal, to octal, and to decimal.
5. Convert the following numbers.
   a. $(\quad)_{16} = (100\,101\,101)_2 = (\quad)_{10}$
   b. $(0.782)_{10} = (\quad)_2 = (\quad)_{16}$
   c. $(47)_{10} = (\quad)_2 = (\quad)_{16}$
   d. $(0.47)_{10} = (\quad)_2 = (\quad)_{16}$
   e. $(51)_{10} = (\quad)_2 = (\quad)_{16}$
   f. $(0.694)_{10} = (\quad)_2 = (\quad)_{16}$
   g. $(\quad)_{16} = (110\,011.111\,010\,110\,110\,1)_2 = (\quad)_8$
   h. $(361.4)_8 = (\quad)_2$
6. Convert 0.4 first to octal and then to binary. Check by converting directly to binary.
7. Prove that the number $\frac{1}{5}$ cannot be represented by a finite expression in the binary system.

8. Prove that a real number has a finite representation in the binary number system if and only if it is of the form $\pm m/2^n$, where n and m are positive integers.

9. Prove that any number that has a finite representation in the binary system must have a finite representation in the decimal system.

10. Explain the algorithm for converting an integer in base 10 to one in base 2, assuming that the calculations will be performed in binary arithmetic. Illustrate by converting 479 to binary.

11. Justify for integers the rule given for the conversion between octal and binary numbers.

12. Find the binary representation of $e \approx 2.718$. Check by reconverting to decimal representation.

13. What is the binary form of $\frac{7}{8}$?

14. What is the binary representation of 592?

15. Write the following numbers in octal.

    a. 27.1

    b. 12.34

    c. 3.14

16. Do you expect your computer to calculate $3 \times \frac{1}{3}$ with infinite precision? What about $2 \times \frac{1}{2}$ or $10 \times \frac{1}{10}$?

17. Justify mathematically the conversion between binary and hexadecimal numbers by regrouping.

COMPUTER PROBLEMS 2.1

1. Show that $e^{\pi\sqrt{163}}$ is incredibly close to being the integer 262 53741 26407 68744. *Hint*: More than 30 decimal digits will be needed to see any difference.

2. Write and test a routine for converting integers into octal and binary forms.

3. (Continuation) Write and test a routine for converting decimal fractions into octal and binary forms.

4. (Continuation) Using the two routines of the preceding problems, write and test a program that reads in decimal numbers and prints out the decimal, octal, and binary representations of these numbers.

5. Read into your computer $x = 1.1$ (base 10) and print it out using several different formats. Explain the results.

2.2 Floating-Point Representation

The standard way to represent a nonnegative real number in decimal form is with an integer part, a fractional part, and a decimal point between them—for example, 37.21829, 0.00227 1828, and 30 00527.11059. Another standard form, often called

normalized scientific notation, is obtained by shifting the decimal point and supplying appropriate powers of 10. Thus, the preceding numbers have alternative representations as

$$37.21829 \qquad = 0.37218\,29 \times 10^2$$
$$0.00227\,1828 \quad = 0.22718\,28 \times 10^{-2}$$
$$30\,00527.11059 = 0.30005\,27110\,59 \times 10^7$$

In normalized scientific notation, the number is represented by a fraction multiplied by 10^n, and the leading digit in the fraction is not zero (except when the number involved *is* zero). Thus, we write 79325 as 0.79325×10^5, not 0.079325×10^6 or 7.9325×10^4 or some other way.

Normalized Floating-Point Representation

In the context of computer science, normalized scientific notation is also called **normalized floating-point representation**. In the decimal system, any real number x (other than zero) can be represented in normalized floating-point form as

$$x = \pm 0.d_1 d_2 d_3 \ldots \times 10^n$$

where $d_1 \neq 0$ and n is an integer (positive, negative, or zero). The numbers $d_1, d_2, \ldots$ are decimal digits 0, 1, 2, 3, 4, 5, 6, 7, 8, and 9.

Stated another way, the real number x, if different from zero, can be represented in normalized floating-point decimal form as

$$x = \pm r \times 10^n \qquad \left(\tfrac{1}{10} \leq r < 1 \right)$$

This representation consists of three parts: a sign that is either $+$ or $-$, a number r in the interval $[\tfrac{1}{10}, 1)$, and an integer power of 10. The number r is called the **normalized mantissa** and n the **exponent**.

The floating-point representation in the binary system is similar to that in the decimal system in several ways. If $x \neq 0$, it can be written as

$$x = \pm q \times 2^m \qquad \left(\tfrac{1}{2} \leq q < 1 \right)$$

The mantissa q would be expressed as a sequence of bits (zeros or ones) in the form $q = (0.b_1 b_2 b_3 \ldots)_2$ with $b_1 \neq 0$. Hence, $b_1 = 1$ and then necessarily $q \geq \tfrac{1}{2}$.

A floating-point number system within a computer is similar to what we have just described with one important difference: Every computer has only a finite word length and a finite total capacity, so only numbers with a finite number of digits can be represented. A number is allotted only one word of storage in the single-precision mode (two or more words in double or extended precision). In either case, the degree of precision is strictly limited. Clearly, irrational numbers cannot be represented, nor can those rational numbers that do not fit the finite format imposed by the computer. Furthermore, numbers may be either too large or too small to be representable. The real numbers representable in a computer are called **machine numbers**.

Since any number used in calculations with a computer system must conform to the format of numbers in that system, it must have a **finite expansion**. Numbers that have a nonterminating expansion cannot be accommodated precisely. Moreover, a number that has a terminating expansion in one base may have a nonterminating expansion in another. A good example of this is the simple fraction $\frac{1}{10}$:

$$\frac{1}{10} = (0.1)_{10} = (0.06314\,6314\,6314\,6314\ldots)_8$$

$$= (0.0\,0011\,0011\,0011\,0011\,0011\,0011\,0011\,0011\ldots)_2$$

The important point here is that in a computer, most numbers cannot be represented exactly.

The effective number system for a computer is *not* a continuum but a rather peculiar discrete set. To illustrate, let us take an extreme example, in which the floating-point numbers must be of the form $x = \pm(0.b_1b_2b_3)_2 \times 2^{\pm m}$, where b_1, b_2, b_3, and m are allowed to have only the value 0 or 1.

EXAMPLE 1   List all the floating-point numbers that can be expressed in the form

$$x = \pm(0.b_1b_2b_3)_2 \times 2^{\pm m} \qquad (m, b_i \in \{0, 1\})$$

Solution   There are two choices for the $\pm$, two choices for b_1, two choices for b_2, two choices for b_3, and three choices for the exponent. Thus, at first one would expect $2 \times 2 \times 2 \times 2 \times 3 = 48$ different numbers. However, there is some duplication. For example, the nonnegative numbers in this system are as follows:

$0.000 \times 2^0 = 0$	$0.000 \times 2^1 = 0$	$0.000 \times 2^{-1} = 0$
$0.001 \times 2^0 = \frac{1}{8}$	$0.001 \times 2^1 = \frac{1}{4}$	$0.001 \times 2^{-1} = \frac{1}{16}$
$0.010 \times 2^0 = \frac{2}{8}$	$0.010 \times 2^1 = \frac{2}{4}$	$0.010 \times 2^{-1} = \frac{2}{16}$
$0.011 \times 2^0 = \frac{3}{8}$	$0.011 \times 2^1 = \frac{3}{4}$	$0.011 \times 2^{-1} = \frac{3}{16}$
$0.100 \times 2^0 = \frac{4}{8}$	$0.100 \times 2^1 = \frac{4}{4}$	$0.100 \times 2^{-1} = \frac{4}{16}$
$0.101 \times 2^0 = \frac{5}{8}$	$0.101 \times 2^1 = \frac{5}{4}$	$0.101 \times 2^{-1} = \frac{5}{16}$
$0.110 \times 2^0 = \frac{6}{8}$	$0.110 \times 2^1 = \frac{6}{4}$	$0.110 \times 2^{-1} = \frac{6}{16}$
$0.111 \times 2^0 = \frac{7}{8}$	$0.111 \times 2^1 = \frac{7}{4}$	$0.111 \times 2^{-1} = \frac{7}{16}$

Altogether there are 31 numbers in the system. The positive numbers obtained are shown on a line in Figure 2.1. Observe that the numbers are symmetrically but unevenly distributed about zero.   □

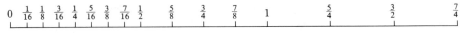

FIGURE 2.1

If, in the course of a computation, a number x is produced of the form $\pm q \times 2^m$ with m outside the computer's permissible range, then we say that an **overflow** or an **underflow** has occurred or that x *is outside the range of the computer*. Generally, an overflow results in a fatal error (or exception), and the normal execution of the program stops. An underflow, however, is usually treated automatically by setting x to zero without any interruption of the program but with a warning message in some computers.

In a computer where floating-point numbers are restricted to the form in the example above, any number closer to zero than $1/16$ would *underflow* to zero, and any number outside the range -1.75 to $+1.75$ would *overflow* to machine infinity.

If, in the preceding example, we allow only *normalized* floating-point numbers, then all our numbers (with the exception of zero) have the form

$$x = \pm(0.1b_2b_3)_2 \times 2^{\pm m}$$

This creates a phenomenon known as *the hole at zero*. Our nonnegative machine numbers are now distributed as in Figure 2.2. There is a relatively wide gap between zero and the smallest positive machine number, which is $(0.100)_2 \times 2^{-1} = \frac{1}{4}$.

Hypothetical `Marc-32` Computer

A computer that operates in single-precision floating-point mode represents numbers as described earlier except for the limitations imposed by the finite word length. Many binary computers have a word length of 32 bits (binary digits). We shall describe a hypothetical machine of this type whose features mimic many workstations and personal computers in widespread use. Because of the 32-bit word length, as much as possible of the normalized floating-point number $\pm q \times 2^m$ must be contained in those 32 bits. One way of allocating the 32 bits is as follows:

sign of q	1 bit
integer $\lvert m \rvert$	8 bits
number q	23 bits

Information on the sign of m is contained in the eight bits allocated for the integer $\lvert m \rvert$. In such a scheme, we can represent real numbers with $\lvert m \rvert$ as large as $2^7 - 1 = 127$. The exponent represents numbers from -127 through 128. Henceforth, we call this fictional computer the `Marc-32`. The `Marc-32` representation given here

FIGURE 2.2

corresponds *closely* to the IEEE floating-point single-precision standard that is used on almost all new computers.

In the normalized representation of a floating-point number, the first bit in the mantissa is *always* 1 so that this bit does not have to be stored. This can be accomplished by shifting the binary point to a "1-plus" form $(1.f)_2$. The mantissa is the rightmost 23 bits and contains f with an understood binary point as in Figure 2.3. So the mantissa *actually* corresponds to 24 binary digits. (Important exceptions are the numbers ± 0.)

In the `Marc-32`, the mantissa is 23 bits, and because $2^{-23} \approx 1.2 \times 10^{-7}$, we infer that in a simple computation approximately six significant decimal digits of accuracy should be obtained in single precision and approximately 15 (30) significant decimal digits in double (quadruple) precision. For integers, the range is from $-(2^{31} - 1)$ to $(2^{31} - 1) = 2147483647$.

We now describe how the hypothetical `Marc-32` represents a machine number of the following form in **standard IEEE floating-point single-precision**

$$(-1)^s \times 2^{e-127} \times (1.f)_2$$

The leftmost bit is used for the sign of the mantissa s with 0 corresponding to $+$ and 1 to $-$. The next eight bits are used to represent the exponent e corresponding to 2^{e-127}, which is interpreted as an *excess-127 code*. Finally, the last 23 bits represent f from the fractional part of the mantissa in the 1-plus form: $(1.f)_2$. Each floating-point single-precision word of the `Marc-32` is partitioned as in Figure 2.3.

We now outline the procedure for determining the representation of a real number x. If x is zero, it is represented by a full word of zero bits with the possible exception of the sign bit. (See Problem **20b**.) For a nonzero x, first assign the sign bit for x and consider $|x|$. Then convert both the integer and fractional parts of $|x|$ from decimal to octal to binary. One-plus normalize $(|x|)_2$ by shifting the binary point so that the first bit to the left of the binary point is a 1 and all bits to the left of this 1 are 0. To compensate for this shift of the binary point, adjust the exponent of 2; that is, multiply by the appropriate power of 2. The 24-bit 1-plus normalized mantissa in binary is thus found. Now the current exponent of 2 should be set equal to $e - 127$ in order to determine e, which is then converted from decimal to octal to binary. The sign bit of the mantissa is combined with $(e)_2$ and $(f)_2$. Finally, write the 32-bit representation of x as eight hexadecimal digits.

The value of e in the `Marc-32` representation of a nonzero floating-point number in single precision is restricted by the inequality

$$0 < e < (11\,111\,111)_2 = 255$$

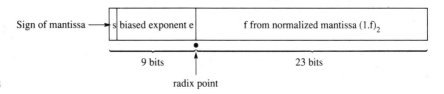

FIGURE 2.3

The values 0 and 255 are reserved for special cases ± 0 and $\pm \infty$, respectively. Hence, the actual exponent of the number is restricted by the inequality

$$-126 \leq e - 127 \leq 127$$

Likewise, we find that the mantissa of each nonzero number is restricted by the inequality

$$1 \leq (1.f)_2 \leq (1.111\ldots 1)_2 = 2 - 2^{-23}$$

The largest number representable in the `Marc-32` is therefore $(2 - 2^{-23})2^{127} \approx 3.4 \times 10^{38}$. The smallest positive number is $2^{-136} \approx 1.2 \times 10^{-38}$.

EXAMPLE 2   Determine the `Marc-32` representation of the decimal number -52.234375.

Solution   Converting the integer part to binary, we have $(52.)_{10} = (64.)_8 = (110\,100.)_2$. Next converting the fractional part, we have $(.234375)_{10} = (.17)_8 = (.001\,111)_2$. Now

$$(52.234375)_{10} = (110\,100.001\,111)_2 = (1.101\,000\,011\,110)_2 \times 2^5$$

is the corresponding 1-plus form in base 2, and $(.101\,000\,011\,110)_2$ is the stored mantissa. Next the exponent is $(5)_{10}$ and, since $e - 127 = 5$, we immediately see that $(132)_{10} = (204)_8 = (10\,000\,100)_2$ is the stored exponent. Thus, the `Marc-32` representation of -52.234375 is

$$[1\,10\,000\,100\,101\,000\,011\,110\,000\,000\,000\,00]_2 =$$

$$[1100\,0010\,0101\,0000\,1111\,0000\,0000\,0000]_2 = [\text{C250F000}]_{16}$$

Here $[\cdots]_k$ is the bit pattern of the `Marc-32` word that represents this floating-point number, which is displayed in in base-k. $\square$

To summarize: If a word in the `Marc-32` contains the bits $(b_1 b_2 \ldots b_{32})$, this word can be interpreted as the real number

$$(-1)^{b_1} \times 2^{(b_2 b_3 \ldots b_9)_2} \times 2^{-127} \times (1.b_{10} b_{11} \ldots b_{32})_2$$

EXAMPLE 3   Determine the decimal numbers that correspond to these `Marc-32` words:

$$[\text{45DE4000}]_{16} \qquad [\text{BA390000}]_{16}$$

Solution   The first number in binary is

$$[0100\,0101\,1101\,1110\,0100\,0000\,0000\,0000]_2$$

The stored exponent is $(10\,001\,011)_2 = (213)_8 = (139)_{10}$, so $139 - 127 = 12$. The mantissa is positive and represents the number

$$(1.101\,111\,001)_2 \times 2^{12} = (1\,101\,111\,001\,000.)_2$$

$$= (15710.)_8$$

$$= 0 \times 1 + 1 \times 8 + 7 \times 8^2 + 5 \times 8^3 + 1 \times 8^4$$

$$= 8(1 + 8(7 + 8(5 + 8(1))))$$

$$= 7112$$

The second word in binary is

$$[1011\,1010\,001\,1001\,0000\,0000\,0000\,0000]_2$$

The exponential part of the word is $(01\,110\,100)_2 = (164)_8 = 116$, so the exponent is $116 - 127 = -11$. The mantissa is negative and corresponds to the following floating-point number:

$$-(1.011\,100\,100)_2 \times 2^{-11} = -(0.000\,000\,000\,010\,111\,001)_2$$

$$= -(0.00027\,1)_8$$

$$= -2 \times 8^{-4} - 7 \times 8^{-5} - 1 \times 8^{-6}$$

$$= -8^{-6}(1 + 8(7 + 8(2)))$$

$$= -\frac{185}{262144} \approx -7.05718\,99 \times 10^{-4} \qquad \square$$

In an actual computer that does not follow the IEEE floating-point standard, the precise manner of allocating the bits in a word and other details can be extremely complicated. For example, the computer might represent negative numbers using a *biased exponent, 1's complement, 2's complement,* or any number of other tricks of the trade that we do not wish to go into here. *Fortunately, one need not know all these details in order to use the computer intelligently.* Nevertheless, in debugging, it may be helpful to understand the representation of numbers in your computer.

Computer Errors in Representing Numbers

We turn now to the errors that can occur when we attempt to represent a given real number x in the computer. We continue to use a model, the `Marc-32` machine, with its 32-bit word length. Suppose first that we let $x = 2^{53\,21697}$ or $x = 2^{-32591}$. The exponents of these numbers far exceed the limitations of the `Marc-32` (as described above). These numbers would overflow and underflow, respectively, and the relative error in replacing x by the closest machine number will be very large. Such numbers are *outside the range* of the `Marc-32`.

Consider next a positive number x in normalized floating-point form:

$$x = q \times 2^m \qquad \left(\tfrac{1}{2} \leqq q < 1, \; -126 \leqq m \leqq 127\right)$$

The process of replacing x by its nearest machine number is called **rounding**, and the error involved is called **roundoff error**. We want to know how large it can be. We suppose that q is expressed in normalized binary notation, so that

$$x = (1.a_1a_2a_3a_4 \ldots a_{23}a_{24}a_{25} \ldots)_2 \times 2^m$$

One nearby machine number can be obtained by *rounding down* or by simply dropping the excess bits $a_{25}a_{26} \ldots$, since only 24 bits have been allocated to q. This machine number is

$$x' = (1.a_1a_2a_3a_4 \ldots a_{23})_2 \times 2^m$$

It lies to the left of x on the real-number axis. Another machine number x'' is just to the right of x on the real axis and is obtained by *rounding up*. It is found by adding one unit to a_{24} in the expression for x'. Thus,

$$x'' = \left[(1.a_1a_2a_3a_4 \ldots a_{23})_2 + 2^{-23} \right] \times 2^m$$

The closer of these machine numbers is the one chosen to represent x.

There are two situations, conveniently illustrated by the simple diagrams in Figure 2.4. If x lies closer to x' than to x'', then

$$|x - x'| \leq \frac{1}{2}|x'' - x'| = 2^{-24+m}$$

In this case, the relative error is bounded as follows:

$$\left| \frac{x - x'}{x} \right| \leq \frac{2^{-24+m}}{(1.a_1a_2a_3 \ldots)_2 \times 2^m} \leq \frac{2^{-24}}{1} = 2^{-24}$$

On the other hand, if x lies closer to x'' than to x', then

$$|x - x''| \leq \frac{1}{2}|x'' - x'|$$

and the same analysis shows that the relative error is no greater than 2^{-24}.

For the Marc-32, seven significant decimal digits should be preserved in most operations because the relative error in approximating a real number (within the range of the machine) by its nearest machine number is not greater than 2^{-24}, which is approximately 0.596×10^{-7}. The number 2^{-24} is the **unit roundoff error** for this machine. Actually, 2^{-23} is *machine epsilon* for a 32-bit binary computer with IEEE standard arithmetic. (See Computer Problem **9**, Section 1.1.)

FIGURE 2.4

We note in passing that when *all* excess digits or bits are discarded, this process is called **chopping**. If the hypothetical `Marc-32` were designed to chop numbers, the bound on the relative roundoff error would be twice as large as above, or 2^{-23}.

We say that a number x is *chopped* to n digits or figures when all digits that follow the nth digit are discarded and none of the remaining n digits is changed. Conversely, x is *rounded* to n digits or figures when x is replaced by an n-digit number that approximates x with minimum error. The question of whether to round up or down an $(n + 1)$-digit decimal number that ends with a 5 is best handled by always selecting the rounded n-digit number with an *even* nth digit.

Suppose that α and β are two numbers, of which one is regarded as an approximation to the other. The **absolute error** of β as an approximation to α is $|\alpha - \beta|$. The **relative error** of β as an approximation to α is $|\alpha - \beta|/|\alpha|$. Notice that in computing the absolute error, the roles of α and β are the same, whereas in computing the relative error, it is essential to distinguish one of the two numbers as *correct*. (Observe that the relative error is undefined in the case $\alpha = 0$.) For practical reasons, the relative error is usually more meaningful than the absolute error. For example, if $\alpha_1 = 1.333$, $\beta_1 = 1.334$, and $\alpha_2 = 0.001$, $\beta_2 = 0.002$, then the absolute error of β_i as an approximation to α_i is the same in both cases—namely, 10^{-3}. However, the relative errors are $\frac{3}{4} \times 10^{-3}$ and 1, respectively. The relative error clearly indicates that β_1 is a good approximation to α_1 but that β_2 is a poor approximation to α_2.

Notation fl(*x*) and Inverse Error Analysis

Next we turn to the errors that are produced in the course of elementary arithmetic operations. To illustrate the principles, suppose that we are working with a five-place decimal machine and wish to add numbers. Two typical machine numbers in normalized floating-point form are

$$x = 0.37218 \times 10^4 \qquad y = 0.71422 \times 10^{-1}$$

Many computers perform arithmetic operations in a double-length work area, so let us assume that our minicomputer will have a ten-place accumulator. The exponent of the smaller number is adjusted so that both exponents are the same; then the numbers are added in the accumulator and the rounded result is placed in a computer word:

$$
\begin{aligned}
x &= 0.37218\,00000 \times 10^4 \\
y &= 0.00000\,71422 \times 10^4 \\
\hline
x + y &= 0.37218\,71422 \times 10^4
\end{aligned}
$$

The nearest machine number is $z = 0.37219 \times 10^4$, and the relative error involved in the machine addition is

$$\frac{|x + y - z|}{|x + y|} = \frac{0.00000\,28578 \times 10^4}{0.37218\,71422 \times 10^4} \approx 0.77 \times 10^{-5}$$

This relative error would be regarded as acceptable on a machine of such low precision.

To facilitate the analysis of such errors, it is convenient to introduce the notation $\mathrm{fl}(x)$ to denote the floating-point machine number that corresponds to the real number x. Of course, the function fl depends on the particular computer involved. The hypothetical five-decimal-digit machine used above would give

$$\mathrm{fl}(0.37218\,71422 \times 10^4) = 0.37219 \times 10^4$$

For the hypothetical $\mathtt{Marc\text{-}32}$ computer, we established previously that if x is any real number within the range of the computer, then

$$\frac{|x - \mathrm{fl}(x)|}{|x|} \leq 2^{-24} \tag{1}$$

This inequality can also be expressed in the more useful form

$$\mathrm{fl}(x) = x(1 + \delta) \qquad \left(|\delta| \leq 2^{-24}\right)$$

To see that these two inequalities are equivalent, simply let $\delta = [\mathrm{fl}(x) - x]/x$. Then by Inequality (1) we have $|\delta| \leq 2^{-24}$; solving for $\mathrm{fl}(x)$ yields $\mathrm{fl}(x) = x\delta + x$.

Now let the symbol $\odot$ denote any one of the arithmetic operations $+$, $-$, $\times$, or $\div$. Suppose the $\mathtt{Marc\text{-}32}$ has been designed so that whenever two *machine* numbers x and y are to be combined arithmetically, the computer will produce $\mathrm{fl}(x \odot y)$ instead of $x \odot y$. We can imagine that $x \odot y$ is first *correctly* formed, normalized, and then rounded to become a machine number. Under this assumption, the relative error (in the $\mathtt{Marc\text{-}32}$) will not exceed 2^{-24} by the previous analysis:

$$\mathrm{fl}(x \odot y) = (x \odot y)(1 + \delta) \qquad \left(|\delta| \leq 2^{-24}\right)$$

Special cases of this are, of course,

$$\mathrm{fl}(x \pm y) = (x \pm y)(1 + \delta)$$

$$\mathrm{fl}(xy) = xy(1 + \delta)$$

$$\mathrm{fl}\left(\frac{x}{y}\right) = \left(\frac{x}{y}\right)(1 + \delta)$$

The assumption we have made about the $\mathtt{Marc\text{-}32}$ is not quite true for a real computer. For example, it is possible for x and y to be machine numbers and for $x \odot y$ to overflow or underflow. Except for this case, the assumption should be realistic for most computing machines.

The equations given above can be written in a variety of ways, some of which suggest alternative interpretations of roundoff. For example,

$$\text{fl}(x + y) = x(1 + \delta) + y(1 + \delta)$$

This says that the result of adding machine numbers x and y is not in general $x + y$ but is the true sum of $x(1 + \delta)$ and $y(1 + \delta)$. We can think of $x(1 + \delta)$ as the result of slightly perturbing x. Thus, the machine version of $x + y$—that is, $\text{fl}(x + y)$—is the *exact* sum of a slightly perturbed x and a slightly perturbed y. The reader can supply similar interpretations in these examples:

$$\text{fl}(x) = x(1 - \delta)$$

$$\text{fl}(xy) = [x(1 + \delta)]y$$

$$\text{fl}\left(\frac{x}{y}\right) = \frac{x(1 + \delta)}{y}$$

$$\text{fl}(xy) = \left(x\sqrt{1 + \delta}\right)\left(y\sqrt{1 + \delta}\right)$$

$$\text{fl}\left(\frac{x}{y}\right) = \frac{x\sqrt{1 + \delta}}{y / \sqrt{1 + \delta}}$$

$$\text{fl}\left(\frac{x}{y}\right) \approx \frac{x}{y(1 - \delta)}$$

This interpretation is an example of **inverse-error analysis**. It attempts to determine what perturbation of the original data would cause the *computer* results to be the exact results for a perturbed problem. In contrast, a **direct-error analysis** attempts to determine how computed answers differ from exact answers based on the same data. In this aspect of numerical analysis, computers have stimulated a new way of looking at computational errors.

EXAMPLE 4   If x, y, and z are machine numbers, what upper bound can be given for the relative roundoff error in computing $z(x + y)$? Use the `Marc-32` as a model.

Solution   In the computer, the calculation of $x + y$ will be done first. This arithmetic operation produces the machine number $\text{fl}(x + y)$, which differs from $x + y$ because of roundoff. By the principles developed above, there is a δ_1 such that

$$\text{fl}(x + y) = (x + y)(1 + \delta_1) \qquad \left(|\delta_1| \leq 2^{-24}\right)$$

Now z is already a machine number. When it multiplies the machine number $\text{fl}(x + y)$, the result is the machine number $\text{fl}[z\,\text{fl}(x + y)]$. This, too, differs from its exact counterpart, and we have, for some δ_2,

$$\text{fl}[z\,\text{fl}(x + y)] = z\,\text{fl}(x + y)(1 + \delta_2) \qquad \left(|\delta_2| \leq 2^{-24}\right)$$

Putting both of our equations together, we have

$$\text{fl}[z\,\text{fl}(x+y)] = z(x+y)(1+\delta_1)(1+\delta_2)$$
$$= z(x+y)(1+\delta_1+\delta_2+\delta_1\delta_2)$$
$$\approx z(x+y)(1+\delta_1+\delta_2)$$
$$= z(x+y)(1+\delta) \qquad \left(|\delta| \leq 2^{-23}\right)$$

In this calculation, $|\delta_1\delta_2| \leq 2^{-48}$, and so we ignore it. Also, we put $\delta = \delta_1 + \delta_2$ and then reason that $|\delta| = |\delta_1 + \delta_2| \leq |\delta_1| + |\delta_2| \leq 2^{-24} + 2^{-24} = 2^{-23}$.   □

PROBLEMS 2.2

1. Generally, when a list of floating-point numbers is added, less roundoff error will occur if the numbers are added in order of increasing magnitude. Give some examples to illustrate this principle.

2. (Continuation) The principle of Problem **1** is not *universally* valid. Consider a decimal machine with two decimal digits allocated to the mantissa. Show that the four numbers 0.25, 0.0034, 0.00051, and 0.061 can be added with less roundoff error if *not* added in ascending order.

3. In the case of machine underflow, what is the relative error involved in replacing a number x by zero?

4. Consider a computer that operates in base β and carries n digits in the mantissa of its floating-point numbers. Show that the rounding of a real number x to the nearest machine number x' involves a relative error of at most $\frac{1}{2}\beta^{1-n}$. *Hint:* Imitate the argument in the text.

5. Consider a decimal machine in which five decimal digits are allocated to the mantissa. Give an example, avoiding overflow or underflow, of a real number x whose closest machine number x' involves the greatest possible relative error.

6. In a five-decimal machine that correctly rounds numbers to the nearest machine number, what real numbers x will have the property: $\text{fl}(1.0 + x) = 1.0$?

7. Consider a computer operating in base β. Suppose that it chops numbers instead of correctly rounding them. If its floating-point numbers have a mantissa of n digits, how large is the relative error in storing a real number in machine format?

8. Which of these are machine numbers on the `Marc-32`?
   a. 10^{403}
   b. $1 + 2^{-32}$
   c. $1/5$
   d. $1/10$
   e. $1/256$

9. In the hypothetical `Marc-32` computer, what is the roundoff error when we represent $2^{-1} + 2^{-25}$ by a machine number? *Note:* This refers to absolute error, not relative error.

10. (Continuation) In the hypothetical `Marc-32` computer, what is the relative roundoff error when we round off $2^{-1} + 2^{-26}$ to get the closest machine number?

11. In the `Marc-32`, the computer word associated with the variable Δ appears as $[7F7FFFFF]_{16}$, which is the largest representable floating-point single-precision number. What is the decimal value of Δ? The variable ε appears as $(00800000)_{16}$, which is the smallest positive number. What is the decimal value of ε?

12. If x is a real number within the range of the `Marc-32` that is rounded and stored, what can happen when x^2 is computed? Explain the difference between $fl[fl(x) fl(x)]$ and $fl(x\,x)$.

13. Determine the representation in the `Marc-32` for the decimal number 2^{-30}.

14. Determine the `Marc-32` representation for 64.015625.

15. In high school, you may have learned that 22/7 was a good approximation to π. Show that 355/113 is a better approximation in terms of both absolute and relative errors. Can you discover a still better fraction that approximates π?

16. Determine the decimal numbers that have the following `Marc-32` representations.

    a. $[3F27E520]_{16}$

    b. $[3BCDCA00]_{16}$

    c. $[BF4F9680]_{16}$

    d. $[CB187ABC]_{16}$

17. What is the representation in the `Marc-32` for -8×2^{-24}?

18. Determine the representation in the `Marc-32` of the following decimal numbers.

    a. $0.5, -0.5$

    b. $0.125, -0.125$

    c. $0.0625, -0.0625$

    d. $0.03125, -0.03125$

19. Determine the decimal numbers that have the following `Marc-32` representations.

    a. $[CA3F2900]_{16}$

    b. $[C705A700]_{16}$

    c. $[494F96A0]_{16}$

    d. $[4B187ABC]_{16}$

    e. $[45223000]_{16}$

    f. $[45607000]_{16}$

    g. $[C553E000]_{16}$

    h. $[437F0000]_{16}$

20. Determine the representation in the `Marc-32` of the following decimal numbers.

    a. $1.0, -1.0$

    b. $+0.0, -0.0$

    c. -9876.54321

    d. 0.234375

    e. 492.78125

    f. 64.37109375

    g. -285.75

    h. 10^{-2}

21. Are these `Marc-32` representations? Why or why not?

    a. $[4BAB2BEB]_{16}$

    b. $[1A1AIA1A]_{16}$

    c. $[FADEDEAD]_{16}$

    d. $[CABE6G94]_{16}$

22. A binary machine that carries 30 bits in the fractional part of each floating-point number is designed to round a number up or down correctly to get the closest floating-point number. What simple upper bound can be given for the relative error in this rounding process?

23. A decimal machine that carries 15 decimal places in its floating-point numbers is designed to chop numbers. If x is a real number in the range of this machine and x' is its machine representation, what upper bound can be given for $|x - x'|/|x|$?

24. If x and y are real numbers within the range of the `Marc-32`, and if xy is also within the range, what relative error can there be in the machine computation of xy? *Hint*: The machine produces $fl[fl(x)\,fl(y)]$.

25. Let x and y be real numbers that are not machine numbers but are within the exponent range of the `Marc-32`. What is the largest possible relative error in the machine representation of $x + y^2$? Include errors made to get the numbers in the machine as well as errors in the arithmetic.

26. Show that if x and y are positive real numbers that have the same first n digits in their decimal representations, then y approximates x with relative error less than 10^{1-n}. Is the converse true?

27. Show that a rough bound on the relative roundoff error when n machine numbers are multiplied in the `Marc-32` is $(n - 1)2^{-24}$.

28. Show that $fl(x + y) = y$ on the `Marc-32` if x and y are positive machine numbers and $x < y \times 2^{-25}$.

29. If 1000 nonzero machine numbers are added in the `Marc-32`, what upper bound can be given for the relative roundoff error in the result? How many decimal digits in the answer can be trusted?

30. Suppose that $x = \sum_{i=1}^{n} a_i 2^{-i}$, where $a_i \in \{-1, 0, 1\}$, is a positive number. Show that x can also be written in the form $\sum_{i=1}^{n} b_i 2^{-i}$, where $b_i \in \{0, 1\}$.

31. If x and y are machine numbers in the `Marc-32` and if $\text{fl}(x/y) = x/[y(1 + \delta)]$, what upper bound can be placed on $|\delta|$?

32. How big is the hole at zero in the `Marc-32`?

33. How many machine numbers are there in the `Marc-32`? (Consider only normalized floating-point numbers.)

34. How many normalized floating-point numbers are available in a binary machine if n bits are allocated to the mantissa and m bits are allocated to the exponent? Assume that two additional bits are used for signs, as in the `Marc-32`.

35. Show by an example that in computer arithmetic $a + (b + c)$ may differ from $(a + b) + c$.

36. Consider a decimal machine in which floating-point numbers have 13 decimal places. Suppose that numbers are correctly rounded up or down to the nearest machine number. Give the best bound for the roundoff error, assuming no underflow or overflow. Use relative error, of course. What if the numbers are always chopped?

37. Consider a computer that uses five-decimal-digit numbers. Let $\text{fl}(x)$ denote the floating-point machine number closest to x. Show that if $x = 0.53214\,87513$ and $y = 0.53213\,04421$, then the operation $\text{fl}(x) - \text{fl}(y)$ involves a large relative error. Compute it.

38. Two numbers x and y that are not machine numbers are read into the `Marc-32`. The machine computes xy^2. What sort of relative error can be expected? Assume no underflow or overflow.

39. Let x, y, and z be three machine numbers in the `Marc-32` computer. By analyzing the relative error in the worst case, determine how much roundoff error should be expected in forming $(xy)z$.

40. Let x and y be machine numbers in the `Marc-32`. What relative roundoff error should be expected in the computation of $x + y$? If x is around 30 and y is around 250, what absolute error should be expected in the computation of $x + y$?

41. A real number x is represented approximately by 0.6032, and we are told that the relative error is 0.1%. What is x? *Note*: There are two answers.

42. What is the relative error involved in rounding 4.9997 to 5.000?

43. Enumerate the set of numbers in the floating-point number system that have binary representations of the form $\pm(0.b_1b_2) \times 2^m$, where:

   a. $m \in \{-1, 0\}$

   b. $m \in \{-1, 1\}$

   c. $m \in \{-1, 0, 1\}$

44. What are the machine numbers immediately to the right and left of 2^m in the `Marc-32`? How far is each from 2^m?

45. Every machine number in the `Marc-32` can be interpreted as the correct machine representation of an entire *interval* of real numbers. Describe this interval for the machine number $q \times 2^m$.

46. Is every machine number on the `Marc-32` the average of two other machine numbers? If not, describe those that are not averages.

COMPUTER PROBLEMS 2.2

1. Use your computer to construct a table of three functions f, g, and h defined as follows. For each integer n in the range 1 to 50, let $f(n) = 1/n$. Then $g(n)$ is computed by adding $f(n)$ to itself $n - 1$ times. Finally, set $h(n) = nf(n)$. We want to see the effects of roundoff error in these computations. Use the function *real(n)* to convert an integer variable n to its real (floating-point) form. Print the table with all the precision of which your computer is capable (in single-precision mode).

2. Print several numbers, both integers and reals, in octal format and try to explain the machine representation used in your computer.

3. The harmonic series $1 + \frac{1}{2} + \frac{1}{3} + \frac{1}{4} + \cdots$ is known to diverge to $+\infty$. The nth partial sum approaches $+\infty$ at the same rate as $\ln(n)$. Euler's constant is defined to be

$$\gamma = \lim_{n \to \infty} \left[\sum_{k=1}^{n} \frac{1}{k} - \ln(n) \right] \approx 0.57721$$

If your computer ran a program for a week based on the pseudocode

real s, x
$x \leftarrow 1.0$
$s \leftarrow 1.0$
repeat
    $x \leftarrow x + 1.0$
    $s \leftarrow s + 1.0/x$
end repeat

what is the largest value of s it would obtain? Write and test a program that uses a loop of 5000 steps to estimate Euler's constant. Print intermediate answers at every 100 steps.

4. Let A denote the set of positive integers whose decimal representation does not contain the digit 0. The sum of the reciprocals of the elements in A is known to be 23.10345. Can you verify this numerically?

5. Write

integer function *ndigit(n, x)*

which returns the nth nonzero digit in the decimal expression for the real number x.

6. Predict and then show what value your computer will print for $\sqrt{2}$ computed in single precision. Repeat for double or extended precision. Explain.

7. Sometimes it is necessary to furnish the unit roundoff error for a routine. This is machine dependent and is defined as the smallest positive number u such that $1 + u > 1$ in the machine. It is usually sufficient to know u within a factor of 2. Write a program to determine u within a factor of 2.

8. Refer to Computer Problem 3. Prove that Euler's constant, γ, can also be represented by

$$\gamma = \lim_{m \to \infty} \left[\sum_{k=1}^{m} \frac{1}{k} - \ln\left(m + \frac{1}{2}\right) \right]$$

Write and test a program that uses $m = 1, 2, 3, \ldots, 5000$ to compute γ by this formula. The convergence should be more rapid than in Computer Problem 3. (See the article by De Temple [1993].)

2.3 Loss of Significance

In this section, we show how loss of significance in subtraction can be reduced or eliminated by various techniques, such as the use of rationalization, Taylor series, trigonometric identities, logarithmic properties, double precision, and/or range reduction. These are some of the techniques that can be used when one wants to guard against the degradation of precision in a calculation.

Significant Digits

We first address the elusive concept of **significant digits** in a number. Suppose that x is a real number expressed in normalized scientific notation in the decimal system:

$$x = \pm q \times 10^n \qquad \left(\tfrac{1}{10} \leqq q < 1\right)$$

For example, x might be

$$x = 0.3721498 \times 10^{-5}$$

The digits $3, 7, 2, 1, 4, 9, 8$ used to express q do not all have the same significance because they represent different powers of 10. Thus, we say that 3 is the *most* significant digit, and the significance of the digits diminishes from left to right. In the example, 8 is the *least* significant digit.

If x is a mathematically exact real number, then its approximate decimal form can be given with as many significant digits as we wish. Thus, we may write

$$\frac{\pi}{10} \approx 0.31415\,92653\,58979$$

and all the digits given are correct. If, however, x is a *measured* quantity, the situation is quite different. Every measured quantity involves an error whose magnitude depends on the nature of the measuring device. Thus, if a meter stick is used, it is not reasonable to measure any length with precision better than 1 millimeter or perhaps 0.1 millimeter. Therefore, the result of measuring, say, a window with a meter stick should not be reported as 0.73594 meter. That would be misleading. Only digits that are believed to be correct or in error by at most a few units should be reported. It is a scientific convention that the least significant digit given in a measured quantity should be in error by at most five units.

Similar remarks pertain to quantities computed from measured quantities. For example, if the side of a square is reported to be $s = 0.736$ meter, then one can assume that the error does not exceed a few units in the third decimal place. The diagonal of that square is then

$$s\sqrt{2} \approx 0.10408\,61182 \times 10^1$$

but should be reported as 0.1041×10^1 or (more conservatively) 0.104×10^1. The infinite precision available in $\sqrt{2}$,

$$\sqrt{2} = 1.41421\,35623\,73095\ldots$$

does *not* convey any more precision to $s\sqrt{2}$ than was already present in s.

Computer-Caused Loss of Significance

Perhaps it is surprising that a loss of significance can occur within the computer. It is essential to understand this process so that blind trust will not be placed in numerical output from a computer. One of the principal causes for a deterioration in precision is the subtraction of one quantity from another nearly equal quantity. This effect is potentially quite serious and can be catastrophic. The more nearly equal the two numbers whose difference is being computed, the more pronounced is the effect.

To illustrate this phenomenon, let us consider the assignment statement

$$y \leftarrow x - \sin(x)$$

and suppose that at some point in a computer program this statement is executed with an x value of $\frac{1}{15}$. Assume further that our computer works with floating-point numbers that have ten decimal digits. Then

$$x \leftarrow 0.66666\,66667 \times 10^{-1}$$

$$\sin(x) \leftarrow 0.66617\,29492 \times 10^{-1}$$

$$x - \sin(x) \leftarrow 0.00049\,37175 \times 10^{-1}$$

$$x - \sin(x) \leftarrow 0.49371\,75000 \times 10^{-4}$$

In the last step, the result has been shifted to normalized floating-point form. Three zeros have then been supplied by the computer in the three *least* significant decimal places. We refer to these as **spurious** zeros; they are *not* significant digits. In fact, the ten-decimal-digit correct value of $\frac{1}{15} - \sin\frac{1}{15}$ is $0.49371\,74327 \times 10^{-4}$. Another way of interpreting this is to note that the final digit in $x - \sin(x)$ is derived from the tenth digits in x and $\sin(x)$. When the eleventh digit in either x or $\sin(x)$ is 5, 6, 7, 8, or 9, the numerical values are rounded up to ten digits so that their tenth digits may be altered by plus one unit. Since these tenth digits may be in error, the final digit in $x - \sin(x)$ may also be in error—which it is.

EXAMPLE 1   If $x = 0.37214\,48693$ and $y = 0.37202\,14371$, what is the relative error in the computation of $x - y$ in a minicomputer that has five decimal digits of accuracy?

Solution   The numbers would first be rounded to $\widetilde{x} = 0.37214$ and $\widetilde{y} = 0.37202$. Then we have $\widetilde{x} - \widetilde{y} = 0.00012$, while the correct answer is $x - y = 0.00012\,34322$. The relative error involved is

$$\frac{|(x - y) - (\widetilde{x} - \widetilde{y})|}{|x - y|} = \frac{0.00000\,34322}{0.00012\,34322} \approx 3 \times 10^{-2}$$

This magnitude of relative error must be judged quite large when compared with the relative error of $\widetilde{x}$ and $\widetilde{y}$. (They cannot exceed $\frac{1}{2} \times 10^{-4}$ by the coarsest estimates, and in this example they are, in fact, approximately 1.3×10^{-5}.)   □

It should be emphasized that this discussion does not pertain to the operation

$$x - y \rightarrow \text{fl}(x - y)$$

but rather to the operation

$$x - y \rightarrow \text{fl}[\text{fl}(x) - \text{fl}(y)]$$

Roundoff error in the former case is governed by the equation

$$\text{fl}(x - y) = (x - y)(1 + \delta)$$

where $|\delta| \leq 2^{-24}$ on the hypothetical `Marc-32` computer, and on a five-decimal-digit computer in the example above $|\delta| \leq \frac{1}{2} \times 10^{-4}$.

In the preceding numerical illustration, we observe that the computed difference of 0.00012 has only two significant figures of accuracy, whereas in general, one expects the numbers and calculations in this minicomputer to have five significant figures of accuracy.

The remedy for this difficulty is first to anticipate that it may occur and then reprogram. The simplest technique may be to carry out part of a computation in double- or extended-precision arithmetic (that means roughly twice as many significant digits), but often a slight change in the formulas is required. Several illustrations of this will be given here, and the reader will find additional ones among the problems.

Consider the previous numerical example, but imagine that double-precision calculations are being used to obtain x, y, and $x - y$. Suppose that single-precision arithmetic is used thereafter. In the computer all ten digits of x, y, and $x - y$ will be retained, but at the end $x - y$ will be rounded to its five-digit form, 0.12343×10^{-3}. This answer has five significant digits of accuracy, as we would like. Of course, the programmer or analyst must know in advance where the double-precision arithmetic will be necessary in the computation. Programming everything in double precision is very costly and wasteful if it is not needed. Another drawback to this approach is that there may be such serious cancellation of significant digits that even double precision will not help.

Theorem on Loss of Precision

Before considering other techniques for avoiding this problem, we ask the following: *Exactly how many significant binary digits are lost in the subtraction $x - y$ when x is close to y?* The closeness of x and y is conveniently measured by $|1 - y/x|$. Here is the result:

THEOREM ON LOSS OF PRECISION

Let x and y be normalized floating-point machine numbers so that $x > y > 0$. If $2^{-p} \leq 1 - y/x \leq 2^{-q}$ for some positive integers p and q, then at most p and at least q significant binary bits are lost in the subtraction $x - y$.

Proof We prove the second part of the theorem and leave the first as an exercise.

Let $x = r \times 2^n$ and $y = s \times 2^m$, where $\frac{1}{2} \leq r, s < 1$. (This is the normalized binary floating-point form.) Since $y < x$, the computer may have to *shift* y before carrying out the subtraction. In any case, y must first be expressed with the same exponent as x. Hence, $y = (s2^{m-n}) \times 2^n$ and

$$x - y = (r - s2^{m-n}) \times 2^n$$

Now the mantissa of this number satisfies

$$r - s2^{m-n} = r\left(1 - \frac{s2^m}{r2^n}\right) = r\left(1 - \frac{y}{x}\right) < 2^{-q}$$

Hence, to normalize the representation of $x - y$, a shift of at least q bits to the left is necessary. Then at least q (spurious) zeros are supplied on the right end of the mantissa. This means that at least q bits of precision have been lost.   ∎

EXAMPLE 2   In the subtraction $37.59362\,1 - 37.58421\,6$, how many bits of significance will be lost?

Solution   Let x denote the first number and y the second. Then

$$1 - \frac{y}{x} = 0.00025\,01754$$

This lies between $2^{-12} = 0.000024\,4$ and $2^{-11} = 0.00048\,8$. Hence, at least 11 but not more that 12 bits are lost.   □

Avoiding Loss of Significance in Subtraction

Now we consider various techniques that can be used to avoid the loss of significance that occurs in subtraction. Consider the function

$$f(x) = \sqrt{x^2 + 1} - 1 \tag{1}$$

whose values may be required for x near zero. Since $\sqrt{x^2 + 1} \approx 1$ when $x \approx 0$, we see that there is a potential loss of significance in the subtraction. However, the function can be rewritten in the form

$$f(x) = (\sqrt{x^2 + 1} - 1)\left(\frac{\sqrt{x^2 + 1} + 1}{\sqrt{x^2 + 1} + 1}\right) = \frac{x^2}{\sqrt{x^2 + 1} + 1} \tag{2}$$

by *rationalizing* the numerator—that is, removing the radical in the numerator. This procedure allows terms to be cancelled and thereby removes the subtraction. For example, if we use five-decimal-digit arithmetic and if $x = 10^{-3}$, then $f(x)$ will be computed incorrectly as 0 by the first formula but as $\frac{1}{2} \times 10^{-6}$ by the second. If we use the first formula together with double precision, the difficulty is ameliorated but not circumvented altogether. For example, in double precision, we have the same problem when $x = 10^{-6}$.

As another example, suppose that the values of

$$f(x) = x - \sin x \tag{3}$$

are required near $x = 0$. A careless programmer may code this function just as indicated in Equation (3), not realizing that serious loss of accuracy will occur. Recall from calculus that

$$\lim_{x \to 0} \frac{\sin x}{x} = 1$$

in order to see that $\sin x \approx x$ when $x \approx 0$. One cure for this problem is to use the Taylor series for $\sin x$:

$$\sin x = x - \frac{x^3}{3!} + \frac{x^5}{5!} - \frac{x^7}{7!} + \cdots$$

This series is known to represent $\sin x$ for all real values of x. For x near zero, it converges quite rapidly. Using this series, we can write the function f as

$$f(x) = x - \left(x - \frac{x^3}{3!} + \frac{x^5}{5!} - \frac{x^7}{7!} - \cdots\right) = \frac{x^3}{3!} - \frac{x^5}{5!} + \frac{x^7}{7!} - \cdots \qquad (4)$$

We see in this equation where the source of the original difficulty arose; namely, for small values of x, the term x in the sine series is much larger than $x^3/3!$ and thus more important. But when $f(x)$ is formed, this dominant x term disappears, leaving only the lesser terms. The series that starts with $x^3/3!$ is very effective for calculating $f(x)$ when x is small.

In this example, further analysis is needed to determine the range in which the Series (4) should be used and the range in which Formula (3) can be used. Using the Theorem on Loss of Precision, we see that the loss of bits in the subtraction of Formula (3) can be limited to at most *one* bit by restricting x so that $\frac{1}{2} \leq 1 - \sin x/x$. (Here we are considering only the case when $\sin x > 0$.) With a calculator, it is easy to see that x must be at least 1.9. Thus, for $|x| < 1.9$ we use the first few terms in the Series (4), and for $|x| \geq 1.9$ we use $f(x) = x - \sin x$. One can verify that for the worst case ($x = 1.9$), ten terms in the series give $f(x)$ with an error of at most 10^{-16}.

To verify these conclusions, the following Maple program was run with three different values (0.4, 1.9, and 3.4) for x.

```
y := sum( (-1)^(k-1)*x^(2*k+1)/(2*k+1)!, k=1..10 );
Digits := 40;
x := 0.4;
z := evalf(y);
t := x - sin(x);
d := z - t;
```

The values obtained for d were 0.2×10^{-31}, 10^{-16}, and 0.6×10^{-10}. This illustrates that using ten terms in the series is more accurate for a value of x below 1.9 and less accurate for one above 1.9.

To construct a function procedure for $f(x)$, notice that the terms in the series can be obtained inductively by the algorithm

$$\begin{cases} t_1 = \dfrac{x^3}{6} \\[2em] t_{n+1} = \dfrac{-t_n x^2}{(2n+2)(2n+3)} \qquad (n \geq 1) \end{cases}$$

Then the partial sums can be obtained inductively by

$$\begin{cases} s_1 = t_1 \\ s_{n+1} = s_n + t_{n+1} \end{cases} \quad (n \geq 1)$$

so that $s_n = \sum_{k=1}^{n} t_k = \sum_{k=1}^{n} (-1)^{k+1} [x^{2k+1}/(2k+1)!]$.
A suitable pseudocode function is given here:

real function $f(x)$
integer parameter $n \leftarrow 10$
integer i
real s, t, x
if $|x| \geq 1.9$ **then**
    $s \leftarrow x - \sin x$
else
    $t \leftarrow x^3/6$
    $s \leftarrow t$
    for $i = 2$ **to** n **do**
        $t \leftarrow -tx^2/[(2i + 2)(2i + 3)]$
        $s \leftarrow s + t$
    end do
end if
$f \leftarrow s$
end function f

EXAMPLE 3   How can accurate values of the function

$$f(x) = e^x - e^{-2x}$$

be computed in the vicinity of $x = 0$?

Solution   Since e^x and e^{-2x} are both equal to 1 when $x = 0$, there will be a loss of significance because of subtraction when x is close to zero. Inserting the appropriate Taylor series, we obtain

$$f(x) = \left(1 + x + \frac{x^2}{2!} + \frac{x^3}{3!} + \cdots\right) - \left(1 - 2x + \frac{4x^2}{2!} - \frac{8x^3}{3!} + \cdots\right)$$

$$= 3x - \frac{3}{2}x^2 + \frac{3}{2}x^3 - \cdots$$

An alternative is to write

$$f(x) = e^{-2x}(e^{3x} - 1)$$

$$= e^{-2x}\left(3x + \frac{9}{2!}x^2 + \frac{27}{3!}x^3 + \cdots\right)$$

By using the Theorem on Loss of Precision, we find that at most one bit is lost in the subtraction $e^x - e^{-2x}$ when $x > 0$ and

$$\frac{1}{2} \leqq 1 - \frac{e^{-2x}}{e^x}$$

This inequality is valid when $x \geq \frac{1}{3} \ln 2 = 0.23105$. Similar reasoning when $x < 0$ shows that for $x \leq -0.23105$ at most one bit is lost. Hence, the series should be used for $|x| < 0.23105$.   □

EXAMPLE 4   Criticize the assignment statement

$$y \leftarrow \cos^2(x) - \sin^2(x)$$

Solution   When $\cos^2(x) - \sin^2(x)$ is computed, there will be a loss of significance at $x = \pi/4$ (and other points). The simple trigonometric identity

$$\cos 2\theta = \cos^2 \theta - \sin^2 \theta$$

should be used. Thus, the assignment statement should be replaced by

$$y \leftarrow \cos(2x)$$ □

EXAMPLE 5   Criticize the assignment statement

$$y \leftarrow \ln(x) - 1$$

Solution   If the expression $\ln x - 1$ is used for x near e, there will be a cancellation of digits and a loss of accuracy. One can use elementary facts about logarithms to overcome the difficulty. Thus, $y = \ln x - 1 = \ln x - \ln e = \ln(x/e)$. Here is a suitable assignment statement:

$$y \leftarrow \ln\left(\frac{x}{e}\right)$$,   □

Range Reduction

Another cause of loss of significant figures is the evaluation of various library functions with very large arguments. This problem is a bit subtler than the ones previously discussed. We illustrate with the sine function.

A basic property of the function $\sin x$ is its periodicity:

$$\sin x = \sin(x + 2n\pi)$$

for all x and for all integer values of n. Because of this relationship, one needs to know only the values of $\sin x$ in some fixed interval of length 2π in order to compute $\sin x$ for arbitrary x. This property is used in the computer evaluation of $\sin x$ and is called **range reduction**.

Suppose now that we want to evaluate $\sin(12532.14)$. By subtracting integer multiples of 2π, we find that this equals $\sin(3.47)$, if we retain only two decimal digits of accuracy. Thus, although our original argument 12532.14 had seven significant figures, the reduced argument has only three. The remaining ones disappeared in the subtraction of 3988π. Since 3.47 has only three significant figures, our computed value of $\sin(12532.14)$ will have *no more than* three significant figures. This decrease in precision is unavoidable if there is no way of increasing the precision of the original argument. If the original argument (12532.14 in this example) can be obtained with more significant figures, these additional figures will be present in the *reduced* argument (3.47 in this example). In some cases, double- or extended-precision programming will help.

EXAMPLE 6 For $\sin x$, how many binary bits of significance are lost in range reduction to the interval $[0, 2\pi)$?

Solution Given an argument $x > 2\pi$, we determine an integer n that satisfies the inequality $0 \leq x - 2n\pi < 2\pi$. Then in evaluating elementary trigonometric functions, we use $f(x) = f(x - 2n\pi)$. In the subtraction $x - 2n\pi$, there will be a loss of significance. By the Theorem on Loss of Precision, at least q bits are lost if

$$1 - \frac{2n\pi}{x} \leq 2^{-q}$$

Since

$$1 - \frac{2n\pi}{x} = \frac{x - 2n\pi}{x} < \frac{2\pi}{x}$$

we conclude that at least q bits are lost if $2\pi/x \leq 2^{-q}$. Stated otherwise, at least q bits are lost if $2^q \leq x/2\pi$. □

PROBLEMS 2.3

1. Indicate how the following formulas may be useful for arranging computations to avoid loss of significant digits.

   a. $\sin x - \sin y = 2 \sin \frac{1}{2}(x - y) \cos \frac{1}{2}(x + y)$

   b. $\log x - \log y = \log(x/y)$

   c. $e^{x-y} = e^x/e^y$

   d. $1 - \cos x = 2 \sin^2(x/2)$

   e. $\arctan x - \arctan y = \arctan\left(\dfrac{x - y}{1 + xy}\right)$

2. What is a good way to compute $\tan x - x$ when x is near zero?

3. How can values of the function $f(x) = \sqrt{x + 4} - 2$ be computed accurately when x is small?

4. Calculate $f(10^{-2})$ for the function $f(x) = e^x - x - 1$. The answer should have five significant figures and can easily be obtained with pencil and paper. Contrast it with the straightforward evaluation of $f(10^{-2})$ using $e^{0.01} \approx 1.0101$.

5. What is a good way to compute values of the function $f(x) = e^x - e$ if full machine precision is needed?

6. What problem could the assignment statement $y \leftarrow 1 - \sin x$ cause? Circumvent it without resorting to a Taylor series if possible.

7. The hyperbolic sine function is defined by $\sinh x = \frac{1}{2}(e^x - e^{-x})$. What drawback could there be in using this formula to obtain values of the function? How can values of $\sinh x$ be computed to full machine precision when $|x| \leq \frac{1}{2}$?

8. On your computer determine the range of x for which $(\sin x)/x \approx 1$ with full machine precision. *Hint*: Use Taylor series.

9. Use of the familiar quadratic formula, $x = (-b \pm \sqrt{b^2 - 4ac})/(2a)$, will cause a problem when the quadratic equation $x^2 - 10^5 x + 1 = 0$ is solved with a machine that carries only eight decimal digits. Investigate the example, observe the difficulty, and propose a remedy. *Hint*: An example in the text is similar.

10. For any $x_0 > -1$, the sequence defined recursively by

$$x_{n+1} = 2^{n+1}\left(\sqrt{1 + 2^{-n}x_n} - 1\right) \qquad (n \geq 0)$$

converges to $\ln(x_0 + 1)$. Arrange this formula in a way that avoids loss of significance.

11. Find ways to compute these functions without serious loss of significant figures.

   a. $e^x - \sin x - \cos x$

   b. $\ln(x) - 1$

   c. $\log x - \log(1/x)$

   d. $x^{-2}(\sin x - e^x + 1)$

   e. $x - \text{arctanh}\, x$

12. Let

$$a(x) = \frac{1 - \cos x}{\sin x} \qquad b(x) = \frac{\sin x}{1 + \cos x} \qquad c(x) = \frac{x}{2} + \frac{x^3}{24}$$

Show that $b(x)$ is identical to $a(x)$ and that $c(x)$ approximates $a(x)$ in a neighborhood of zero.

13. Determine the first two nonzero terms in the expansion about zero for the function $f(x) = (\tan x - \sin x)/(x - \sqrt{1 + x^2})$. Give an approximate value for $f(0.0125)$.

14. Find a method for computing $(\sinh x - \tanh x)/x$ that avoids loss of significance when x is small. Find appropriate identities to solve this problem without using Taylor series.

15. Find a way to calculate accurate values for

$$f(x) = \frac{\sqrt{1 + x^2} - 1}{x^2} - \frac{x^2 \sin x}{x - \tan x}$$

Determine $\lim_{x \to 0} f(x)$.

16. For some values of x, the assignment statement $y \leftarrow 1 - \cos x$ involves a difficulty. What is it, what values of x are involved, and what remedy do you propose?

17. For some values of x, the function $f(x) = \sqrt{x^2 + 1} - x$ cannot be accurately computed by using this formula. Explain and find a way around the difficulty.

18. The inverse hyperbolic sine is given by $f(x) = \ln(x + \sqrt{x^2 + 1})$. Show how to avoid loss of significance in computing $f(x)$ when x is negative. *Hint*: Find and exploit the relationship between $f(x)$ and $f(-x)$.

19. On most computers a highly accurate routine for $\cos x$ is provided. It is proposed to base a routine for $\sin x$ on the fundamental formula $\sin x = \pm\sqrt{1 - \cos^2 x}$. From the standpoint of precision (not efficiency), what problems do you foresee and how can they be avoided if we insist on using the routine for $\cos x$?

20. Criticize and recode the assignment statement $z \leftarrow \sqrt{x^4 + 4} - 2$ assuming that z will sometimes be needed for an x close to zero.

21. How can values of the function $f(x) = \sqrt{x + 2} - \sqrt{x}$ be computed accurately when x is large?

22. Write a function that computes accurate values of $f(x) = \sqrt[4]{x + 4} - \sqrt[4]{x}$ for positive x.

23. Find a way to calculate $f(x) = (\cos x - e^{-x})/\sin x$ correctly. Determine $f(0.008)$ correctly to ten decimal places (rounded).

24. Without using series, how could the function

$$f(x) = \frac{\sin x}{x - \sqrt{x^2 - 1}}$$

be computed to avoid loss of significance?

25. Write a function procedure that returns accurate values of the hyperbolic tangent function

$$\tanh x = \frac{e^x - e^{-x}}{e^x + e^{-x}}$$

for all values of x. Notice the difficulty when $|x| < \frac{1}{2}$.

26. Find a good way to compute $\sin x + \cos x - 1$ for x near zero.

27. Find a good way to compute $\arctan x - x$ for x near zero.

28. Find a good bound for $|\sin x - x|$ using Taylor series and assuming $|x| < \frac{1}{10}$.

29. How would you compute $(e^{2x} - 1)/(2x)$ to avoid loss of significance near zero?

30. When accurate values for the roots of a quadratic equation are desired, some loss of significance may occur if $b^2 \approx 4ac$. What (if anything) can be done to overcome this when writing a computer routine?

31. Let x and y be two normalized binary floating-point machine numbers. Assume that $x = q \times 2^n$, $y = r \times 2^{n-1}$, $\frac{1}{2} \le r, q < 1$, and $2q - 1 \ge r$. How much loss of significance occurs in subtracting $x - y$? Answer the same question when $2q - 1 < r$. Observe that the Theorem on Loss of Precision is not strong enough to solve this problem precisely.

32. Prove the first part of the Theorem on Loss of Precision.

33. Show that if x is a machine number on the `Marc-32` that satisfies the inequality $x > \pi 2^{25}$, then $\sin x$ will be computed with *no* significant digits.

34. Let x and y be two positive normalized floating-point machine numbers in the `Marc-32`. Let $x = q \times 2^m$ and $y = r \times 2^n$ with $\frac{1}{2} \le r, q < 1$. Show that if $n = m$, then at least one bit of significance is lost in the subtraction $x - y$.

35. Refer to the discussion of the function $f(x) = x - \sin x$ given in the text. Show that when $0 < x < 1.9$, there will be no undue loss of significance from subtraction in Equation (2).

36. Discuss the problem of computing $\tan(10^{100})$. (See Gleick [1992], p. 178.)

COMPUTER PROBLEMS 2.3

1. Write code using double or extended precision to evaluate $f(x) = \cos(10^4 x)$ on the interval $[0, 1]$. Determine how many significant figures the values of $f(x)$ will have.

2. Write a procedure to compute $f(x) = \sin x - 1 + \cos x$. The routine should produce nearly full machine precision for all x in the interval $[0, \pi/4]$. *Hint:* The trigonometric identity $\sin^2 \theta = \frac{1}{2}(1 - \cos 2\theta)$ may be useful.

3. Write a procedure to compute $f(x, y) = \int_1^x t^y \, dt$ for arbitrary x and y. Note the exceptional case $y = -1$ and the numerical problem *near* the exceptional case.

4. Suppose that we wish to evaluate the function $f(x) = (x - \sin x)/x^3$ for values of x close to zero.

    a. Write a routine for this function. Evaluate $f(x)$ sixteen times. Initially, let $x \leftarrow 1$ and then let $x \leftarrow \frac{1}{10} x$ fifteen times. Explain the results. *Note:* L'Hôpital's rule indicates that $f(x)$ should tend to $\frac{1}{6}$. Test this code.

    b. Write a function procedure that produces more accurate values of $f(x)$ for all values of x. Test this code.

5. Write a program to print a table of the function $f(x) = 5 - \sqrt{25 + x^2}$ for $x = 0$ to 1 with steps of 0.01. Be sure that your program yields full machine precision but do not program the problem in double precision. Explain the results.

6. Quite important in many numerical calculations is the accurate computation of the absolute value $|z|$ of a complex number $z = a + bi$. Design and carry out a computer experiment to compare the following three schemes:

$$|z| = (a^2 + b^2)^{1/2}$$

$$|z| = v\left[1 + \left(\frac{w}{v}\right)^2\right]^{1/2}$$

$$|z| = 2v\left[\frac{1}{4} + \left(\frac{w}{2v}\right)^2\right]^{1/2}$$

where $v = \max\{|a|, |b|\}$ and $w = \min\{|a|, |b|\}$. Use very small and large numbers for the experiment.

7. For what range of x is the approximation $(e^x - 1)/2x \approx 0.5$ correct to 15 decimal digits of accuracy? Using this information, write a function procedure for $(e^x - 1)/2x$, producing 15 decimals of accuracy throughout the interval $[-10, 10]$.

8. In the theory of Fourier series, some numbers known as **Lebesgue constants** play a role. A formula for them is

$$\rho_n = \frac{1}{2n + 1} + \frac{2}{\pi} \sum_{k=1}^{n} \frac{1}{k} \tan \frac{\pi k}{2n + 1}$$

Write and run a program to compute $\rho_1, \rho_2, \ldots, \rho_{100}$ with eight decimal digits of accuracy. Then test the validity of the inequality

$$0 \leq \frac{4}{\pi^2} \ln(2n + 1) + 1 - \rho_n \leq 0.0106$$

9. Write a routine for computing the two roots x_1 and x_2 of the quadratic equation $f(x) = ax^2 + bx + c = 0$ with real constants a, b, and c and for evaluating $f(x_1)$ and $f(x_2)$. Use formulas that reduce roundoff errors and write efficient code. Test your routine on the following (a, b, c) values: $(0, 0, 1)$; $(0, 1, 0)$; $(1, 0, 0)$; $(0, 0, 0)$; $(1, 1, 0)$; $(2, 10, 1)$; $(1, -4, 3.99999)$; $(1, -8.01, 16.004)$; $(2 \times 10^{17}, 10^{18}, 10^{17})$; and $(10^{-17}, -10^{17}, 10^{17})$.

10. (Continuation) Write and test a routine for solving a quadratic equation that may have complex roots.

11. Alter the code in the text for computing $x - \sin x$ by using nested multiplication to evaluate the series.

12. Write a routine for the function $f(x) = e^x - e^{-2x}$ using the examples in the text for guidance.

13. (Double or extended precision) Compute in double or extended precision the following number:

$$x = \left[\frac{1}{\pi} \ln(6\,40320^3 + 744) \right]^2$$

What is the point of this problem? (See Good [1972].)

14. Write a routine that computes e^x by summing n terms of the Taylor series until the $n + 1$st term t is such that $|t| < \varepsilon = 10^{-6}$. Use the reciprocal of e^x for negative values of x. Test on the following data: 0, +1, −1, 0.5, −0.123, −25.5, −1776, 3.14159. Compute the relative error, the absolute error, and n for each case, using the exponential function on your computer system for the exact value. Sum no more than 25 terms.

15. (Continuation) The computation of e^x can be reduced to computing e^u for $|u| < (\ln 2)/2$ only. This algorithm removes powers of 2 and computes e^u in a range where the series converges very rapidly. It is given by

$$e^x = 2^m e^u$$

where m and u are computed by the steps

$$z \leftarrow x / \ln 2$$

$$m \leftarrow \text{integer } (z \pm \tfrac{1}{2})$$

$$w \leftarrow z - m$$

$$u \leftarrow w \ln 2$$

Here the minus sign is used if $x < 0$ because $z < 0$. Incorporate this range reduction technique into the code.

16. (Continuation) Write a routine that uses range reduction $e^x = 2^m e^u$ and computes e^u from the even part of the *Gaussian continued fraction*; that is,

$$e^u = \frac{s + u}{s - u} \quad \text{where} \quad s = 2 + u^2 \left(\frac{2520 + 28u^2}{15120 + 420u^2 + u^4} \right)$$

Test on the data given in Computer Problem **14**. *Note*: Computer Problems **16–21** contain rather complicated algorithms for various intrinsic functions that correspond to those actually used on a large mainframe computer system. Descriptions of these and other similar library functions are frequently found in the documentation material or on-line manual files of your computer system.

17. Write a routine to compute $\sin x$ for x in radians as follows. First, using properties of the sine function, reduce the range so that $-\pi/2 \leq x \leq \pi/2$. Then if $|x| < 10^{-8}$, set $\sin x \approx x$; if $|x| > \pi/6$, set $u = x/3$, compute $\sin u$ by the formula below, and then set $\sin x \approx [3 - 4 \sin^2 u] \sin u$; if $|x| \leq \pi/6$, set $u = x$

and compute sin u as follows:

$$\sin u \approx u \left[\frac{1 - \left(\dfrac{3\,25523}{22\,83996}\right)u^2 + \left(\dfrac{34911}{76\,13320}\right)u^4 - \left(\dfrac{4\,79249}{1\,15113\,39840}\right)u^6}{1 + \left(\dfrac{18381}{7\,61332}\right)u^2 + \left(\dfrac{1261}{45\,67992}\right)u^4 + \left(\dfrac{2623}{1644\,77120}\right)u^6} \right]$$

Try to determine whether the sine function on your computer system uses this algorithm. *Note*: This is the Padé rational approximation for sine.

18. Write a routine to compute the natural logarithm by the algorithm outlined here based on *telescoped rational* and *Gaussian continued fractions* for ln x and test for several values of x. First check whether $x = 1$ and return zero if so. Reduce the range of x by determining n and r such that $x = r \times 2^n$ with $\frac{1}{2} \leq r < 1$. Next, set $u = (r - \sqrt{2}/2)(r + \sqrt{2}/2)$ and compute $\ln[(1 + u)/(1 - u)]$ by the approximation

$$\ln\left(\frac{1 + u}{1 - u}\right) \approx u\left(\frac{20790 - 21545.27u^2 + 4223.9187u^4}{10395 - 14237.635u^2 + 4778.8377u^4 - 230.41913u^6}\right)$$

which is valid for $|u| < 3 - 2\sqrt{2}$. Finally, set

$$\ln x \approx \left(n - \frac{1}{2}\right)\ln 2 + \ln\left[\frac{1 + u}{1 - u}\right]$$

19. Write a routine to compute the tangent of x in radians, using the algorithm below. Test the resulting routine over a range of values of x. First, the argument x is reduced to $|x| \leq \pi/2$ by adding or subtracting multiples of π. If we have $0 \leq |x| \leq 1.7 \times 10^{-9}$, set $\tan x \approx x$. If $|x| > \pi/4$, set $u = \pi/2 - x$; otherwise, set $u = x$. Now compute the approximation

$$\tan u \approx u\left(\frac{1\,35135 - 17336.106u^2 + 379.23564u^4 - 1.01186\,25u^6}{1\,35135 - 62381.106u^2 + 3154.9377u^4 + 28.17694u^6}\right)$$

Finally, if $|x| > \pi/4$, set $\tan x \approx 1/\tan u$; if $|x| \leq \pi/4$, set $\tan x \approx \tan u$. *Note*: This algorithm is obtained from the *telescoped rational* and *Gaussian continued fraction* for the tangent function.

20. Write a routine to compute arcsin x based on the following algorithm, using telescoped polynomials for the arcsine. If $|x| < 10^{-8}$, set arcsin $x \approx x$. Otherwise, if $0 \leq x \leq \frac{1}{2}$, set $u = x$, $a = 0$, and $b = 1$; if $\frac{1}{2} < x \leq \frac{1}{2}\sqrt{3}$, set $u = 2x^2 - 1$, $a = \pi/4$, and $b = \frac{1}{2}$; if $\frac{1}{2}\sqrt{3} < x \leq \frac{1}{2}\sqrt{2 + \sqrt{3}}$, set $u = 8x^4 - 8x^2 + 1$, $a = 3\pi/8$, and $b = \frac{1}{4}$; if $\frac{1}{2}\sqrt{2 + \sqrt{3}} < x \leq 1$, set $u = \sqrt{\frac{1}{2}(1 - x)}$, $a = \pi/2$, and $b = -2$. Now compute the approximation

$$\arcsin u \approx u\left(1.0 + \frac{1}{6}u^2 + 0.075u^4 + 0.04464\,286u^6 + 0.03038\,182u^8\right.$$

$$+ 0.02237\,5u^{10} + 0.01731\,276u^{12} + 0.01433\,124u^{14}$$

$$+ 0.00934\,2806u^{16} + 0.01835\,667u^{18} - 0.01186\,224u^{20}$$

$$\left. + 0.03162\,712u^{22}\right)$$

Finally, set $\arcsin x \approx a + b\arcsin u$. Test this routine for various values of x.

21. Write and test a routine to compute $\arctan x$ for x in radians as follows. If $0 \le x \le 1.7 \times 10^{-9}$, set $\arctan x \approx x$. If $1.7 \times 10^{-9} < x \le 2 \times 10^{-2}$, use the series approximation

$$\arctan x \approx x - \frac{x^3}{3} + \frac{x^5}{5} - \frac{x^7}{7}$$

Otherwise, set $y = x$, $a = 0$, and $b = 1$ if $0 \le x \le 1$; set $y = 1/x$, $a = \pi/2$, and $b = -1$ if $1 < x$. Then set $c = \pi/16$ and $d = \tan c$ if $0 \le y \le \sqrt{2} - 1$ and $c = 3\pi/16$ and $d = \tan c$ if $\sqrt{2} - 1 < y \le 1$. Compute $u = (y - d)/(1 + dy)$ and the approximation

$$\arctan u \approx u\left(\frac{1\,35135 + 1\,71962.46u^2 + 52490.4832u^4 + 2218.1u^6}{1\,35135 + 2\,17007.46u^2 + 97799.3033u^4 + 10721.3745u^6}\right)$$

Finally, set $\arctan x \approx a + b(c + \arctan u)$. *Note:* This algorithm uses *telescoped rational* and *Gaussian continued fractions.*

22. A fast algorithm for computing $\arctan x$ to n-bit precision for x in the interval $(0, 1]$ is as follows: Set $a = 2^{-n/2}$, $b = x/(1 + \sqrt{1 + x^2})$, $c = 1$, and $d = 1$. Then repeatedly update these variables by these formulas (in order from top to bottom):

real a, b, c, d

$\vdots$

$$c \leftarrow \frac{2c}{1 + a}$$

$$d \leftarrow \frac{2ab}{1 + b^2}$$

$$d \leftarrow \frac{d}{1 + \sqrt{1 - d^2}}$$

$$d \leftarrow \frac{b + d}{1 - bd}$$

$$\vdots$$

$$c \leftarrow \frac{2c}{1 + a}$$

$$b \leftarrow \frac{d}{1 + \sqrt{1 + d^2}}$$

$$a \leftarrow \frac{2\sqrt{a}}{1 + a}$$

After each sweep, print $f = c \ln[(1 + b)/(1 - b)]$. Stop when $1 - a \leq 2^{-n}$. Write a double-precision routine to implement this algorithm and test it for various values of x. Compare the results to those obtained from the arctangent function on your computer system. *Note*: This fast multiple-precision algorithm depends on the theory of *elliptic integrals*, using the arithmetic-geometric mean iteration and *ascending Landen transformations*. Other fast algorithms for trigonometric functions are discussed in Brent [1976].

3 LOCATING ROOTS OF EQUATIONS

An electric power cable is suspended (at points of equal height) from two towers that are 100 meters apart. The cable is allowed to dip 10 meters in the middle. How long is the cable?

It is known that the curve assumed by a suspended cable is a **catenary**. When the y-axis passes through the lowest point, we can assume an equation of the form $y = \lambda \cosh(x/\lambda)$. Here λ is a parameter to be determined. The conditions of the problem are that $y(50) = y(0) + 10$. Hence,

$$\lambda \cosh \frac{50}{\lambda} = \lambda + 10$$

From this equation λ can be determined by the methods of this chapter. The result is $\lambda = 126.632$. From the equation of the catenary, the arc length is easily computed by a standard method of calculus to be 102.619 meters.

Let f be a real-valued function of a real variable. Any real number r for which $f(r) = 0$ is called a **root** of that equation or a **zero** of f. For example, the function

$$f(x) = 6x^2 - 7x + 2$$

has $\frac{1}{2}$ and $\frac{2}{3}$ as zeros, as can be verified by direct substitution or by writing f in its factored form

$$f(x) = (2x - 1)(3x - 2)$$

For another example, the function

$$g(x) = \cos 3x - \cos 7x$$

has not only the obvious zero $x = 0$ but every integer multiple of $\pi/5$ and of $\pi/2$ as well, which we discover by applying the trigonometric identity

$$\cos A - \cos B = 2 \sin \frac{1}{2}(A + B) \sin \frac{1}{2}(B - A)$$

Why is locating roots important? Frequently the solution to a scientific problem is a number about which we have little information other than that it satisfies some equation. Since every equation can be written so that a function stands on one side and zero on the other, the desired number must be a zero of the function. Thus, if we possess an arsenal of methods for locating zeros of functions, we shall be able to solve such problems.

We illustrate this claim by use of a specific engineering problem with a solution that is the root of an equation. In a certain electrical circuit, the voltage V and current I are related by two equations of the form

$$\begin{cases} I = a(e^{bV} - 1) \\ c = dI + V \end{cases}$$

in which a, b, c, and d are constants. When these equations are combined by eliminating I between them, the result is a single equation:

$$c = ad(e^{bV} - 1) + V$$

In a concrete case, this might reduce to

$$12 = 14.3(e^{2V} - 1) + V$$

and the solution of this is required. (It turns out that $V \approx 0.299$ in this case.)

In some problems where a root of an equation is sought, we can perform the required calculation with a hand calculator. But how can we locate zeros of complicated functions such as these:

$$f(x) = 3.24x^8 - 2.42x^7 + 10.34x^6 + 11.01x^2 + 47.98$$
$$g(x) = 2^{x^2} - 10x + 1$$
$$h(x) = \cosh\left(\sqrt{x^2 + 1} - e^x\right) + \log|\sin x|$$

What is needed is a general numerical method that does not depend on special properties of our functions. Of course, continuity and differentiability are special properties, but they are common attributes of functions usually encountered. The sort of special property that we probably *cannot* easily exploit in general-purpose codes is typified by the trigonometric identity mentioned previously.

Myriads of methods are available for locating zeros of functions, and three of the most useful have been selected for study. These are the bisection method, Newton's method, and the secant method.

3.1   Bisection Method

Let f be a function that has values of opposite sign at the two ends of an interval. Suppose also that f is continuous on that interval. To fix the notation, let $a < b$ and $f(a)f(b) < 0$. It then follows that f has a root in the interval (a, b). In other words, there must exist a number r that satisfies the two conditions $a < r < b$ and $f(r) = 0$. How is this conclusion reached? One must recall the **Intermediate-Value Theorem of Continuous Functions**.* If x traverses an interval $[a, b]$, then the values of $f(x)$ completely fill out the interval between $f(a)$ and $f(b)$. No intermediate values can be skipped. So our function f must take on the value zero somewhere in the interval (a, b) because $f(a)$ and $f(b)$ are of opposite signs.

Bisection Algorithm and Pseudocode

The bisection method and its algorithm exploit this property of continuous functions. At each step in this algorithm, we have an interval $[a, b]$ and the values $u = f(a)$ and $v = f(b)$. The numbers u and v satisfy $uv < 0$. Next, we construct the midpoint of the interval, $c = \frac{1}{2}(a + b)$, and compute $w = f(c)$. It can happen fortuitously that $f(c) = 0$. If so, the objective of the algorithm has been fulfilled. In the usual case, $w \neq 0$ and either $wu < 0$ or $wv < 0$. (Why?) If $wu < 0$, we can be sure that a root of f exists in the interval $[a, c]$. Consequently, we store the value of c in b and w in v. If $wu > 0$, then we cannot be sure that f has a root in $[a, c]$, but since $wv < 0$, f must have a root in $[c, b]$. So in this case we store the value of c in a and w in u. In either case, the situation at the end of this step is just like that at the beginning except that the final interval $[a, b]$ is half as large as the initial interval. This step can now be repeated until the interval is satisfactorily small, say $|b - a| < \frac{1}{2} \times 10^{-5}$. At the end, the best estimate of the root would be $(a + b)/2$.

Now let us construct pseudocode to carry out this procedure. We shall not try to create a piece of high-quality software with many "bells and whistles," but we shall write the pseudocode in the form of a procedure for general use. This will afford the reader an opportunity to review how a main program and one or more procedures can be connected.

The procedure to be constructed will operate on an arbitrary function f. An interval $[a, b]$ is also specified, and the number of steps to be taken, m, is given. Pseudocode to perform m steps in the bisection algorithm follows:

---

*A formal statement of the **Intermediate-Value Theorem of Continuous Functions** is as follows: If the function f is continuous on the closed interval $[a, b]$, and if $f(a) \leqq y \leqq f(b)$ or $f(b) \leqq y \leqq f(a)$, then there exists a point c such that $a \leqq c \leqq b$ and $f(c) = y$.

```
procedure bisection (f, a, b, m)
integer n, m
real a, b, c, u, v, w
u ← f(a)
v ← f(b)
output a, b, u, v
for n = 0 to m do
   c ← (a + b)/2
   w ← f(c)
   output n, c, w
   if wv = 0 then
      exit loop
   else
      if wv < 0 then
         b ← c
         v ← w
      else
         a ← c
         u ← w
      end if
   end if
end do
end procedure bisection
```

This code follows exactly the description of the algorithm given previously. The reader should trace the steps in the routine to see that it does what is claimed.

To illustrate some techniques of structured programming and some other alternatives, a second version of the same procedure is now offered:

```
procedure bisection (f, a, b, m)
integer n, m
real a, b, c, fa, fb, fc, error
fa ← f(a)
fb ← f(b)
if sign(fa) = sign(fb) then
   output "function has same signs at a and b"
   output a, b, fa, fb
else
   error ← b − a
   for n = 0 to m do
      error ← error/2
      c ← a + error
      fc ← f(c)
      if sign(fa) ≠ sign(fc) then
         b ← c
         fb ← fc
```

```
        else
            a ← c
            fa ← fc
        end if
        output n, c, fc, error
    end do
end if
end function bisection
```

As a general rule, in programming routines to locate the roots of arbitrary functions, unnecessary evaluations of the function should be avoided because a given function may be costly to evaluate in terms of computer time. Thus, any value of the function that may be needed later should be stored rather than recomputed. A careless programming of the bisection method might violate this desideratum.

As another illustration, we show how this routine can be written as a recursive procedure:

```
recursive procedure bisection(f, a, b, fa, fb, m, n)
integer m, n
real a, b, c, fa, fb, fc, error
if sign(fa) = sign(fb) then
    output "function has same signs at a and b"
    output a, b, fa, fb
else
    error ← (b − a)/2
    c ← a + error
    fc ← f(c)
    output n, c, fc, error
    n ← n + 1
    if n ≦ m then
        if sign(fa) ≠ sign(fc) then
            call bisection(f, a, c, fa, fc, m, n)
        else
            call bisection(f, c, b, fc, fb, m, n)
        end if
    end if
end if
end function bisection
```

Use of Pseudocode

Now we want to illustrate how the bisection pseudocode can be used. Suppose that we have two functions, and for each we seek a zero in a specified interval:

$$f(x) = x^3 - 3x + 1 \quad \text{on } [0, 1]$$
$$g(x) = x^3 - 2\sin x \quad \text{on } [0.5, 2]$$

First, we write two procedure functions to compute $f(x)$ and $g(x)$. Then we input the initial intervals and the number of steps to be performed in a main program. Since this is a rather simple example, this information could be assigned directly in the main program or by way of data statements, parameter statements, or some other method rather than being read into the program. Also, depending on the computer language being used, an external or interface statement is needed to tell the compiler that the parameter f in the bisection procedure is *not* an ordinary variable with numerical values but the name of a function procedure defined externally to the main program. In this example, there would be two of these function procedures and two calls to the bisection procedure.

A call program or main program that calls the recursive bisection routine might be written as follows:

program *main*
integer parameter $m \leftarrow 20$
integer n
real a, b, fa, fb, ga, gb
interface external function f, g
$a \leftarrow 0.0$
$b \leftarrow 1.0$
$fa \leftarrow f(a)$
$fb \leftarrow f(b)$
output a, b, fa, fb
$n \leftarrow 0$
call $bisection(f, a, b, fa, fb, m, n)$
$a \leftarrow 0.5$
$b \leftarrow 2.0$
$ga \leftarrow g(a)$
$gb \leftarrow g(b)$
output a, b, ga, gb
$n \leftarrow 0$
call $bisection(g, a, b, ga, gb, m, n)$
end *main*

real function $f(x)$
real x
$f \leftarrow x^3 - 3x + 1$
end function f

real function $g(x)$
real x
$g \leftarrow x^3 - 2\sin x$
end function g

The computer results for the iterative steps of the bisection method for $f(x)$ and $g(x)$ are as follows:

n	c_n	$f(c_n)$	error
0	0.5	−0.375	0.5
1	0.25	0.266	0.25
2	0.375	-7.23×10^{-2}	0.125
3	0.3125	9.30×10^{-2}	6.25×10^{-2}
4	0.34375	9.37×10^{-3}	3.125×10^{-2}
⋮			
19	0.34729 67	-9.54×10^{-7}	9.54×10^{-7}
20	0.34729 62	3.58×10^{-7}	4.77×10^{-7}

n	c_n	$g(c_n)$	error
0	1.25	5.52×10^{-2}	0.75
1	0.875	−0.865	0.375
2	1.0625	−0.548	0.188
3	1.15625	−0.285	9.38×10^{-2}
4	1.20312 5	−0.125	4.69×10^{-2}
⋮			
19	1.23618 27	-4.88×10^{-6}	1.43×10^{-6}
20	1.23618 34	-2.15×10^{-6}	7.15×10^{-7}

Using the Maple commands

```
fsolve(x^3 - 3*x + 1, x, 0..1);
fsolve(x^3 - 2*sin(x), x, 0.5..2);
```

we find the desired roots of f and g to be 0.34729 63553 and 1.23618 3928, respectively.

Convergence Analysis

Now let us investigate the *accuracy* with which the bisection method determines a root of a function. Suppose that f is a continuous function that takes values of opposite sign at the ends of an interval $[a_0, b_0]$. Then there is a root r in $[a_0, b_0]$, and if we use the midpoint $c_0 = (a_0 + b_0)/2$ as our estimate of r, we have

$$|r - c_0| \leq \frac{b_0 - a_0}{2}$$

as illustrated in Figure 3.1.

If the bisection algorithm is now applied, and if the computed quantities are denoted by $a_0, b_0, c_0, a_1, b_1, c_1$ and so on, then by the same reasoning,

$$|r - c_n| \leq \frac{b_n - a_n}{2} \qquad (n \geq 0)$$

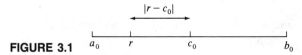

FIGURE 3.1

Since the widths of the intervals are divided by 2 in each step, we conclude that

$$|r - c_n| \leq \frac{b_0 - a_0}{2^{n+1}} \tag{1}$$

To summarize:

THEOREM ON CONVERGENCE OF BISECTION METHOD

If the bisection algorithm is applied to a continuous function f on an interval $[a, b]$, where $f(a)f(b) < 0$, then, after n steps, an approximate root will have been computed with error of at most $(b - a)/2^{n+1}$.

If an error tolerance has been prescribed in advance, it is possible to determine the number of steps required in the bisection method. Suppose that we want $|r - c_n| < \varepsilon$. Then it is necessary to solve the following inequality for n:

$$\frac{b - a}{2^{n+1}} < \varepsilon$$

By taking logarithms (with any convenient base), one obtains

$$n > \frac{\log(b - a) - \log(2\varepsilon)}{\log 2} \tag{2}$$

EXAMPLE 1   How many steps of the bisection algorithm are needed to compute a root of f to full machine precision on the `Marc-32` if $a = 16$ and $b = 17$?

Solution   The root is between the two binary numbers $a = (10\,000.0)_2$ and $b = (10\,001.0)_2$. Thus, we already know five of the binary digits in the answer. Since the `Marc-32` has a 24-bit mantissa, we can expect the answer to have an accuracy of only $\varepsilon = 2^{-20}$. From the above,

$$n > \frac{\log 1 - \log 2^{-19}}{\log 2}$$

Using a basic property of logarithms ($\log x^y = y \log x$), we find that $n \geq 19$. In this example, each step of the algorithm determines the root with one additional binary digit of precision. ☐

A sequence $\{x_n\}$ is said to have **linear convergence** to a limit x if there is a constant C in the interval $[0, 1)$ such that

$$|x_{n+1} - x| \leq C |x_n - x| \qquad (n \geq 1) \tag{3}$$

If this inequality is true for all n, then

$$|x_{n+1} - x| \leq C |x_n - x| \leq C^2 |x_{n-1} - x| \leq \cdots \leq C^n |x_1 - x|$$

Thus, it is a consequence of linear convergence that

$$|x_{n+1} - x| \leq AC^n \qquad (0 \leq C < 1) \tag{4}$$

The sequence produced by the bisection method obeys Inequality (4), as we see from Equation (1). However, the sequence need not obey Inequality (3). Problems **19** and **21** pertain to this assertion.

PROBLEMS 3.1

1. Prove inequality (1).

2. If $a = 0.1$ and $b = 1.0$, how many steps of the bisection method are needed to determine the root with an error of at most $\frac{1}{2} \times 10^{-8}$?

3. Find all the roots of the function $f(x) = \cos x - \cos 3x$.

4. Give a graphical demonstration that the equation $\tan x = x$ has infinitely many roots. Determine one root precisely and another approximately by using a graph.

5. Demonstrate graphically that the equation $50\pi + \sin x = 100 \arctan x$ has infinitely many roots.

6. Find the root or roots of $\ln[(1 + x)/(1 - x^2)] = 0$.

7. If f has an inverse, then the equation $f(x) = 0$ can be solved by simply writing $x = f^{-1}(0)$. Does this remark eliminate the problem of finding roots of equations? Illustrate with $\sin x = 1/\pi$.

8. Find where the graphs of $y = 3x$ and $y = e^x$ intersect by finding roots of $e^x - 3x = 0$ correct to four decimal digits.

9. How many binary digits of precision are gained in each step of the bisection method? How many steps are required for each decimal digit of precision?

10. Try to devise a stopping criterion for the bisection method to guarantee that the root is determined with *relative* error at most ε.

11. Denote the successive intervals that arise in the bisection method by $[a_0, b_0]$, $[a_1, b_1]$, $[a_2, b_2]$, and so on.

    a. Show that $a_0 \leq a_1 \leq a_2 \leq \cdots$ and that $b_0 \geq b_1 \geq b_2 \geq \cdots$.

    b. Show that $b_n - a_n = 2^{-n}(b_0 - a_0)$.

    c. Show that, for all n, $a_n b_n + a_{n-1} b_{n-1} = a_{n-1} b_n + a_n b_{n-1}$.

12. (Continuation) Using the notation of Problem **11**, let $c_n = (a_n + b_n)/2$. Show that

$$\lim_{n \to \infty} c_n = \lim_{n \to \infty} a_n = \lim_{n \to \infty} b_n$$

13. (Continuation) Consider the bisection method with the initial interval $[a_0, b_0]$. Show that after ten steps with this method,

$$\left|\tfrac{1}{2}(a_{10} + b_{10}) - \tfrac{1}{2}(a_9 + b_9)\right| = 2^{-11}(b_0 - a_0)$$

Also, determine how many steps are required to guarantee an approximation of a root to six decimal places (rounded).

14. By graphical methods, locate approximations to all roots of the nonlinear equation $\ln(x + 1) + \tan(2x) = 0$.

15. If the bisection method generates intervals $[a_0, b_0]$, $[a_1, b_1]$, and so on, which of these inequalities are true for the root r that is being calculated? Give proofs or counterexamples in each case.

   a. $|r - a_n| \leq 2|r - b_n|$

   b. $|r - a_n| \leq 2^{-n-1}(b_0 - a_0)$

   c. $\left|r - \tfrac{1}{2}(a_n + b_n)\right| \leq 2^{-n-2}(b_0 - a_0)$

   d. $0 \leq r - a_n \leq 2^{-n}(b_0 - a_0)$

   e. $|r - b_n| \leq 2^{-n-1}(b_0 - a_0)$

16. If the bisection method is applied with starting interval $[a, a + 1]$ and $a \geq 2^m$, where $m \geq 0$, what is the correct number of steps to compute the root with full machine precision on the `Marc-32`?

17. If the bisection method is applied with starting interval $[2^m, 2^{m+1}]$, where m is a positive or negative integer, how many steps should be taken to compute the root to full machine precision on the `Marc-32`?

18. Every polynomial of degree n has n zeros (counting multiplicities) in the complex plane. Does every real polynomial have n real zeros? Does every "polynomial of infinite degree" $f(x) = \sum_{n=0}^{\infty} a_n x^n$ have infinitely many zeros?

19. Using the notation of the text, determine which of these assertions are true and which are generally false.

   a. $|r - c_n| < |r - c_{n-1}|$

   b. $a_n \leq r \leq c_n$

   c. $c_n \leq r \leq b_n$

   d. $|r - a_n| \leq 2^{-n}$

   e. $|r - b_n| \leq 2^{-n}(b_0 - a_0)$

20. Prove that $|c_n - c_{n+1}| = 2^{-n-2}(b_0 - a_0)$

21. Give an example of a function for which the bisection method does *not* converge linearly.

22. Can it happen that $a_0 = a_1 = a_2 = \cdots$? Refer to Problem **11** for notation.

23. Draw a graph of a function that is discontinuous yet the bisection method converges. Repeat, getting a function for which it diverges.

1. Use the bisection method to determine roots of these functions on the intervals indicated. Process all three functions in *one* computer run.

$$f(x) = x^3 + 3x - 1 \qquad \text{on } [0, 1]$$

$$g(x) = x^3 - 2\sin x \qquad \text{on } [0.5, 2]$$

$$h(x) = x + 10 - x\cosh(50/x) \qquad \text{on } [120, 130]$$

Find each root to full machine precision. Use the correct number of steps, at least approximately.

2. Write a new procedure for the bisection method that takes a number ε as input and then carries out enough steps so that $|r - c_n| < \varepsilon$. (See text for definitions.) The routine should print at each step a, b, c, $f(c)$, and $(b - a)/2$. Include a test to ensure that at each step $f(a)f(b) < 0$. Do this by testing $\text{sign}(u) \neq \text{sign}(v)$, where $u = f(a)$ and $v = f(b)$. If u and v are close to zero, uv may underflow. The suggested test involving sign would avoid this difficulty. Make any other modifications that you think will improve or enhance the code. Test your routine on $f(x) = x^3 + 2x^2 + 10x - 20$, with $a = 1$ and $b = 2$. The zero is 1.368808108. In programming this polynomial function, use nested multiplication.

3. Write a program to find a zero of a function f in the following way: In each step, an interval $[a, b]$ is given and $f(a)f(b) < 0$. Then c is computed as the root of the linear function that agrees with f at a and b. We retain either $[a, c]$ or $[c, b]$, depending on whether $f(a)f(c) < 0$ or $f(c)f(b) < 0$. Test your program on several functions.

4. Select a routine from your program library to solve polynomial equations and use it to find the roots of the equation

$$x^8 - 36x^7 + 546x^6 - 4536x^5 + 22449x^4 - 67284x^3$$
$$+ 118124x^2 - 109584x + 40320 = 0$$

The correct roots are the integers $1, 2, \ldots, 8$. Next, solve the same equation when the coefficient of x^7 is changed to -37. Observe how a minor perturbation in the coefficients can cause massive changes in the roots. Thus, the roots are **unstable** functions of the coefficients. (Be sure to program the problem to allow for complex roots.)

5. Using the bisection method, determine the point of intersection of the curves given by $y = x^3 - 2x + 1$ and $y = x^2$.

6. Find a root of the following equation in the interval $[0, 1]$ by using the bisection method: $9x^4 + 18x^3 + 38x^2 - 57x + 14 = 0$.

7. Find a root of the equation $\tan x = x$ on the interval $[1, 2]$ by using the bisection method.

8. Find a root of the equation $6(e^x - x) = 6 + 3x^2 + 2x^3$ between -1 and $+1$ using the bisection method.

9. A circular metal shaft is being used to transmit power. It is known that at a certain critical angular velocity ω, any jarring of the shaft during rotation will cause the shaft to deform or buckle. This is a dangerous situation because the shaft might shatter under the increased centrifugal force. To find this critical velocity ω, we must first compute a number x that satisfies the equation

$$\tan x + \tanh x = 0$$

This number is then used in a formula to obtain ω. Solve for x ($x > 0$).

10. Use the bisection method to solve the problem that begins this chapter.

11. Code and test the recursive procedure for the bisection method.

3.2 Newton's Method

The procedure known as Newton's method is also called the **Newton-Raphson iteration**. It has a more general form than the one seen here, and the more general form can be used to find roots of systems of equations. Indeed, it is one of the more important procedures in numerical analysis, and its applicability extends to differential equations and integral equations. Here it is being applied to a single equation of the form $f(x) = 0$. As before, we seek one or more points at which the value of the function f is zero.

Interpretations of Newton's Method

In Newton's method, it is assumed at once that the function f is differentiable. This implies that the graph of f has a definite *slope* at each point and hence a unique tangent line. Now let us pursue the following simple idea. At a certain point $(x_0, f(x_0))$ on the graph of f there is a tangent, which is a rather good approximation to the curve in the vicinity of that point. Analytically, it means that the linear function

$$l(x) = f'(x_0)(x - x_0) + f(x_0)$$

is close to the given function f near x_0. At x_0 the two functions l and f agree. We take the zero of l as an approximation to the zero of f. The zero of l is easily found:

$$x_1 = x_0 - \frac{f(x_0)}{f'(x_0)}$$

Thus, starting with point x_0 (which we may interpret as an approximation to the root sought), we pass to a new point x_1 obtained from the preceding formula. Naturally the process can be repeated (iterated) to produce a sequence of points:

$$x_2 = x_1 - \frac{f(x_1)}{f'(x_1)}$$

$$x_3 = x_2 - \frac{f(x_2)}{f'(x_2)}$$

etc.

Under favorable conditions, the sequence of points will approach a zero of f, say r.

The geometry of Newton's method is shown in Figure 3.2. The line $y = l(x)$ is tangent to the curve $y = f(x)$. It intersects the x-axis at a point x_1. The slope of $l(x)$ is $f'(x_0)$.

There are other ways of interpreting Newton's method. Suppose again that x_0 is an initial approximation to a root of f. We ask: *What* **correction** *h should be added to x_0 in order to obtain the root precisely?* Obviously we want

$$f(x_0 + h) = 0$$

If f is a sufficiently well-behaved function, it will have a Taylor series at x_0 [see Equation (11) in Section 1.2]. Thus, we could write

$$f(x_0) + hf'(x_0) + \frac{h^2}{2}f''(x_0) + \cdots = 0$$

Determining h from this equation is, of course, not easy. Therefore, we give up the expectation of arriving at the true root in one step and seek only an approximation to h. This can be obtained by ignoring all but the first two terms in the series:

$$f(x_0) + hf'(x_0) = 0$$

The h that solves this is *not* the h that solves $f(x + h) = 0$, but it is the easily computed number

$$h = -\frac{f(x_0)}{f'(x_0)}$$

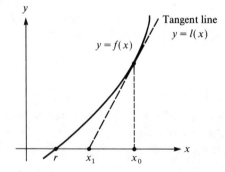

FIGURE 3.2
Newton's method

Our new approximation is then

$$x_1 = x_0 + h$$

and the process can be repeated. In retrospect, we see that the Taylor series was not needed after all because we used only the first two terms. In the analysis to be given later, it is assumed that f'' is continuous in a neighborhood of the root. This assumption enables us to estimate the errors in the process.

If Newton's method is described in terms of a sequence $x_0, x_1, \ldots$, then the following **recursive** or **inductive** definition applies:

$$x_{n+1} = x_n - \frac{f(x_n)}{f'(x_n)}$$

Naturally, the interesting question is whether

$$\lim_{n \to \infty} x_n = r$$

EXAMPLE 1   If $f(x) = x^3 - x + 1$ and $x_0 = 1$, what are x_1 and x_2 in the Newton iteration?

Solution   From the basic formula, $x_1 = x_0 - f(x_0)/f'(x_0)$. Now $f'(x) = 3x^2 - 1$, and so $f'(1) = 2$. Also $f(1) = 1$. Hence, we have $x_1 = 1 - \frac{1}{2} = \frac{1}{2}$. Similarly, $f(\frac{1}{2}) = \frac{5}{8}$, $f'(\frac{1}{2}) = -\frac{1}{4}$, and $x_2 = 3$.   □

Pseudocode

The pseudocode for Newton's method can be written as follows:

```
procedure newton (f, f', x, m)
integer n, m
real x, fx
interface external function f, f'
fx ← f(x)
output 0, x, fx
for n = 1 to m do
  x ← x − fx/f'(x)
  fx ← f(x)
  output n, x, fx
end do
end procedure newton
```

Here the procedure functions for $f(x)$ and $f'(x)$ must be supplied.

Illustration

Now we illustrate Newton's method by locating a root of $x^3 + x = 2x^2 + 3$. We apply the method to the function $f(x) = x^3 - 2x^2 + x - 3$, starting with $x_0 = 4$.

Of course, $f'(x) = 3x^2 - 4x + 1$, and these two functions should be arranged in nested form for efficiency:

$$f(x) = ((x - 2)x + 1)x - 3$$

$$f'(x) = (3x - 4)x + 1$$

To see in greater detail the rapid convergence of Newton's method, we use arithmetic with double the normal precision in the program and obtain the following results:

n	x_n	$f(x_n)$
0	4.0	33.0
1	3.0	9.0
2	2.4375	2.037
3	2.21303 27224 73144 5	0.256
4	2.17555 49386 14368 4	6.463×10^{-3}
5	2.17456 01006 55071 4	4.479×10^{-6}
6	2.17455 94102 93284 1	1.973×10^{-12}

Notice the doubling of the accuracy in $f(x)$ (and also in x) until the maximum precision of the computer is encountered. Figure 3.3 is a computer plot of three iterations of Newton's method for this sample problem.

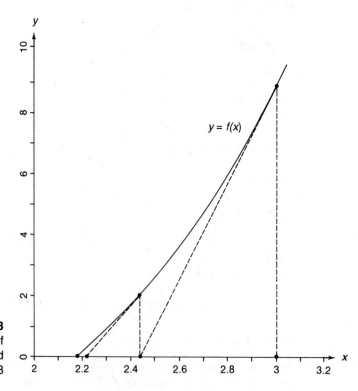

FIGURE 3.3
Three steps of
Newton's method
$f(x) = x^3 - 2x^2 + x - 3$

Next we use Maple to compute ten iterates in the above example to 100 decimal places.

```
Digits := 100;
x := 4.0;
f := ((x - 2.0)*x + 1.0)*x - 3.0;
for n from 1 by 1 to 10 do
x := x - f/((3.0*x - 4.0)*x + 1.0);
f := ((x - 2.0)*x + 1.0)*x - 3.0;
e := evalf(\kf f,2);
od;
```

After nine iterations, the value of x is

> 2.17455 94102 92980 07420 23189 88695 65392 56759 48725 33708
> 24983 36733 92030 23647 64792 75760 66115 28969 38832 0640

The values of the function at each iteration are $9.0, 2.0, 0.26, 0.0065, 0.45 \times 10^{-5}$, $0.22 \times 10^{-11}, 0.50 \times 10^{-24}, 0.27 \times 10^{-49}, 0.1 \times 10^{-98}$, and 0.1×10^{-98}. Again we see the number of digits of accuracy in Newton's method doubling (approximately) with each iteration once they are sufficiently close to the root. One can also use Maple to discover that this root is exactly

$$\sqrt[3]{\frac{79}{54} + \frac{1}{6}\sqrt{77}} + \frac{1}{9\sqrt[3]{\frac{79}{54} + \frac{1}{6}\sqrt{77}}} + \frac{2}{3}$$

Convergence Analysis

Anyone who has experimented with Newton's method—for instance, by working one of the problems—will have observed the remarkable rapidity in the convergence of the sequence to the root. This phenomenon is also noticeable in the example just given. Indeed, the number of correct figures in the answer is nearly *doubled* at each successive step. Thus in the example above, we have first 0 and then $1, 2, 3, 6, 12, 24, \ldots$ accurate digits from each Newton iteration. Five or six steps of Newton's method often suffice to yield full machine precision in the determination of a root. There is a theoretical basis for this dramatic performance, as we shall now see.

Let the function f, whose zero we seek, possess two continuous derivatives f' and f'', and let r be a zero of f. Assume further that r is a *simple* zero; that is, $f'(r) \neq 0$. Then Newton's method, if started sufficiently close to r, converges *quadratically* to r. This means that the errors in successive steps obey an inequality of the form

$$|x_{n+1} - r| \leq c |x_n - r|^2$$

We shall establish this fact presently, but first, an informal interpretation of the inequality may be helpful.

Suppose, for simplicity, that $c = 1$. Suppose also that x_n is an estimate of the root r that differs from it by at most one unit in the kth decimal place. This means that

$$|x_n - r| \leq 10^{-k}$$

The inequality above then implies that

$$|x_{n+1} - r| \leq 10^{-2k}$$

In other words, x_{n+1} differs from r by at most one unit in the $(2k)$th decimal place. So x_{n+1} has approximately twice as many correct digits as x_n! This is the doubling of significant digits alluded to previously.

To establish the quadratic convergence of Newton's method, let $e_n = x_n - r$. The formula that defines the sequence $\{x_n\}$ then gives

$$e_{n+1} = x_{n+1} - r = x_n - \frac{f(x_n)}{f'(x_n)} - r = e_n - \frac{f(x_n)}{f'(x_n)} = \frac{e_n f'(x_n) - f(x_n)}{f'(x_n)}$$

By Taylor's Theorem (see Section 1.2), there exists a point ξ_n situated between x_n and r for which

$$0 = f(r) = f(x_n - e_n) = f(x_n) - e_n f'(x_n) + \frac{1}{2} e_n^2 f''(\xi_n)$$

(The subscript on ξ_n emphasizes the dependence on x_n.) This last equation can be rearranged to read

$$e_n f'(x_n) - f(x_n) = \frac{1}{2} e_n^2 f''(\xi_n)$$

and if this is used in the previous equation for e_{n+1}, the result is

$$e_{n+1} = \frac{1}{2} \left(\frac{f''(\xi_n)}{f'(x_n)} \right) e_n^2 \tag{1}$$

This is, at least qualitatively, the sort of equation we want. Continuing the analysis, we define a function

$$c(\delta) = \frac{1}{2} \frac{\max\limits_{|x-r| \leq \delta} |f''(x)|}{\min\limits_{|x-r| \leq \delta} |f'(x)|} \qquad (\delta > 0) \tag{2}$$

By virtue of this definition, we can assert that, for any two points x and ξ within distance δ of the root r, the inequality $\frac{1}{2}|f''(\xi)/f'(x)| \leq c(\delta)$ is true. Now select δ so small that $\delta c(\delta) < 1$. This is possible because as δ approaches 0, $c(\delta)$ converges to $\frac{1}{2}|f''(r)/f'(r)|$, and so $\delta c(\delta)$ converges to 0. Recall that we assumed $f'(r) \neq 0$. Let $\rho = \delta c(\delta)$. In the remainder of this argument, we hold δ, $c(\delta)$, and ρ fixed with $\rho < 1$.

Suppose now that some iterate x_n lies within distance δ from the root r. From Equation (1), we have

$$|e_n| = |x_n - r| \leq \delta \qquad \text{and} \qquad |\xi_n - r| \leq \delta$$

By the definition of $c(\delta)$, it follows that $\frac{1}{2}|f''(\xi_n)|/|f'(x_n)| \leq c(\delta)$. From Equation (1), we now have

$$|e_{n+1}| = \frac{1}{2}\left|\frac{f''(\xi_n)}{f'(x_n)}\right| e_n^2 \leq c(\delta)e_n^2 \leq \delta c(\delta)|e_n| = \rho|e_n|$$

Consequently, x_{n+1} is also within distance δ of r because

$$|x_{n+1} - r| = |e_{n+1}| \leq \rho|e_n| \leq |e_n| \leq \delta$$

If the initial point x_0 is chosen within distance δ of r, then

$$|e_n| \leq \rho|e_{n-1}| \leq \rho^2|e_{n-1}| \leq \cdots \leq \rho^n|e_0|$$

Since $0 < \rho < 1$, $\lim_{n\to\infty} \rho^n = 0$ and $\lim_{n\to\infty} e_n = 0$. In other words, $\lim_{n\to\infty} x_n = r$. In this process, $|e_{n+1}| \leq c(\delta)e_n^2$.

To summarize:

THEOREM ON NEWTON'S METHOD

If f, f', and f'' are continuous in a neighborhood of a root r of f and if $f'(r) \neq 0$, then there is a positive δ with the following property: If the initial point in Newton's method satisfies $|x_0 - r| \leq \delta$, then all subsequent points x_n satisfy the same inequality, converge to r, and do so quadratically; that is,

$$|x_{n+1} - r| \leq c(\delta)|x_n - r|^2$$

where $c(\delta)$ is given by Equation (2).

In the use of Newton's method, consideration must be given to the proper choice of a starting point. Usually one must have some insight into the shape of the graph of the function. Sometimes a coarse graph is adequate, but in other cases a step-by-step evaluation of the function at various points may be necessary to find a point near the root. Often the bisection method is used initially to obtain a suitable starting point, and Newton's method is used to improve the precision.

Although Newton's method is truly a marvelous invention, its convergence depends upon hypotheses that are difficult to verify *a priori*. Some graphical examples will show what can happen. In Figure 3.4(a), the tangent to the graph of the function f at x_0 intersects the x-axis at a point remote from the root r, and successive points in Newton's iteration *recede* from r instead of converging to r. The difficulty can be ascribed to a poor choice of the initial point x_0; it is *not* sufficiently close to r. In Figure 3.4(b), the tangent to the curve is parallel to the x-axis and $x_1 = \pm\infty$ or it is assigned the value of machine infinity in a computer. In Figure 3.4(c), the iteration values *cycle* because $x_2 = x_0$. In a computer, roundoff errors or limited precision may eventually cause this situation to become unbalanced so that the iterates either spiral inward and converge or outward and diverge.

The analysis that establishes the quadratic convergence discloses another troublesome hypothesis—namely, that $f'(r) \neq 0$. If $f'(r) = 0$, then r is a zero of f and f'. Such a zero is termed a **multiple** zero of f—in this case, at least a double zero. Newton's iteration for a multiple zero converges only linearly! Ordinarily, one would not know in advance that the zero sought was a multiple zero. If one knew this, however, Newton's method could be accelerated by modifying the equation to read

$$x_{n+1} = x_n - m\frac{f(x_n)}{f'(x_n)}$$

in which m is the **multiplicity** of the zero in question. The multiplicity is the least m such that $f^{(k)}(r) = 0$ for $0 \leq k < m$, but $f^{(m)}(r) \neq 0$. (See Problem **23**.)

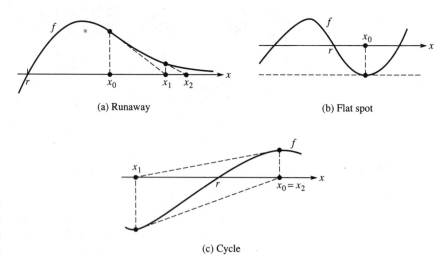

(a) Runaway

(b) Flat spot

(c) Cycle

FIGURE 3.4
Failure of Newton's
method due to
bad starting points

**PROBLEMS
3.2**

1. Verify that when Newton's method is used to compute $\sqrt{R}$ (by solving the equation $x^2 = R$), the sequence of iterates is defined by $x_{n+1} = (x_n + R/x_n)/2$.

2. (Continuation) Show that if the sequence $\{x_n\}$ is defined as in Problem 1, then $x_{n+1}^2 - R = [(x_n^2 - R)/2x_n]^2$. Interpret this equation in terms of quadratic convergence.

3. Write Newton's method in simplified form for determining the reciprocal of the square root of a positive number. Perform two iterations to approximate $1/\pm\sqrt{5}$, starting with $x_0 = 1$ and $x_0 = -1$.

4. Two of the four zeros of $x^4 + 2x^3 - 7x^2 + 3$ are positive. Find them by Newton's method, correct to two significant figures.

5. Derive a formula for Newton's method for the function $F(x) = f(x)/f'(x)$, where $f(x)$ is a function with simple zeros that is three times continuously differentiable. Show that the convergence of the resulting method to any zero r of $f(x)$ is at least quadratic. *Hint:* Apply the result in the text to F, making sure that F has the required properties.

6. The equation $x - Rx^{-1} = 0$ has $x = \pm R^{1/2}$ for its solution. Establish Newton's iterative scheme, in simplified form, for this situation. Carry out five steps for $R = 25$ and $x_0 = 1$.

7. The Taylor series for a function f looks like this:

$$f(x + h) = f(x) + hf'(x) + \frac{h^2}{2}f''(x) + \frac{h^3}{6}f'''(x) + \cdots$$

Suppose that $f(x)$, $f'(x)$, and $f''(x)$ are easily computed. Derive an algorithm like Newton's method that uses three terms in the Taylor series. The algorithm should take as input an approximation to the root and produce as output a better approximation to the root. Show that the method is cubically convergent.

8. Using a calculator, observe the sluggishness with which Newton's method converges in the case of $f(x) = (x - 1)^m$ with $m = 8$ or 12. Reconcile this with the theory. Use $x_0 = 1.1$.

9. What linear function $y = ax + b$ approximates $f(x) = \sin x$ best in the vicinity of $x = \pi/4$? How does this problem relate to Newton's method?

10. In Problems 7 to 9 of Section 1.2, several methods are suggested for computing $\ln 2$. Compare them with the use of Newton's method applied to the equation $e^x = 2$.

11. Define a sequence $x_{n+1} = x_n - \tan x_n$ with $x_0 = 3$. What is $\lim_{n\to\infty} x_n$?

12. The iteration formula $x_{n+1} = x_n - (\cos x_n)(\sin x_n) + R\cos^2 x_n$, where R is a positive constant, was obtained by applying Newton's method to some function $f(x)$. What was $f(x)$? What can this formula be used for?

13. **a.** Establish Newton's iterative scheme in simplified form, not involving the reciprocal of x, for the function $f(x) = xR - x^{-1}$.

    b. Carry out several steps of this procedure using $R = 4$.

14. To avoid computing the derivative at each step in Newton's method, it has been proposed to replace $f'(x_n)$ by $f'(x_0)$. Derive the rate of convergence for this method.

15. Consider the following procedures
   a. $x_{n+1} = \frac{1}{3}\left(2x_n - R/x_n^2\right)$
   b. $x_{n+1} = \frac{1}{2}x_n + 1/x_n$
   Do they converge for $x_0 \neq 0$? If so, to what values?

16. Each of the following functions has $\sqrt[3]{R}$ as a zero for any positive real number R. Determine the formulas for Newton's method for each and any necessary restrictions on the choice for x_0.
   a. $a(x) = x^3 - R$
   b. $b(x) = 1/x^3 - 1/R$
   c. $c(x) = x^2 - R/x$
   d. $d(x) = x - R/x^2$
   e. $e(x) = 1 - R/x^3$
   f. $f(x) = 1/x - x^2/R$
   g. $g(x) = 1/x^2 - x/R$
   h. $h(x) = 1 - x^3/R$

17. Determine the formulas for Newton's method for finding a root of the function $f(x) = x - e/x$. What is the behavior of the iterates?

18. If Newton's method is used on $f(x) = x^3 - x + 1$ starting with $x_0 = 1$, what will x_1 be?

19. Locate the root of $f(x) = e^{-x} - \cos x$ that is nearest $\pi/2$.

20. If Newton's method is used on $f(x) = x^5 - x^3 + 3$ and if $x_n = 1$, what is x_{n+1}?

21. Determine Newton's iteration formula for computing the cube root of N/M for nonzero integers N and M.

22. Refer to the discussion of Newton's method and establish that

$$\lim_{n \to \infty}(e_{n+1}e_n^{-2}) = \frac{1}{2}\left[\frac{f''(r)}{f'(r)}\right]$$

How can this be used in a practical case to test whether the convergence is quadratic? Devise an example in which r, $f'(r)$, and $f''(r)$ are all known, and test numerically the convergence of $e_{n+1}e_n^{-2}$.

23. Show that in the case of a zero of multiplicity m, the modified Newton's method

$$x_{n+1} = x_n - m\frac{f(x_n)}{f'(x_n)}$$

is quadratically convergent. *Hint:* Use Taylor series for $f(r + e_n)$ and $f'(r + e_n)$.

24. A method due to Steffensen for solving the equation $f(x) = 0$ uses the formula

$$x_{n+1} = x_n - \frac{f(x_n)}{g(x_n)}$$

in which $g(x) = \{f[x + f(x)] - f(x)\}/f(x)$. It is quadratically convergent, like Newton's method. How many function evaluations are necessary per step? Using Taylor series, show that $g(x) \approx f'(x)$ if $f(x)$ is small and thus relate Steffensen's iteration to Newton's. What advantage does Steffensen's have? Establish the quadratic convergence.

25. What happens if the Newton iteration is applied to the function $f(x) = \arctan x$ with $x_0 = 2$? For what starting values will Newton's method converge? (See Computer Problem **20**.)

26. Newton's method can be interpreted as follows: Suppose that $f(x + h) = 0$. Then $f'(x) \approx [f(x + h) - f(x)]/h = -f(x)/h$. Continue this argument.

27. For what starting values will Newton's method converge if $f(x) = x^2/(1 + x^2)$?

28. Starting at $x = 3$, $x < 3$, or $x > 3$, analyze what happens when Newton's method is applied to the function $f(x) = 2x^3 - 9x^2 + 12x + 15$.

29. (Continuation) Repeat Computer Problem **28** for $f(x) = \sqrt{|x|}$, starting with $x < 0$ or $x > 0$.

30. In order to determine $x = \sqrt[3]{R}$, we can solve the equation $x^3 = R$ by Newton's method. Write the loop that carries out this process, starting from the initial approximation $x_0 = R$.

31. The reciprocal of a number R can be computed without division by the formula

$$x_{n+1} = x_n(2 - x_n R)$$

Establish this relation by applying Newton's method to some $f(x)$. Beginning with $x_0 = 0.2$, compute the reciprocal of 4 correct to six decimal digits or more by this rule. Tabulate the error at each step and observe the quadratic convergence.

32. Newton's method for finding $\sqrt{R}$ is

$$x_{n+1} = \frac{1}{2}\left[x_n + \left(\frac{R}{x_n}\right)\right]$$

Perform three iterations of this scheme for computing $\sqrt{2}$, starting with $x_0 = 1$, and of the bisection method for $\sqrt{2}$, starting with interval $[1, 2]$. How many iterations are needed for each method in order to obtain 10^{-6} accuracy?

33. (Continuation) Newton's method for finding $\sqrt{R}$, where $R = AB$, gives this approximation:

$$\sqrt{AB} \approx \frac{A + B}{4} + \frac{AB}{A + B}$$

Show that if $x_0 = A$ or B, then two iterations of Newton's method are needed to obtain this approximation, whereas if $x_0 = \frac{1}{2}(A + B)$, then only one iteration is needed.

34. Consider the algorithm of which *one* step consists of two steps of Newton's method. What is its order of convergence?

35. Show that Newton's method applied to $x^m - R$ and to $1 - (R/x^m)$ for determining $\sqrt[m]{R}$ results in two similar yet different iterative formulas. Here $R > 0, m \geq 2$. Which is better and why?

36. A proposed generalization of Newton's method is

$$x_{n+1} = x_x - \omega \frac{f(x_n)}{f'(x_n)}$$

where the constant ω is an acceleration factor chosen to increase the rate of convergence. For what range of values of ω is a simple root r of $f(x)$ a **point of attraction**; that is, $|g'(r)| < 1$, where $g(x) = x - \omega f(x)/f'(x)$? This method is quadratically convergent *only* if $\omega = 1$ because $g'(r) \neq 0$ when $\omega \neq 1$.

37. Suppose that r is a double root of $f(x) = 0$; that is, $f(r) = f'(r) = 0$ but $f''(r) \neq 0$, and suppose that f and all derivatives up to and including the second are continuous in some neighborhood of r. Show that $e_{n+1} \approx \frac{1}{2}e_n$ for Newton's method and thereby conclude that the rate of convergence is *linear* near a double root. (If the root has multiplicity m, then $e_{n+1} \approx [(m-1)/m]e_n$.)

38. (Simultaneous nonlinear equations) Using the Taylor series in two variables (x, y) of the form

$$f(x + h, y + k) = f(x, y) + hf_x(x, y) + kf_y(x, y) + \cdots$$

where $f_x = \partial f/\partial x$ and $f_y = \partial f/\partial y$, establish that Newton's method for solving the two simultaneous nonlinear equations

$$\begin{cases} f(x, y) = 0 \\ g(x, y) = 0 \end{cases}$$

can be described with the formulas

$$x_{n+1} = x_n - \frac{fg_y - gf_y}{f_xg_y - g_xf_y}$$

$$y_{n+1} = y_n - \frac{f_xg - g_xf}{f_xg_y - g_xf_y}$$

Here the functions f, f_x, and so on are evaluated at (x_n, y_n).

39. Newton's method can be defined for the equation $f(z) = g(x, y) + ih(x, y)$, where $f(z)$ is an analytic function of the complex variable $z = x + iy$ (x and y real) and $g(x, y)$ and $h(x, y)$ are real functions for all x, y. The derivative

$f'(z)$ is given by $f'(z) = g_x + ih_x = h_y - ig_y$ because the **Cauchy-Riemann equations** $g_x = h_y$ and $h_x = -g_y$ hold. Here $g_x = \partial g / \partial x$, $g_y = \partial g / \partial y$, and so on. Show that Newton's method

$$z_{n+1} = z_n - \frac{f(z_n)}{f'(z_n)}$$

can be written in the form

$$x_{n+1} = x_n - \frac{gh_y - hg_y}{g_x h_y - g_y h_x}$$

$$y_{n+1} = y_n - \frac{hg_x - gh_x}{g_x h_y - g_y h_x}$$

Here all functions are evaluated at $z_n = x_n + iy_n$.

40. Using the idea of Problem **34**, show how we can easily create methods of arbitrarily high order for solving $f(x) = 0$. Why is the order of a method not the only criterion that should be considered in assessing its merits?

41. Using a hand calculator, carry out three iterations of Newton's method using $x_0 = 1$ and $f(x) = 3x^3 + x^2 - 15x + 3$.

COMPUTER PROBLEMS 3.2

1. Write

  procedure *newton*(f, f', x, m)

which carries out m steps of Newton's method using an external function f, its derivative f', and a starting point x. Test your code on these examples, using a single computer run: $f(t) = \tan t - t$ with $x_0 = 7$, and $g(t) = e^t - \sqrt{t + 9}$ with $x_0 = 2$. Print each iterate and its accompanying function value.

2. Write a simple, self-contained program to apply Newton's method to the equation $x^3 + 2x^2 + 10x = 20$, starting with $x_0 = 2$. Evaluate the appropriate $f(x)$ and $f'(x)$ using nested multiplication. Stop the computation when two successive points differ by $\frac{1}{2} \times 10^{-5}$ or some other convenient tolerance close to your machine's capability. Print all intermediate points and function values. Put an upper limit of ten on the number of steps.

3. (Continuation) Repeat Computer Problem **2** using double precision and more steps.

4. Find the root of the equation

$$2x(1 - x^2 + x) \ln x = x^2 - 1$$

in the interval $[0, 1]$ by Newton's method using double precision. Make a table that shows the number of correct digits in each step.

5. In 1685 John Wallis published a book called *Algebra*, in which he described a method devised by Newton for solving equations. In slightly modified form, this method was also published by Joseph Raphson in 1690. This form is the one now commonly called Newton's method or the Newton-Raphson method. Newton himself discussed the method in 1669 and illustrated it with the equation $x^3 - 2x - 5 = 0$. Wallis used the same example. Find a root of this equation in double precision, thus continuing the tradition that every numerical analysis student should solve this venerable equation.

6. In celestial mechanics, Kepler's equation is important. It reads $x = y - \varepsilon \sin y$, in which x is a planet's mean anomaly, y its eccentric anomaly, and ε the eccentricity of its orbit. Taking $\varepsilon = 0.9$, construct a table of y for 30 equally spaced values of x in the interval $0 \leq x \leq \pi$. Use Newton's method to obtain each value of y. The y corresponding to an x can be used as the starting point for the iteration when x is changed slightly.

7. In Newton's method, we progress in each step from a given point x to a new point $x - h$, where $h = f(x)/f'(x)$. A refinement that is easily programmed is this: If $|f(x - h)|$ is not smaller than $|f(x)|$, then reject this value of h and use $h/2$ instead. Test this refinement. See Problem **25** in the preceding set.

8. Write a brief program to compute a root of the equation $x^3 = x^2 + x + 1$, using Newton's method. Be careful to select a suitable starting value.

9. Find the root of the equation $5(3x^4 - 6x^2 + 1) = 2(3x^5 - 5x^3)$ that lies in the interval $[0, 1]$ by using Newton's method and a short program.

10. For each equation, write a brief program to compute and print eight steps of Newton's method for finding a positive root.

   a. $x = 2 \sin x$

   b. $x^3 = \sin x + 7$

   c. $x^5 + x^2 = 1 + 7x^3$ for $x \geq 2$

   d. $\sin x = 1 - x$

11. On a certain modern computer, floating-point numbers have a 48-bit mantissa. Moreover, floating-point hardware can perform addition, subtraction, multiplication, and reciprocation, but not division. Unfortunately, the reciprocation hardware produces a result accurate to only 30 bits, whereas the other operations produce results accurate to full floating-point precision.

   a. Show that Newton's method can be used to solve $f(x) = 1 - 1/(ax)$ for an approximation to $1/a$, which is accurate to full floating-point precision. How many iterations are required?

   b. Show how to obtain an approximation to b/a that is accurate to full floating-point precision.

12. Solve this pair of simultaneous nonlinear equations by first eliminating y and then solving the resulting equation in x by Newton's method. Start with $x_0 = 1.0$.

$$\begin{cases} x^3 - 2xy + y^7 - 4x^3y = 5 \\ y \sin x + 3x^2y + \tan x = 4 \end{cases}$$

13. In the Newton method for finding a root r of $f(x) = 0$, we start with x_0 and compute the sequence $x_1, x_2, \ldots$ using the formula $x_{n+1} = x_n - f(x_n)/f'(x_n)$. To avoid computing the derivative at each step, it has been proposed to replace $f'(x_n)$ with $f'(x_0)$ in all steps. It has also been suggested that the derivative in Newton's formula be computed only every other step. This method is given by

$$\begin{cases} x_{2n+1} = x_{2n} - \dfrac{f(x_{2n})}{f'(x_{2n})} \\ x_{2n+2} = x_{2n+1} - \dfrac{f(x_{2n+1})}{f'(x_{2n})} \end{cases}$$

Numerically compare both proposed methods to Newton's method for several simple functions that have known roots. Print the error of each method on every iteration in order to monitor the convergence. How well do the proposed methods work?

14. Write

    real function *sqrtf*(x)

    to compute the square root of a real argument x by the following algorithm: First, reduce the range of x by finding a real number r and an integer m such that $x = 2^{2m} r$ with $\frac{1}{4} \leq r < 1$. Next, compute x_2 by using two iterations of Newton's method given by

$$x_{n+1} = \frac{1}{2} \left(x_n + \frac{r}{x_n} \right)$$

    with the special initial approximation

$$x_0 = 1.27235\,367 + 0.24269\,3281 r - \frac{1.02966\,039}{1 + r}$$

    Then set $\sqrt{x} \approx 2^m x_2$. Test this algorithm on various values of x. Obtain a listing of the code for the square-root function on your computer system. By reading the comments, try to determine what algorithm it uses.

15. Write

    real function *cuberoot*(x)

    to compute the cube root of a real argument x by the following procedure: First, determine a real number r and an integer m such that $x = r2^{3m}$ with $\frac{1}{8} \leq r < 1$. Compute x_4 using four iterations of Newton's method:

$$x_{n+1} = \frac{2}{3} \left(x_n + \frac{r}{2x_n^2} \right)$$

with the special starting value

$$x_0 = 2.50292\,6 - \frac{8.04512\,5(r + 0.38775\,52)}{(r + 4.61224\,4)(r + 0.38775\,52) - 0.35984\,96}$$

Then set $\sqrt[3]{x} \approx 2^m x_4$. Test this algorithm on a variety of x values.

16. Would you like to see the number $0.55887\,766$ come out of a calculation? Take three steps in Newton's method on $10 + x^3 - 12\cos x = 0$ starting with $x_0 = 1$.

17. Write a short program to solve for a root of the equation $e^{-x^2} = \cos x + 1$ on $[0, 4]$. What happens in Newton's method if we start with $x_0 = 0$ or with $x_0 = 1$?

18. Find the root of the equation $\frac{1}{2}x^2 + x + 1 - e^x = 0$ by Newton's method, starting with $x_0 = 1$, and account for the slow convergence.

19. Find the zero of the function $f(x) = x - \tan x$ that is closest to 99 (radians) by both the bisection method and Newton's method. *Hint*: Extremely accurate starting values are needed for this function. Use the computer to construct a table of values of $f(x)$ around 99 to determine the nature of this function.

20. Using the bisection method, find the positive root of $2x(1 + x^2)^{-1} = \arctan x$. Using the root as x_0, apply Newton's method to the function $\arctan x$. Interpret the results.

21. If the root r of $f(x) = 0$ is a double root, then Newton's method can be accelerated by using

$$x_{n+1} = x_n - 2\frac{f(x_n)}{f'(x_n)}$$

Numerically compare the convergence of this scheme with Newton's method on a function with a known double root.

22. Program and test Steffensen's method, as described in Problem **24** of the preceding set.

23. Using Problem **38**, find a root of the nonlinear system

$$\begin{cases} 4y^2 + 4y + 52x - 19 = 0 \\ 169x^2 + 3y^2 + 111x - 10y - 10 = 0 \end{cases}$$

Use Newton's method and start at $(-0.01, -0.01)$.

24. Find a root of the nonlinear system using Problem **38**.

$$\begin{cases} \sin(x + y) = e^{x-y} \\ \cos(x + 6) = x^2 y^2 \end{cases}$$

25. Using Problem **39**, find a complex root of each of the following.

a. $z^3 - z - 1 = 0$

b. $z^4 - 2z^3 - 2iz^2 + 4iz = 0$

c. $2z^3 - 6(1 + i)z^2 - 6(1 - i) = 0$

d. $z = e^z$

Hint: For part **d**, use Euler's relation $e^{iy} = \cos y + i \sin y$.

26. Write and test a recursive procedure for Newton's method.

27. Using $f(x) = x^5 - 9x^4 - x^3 + 17x^2 - 8x - 8$ and $x_0 = 0$, study and explain the behavior of Newton's method. *Hint*: The iterates are initially cyclic.

3.3 Secant Method

Interpretations of Secant Method

We now consider a general-purpose procedure that converges almost as fast as Newton's method. This method mimics Newton's method but avoids the calculation of derivatives. Recall that Newton's iteration defines x_{n+1} in terms of x_n via the formula

$$x_{n+1} = x_n - \frac{f(x_n)}{f'(x_n)} \tag{1}$$

In the secant method, we replace $f'(x_n)$ in Formula (1) by an approximation that is easily computed. Since the derivative is *defined* by

$$f'(x) = \lim_{h \to 0} \frac{f(x + h) - f(x)}{h}$$

we can say that for small h,

$$f'(x) \approx \frac{f(x + h) - f(x)}{h}$$

In particular, if $x = x_n$ and $h = x_{n-1} - x_n$, we have

$$f'(x_n) \approx \frac{f(x_{n-1}) - f(x_n)}{x_{n-1} - x_n} \tag{2}$$

When this is used in Equation (1), the result defines the **secant method**:

$$x_{n+1} = x_n - f(x_n) \left(\frac{x_n - x_{n-1}}{f(x_n) - f(x_{n-1})} \right) \tag{3}$$

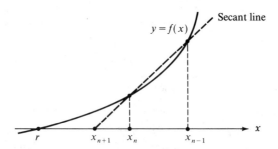

FIGURE 3.5
Secant method

The name of the method is taken from the fact that the right member of Equation (2) is the slope of a *secant* line to the graph of f (see Figure 3.5). Of course, the left member is the slope of a *tangent* line to the graph of f.

A few remarks about Equation (3) are in order. Clearly, x_{n+1} depends on *two* previous elements of the sequence. So in order to start, two points (x_0 and x_1) must be provided. Equation (3) can then generate $x_2, x_3, \ldots$. In programming the secant method, calculate and test the quantity $f(x_n) - f(x_{n-1})$. If it is nearly zero, an overflow can occur in Equation (3). Of course, if the method is succeeding, the points x_n will be approaching a zero of f, so $f(x_n)$ will be converging to zero. (We are assuming that f is continuous.) Also, $f(x_{n-1})$ will be converging to zero, and, *a fortiori*, $f(x_n) - f(x_{n-1})$ will approach zero. If the terms $f(x_n)$ and $f(x_{n-1})$ have the same sign, additional significant digits are canceled in the subtraction. So we could perhaps halt the iteration when $|f(x_n) - f(x_{n-1})| \leq \varepsilon |f(x_n)|$ for some specified tolerance ε, such as $\frac{1}{2} \times 10^{-5}$.

Secant Algorithm

The pseudocode for m steps of the secant method applied to the function f starting with the interval $[a, b] = [x_0, x_1]$ can be written as follows:

procedure *secant* (f, a, b, m)
integer n, m
real $a, b, fa, fb, temp$
interface external function f
$fa \leftarrow f(a)$
$fb \leftarrow f(b)$
output $0, a, fa$
output $1, b, fb$
for $n = 2$ **to** m **do**
  if $|fa| > |fb|$ **then**
    $a \leftrightarrow b$
    $fa \leftrightarrow fb$
  end if
  $temp \leftarrow (b - a)/(fb - fa)$
  $b \leftarrow a$
  $fb \leftarrow fa$

$$a \leftarrow a - (fa)(temp)$$
$$fa \leftarrow f(a)$$
output n, a, fa
end do
end procedure *secant*

The endpoints $[a, b]$ are interchanged, if necessary, in order to keep $|f(a)| \leq |f(b)|$. Consequently, the absolute values of the function are nonincreasing; thus we have $|f(x_n)| \geq |f(x_{n+1})|$ for $n \geq 1$.

EXAMPLE 1   If the secant method is used on $f(x) = x^5 + x^3 + 3$ with $x_0 = 1$ and $x_1 = -1$, what is x_8?

Solution   The output from the computer program corresponding to the pseudocode for the secant method is as follows. (We used a machine similar to the `Marc-32`.)

n	x_n	$f(x_n)$
0	-1.0	1.0
1	1.0	5.0
2	-1.5	-7.99
3	-1.05575	0.512
4	-1.11416	-9.991×10^{-2}
5	-1.10462	7.593×10^{-3}
6	-1.10529	1.010×10^{-4}
7	-1.10530	4.768×10^{-7}
8	-1.10530	4.768×10^{-7}

Using the following Maple statement

```
fsolve(x^5 + x^3 + 3, x, complex);
```

we find that there is one real root at -1.105298546 and there are two complex-conjugate pairs of roots. $\square$

Convergence Analysis

The advantages of the secant method are that (after the first step) only one function evaluation is required per step (in contrast to Newton's iteration, which requires two) and that it is almost as rapidly convergent. It can be shown that the errors obey an equation of the form

$$e_{n+1} = \frac{1}{2} \left(\frac{f''(\xi_n)}{f'(\zeta_n)} \right) e_n e_{n-1} \approx \frac{1}{2} \left(\frac{f''(r)}{f'(r)} \right) e_n e_{n-1} \tag{4}$$

where ξ_n and ζ_n are in the smallest interval that contains r, x_n, and x_{n-1}. Thus, the ratio $e_{n+1}(e_n e_{n-1})^{-1}$ converges to $\frac{1}{2} f''(r)/f'(r)$. The rapidity of convergence of this method is, in general, between those for bisection and for Newton's method.

To prove Equation (4), we begin with the definition of the secant method in Equation (3) and the error

$$e_{n+1} = x_{n+1} - r$$

$$= \frac{f(x_n)x_{n-1} - f(x_{n-1})x_n}{f(x_n) - f(x_{n-1})} - r$$

$$= \frac{f(x_n)e_{n-1} - f(x_{n-1})e_n}{f(x_n) - f(x_{n-1})}$$

$$= \left[\frac{x_n - x_{n-1}}{f(x_n) - f(x_{n-1})} \right] \left[\frac{\dfrac{f(x_n)}{e_n} - \dfrac{f(x_{n-1})}{e_{n-1}}}{x_n - x_{n-1}} \right] e_n e_{n-1} \qquad (5)$$

By Taylor's Theorem,

$$f(x_n) = f(r + e_n) = f(r) + e_n f'(r) + \frac{1}{2}e_n^2 f''(r) + \mathcal{O}(e_n^3)$$

Since $f(r) = 0$, this gives us

$$\frac{f(x_n)}{e_n} = f'(r) + \frac{1}{2}e_n f''(r) + \mathcal{O}(e_n^2)$$

Changing the index to $n - 1$ yields

$$\frac{f(x_{n-1})}{e_{n-1}} = f'(r) + \frac{1}{2}e_{n-1} f''(r) + \mathcal{O}(e_{n-1}^2)$$

By subtraction between these equations, we arrive at

$$\frac{f(x_n)}{e_n} - \frac{f(x_{n-1})}{e_{n-1}} = \frac{1}{2}(e_n - e_{n-1})f''(r) + \mathcal{O}(e_{n-1}^2)$$

Since $x_n - x_{n-1} = e_n - e_{n-1}$, we reach the equation

$$\frac{\dfrac{f(x_n)}{e_n} - \dfrac{f(x_{n-1})}{e_{n-1}}}{x_n - x_{n-1}} \approx \frac{1}{2}f''(r)$$

The first bracketed expression in Equation (5) can be written as

$$\frac{x_n - x_{n-1}}{f(x_n) - f(x_{n-1})} \approx \frac{1}{f'(r)}$$

Hence, we have shown part of Equation (4).

We can establish the other part of Equation (4) from Equation (5) by using some results proved in Chapter 4:

$$e_{n+1} = \frac{1}{f[x_n, x_{n-1}]} \left[\frac{\dfrac{f(x_n) - f(r)}{x_n - r} - \dfrac{f(x_{n-1}) - f(r)}{x_{n-1} - r}}{x_n - x_{n-1}} \right] e_n e_{n-1}$$

$$= \frac{1}{f[x_n, x_{n-1}]} \left[\frac{f[x_n, r] - f[r, x_{n-1}]}{x_n - x_{n-1}} \right] e_n e_{n-1}$$

$$= \frac{1}{f[x_n, x_{n-1}]} f[x_n, r, x_{n-1}] e_n e_{n-1}$$

$$= \frac{1}{f'(\zeta_n)} \left[\frac{1}{2} f''(\xi_n) \right] e_n e_{n-1}$$

Here we have used the notation of divided differences, as in Section 4.1. We used their recursive property and a theorem in Section 4.2 that relates divided differences to derivatives.

From Equation (4), we can now derive the order of convergence, which is

$$|e_{n+1}| \leq C |e_n|^\alpha \quad \text{where} \quad \alpha = \frac{1}{2}\left(1 + \sqrt{5}\right) \approx 1.62 \tag{6}$$

Since $\alpha > 1$, we say that the convergence is **superlinear**. Let $c = c(\delta)$ be defined as in Equation (2) of the preceding section. If $|x_n - r| \leq \delta$ and $|x_{n-1} - r| \leq \delta$, then Equation (4) yields

$$|e_{n+1}| \leq c |e_n| |e_{n-1}|$$

Suppose that the initial points x_0 and x_1 are sufficiently close to r so that $c |e_0| \leq D$ and $c |e_1| \leq D$ for some $D < 1$. Then

$$c |e_2| \leq c |e_1| \, c |e_0| \leq D^2$$
$$c |e_3| \leq c |e_2| \, c |e_1| \leq D^3$$
$$c |e_4| \leq c |e_3| \, c |e_2| \leq D^5$$
$$c |e_5| \leq c |e_4| \, c |e_3| \leq D^8$$

etc.

In general, we have

$$c |e_n| \leq D^{\lambda_n}$$

where inductively

$$\begin{cases} \lambda_0 = \lambda_1 = 1 \\ \lambda_n = \lambda_{n-1} + \lambda_{n-2} \end{cases} \tag{7}$$

These are the famous **Fibonacci** numbers $(1, 1, 2, 3, 5, 8, \ldots)$. Since $D < 1$ and $\lambda_n \rightarrow \infty$, we conclude that $e_n \rightarrow 0$. The Fibonacci sequence has the explicit form

$$\lambda_n = \frac{1}{\sqrt{5}} \left(\alpha^{n+1} + \beta^{n+1} \right) \tag{8}$$

where $\alpha = \frac{1}{2}(1 + \sqrt{5})$ and $\beta = \frac{1}{2}(1 - \sqrt{5})$. From a previous equation, we now have

$$\begin{aligned}
|e_{n+1}| &\leq c |e_n| |e_{n-1}| \\
&= c |e_n|^\alpha |e_n|^{1-\alpha} |e_{n-1}| \\
&\leq c |e_n|^\alpha \left(\frac{D^{\lambda_n}}{c} \right)^{1-\alpha} \left(\frac{D^{\lambda_{n-1}}}{c} \right) \\
&= |e_n|^\alpha c^{\alpha-1} D^{\lambda_n(1-\alpha) + \lambda_{n-1}} \\
&= |e_n|^\alpha c^{\alpha-1} D^{\lambda_{n+1} - \alpha \lambda_n}
\end{aligned}$$

In the last line we used the relation (7). Now $\lambda_{n+1} - \alpha \lambda_n$ converges to zero (see Problem **15**). Hence, $c^{\alpha-1} D^{\lambda_{n+1} - \alpha \lambda_n}$ is bounded, say by C, as a function of n. Hence, Equation (6) is established.

Summary

In this chapter, three methods for solving an equation $f(x) = 0$ have been presented. The bisection method is reliable but slow. Newton's method is fast, but often only near the root. The secant method is nearly as fast as Newton's method and does not require a programming or even a knowledge of the derivative f'. The user of the bisection method must provide two points at which the signs of $f(x)$ differ, and the function f need only be continuous. In using Newton's method, one must specify a starting point near the root, and f must be differentiable. The secant method also requires good starting points. Newton's procedure can be interpreted as the repetition of a two-step procedure summarized by the prescription *linearize and solve*. This strategy is applicable in many other numerical problems, and its importance cannot be overemphasized.

**PROBLEMS
3.3**

1. What is the appropriate formula for finding square roots using the secant method? (Refer to Problem **1** in Section 3.2.)

2. The formula for the secant method can also be written as

$$x_{n+1} = \frac{x_{n-1}f(x_n) - x_n f(x_{n-1})}{f(x_n) - f(x_{n-1})}$$

Establish this, and explain why it is inferior to Equation (3) in a computer program.

3. Show that if the iterates in Newton's method converge to a point r for which $f'(r) \neq 0$, then $f(r) = 0$. Establish the same assertion for the secant method. *Hint*: In the latter, the Mean-Value Theorem of Differential Calculus is useful. This is the case $n = 0$ in Taylor's Theorem.

4. A method of finding a zero of a given function f proceeds as follows. Two initial approximations x_0 and x_1 to the zero are chosen, the value of x_0 is fixed, and successive iterations are given by

$$x_{n+1} = x_n - \left(\frac{x_n - x_0}{f(x_n) - f(x_0)}\right) f(x_n)$$

This process will converge to a zero of f under certain conditions. Show that the rate of convergence to a simple zero is *linear*.

5. If $x_{n+1} = x_n + (2 - e^{x_n})(x_n - x_{n-1})/(e^{x_n} - e^{x_{n-1}})$ with $x_0 = 0$ and $x_1 = 1$, what is $\lim_{n \to \infty} x_n$?

6. Using the bisection method, Newton's method, and the secant method, find the largest positive root correct to three decimal places of $x^3 - 5x + 3 = 0$. (All roots are in $[-3, +3]$.)

7. Calculate an approximate value for $4^{3/4}$ using one step of the secant method with $x_0 = 3$ and $x_1 = 2$.

8. If we use the secant method on $f(x) = x^3 - 2x + 2$ starting with $x_0 = 0$ and $x_1 = 1$, what is x_2?

9. If the secant method is used on $f(x) = x^5 + x^3 + 3$ and if $x_{n-2} = 0$ and $x_{n-1} = 1$, what is x_n?

10. Test the following sequences for different types of convergence (i.e., linear, superlinear, or quadratic), where $n = 1, 2, 3 \ldots$.

    a. $x_n = n^{-2}$

    b. $x_n = 2^{-n}$

    c. $x_n = 2^{-a_n}$ with $a_0 = a_1 = 1$ and $a_{n+1} = a_n + a_{n-1}$ for $n \geq 2$

    d. $x_n = 2^{-2^n}$

11. This problem and the next three deal with the method of **functional iteration**. A **fixed point** of a function f is an x such that $f(x) = x$. The method of functional iteration is as follows: Starting with any x_0, we define $x_{n+1} = f(x_n)$, where $n = 0, 1, 2, \ldots$. Show that if f is continuous and if the sequence $\{x_n\}$ converges, then its limit is a fixed point of f.

12. (Continuation) Show that if f is a function defined on the whole real line whose derivative satisfies $|f'(x)| \leq c$ with a constant c less than 1, then the method of functional iteration produces a fixed point of f. *Hint*: In establishing this, the Mean-Value Theorem from Section 1.2 is helpful.

13. (Continuation) With a calculator, try the method of functional iteration with $f(x) = x/2 + 1/x$, taking $x_0 = 1$. What is the limit of the resulting sequence?

14. (Continuation) Using Problem **11**, show that the equation $10 - 2x + \sin x = 0$ has a root. Locate the root approximately by drawing a graph. Starting with your approximate root, use functional iteration to obtain the root accurately by using a calculator. *Hint*: Write the equation in the form $x = 5 + \frac{1}{2} \sin x$.

15. Prove that in the analysis of the secant method, $\lambda_{n+1} - \alpha \lambda_n$ converges to zero as $n \to \infty$.

COMPUTER PROBLEMS 3.3

1. Write a simple program to compare the secant method with Newton's method for finding a root of each function.

   a. $x^3 - 3x + 1$ with starting point $x_0 = 2$

   b. $x^3 - 2 \sin x$ with starting point $x_0 = \frac{1}{2}$ (x is in radians.)

   Use the x_1 value from Newton's method as the second starting point for the secant method. Print out each iteration for both methods.

2. Write a simple program to find the root of $f(x) = x^3 + 2x^2 + 10x - 20$ using the secant method with starting values $x_0 = 2$ and $x_1 = 1$. Let it run at most 20 steps, but stop the iteration when $f(x_n) - f(x_{n-1})$ is very small. Compare the number of steps needed here to the number needed in Newton's method. Is the convergence quadratic?

3. Test the secant method on the set of functions $f_k(x) = 2e^{-k}x + 1 - 3e^{-kx}$ for $k = 1, 2, 3, \ldots, 10$. Use the starting points 0 and 1 in each case.

4. An example by Wilkinson [1963] shows that minute alterations in the coefficients of a polynomial may have massive effects on the roots. Let

$$f(x) = (x - 1)(x - 2) \cdots (x - 20)$$

The zeros of f are, of course, the integers $1, 2, \ldots, 20$. Try to determine what happens to the zero $r = 20$ when the function is altered to $f(x) - 10^{-8}x^{19}$. *Hint*: The secant method in double precision will locate a zero in the interval $[20, 21]$.

5. Test the secant method on an example in which r, $f'(r)$, and $f''(r)$ are known in advance. Monitor the ratios $e_{n+1}/(e_n e_{n-1})$ to see whether they converge to $\frac{1}{2}f''(r)/f'(r)$. The function $f(x) = \arctan x$ is suitable for this experiment.

6. Use the secant method to find the zero near -0.5 of $f(x) = e^x - 3x^2$. This function also has a zero near 4. Find this positive zero by Newton's method.

7. Write

procedure *secant*($f, x1, x2, epsi, delta, maxf, x, ierr$)

which uses the secant method to solve $f(x) = 0$. The input parameters are as follows: f is the name of the given function; $x1$ and $x2$ are the initial estimates of the solution; *epsi* is a positive tolerance such that the iteration stops if the difference between two consecutive iterates is smaller than this value; *delta* is a positive tolerance such that the iteration stops if a function value is smaller in magnitude than this value; and *maxf* is a positive integer bounding the number of evaluations of the function allowed. The output parameters are as follows: x is the final estimate of the solution, and *ierr* is an integer error flag that indicates whether a tolerance test was violated. Test this routine using the function of Computer Problem **6**. Print the final estimate of the solution and the value of the function at this point.

8. Using a function of your choice, verify numerically that the iterative method

$$x_{n+1} = x_n - \frac{f(x_n)}{\sqrt{[f'(x_n)]^2 - f(x_n)f''(x_n)}}$$

is cubically convergent at a simple root but only linearly convergent at a multiple root.

9. Test numerically whether Olver's method, given by

$$x_{n+1} = x_n - \frac{f(x_n)}{f'(x_n)} - \frac{1}{2}\frac{f''(x_n)}{f'(x_n)}\left[\frac{f(x_n)}{f'(x_n)}\right]^2$$

is cubically convergent to a root of f. Try to establish that it is.

10. (Continuation) Repeat Computer Problem **9** for Halley's method

$$x_{n+1} = x_n - \frac{1}{a_n}$$

with

$$a_n = \frac{f'(x_n)}{f(x_n)} - \frac{1}{2}\left[\frac{f''(x_n)}{f'(x_n)}\right]$$

11. (Moler-Morrison algorithm) Computing an approximation for $\sqrt{x^2 + y^2}$ does not require square roots. It can be done as follows:

real function $f(x, y)$
integer n
real a, b, c, x, y

$$f \leftarrow \max\{|x|, |y|\}$$
$$a \leftarrow \min\{|x|, |y|\}$$
for $n = 1$ **to** 3 **do**
$$b \leftarrow (a/f)^2$$
$$c \leftarrow b/(4 + b)$$
$$f \leftarrow f + 2cf$$
$$a \leftarrow ca$$
end do
end function f

Test the algorithm on some simple cases such as $(x, y) = (3, 4)$, $(-5, 12)$, and $(7, -24)$. Then write a routine that uses the function $f(x, y)$ for approximating the Euclidean norm of a vector $x = (x_1, x_2, \ldots, x_n)$; that is, the nonnegative number $\|x\| = (x_1^2 + x_2^2 + \cdots + x_n^2)^{1/2}$.

12. Find a zero of one of the functions given in the introduction of this chapter using one of the methods introduced in this chapter.

13. Write and test a recursive procedure for the secant method.

14. Rerun the example in this section with $x_0 = 0$ and $x_1 = 1$. Explain any unusual results.

4 INTERPOLATION AND NUMERICAL DIFFERENTIATION

As a result of an expensive and laborious calculation, the following values of a function have been obtained:

x	y
0.924	−0.00851 37255 5933
0.928	−0.00387 71887 1002
0.932	0.00080 22122 9861
0.936	0.00552 49809 8693
0.940	0.01029 16304 4063

From this information, what value of x corresponds to $y = 0$?

Let $(x_0, y_0), (x_1, y_1), \ldots, (x_4, y_4)$ be the five number pairs in the table. Using the methods of this chapter, one can easily construct a polynomial p of degree 4 such that $p(y_i) = x_i$ for $0 \leq i \leq 4$. In the neighborhood of these five points, we can assume that the equation $x = p(y)$ is approximately true. The x value wanted is then $x = p(0)$. When this calculation is carried out, $x \approx 0.93132$. (This is an example of *inverse* interpolation.)

We pose three problems concerning the representation of functions to give an indication of the subject matter that we present in this chapter, Chapter 7 (on splines), and Chapter 10 (on least squares).

First, suppose that we have the following table of numerical values of a function:

x	x_0	x_1	$\cdots$	x_n
y	y_0	y_1	$\cdots$	y_n

Is it possible to find a simple and convenient formula that reproduces the given points exactly?

The second problem is similar, but it is assumed that the given table of numerical values is contaminated by errors, as might occur if the values came from a physical experiment. Now we ask for a formula that represents the data (approximately) and, if possible, filters out the errors.

As a third problem, a function f is given, perhaps in the form of a computer procedure, but it is an expensive function to evaluate. In this case, we ask for another function g that is simpler to evaluate and produces a reasonable approximation to f.

In all of these problems, a simple function p can be obtained that represents or approximates the given table or function f. The representation p can always be taken to be a polynomial, although many other types of simple functions can also be used. Once a simple function p has been obtained, it can be used in place of f in many situations. For example, the integral of f could be estimated by the integral of p, and the latter should generally be easier to evaluate.

In many situations, a polynomial solution to the problems outlined above will be unsatisfactory from a practical point of view, and other classes of functions must be considered. In this book, one other class of versatile functions is discussed: the **spline** functions (see Chapter 7). The present chapter concerns polynomials exclusively, and Chapter 10 discusses general linear families of functions, with splines and polynomials being important examples.

The obvious way in which a polynomial can *fail* as a practical solution to one of the preceding problems is that its degree may be unreasonably high. For instance, if the table considered contains 1000 entries, a polynomial of degree 999 may be required to represent it. Polynomials also may have the surprising defect of being highly **oscillatory**. If the table is precisely represented by a polynomial p, then $p(x_i) = y_i$ for $0 \leqq i \leqq n$. For points other than the given x_i, however, $p(x)$ may be a very poor representation of the function from which the table arose. The example in Section 4.2 involving the Runge function illustrates this phenomenon.

4.1 Polynomial Interpolation

Existence of Interpolating Polynomial

We begin again with a table of values:

x	x_0	x_1	$\cdots$	x_n
y	y_0	y_1	$\cdots$	y_n

and assume that the x_i's form a set of $n + 1$ distinct points. The table represents $n + 1$ points in the Cartesian plane, and we want to find a polynomial curve that passes through all points. Thus, we seek to determine a polynomial, which is defined for *all* x, that takes on the corresponding values of y_i for each of the $n + 1$ distinct x_i's in this table. A polynomial p for which $p(x_i) = y_i$ when $0 \leqq i \leqq n$ is said to **interpolate** the table. The points x_i are called **nodes**. Generally, a different polynomial is required for different-sized tables; that is, the desired polynomial depends on n.

Consider the first and simplest case, $n = 0$. Here a constant function solves the problem. That is, the polynomial p of degree 0 defined by the equation $p(x) = y_0$ reproduces the one-node table.

The next simplest case is when $n = 1$. Since a straight line can be passed through two points, a linear function is capable of solving the problem. Explicitly, the polynomial p defined by

$$p(x) = y_0 + \left(\frac{y_1 - y_0}{x_1 - x_0}\right)(x - x_0)$$

is of first degree (at most) and reproduces the table. That means (in this case) that $p(x_0) = y_0$ and $p(x_1) = y_1$, as is easily verified. This p is used for *linear interpolation*.

EXAMPLE 1   Find the polynomial of least degree that interpolates this table:

x	1.4	1.25
y	3.7	3.9

Solution   By the equation above, the polynomial sought is

$$p(x) = 3.7 + \frac{3.9 - 3.7}{1.25 - 1.4}(x - 1.4)$$

$$= 3.7 - \frac{4}{3}(x - 1.4) \qquad \qquad \square$$

Now suppose that we have succeeded in finding a polynomial p that reproduces *part* of the table. Assume, say, that $p(x_i) = y_i$ for $0 \le i \le k$. We shall attempt to add to p another term that will enable the new polynomial to reproduce one more entry in the table. We consider

$$p(x) + c(x - x_0)(x - x_1) \cdots (x - x_k)$$

where c is a constant to be determined. This is surely a polynomial. It also reproduces the first k points in the table because p itself does so, and the added portion takes the value 0 at each of the points $x_0, x_1, \ldots, x_k$. (Its form is chosen for precisely this reason.) Now we adjust the parameter c so that the new polynomial takes the value y_{k+1} at x_{k+1}. Imposing this condition, we obtain

$$p(x_{k+1}) + c(x_{k+1} - x_0)(x_{k+1} - x_1) \cdots (x_{k+1} - x_k) = y_{k+1}$$

The proper value of c *can* be obtained from this equation because none of the factors $x_{k+1} - x_i$, for $0 \le i \le k$, can be zero. Remember our original assumption that the x_i's are all distinct.

This analysis is an example of inductive reasoning. We have shown that the process can be started and that it can be continued. Hence, the following formal statement has been partially justified:

THEOREM ON EXISTENCE OF POLYNOMIAL INTERPOLATION

If points $x_0, x_1, \ldots, x_n$ are distinct, then for arbitrary real values $y_0, y_1, \ldots, y_n$, there is a unique polynomial p of degree $\leq n$ such that $p(x_i) = y_i$ for $0 \leq i \leq n$.

Two parts of this formal statement must still be established. First, the degree of the polynomial increases by at most 1 in each step of the inductive argument. At the beginning the degree was 0, so at the end the degree is at most n.

Second, we establish the uniqueness of the polynomial p. Suppose that another polynomial q claims to accomplish what p does; that is, q is also of degree at most n and satisfies $q(x_i) = y_i$ for $0 \leq i \leq n$. Then the polynomial $p - q$ is of degree at most n and takes the value 0 at $x_0, x_1, \ldots, x_n$. Recall, however, that a *nonzero* polynomial of degree n can have at most n roots. We conclude that $p = q$, which establishes the uniqueness of p.

Newton Interpolating Polynomial

The preceding discussion provides a method for constructing an interpolating polynomial. The method is known as the *Newton algorithm*, and the resulting polynomial is the *Newton interpolating polynomial*.

EXAMPLE 2   Using the Newton algorithm, find the interpolating polynomial of least degree for this table:

x	0	1	-1	2	-2
y	-5	-3	-15	39	-9

Solution   In the construction, five successive polynomials will appear; these are labeled p_0, p_1, p_2, p_3, and p_4. The polynomial p_0 is defined to be

$$p_0(x) = -5$$

The polynomial p_1 has the form

$$p_1(x) = p_0(x) + c(x - x_0) = -5 + c(x - 0)$$

The interpolation condition placed on p_1 is that $p_1(1) = -3$. Therefore, we have $-5 + c(1 - 0) = -3$. Hence, $c = 2$, and p_1 is

$$p_1(x) = -5 + 2x$$

The polynomial p_2 has the form

$$p_2(x) = p_1(x) + c(x - x_0)(x - x_1)$$
$$= -5 + 2x + cx(x - 1)$$

The interpolation condition placed on p_2 is that $p_2(-1) = -15$. Hence, we have $-5 + 2(-1) + c(-1)(-1 - 1) = -15$. This yields $c = -4$, and so

$$p_2(x) = -5 + 2x - 4x(x - 1)$$

The remaining steps are similar, and the final result is the Newton interpolating polynomial

$$p_4(x) = -5 + 2x - 4x(x - 1) + 8x(x - 1)(x + 1)$$
$$+ 3x(x - 1)(x + 1)(x - 2)$$

The Maple statement

```
interp([0, 1, -1, 2, -2], [-5, -3, -15, 39, -9], x);
```

results in the polynomial

$$p_4(x) = 3x^4 + 2x^3 - 7x^2 + 4x - 5 \qquad \square$$

Later we shall develop a better algorithm for constructing the Newton interpolating polynomial. Nevertheless, the method just explained is a systematic one and involves very little computation. An important feature to notice is that each new polynomial in the algorithm is obtained from its predecessor by adding a new term. Thus, at the end, the final polynomial exhibits all the previous polynomials as constituents.

Nested Form

Before continuing, let us rewrite the Newton interpolating polynomial for efficient evaluation.

EXAMPLE 3   Write the polynomial p_4 of the preceding example in *nested* form and use it to evaluate $p_4(3)$.

Solution   We write p_4 as

$$p_4(x) = -5 + x(2 + (x - 1)(-4 + (x + 1)(8 + (x - 2)3)))$$
$$= (((3(x - 2) + 8)(x + 1) - 4)(x - 1) + 2)x - 5$$

Therefore,

$$p_4(3) = (((3(3 - 2) + 8)(3 + 1) - 4)(3 - 1) + 2)3 - 5$$

$$= 241 \qquad \qquad \square$$

This form is obtained by systematic factoring of the original polynomial. It is known as the **nested** form and its evaluation is by **nested multiplication**.

To describe nested multiplication in a formal way (so that it can be translated into a pseudocode), consider a general polynomial in the Newton form. It might be

$$p(x) = a_0 + a_1[(x - x_0)] + a_2[(x - x_0)(x - x_1)] + \cdots$$
$$+ a_n[(x - x_0)(x - x_1) \cdots (x - x_{n-1})]$$

This can be written succinctly as

$$p(x) = a_0 + \sum_{i=1}^{n} a_i \left[\prod_{j=0}^{i-1} (x - x_j) \right] \qquad (1)$$

where the standard product notation has been used. The nested form of $p(x)$ is

$$p(x) = a_0 + (x - x_0)(a_1 + (x - x_1)(a_2 + \cdots + (x - x_{n-1})a_n)) \cdots))$$
$$= (\cdots ((a_n(x - x_{n-1}) + a_{n-1})(x - x_{n-2}) + a_{n-2}) \cdots)(x - x_0) + a_0$$

In evaluating $p(t)$ for a given value of t, we naturally start with the innermost parentheses, forming successively the following quantities:

$$v_0 = a_n$$

$$v_1 = v_0(t - x_{n-1}) + a_{n-1}$$

$$v_2 = v_1(t - x_{n-2}) + a_{n-2}$$

$$\vdots$$

$$v_n = v_{n-1}(t - x_0) + a_0$$

The quantity v_n is now $p(t)$. In the following pseudocode, a subscripted variable is not needed for v_i. Instead, we can write

real array $(a_i)_{0:n}, (x_i)_{0:n}$
integer i, n
real t, v

$\vdots$

$v \leftarrow a_n$

for $i = n - 1$ **to** 0 **step** -1 **do**
    $v \leftarrow v(t - x_i) + a_i$
end do

Here the array $(a_i)_{0:n}$ contains the $n + 1$ coefficients of the Newton interpolating polynomial (1) of degree at most n, and the array $(x_i)_{0:n}$ contains the $n + 1$ nodes x_i.

Calculating Coefficients a_i Using Divided Differences

We turn now to the problem of determining the coefficients $a_0, a_1, \ldots, a_n$ efficiently. Again we start with a table of values of a function f:

x	x_0	x_1	x_2	$\cdots$	x_n
$f(x)$	$f(x_0)$	$f(x_1)$	$f(x_2)$	$\cdots$	$f(x_n)$

The points $x_0, x_1, \ldots, x_n$ are assumed to be distinct, but no assumption is made about their positions on the real line.

Previously, we established that for each $n = 0, 1, \ldots$, there exists a unique polynomial p_n such that:

1. The degree of p_n is at most n.

2. $p_n(x_i) = f(x_i)$ for $i = 0, 1, \ldots, n$.

It was shown that p_n can be expressed in the Newton form:

$$p_n(x) = a_0 + a_1(x - x_0) + a_2(x - x_0)(x - x_1) + \cdots + a_n(x - x_0) \cdots (x - x_{n-1})$$

The compact form of this equation is

$$p_n(x) = \sum_{i=0}^{n} a_i \prod_{j=0}^{i-1} (x - x_j) \tag{2}$$

in which $\prod_{j=0}^{-1}(x - x_j)$ is interpreted to be 1. A crucial observation about p_n is that the coefficients $a_0, a_1, \ldots$ do not depend on n. In other words, p_n is obtained from p_{n-1} by adding one more term, without altering the coefficients already present in p_{n-1} itself. This is because we began with the hope that p_n could be expressed in the form

$$p_n(x) = p_{n-1}(x) + a_n(x - x_0) \cdots (x - x_{n-1})$$

and discovered that it was indeed possible.

A way of systematically determining the unknown coefficients $a_0, a_1, \ldots, a_n$ is to set x equal in turn to $x_0, x_1, \ldots, x_n$ in Equation (2) and to write down the resulting equations:

$$\begin{cases} f(x_0) = a_0 \\ f(x_1) = a_0 + a_1(x_1 - x_0) \\ f(x_2) = a_0 + a_1(x_2 - x_0) + a_2(x_2 - x_0)(x_2 - x_1) \\ \text{etc.} \end{cases} \tag{3}$$

The compact form of Equations (3) is

$$f(x_k) = \sum_{i=0}^{k} a_i \prod_{j=0}^{i-1}(x_k - x_j) \qquad (0 \le k \le n) \tag{4}$$

Equation (3) can be solved for the a_i's in turn, starting with a_0. Then we see that a_0 depends on $f(x_0)$, that a_1 depends on $f(x_0)$ and $f(x_1)$, and so on. In general, a_k depends on $f(x_0), f(x_1), \dots, f(x_k)$. In other words, a_k depends on f at $x_0, x_1, \dots, x_k$. The traditional notation is

$$a_k = f[x_0, x_1, \dots, x_k] \tag{5}$$

This equation defines $f[x_0, x_1, \dots, x_k]$. The quantities $f[x_0, x_1, \dots, x_k]$ are called the **divided differences** of f. Notice also that the coefficients $a_0, a_1, \dots, a_k$ are *uniquely* determined by System (3). Indeed, there is no possible choice for a_0 other than $a_0 = f(x_0)$. Then there is no choice for a_1 other than $[f(x_1) - a_0]/(x_1 - x_0)$, and so on.

EXAMPLE 4   For the table

x	1	-4	0
$f(x)$	3	13	-23

determine the quantities $f[x_0]$, $f[x_0, x_1]$, and $f[x_0, x_1, x_2]$.

Solution   We write out the system of equations (3) for this concrete case:

$$\begin{cases} 3 = a_0 \\ 13 = a_0 + a_1(-5) \\ -23 = a_0 + a_1(-1) + a_2(-1)(4) \end{cases}$$

The solution is $a_0 = 3$, $a_1 = -2$, and $a_2 = 7$. Hence, for this function, $f[1] = 3$, $f[1, -4] = -2$, and $f[1, -4, 0] = 7$. ∎

With this new notation, the **Newton interpolating polynomial** takes the form

$$p_n(x) = \sum_{i=0}^{n} \left\{ f[x_0, x_1, \dots, x_i] \prod_{j=0}^{i-1}(x - x_j) \right\} \tag{6}$$

with the usual convention that $\prod_{j=0}^{-1}(x - x_j) = 1$. Notice that the coefficient of x^n in p_n is $f[x_0, x_1, \ldots, x_n]$ because the term x^n occurs only in $\prod_{j=0}^{n-1}(x - x_j)$. It follows that if f is a polynomial of degree $\leqq n - 1$, then $f[x_0, x_1, \ldots, x_n] = 0$.

We return to the question of how to compute the coefficients $f[x_0, x_1, \ldots, x_k]$. From System (3) or (4), it is evident that this computation can be performed *recursively*. We simply solve Equation (4) for a_k as follows:

$$f(x_k) = a_k \prod_{j=0}^{k-1}(x_k - x_j) + \sum_{i=0}^{k-1} a_i \prod_{j=0}^{i-1}(x_k - x_j)$$

and

$$a_k = \frac{f(x_k) - \sum_{i=0}^{k-1} a_i \prod_{j=0}^{i-1}(x_k - x_j)}{\prod_{j=0}^{k-1}(x_k - x_j)}$$

Using Equation (5), we have

$$f[x_0, x_1, \ldots, x_k] = \frac{f(x_k) - \sum_{i=0}^{k-1} f[x_0, x_1, \ldots, x_i] \prod_{j=0}^{i-1}(x_k - x_j)}{\prod_{j=0}^{k-1}(x_k - x_j)} \tag{7}$$

Thus, one possible algorithm for computing the divided differences of f is:

1. Set $f[x_0] = f(x_0)$.

2. For $k = 1, 2, \ldots, n$, compute $f[x_0, x_1, \ldots, x_k]$ by Equation (7). $\tag{8}$

EXAMPLE 5 Using Algorithm (8), write out the formulas for $f[x_0]$, $f[x_0, x_1]$, $f[x_0, x_1, x_2]$, and $f[x_0, x_1, x_2, x_3]$.

Solution $$f[x_0] = f(x_0)$$

$$f[x_0, x_1] = \frac{f(x_1) - f[x_0]}{x_1 - x_0}$$

$$f[x_0, x_1, x_2] = \frac{f(x_2) - f[x_0] - f[x_0, x_1](x_2 - x_0)}{(x_2 - x_0)(x_2 - x_1)}$$

$$f[x_0, x_1, x_2, x_3] = \frac{f(x_3) - f[x_0] - f[x_0, x_1](x_3 - x_0) - f[x_0, x_1, x_2](x_3 - x_0)(x_3 - x_1)}{(x_3 - x_0)(x_3 - x_1)(x_3 - x_2)}$$

$\square$

Algorithm (8) is easily programmed and is capable of computing the divided differences $f[x_0], f[x_0, x_1], \ldots, f[x_0, x_1, \ldots, x_n]$ at the cost of $\frac{1}{2}n(3n + 1)$ additions/subtractions, $(n - 1)(n - 2)$ multiplications, and n divisions excluding arithmetic operations on the indices. A more refined method is now presented for which the pseudocode requires only three statements (!) and costs only $\frac{1}{2}n(n + 1)$ divisions and $n(n + 1)$ additions.

At the heart of the new method is the following remarkable theorem:

THEOREM ON RECURSIVE PROPERTY OF DIVIDED DIFFERENCES

The divided differences obey the formula

$$f[x_0, x_1, \ldots, x_k] = \frac{f[x_1, x_2, \ldots, x_k] - f[x_0, x_1, \ldots, x_{k-1}]}{x_k - x_0} \qquad (9)$$

Proof   Since $f[x_0, x_1, \ldots, x_k]$ was defined to be equal to the coefficient a_k in the Newton interpolating polynomial p_k of Equation (2), we can say that $f[x_0, x_1, \ldots, x_k]$ is the coefficient of x^k in the polynomial p_k of degree $\leq k$, which interpolates f at $x_0, x_1, \ldots, x_k$. Similarly, $f[x_1, x_2, \ldots, x_k]$ is the coefficient of x^{k-1} in the polynomial q of degree $\leq k - 1$, which interpolates f at $x_1, x_2, \ldots, x_k$. Likewise, $f[x_0, x_1, \ldots, x_{k-1}]$ is the coefficient of x^{k-1} in the polynomial p_{k-1} of degree $\leq k - 1$, which interpolates f at $x_0, x_1, \ldots, x_{k-1}$. The three polynomials p_k, q, and p_{k-1} are intimately related. In fact,

$$p_k(x) = q(x) + \frac{x - x_k}{x_k - x_0}[q(x) - p_{k-1}(x)] \qquad (10)$$

To establish Equation (10), observe that the right side is a polynomial of degree at most k. Evaluating it at x_0 gives $f(x_0)$:

$$q(x_0) + \frac{x_0 - x_k}{x_k - x_0}[q(x_0) - p_{k-1}(x_0)] = q(x_0) - [q(x_0) - p_{k-1}(x_0)]$$

$$= p_{k-1}(x_0) = f(x_0)$$

Evaluating it at x_i, for $1 \leq i \leq k - 1$, results in $f(x_i)$:

$$q(x_i) + \frac{x_i - x_k}{x_k - x_0}[q(x_i) - p_{k-1}(x_i)] = f(x_i) + \frac{x_i - x_k}{x_k - x_0}[f(x_i) - f(x_i)]$$

$$= f(x_i)$$

Similarly at x_k, we get $f(x_k)$:

$$q(x_k) + \frac{x_k - x_k}{x_k - x_0}[q(x_k) - p_{k-1}(x_k)] = q(x_k) = f(x_k)$$

By the uniqueness of interpolating polynomials, the right side of Equation (10) must be $p_k(x)$, and Equation (10) is established.

Completing the argument to justify Equation (9), we take the coefficient of x^k on both sides of Equation (10). The result is Equation (9). Indeed, $f[x_1, x_2, \ldots, x_k]$ is the coefficient of x^{k-1} in q, and $f[x_0, x_1, \ldots, x_{k-1}]$ is the coefficient of x^{k-1} in p_{k-1}. ∎

Notice that $f[x_0, x_1, \ldots, x_k]$ is not changed if the nodes $x_0, x_1, \ldots, x_k$ are permuted: thus, for example, $f[x_0, x_1, x_2] = f[x_1, x_2, x_0]$. The reason is that $f[x_0, x_1, x_2]$ is the coefficient of x^2 in the quadratic polynomial interpolating f at x_0, x_1, x_2, whereas $f[x_1, x_2, x_0]$ is the coefficient of x^2 in the quadratic polynomial interpolating f at x_1, x_2, x_0. These two polynomials are, of course, the same. A formal statement in mathematical language is as follows:

INVARIANCE THEOREM

The divided difference $f[x_0, x_1, \ldots, x_k]$ is invariant under all permutations of the arguments $x_0, x_1, \ldots, x_k$.

Since the variables $x_0, x_1, \ldots, x_k$ and k are arbitrary, the recursive Formula (9) can also be written

$$f[x_i, x_{i+1}, \ldots, x_{j-1}, x_j] = \frac{f[x_{i+1}, x_{i+2}, \ldots, x_j] - f[x_i, x_{i+1}, \ldots, x_{j-1}]}{x_j - x_i} \tag{11}$$

The first three divided differences are thus

$$f[x_i] = f(x_i)$$

$$f[x_i, x_{i+1}] = \frac{f[x_{i+1}] - f[x_i]}{x_{i+1} - x_i}$$

$$f[x_i, x_{i+1}, x_{i+2}] = \frac{f[x_{i+1}, x_{i+2}] - f[x_i, x_{i+1}]}{x_{i+2} - x_i}$$

Using Formula (11), we can construct a divided-difference table for a function f. It is customary to arrange it as follows (here $n = 3$):

x	$f[\ ]$	$f[\ ,\ ]$	$f[\ ,\ ,\ ]$	$f[\ ,\ ,\ ,\ ]$
x_0	$f[x_0]$			
		$f[x_0, x_1]$		
x_1	$f[x_1]$		$f[x_0, x_1, x_2]$	
		$f[x_1, x_2]$		$f[x_0, x_1, x_2, x_3]$
x_2	$f[x_2]$		$f[x_1, x_2, x_3]$	
		$f[x_2, x_3]$		
x_3	$f[x_3]$			

In the table, the coefficients along the top diagonal are the ones needed to form the Newton interpolating polynomial (6).

EXAMPLE 6   Construct a divided-difference diagram for the function f given in the following table, and write out the Newton interpolating polynomial.

x	1	3/2	0	2
$f(x)$	3	13/4	3	5/3

Solution   The first entry is $f[x_0, x_1] = (\frac{13}{4} - 3)/(\frac{3}{2} - 1) = \frac{1}{2}$. After completion of column 3, the first entry in column 4 is

$$f[x_0, x_1, x_2] = \frac{f[x_1, x_2] - f[x_0, x_1]}{x_2 - x_0} = \frac{1/6 - 1/2}{0 - 1} = \frac{1}{3}$$

The complete diagram is

x	$f[\,]$	$f[\,,\,]$	$f[\,,\,,\,]$	$f[\,,\,,\,,\,]$
1	3			
		1/2		
3/2	13/4		1/3	
		1/6		-2
0	3		$-5/3$	
		$-2/3$		
2	5/3			

Thus,

$$p_3(x) = 3 + \frac{1}{2}(x - 1) + \frac{1}{3}(x - 1)\left(x - \frac{3}{2}\right) - 2(x - 1)\left(x - \frac{3}{2}\right)x$$

The Maple statement

```
interp([1, 3/2, 0, 2], [3, 13/4, 3, 5/3], x);
```

gives us the polynomial

$$p_3(x) = -2x^3 + \frac{16}{3}x^2 - \frac{10}{3}x + 3 \qquad \square$$

Algorithms and Pseudocode

Turning next to algorithms, suppose that a table for f is given at points $x_0, x_1, \ldots, x_n$ and that all the divided differences $a_{ij} \equiv f[x_i, x_{i+1}, \ldots, x_j]$ are to be computed. The following segment of pseudocode accomplishes this:

real array $(a_{ij})_{0:n \times 0:n}, (x_i)_{0:n}$
integer i, j, n

$\vdots$

for $i = 0$ **to** n **do**
   $a_{i0} \leftarrow f(x_i)$
end do
for $j = 1$ **to** n **do**
  for $i = 0$ **to** $n - j$ **do**
    $a_{ij} \leftarrow (a_{i+1,j-1} - a_{i,j-1})/(x_{i+j} - x_i)$
  end do
end do

Observe that the coefficients of the interpolating polynomial (2) are stored in the first row of the array $(a_{ij})_{0:n,0:n}$.

If the divided differences are being computed for use only in constructing the Newton interpolation polynomial

$$p_n(x) = \sum_{i=0}^{n} a_i \prod_{j=0}^{i-1} (x - x_j)$$

where $a_i = f[x_0, x_1, \ldots, x_i]$, there is no need to store all of them. Only $f[x_0]$, $f[x_0, x_1], \ldots, f[x_0, x_1, \ldots, x_n]$ need to be stored.

When a one-dimensional array $(a_i)_{0:n}$ is used, the divided differences can be overwritten each time from the last storage location backward so that, finally, only the desired coefficients remain. In this case, the amount of computing is the same as in the preceding case, but the storage requirements are less. (Why?) Here is a pseudocode to do this:

real array $(a_i)_{0:n}, (x_i)_{0:n}$
integer i, j, n

$\vdots$

for $i = 0$ **to** n **do**
   $a_i \leftarrow f(x_i)$
end do
for $j = 1$ **to** n **do**
  for $i = n$ **to** j **step** -1 **do**
    $a_i \leftarrow (a_i - a_{i-1})/(x_i - x_{i-j})$
  end do
end do

This algorithm is more intricate, and the reader is invited to verify it—say, in the case $n = 3$.

For the numerical experiments suggested in the computer problems, the following two procedures should be satisfactory. The first is called *coef*. It requires as input the number of points $n + 1$ in the table and the tabular values in the arrays (x_i) and (y_i). The procedure then computes the coefficients required in the Newton interpolating polynomial, storing them in the array (a_i).

procedure $coef(n, (x_i), (y_i), (a_i))$
real array $(x_i)_{0:n}, (y_i)_{0:n}, (a_i)_{0:n}$
integer i, j, n
for $i = 0$ **to** n **do**
   $a_i \leftarrow y_i$
end do
for $j = 1$ **to** n **do**
   for $i = n$ **to** j **step** -1 **do**
      $a_i \leftarrow (a_i - a_{i-1})/(x_i - x_{i-j})$
   end do
end do
end procedure *coef*

The second procedure is function *eval*. It requires as input the array (x_i) from the original table and the array (a_i), which is *output* from *coef*. The array (a_i) contains the coefficients for the Newton interpolation polynomial. Finally, as input, a single real value for t is given. The function then returns the value of the interpolating polynomial at t.

real function $eval(n, (x_i), (a_i), t)$
real array $(x_i)_{0:n}, (a_i)_{0:n}$
integer i, n
real $t, temp$
$temp \leftarrow a_n$
for $i = n - 1$ **to** 0 **step** -1 **do**
   $temp \leftarrow (temp)(t - x_i) + a_i$
end do
$eval \leftarrow temp$
end function *eval*

Since the coefficients of the interpolating polynomial need be computed only once, we call *coef* first, and then all subsequent calls for evaluating this polynomial are accomplished with *eval*. Notice that only the t argument should be changed between successive calls to this function.

EXAMPLE 7   Write pseudocode that determines the Newton interpolating polynomial p for $\sin x$ at ten equidistant points in the interval $[0, 1.6875]$. The code should print the value of $\sin x - p(x)$ at 37 equally spaced points in the same interval.

Solution   If we take ten points, including the ends of the interval, then we create nine subintervals, each of length $h = 0.1875$. The points are then $x_i = ih$ for $i = 0, 1, \ldots, 9$. After obtaining the polynomial, we evaluate $\sin x - p(x)$ at 37 points (called t in the pseudocode). These are $t_i = ih/4$ for $i = 0, 1, \ldots, 36$. Here is a suitable main pseudocode that calls the procedures *coef* and *eval* previously given.

program *main*
integer parameter $n \leftarrow 9$
real array $(x_i)_{0:n}, (y_i)_{0:n}, (a_i)_{0:n}$
integer i, n
real h
$h \leftarrow 1.6875/n$
for $i = 0$ **to** n **do**
  $x_i \leftarrow ih$
  $y_i \leftarrow \sin(x_i)$
end do
call $coef(n, (x_i), (y_i), (a_i))$
output (a_i)
for $j = 0$ **to** $4n$ **do**
  $t \leftarrow jh/4$
  $d \leftarrow \sin(t) - eval(n, (x_i)_n, (a_i)_n, t)$
  output i, t, d
end do
end program *main*

We have not shown the computer output from this numerical experiment. The first coefficient in the Newton interpolating polynomial is 0 (why?), and the others range in magnitude from approximately 0.99 to 0.18×10^{-5}. The deviation between $\sin x$ and $p(x)$ is practically zero at each interpolation node. (Because of roundoff errors, they are not precisely zero.) Between the nodes, the deviations printed in the computer program are all less than $\frac{1}{2} \times 10^{-9}$ in magnitude. From the computer output, a typical result is $i = 20, t = .9375000$, and $d = -5.961E-08$, which shows the value and deviation at the 20th point.   □

Inverse Interpolation

A process called **inverse interpolation** is often used to solve equations of the form $f(x) = 0$. Suppose that values $y_i = f(x_i)$ have been computed at $x_0, x_1, \ldots, x_n$. Using the table

y	y_0	y_1	$\cdots$	y_n
x	x_0	x_1	$\cdots$	x_n

we form the interpolation polynomial

$$p_n(y) = \sum_{i=0}^{n} f[y_0, y_1, \ldots, y_i] \prod_{j=0}^{i-1} (y - y_j)$$

Now evaluating p_n at 0, we have an approximation $\widehat{x} = p_n(0)$ such that $f(\widehat{x}) \approx 0$. Procedures *coef* and *eval* can be used to carry out the inverse interpolation by reversing the arguments x and y in the calling sequence for *coef* and using $t = 0.0$ in function *eval*.

Lagrange Interpolating Polynomial

Interpolating polynomials can be written in a variety of forms, and among these the Newton form is probably the most convenient and efficient. Conceptually, however, the form known by the name of **Lagrange** has several advantages. Suppose that we wish to interpolate arbitrary functions at a set of fixed nodes $x_0, x_1, \ldots, x_n$. We first define a system of $n + 1$ special polynomials of degree n known as **cardinal functions** in interpolation theory. These are denoted by $\ell_0, \ell_1, \ldots, \ell_n$ and have the property

$$\ell_i(x_j) = \delta_{ij} = \begin{cases} 0 & \text{if } i \neq j \\ 1 & \text{if } i = j \end{cases}$$

Once these are available, we can interpolate *any* function f by the formula of Lagrange:

$$p_n(x) = \sum_{i=0}^{n} \ell_i(x) f(x_i) \tag{12}$$

This function p_n, being a linear combination of the polynomials ℓ_i, is itself a polynomial of degree $\leq n$. Furthermore, when we evaluate p_n at x_j, we get $f(x_j)$:

$$p_n(x_j) = \sum_{i=0}^{n} \ell_i(x_j) f(x_i) = \ell_j(x_j) f(x_j) = f(x_j)$$

Thus, p_n is the interpolating polynomial for the function f at nodes $x_0, x_1, \ldots, x_n$. It remains now only to write the formula for ℓ_i, which is

$$\ell_i(x) = \prod_{\substack{j \neq i \\ j=0}}^{n} \left(\frac{x - x_j}{x_i - x_j} \right) \qquad (0 \leq i \leq n) \tag{13}$$

This formula indicates that $\ell_i(x)$ is the product of n linear factors:

$$\ell_i(x) = \left(\frac{x - x_0}{x_i - x_0} \right) \left(\frac{x - x_1}{x_i - x_1} \right) \cdots \left(\frac{x - x_{i-1}}{x_i - x_{i-1}} \right) \left(\frac{x - x_{i+1}}{x_i - x_{i+1}} \right) \cdots \left(\frac{x - x_n}{x_i - x_n} \right)$$

(The denominators are just numbers; the variable x occurs only in the numerators.) Thus, ℓ_i is a polynomial of degree n. Notice that when $\ell_i(x)$ is evaluated at $x = x_i$, each factor in the preceding equation becomes 1. Hence, $\ell_i(x_i) = 1$. But when $\ell_i(x)$

is evaluated at any *other* node, say x_j, one of the factors in the above equation will be 0, and $\ell_i(x_j) = 0$.

EXAMPLE 8   Write out the cardinal functions appropriate to the problem of interpolating the following table, and give the Lagrange interpolating polynomial:

x	1/3	1/4	1
$f(x)$	2	−1	7

Solution   Using Equation (13), we have

$$\ell_0(x) = \frac{(x - 1/4)(x - 1)}{(1/3 - 1/4)(1/3 - 1)} = -18\left(x - \frac{1}{4}\right)(x - 1)$$

$$\ell_1(x) = \frac{(x - 1/3)(x - 1)}{(1/4 - 1/3)(1/4 - 1)} = 16\left(x - \frac{1}{3}\right)(x - 1)$$

$$\ell_2(x) = \frac{(x - 1/3)(x - 1/4)}{(1 - 1/3)(1 - 1/4)} = 2\left(x - \frac{1}{3}\right)\left(x - \frac{1}{4}\right)$$

Therefore, the interpolating polynomial in Lagrange's form is

$$p_2(x) = -36\left(x - \frac{1}{4}\right)(x - 1) - 16\left(x - \frac{1}{3}\right)(x - 1) + 14\left(x - \frac{1}{3}\right)\left(x - \frac{1}{4}\right)$$

The nested Newton form

$$p_2(x) = 2 + \left(x - \frac{1}{3}\right)\left(36 + \left(x - \frac{1}{4}\right)(-38)\right)$$

involves the fewest arithmetic operations and is recommended for computing $p_2(x)$. However, the Lagrange form might be slightly more accurate in computing $p_2(x)$ in the vicinity of the nodes $\frac{1}{3}$, $\frac{1}{4}$, and 1 but at a higher cost.   ◻

PROBLEMS 4.1

1. Complete the following divided-difference tables and use them to obtain polynomials of degree 3 that interpolate the function values indicated:

a.

x	$f[\ ]$	$f[\ ,\ ]$	$f[\ ,\ ,\ ]$	$f[\ ,\ ,\ ,\ ]$
−1	2			
1	−4		2	
3	6			
		2		
5	10			

b.

x	$f[\ ]$	$f[\ ,\ ]$	$f[\ ,\ ,\ ]$	$f[\ ,\ ,\ ]$
-1	2			
1	-4			
3	46			
		53.5		
4	99.5			

Write the final polynomials in a form most efficient for computing.

2. Find an interpolating polynomial for this table.

x	1	2	2.5	3	4
y	-1	$-1/3$	$3/32$	$4/3$	25

3. Given the data

x	0	1	2	4	6
$f(x)$	1	9	23	93	259

do the following.

a. Construct the divided-difference table.

b. Using Newton's interpolation polynomial, find an approximation to $f(4.2)$.

4. a. Construct Newton's interpolation polynomial for the data.

x	0	2	3	4
y	7	11	28	63

b. Without simplifying it, write the polynomial obtained in nested form for easy evaluation.

5. From census data, the approximate population of the United States was 150.7 million in 1950, 179.3 million in 1960, 203.3 million in 1970, 226.5 million in 1980, and 249.6 million in 1990. Using Newton's interpolation polynomial for these data, find an approximate value for the population in 2000. Then use the polynomial to estimate the population in 1920 based on these data. What conclusion should be drawn?

6. The polynomial $p(x) = x^4 - x^3 + x^2 - x + 1$ has the values shown.

x	-2	-1	0	1	2	3
$p(x)$	31	5	1	1	11	61

Find a polynomial q that takes these values:

x	-2	-1	0	1	2	3
$q(x)$	31	5	1	1	11	30

Hint: This can be done with little work.

7. Use the divided-difference method to obtain a polynomial of least degree that fits the values shown.

   a.

x	0	1	2	−1	3
y	−1	−1	−1	−7	5

   b.

x	1	3	−2	4	5
y	2	6	−1	−4	2

8. Find the interpolating polynomial for these data.

x	1.0	2.0	2.5	3.0	4.0
f(x)	−1.5	−0.5	0.0	0.5	1.5

9. It is suspected that the table

x	−2	−1	0	1	2	3
y	1	4	11	16	13	−4

   comes from a cubic polynomial. How can this be tested? Explain.

10. There exists a unique polynomial $p(x)$ of degree 2 or less such that $p(0) = 0$, $p(1) = 1$, and $p'(\alpha) = 2$ for any value of α between 0 and 1 (inclusive), except one value of α, say α_0. Determine α_0 and give this polynomial for $\alpha \neq \alpha_0$.

11. Determine by two methods the polynomial of degree 2 or less whose graph passes through the points $(0, 1.1)$, $(1, 2)$, and $(2, 4.2)$. Verify that they are the same.

12. Develop the divided-difference table from the given data. Write down the interpolating polynomial and rearrange it for fast computation without simplifying.

x	0	1	3	2	5
f(x)	2	1	5	6	−183

    Checkpoint: $f[1, 3, 2, 5] = -7$.

13. Let $f(x) = x^3 + 2x^2 + x + 1$. Find the polynomial of degree 4 that interpolates the values of f at $x = -2, -1, 0, 1, 2$. Find the polynomial of degree 2 that interpolates the values of f at $x = -1, 0, 1$.

14. Without using a divided-difference table, derive and simplify the polynomial of least degree that assumes these values.

x	−2	−1	0	1	2
y	2	14	4	2	2

15. (Continuation) Find a polynomial that takes the values shown in Problem **14** and has at $x = 3$ the value 10. *Hint*: Add a suitable polynomial to the $p(x)$ of Problem **14**.

16. Find a polynomial of least degree that takes these values.

x	1.73	1.82	2.61	5.22	8.26
y	0	0	7.8	0	0

Hint: Rearrange the table so that the nonzero value of y is the *last* entry or think of some better way.

17. Form a divided-difference table for the following and explain what happened.

x	1	2	3	1
y	3	5	5	7

18. Simple polynomial interpolation in two dimensions is not always possible. For example, suppose that the following data are to be represented by a polynomial of first degree in x and y, $p(t) = a + bx + cy$, where $t = (x, y)$:

t	(1, 1)	(3, 2)	(5, 3)
$f(t)$	3	2	6

Show that it is not possible.

19. Consider a function $f(x)$ such that $f(2) = 1.5713$, $f(3) = 1.5719$, $f(5) = 1.5738$, and $f(6) = 1.5751$. Estimate $f(4)$ using a second-degree interpolating polynomial and a third-degree polynomial. Round the final results off to four decimal places. Is there any advantage here in using a third-degree polynomial?

20. Use inverse interpolation to find an approximate value of x such that $f(x) = 0$ given the following table of values for f.

x	-2	-1	1	2	3
$f(x)$	-31	5	1	11	61

21. Find a polynomial $p(x)$ of degree at most 3 such that $p(0) = 1$, $p(1) = 0$, $p'(0) = 0$, and $p'(-1) = -1$.

22. From a table of logarithms we obtain the following values of $\log x$ at the indicated tabular points:

x	1	1.5	2	3	3.5	4
$\log x$	0	0.17609	0.30103	0.47712	0.54407	0.60206

Form a divided-difference table based on these values. Interpolate for $\log 2.4$ and $\log 1.2$ using third-degree interpolation polynomials in Newton form.

23. Show that the divided differences are linear maps; that is,

$$(\alpha f + \beta g)[x_0, x_1, \ldots, x_n] = \alpha f[x_0, x_1, \ldots, x_n] + \beta g[x_0, x_1, \ldots, x_n]$$

Hint: Use induction or use Problem **29**.

24. Show that another form for the polynomial p_n of degree at most n that takes values $y_0, y_1, \ldots, y_n$ at abscissas $x_0, x_1, \ldots, x_n$ is

$$\sum_{i=0}^{n} f[x_n, x_{n-1}, \ldots, x_{n-i}] \prod_{j=0}^{i-1} (x - x_{n-j})$$

25. Use the Lagrange interpolation process to obtain a polynomial of least degree that assumes these values.

x	0	2	3	4
y	7	11	28	63

26. (Continuation) Rearrange the points in the table of Problem **25** and find the Newton interpolating polynomial. Show that the polynomials obtained are identical, although their forms may differ.

27. For the four interpolation nodes $-1, 1, 3, 4$, what are the ℓ_i functions (13) required in the Lagrange interpolation procedure? Draw the graphs of these four functions to show their essential properties.

28. Use the uniqueness of the interpolating polynomial to verify that

$$\sum_{i=0}^{n} f(x_i)\ell_i(x) = \sum_{i=0}^{n} f[x_0, x_1, \ldots, x_i] \prod_{j=0}^{i-1}(x - x_j)$$

29. (Continuation) Show that the following explicit formula is valid for divided differences:

$$f[x_0, x_1, \ldots, x_n] = \sum_{i=0}^{n} f(x_i) \prod_{\substack{j \neq i \\ j=0}}^{n}(x_i - x_j)^{-1}$$

Hint: If two polynomials are equal, the coefficients of x^n in each are equal.

30. Verify directly that

$$\sum_{i=0}^{n} \ell_i(x) = 1$$

for the case $n = 1$. Then establish the result for arbitrary values of n.

31. Write the Lagrange form (12) of the interpolating polynomial of degree ≤ 2 that interpolates $f(x)$ at x_0, x_1, and x_2, where $x_0 < x_1 < x_2$.

32. (Continuation) Write the Newton form of the interpolating polynomial $p_2(x)$ and show that it is equivalent to the Lagrange form.

33. (Continuation) Show directly that

$$p_2''(x) = 2f[x_0, x_1, x_2]$$

34. (Continuation) Show directly for uniform spacing $h = x_1 - x_0 = x_2 - x_1$ that

$$f[x_0, x_1] = \frac{\Delta f_0}{h} \quad \text{and} \quad f[x_0, x_1, x_2] = \frac{\Delta^2 f_0}{2h^2}$$

where $\Delta f_i = f_{i+1} - f_i$, $\Delta^2 f_i = \Delta f_{i+1} - \Delta f_i$, and $f_i = f(x_i)$.

35. (Continuation) Establish a **Newton forward-difference interpolating polynomial** for uniform spacing

$$p_2(x) = f_0 + \binom{s}{1}\Delta f_0 + \binom{s}{2}\Delta^2 f_0$$

where $x = x_0 + sh$. Here $\binom{s}{m}$ is the binomial coefficient $[s!]/[(s-m)!m!]$ and $s!/(s-m)! = s(s-1)(s-2)\cdots(s-m+1)$ because s can be any real number and $m!$ has the usual definition because m is an integer.

36. (Continuation) From the following table of values of $\ln x$, interpolate $\ln 2.352$ and $\ln 2.387$ using the forward-difference forms of Newton's interpolating polynomial:

x	$f(x)$	Δf	$\Delta^2 f$
2.35	0.85442		
		0.00424	
2.36	0.85866		−0.00001
		0.00423	
2.37	0.86289		−0.00002
		0.00421	
2.38	0.86710		−0.00002
		0.00419	
2.39	0.87129		

Using the correctly rounded values $\ln 2.352 \approx 0.85527$ and $\ln 2.387 \approx 0.87004$, show that the forward-difference formula is more accurate near the top of the table than it is near the bottom.

37. Count the number of multiplications, divisions, and additions/subtractions in the generation of the divided-difference table that has $n + 1$ points.

38. Verify directly that for any three distinct points $x_0, x_1,$ and x_2,

$$f[x_0, x_1, x_2] = f[x_2, x_0, x_1] = f[x_1, x_2, x_0]$$

Compare this argument to the one in the text.

39. Let p be a polynomial of degree n. What is $p[x_0, x_1, \ldots, x_{n+1}]$?

40. Show that if f is continuously differentiable on the interval $[x_0, x_1]$, then $f[x_0, x_1] = f'(c)$ for some c in (x_0, x_1).

41. If f is a polynomial of degree n, show that in a divided-difference table for f, the nth column has a single constant value. This is the column that contains entries $f[x_i, x_{i+1}, \ldots, x_{i+n}]$.

42. Determine whether the following assertion is true or false. If $x_0, x_1, \ldots, x_n$ are distinct, then for arbitrary real values $y_0, y_1, \ldots, y_n$, there is a unique polynomial p_{n+1} of degree $\leqq n + 1$ such that $p_{n+1}(x_i) = y_i$ for all $i = 0, 1, \ldots, n$.

43. Show that if a function g interpolates the function f at $x_0, x_1, \ldots, x_{n-1}$ and h interpolates f at $x_1, x_2, \ldots, x_n$, then

$$g(x) + \frac{x_0 - x}{x_n - x_0}[g(x) - h(x)]$$

interpolates f at $x_0, x_1, \ldots, x_n$.

44. (Vandermonde determinant) Using $f_i = f(x_i)$, show the following.

a. $f[x_0, x_1] = \dfrac{\begin{vmatrix} 1 & f_0 \\ 1 & f_1 \end{vmatrix}}{\begin{vmatrix} 1 & x_0 \\ 1 & x_1 \end{vmatrix}}$

b. $f[x_0, x_1, x_2] = \dfrac{\begin{vmatrix} 1 & x_0 & f_0 \\ 1 & x_1 & f_1 \\ 1 & x_2 & f_2 \end{vmatrix}}{\begin{vmatrix} 1 & x_0 & x_0^2 \\ 1 & x_1 & x_1^2 \\ 1 & x_2 & x_2^2 \end{vmatrix}}$

45. Verify that the polynomials

$$p(x) = 5x^3 - 27x^2 + 45x - 21 \qquad q(x) = x^4 - 5x^3 + 8x^2 - 5x + 3$$

interpolate the data

x	1	2	3	4
y	2	1	6	47

and explain why this does not violate the uniqueness part of the Existence Theorem for Polynomial Interpolation.

46. Verify that the polynomials

$$p(x) = 3 + 2(x - 1) + 4(x - 1)(x + 2) \qquad q(x) = 4x^2 + 6x - 7$$

are both interpolating polynomials for the given table, and explain why this does not violate the existence theorem for polynomial interpolation.

x	1	-2	0
y	3	-3	-7

COMPUTER PROBLEMS 4.1

1. Find the polynomial of degree 10 that interpolates the function $\arctan x$ at 11 equally spaced points in the interval $[1, 6]$. Print the coefficients in the Newton form of the polynomial. Compute and print the difference between the polynomial and the function at 33 equally spaced points in the interval $[0, 8]$. What conclusion can be drawn?

2. Test the procedure given in the text for determining the Newton interpolating polynomial. For example, consider this table:

x	1	2	3	-4	5
y	2	48	272	1182	2262

Find the interpolating polynomial and verify that $p(-1) = 12$.

3. Write a simple program using procedure *coef* that interpolates e^x by a polynomial of degree 10 on $[0, 2]$ and then compares the polynomial to exp at 100 points.

4. Use as input data to procedure *coef* the annual rainfall in your town for each of the last 5 years. Using function *eval*, predict the rainfall for this year. Is the answer reasonable?

5. A table of values of a function f is given at the points $x_i = i/10$ for $0 \leq i \leq 100$. In order to obtain a graph of f with the aid of an automatic plotter, the values of f are required at the points $z_i = i/20$ for $0 \leq i \leq 200$. Write a procedure to do this, using a cubic interpolating polynomial with nodes x_i, x_{i+1}, x_{i+2}, and x_{i+3} to compute f at $\frac{1}{2}(x_{i+1} + x_{i+2})$. For z_1 and z_{199}, use the cubic polynomial associated with z_3 and z_{197}, respectively. Compare this routine to *coef* for a given function.

6. Write routines analogous to *coef* and *eval* using the Lagrange form of the interpolation polynomial. Test on the example given in this section at 20 points with $h/2$. Does the Lagrange form have any advantage over the Newton form?

7. (Continuation) Design and carry out a numerical experiment to compare the accuracy of the Newton and Lagrange interpolation polynomials at values throughout the interval $[x_0, x_n]$. *Hint*: See the last example in this section.

8. Rewrite and test routines *coef* and *eval* so that the array (a_i) is not used. *Hint*: When the elements in the array (y_i) are no longer needed, store the divided differences in their places.

9. Write a procedure just for carrying out inverse interpolation to solve equations of the form $f(x) = 0$. Test it on the leading example at the beginning of this chapter.

4.2 Errors in Polynomial Interpolation

When a function f is approximated on an interval $[a, b]$ by means of an interpolating polynomial p, the discrepancy between f and p will (theoretically) be zero at each node of interpolation. A natural expectation is that the function f will be well approximated at all intermediate points and that, as the number of nodes increases, this agreement will become better and better.

Examples

In the history of numerical mathematics, a severe shock occurred when it was realized that this expectation was ill-founded. Of course, if the function being approximated is not required to be continuous, then there may be no agreement at all between $p(x)$ and $f(x)$ except at the nodes. As a pathological example, consider the so-called **Dirichlet function** f, defined to be 1 at each irrational point and 0 at each rational point. If we choose nodes that are rational numbers, then $p(x) \equiv 0$ and $f(x) - p(x) = 0$ for all rational values of x, but $f(x) - p(x) = 1$ for all irrational values of x.

However, if the function f is well behaved, can we not assume that $|f(x) - p(x)|$ will be small when the number of interpolating nodes is large? The answer is still *no*, even for functions that possess continuous derivatives of all orders on the interval! A specific example of this remarkable phenomenon is provided by the **Runge function**:

$$f(x) = (1 + x^2)^{-1} \tag{1}$$

on the interval $[-5, 5]$. Let p_n be the polynomial that interpolates this function at $n + 1$ equally spaced points on the interval $[-5, 5]$, including the endpoints. Then

$$\lim_{n \to \infty} \max_{-5 \le x \le 5} |f(x) - p_n(x)| = \infty$$

Thus, the effect of requiring the agreement of f and p_n at more and more points is to *increase* the error at nonnodal points, and the error actually increases beyond all bounds!

The moral of this example, then, is that polynomial interpolation of high degree with many nodes is a risky operation; the resulting polynomials may be very unsatisfactory as representations of functions unless the set of nodes is chosen very carefully.

The reader can easily observe the phenomenon just described by using the pseudocodes already developed in this chapter. See Computer Problem **1** for a suggested numerical experiment. In a more advanced study of this topic, it would be shown that the divergence of the polynomials can often be ascribed to the fact that the nodes are equally spaced. Again, contrary to intuition, equally distributed nodes are a very poor choice in interpolation. A much better choice for $n + 1$ nodes in $[-1, 1]$ is the set of **Chebyshev nodes**:

$$x_i = \cos\left[\left(\frac{i}{n}\right)\pi\right] \qquad (0 \le i \le n)$$

The corresponding set of nodes on an arbitrary interval $[a, b]$ would be derived from a simple mapping to obtain:

$$x_i = \frac{1}{2}(a + b) + \frac{1}{2}(b - a)\cos\left[\left(\frac{i}{n}\right)\pi\right] \qquad (0 \le i \le n)$$

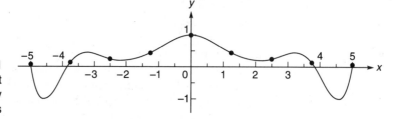

FIGURE 4.1

Polynomial interpolant with nine equally spaced nodes

A simple graph illustrates this phenomenon best. Again, consider Equation (1) on the interval $[-5, 5]$. First, we select nine equally spaced nodes and use routines *coef* and *eval* with an automatic plotter to graph p_8. As shown in Figure 4.1, the resulting curve assumes negative values, which, of course, $f(x)$ does not have! Adding more equally spaced nodes—and thereby obtaining a higher-degree polynomial—only makes matters worse with wilder oscillations. In Figure 4.2, nine Chebyshev nodes are used, and the resulting polynomial curve is smoother. However, cubic splines (discussed in Chapter 7) produce an even better curve fit.

Theorems on Interpolation Errors

It is possible to assess the errors of interpolation by means of a formula that involves the $(n + 1)$st derivative of the function being interpolated. Here is the formal statement:

INTERPOLATION ERRORS THEOREM 1

If p is the polynomial of degree at most n that interpolates f at the $n + 1$ distinct nodes $x_0, x_1, \ldots, x_n$ belonging to an interval $[a, b]$ and if $f^{(n+1)}$ is continuous, then for each x in $[a, b]$, there is a ξ in (a, b) for which

$$f(x) - p(x) = \frac{1}{(n + 1)!} f^{(n+1)}(\xi) \prod_{i=0}^{n} (x - x_i) \tag{2}$$

Proof   Observe first that Equation (2) is obviously valid if x is one of the nodes x_i because then both sides of the equation reduce to zero. If x is not a node, let it be fixed in the remainder of the discussion, and define

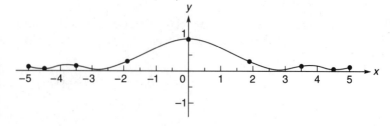

FIGURE 4.2

Polynomial interpolant with nine Chebyshev nodes

$$w(t) = \prod_{i=0}^{n}(t - x_i) \qquad \text{(polynomial in the variable } t\text{)}$$

$$c = \frac{f(x) - p(x)}{w(x)} \qquad \text{(constant)} \tag{3}$$

$$\phi(t) = f(t) - p(t) - cw(t) \qquad \text{(function in the variable } t\text{)}$$

Observe that c is well defined because $w(x) \neq 0$ (x is not a node). Note also that ϕ takes the value 0 at the $n + 2$ points $x_0, x_1, \ldots, x_n$, and x. Now invoke **Rolle's Theorem**,* which states that between any two roots of ϕ there must occur a root of ϕ'. Thus, ϕ' has at least $n + 1$ roots. By similar reasoning, ϕ'' has at least n roots, ϕ''' has at least $n - 1$ roots, and so on. Finally, it can be inferred that $\phi^{(n+1)}$ must have at least one root. Let ξ be a root of $\phi^{(n+1)}$. All the roots being counted in this argument are in (a, b). Thus,

$$0 = \phi^{(n+1)}(\xi) = f^{(n+1)}(\xi) - p^{(n+1)}(\xi) - cw^{(n+1)}(\xi)$$

In this equation, $p^{(n+1)}(\xi) = 0$ because p is a polynomial of degree $\leq n$. Also, $w^{(n+1)}(\xi) = (n + 1)!$ because $w(t) = t^{n+1} +$ (lower-order terms in t). Thus, we have

$$0 = f^{(n+1)}(\xi) - c(n + 1)! = f^{(n+1)}(\xi) - \frac{(n + 1)!}{w(x)}[f(x) - p(x)]$$

This equation is a rearrangement of Equation (2).   ∎

A special case that often arises is the one in which the interpolation nodes are equally spaced. Suppose that $x_i = a + ih$ for $i = 0, 1, \ldots, n$ and that $h = (b - a)/n$. Then we can show that for any $x \in [a, b]$,

$$\prod_{i=0}^{n} |x - x_i| \leq \frac{1}{4} h^{n+1} n! \tag{4}$$

To establish this inequality, fix x and select j so that $x_j \leq x \leq x_{j+1}$. It is an exercise in calculus (Problem **1**) to show that

$$|x - x_j||x - x_{j+1}| \leq \frac{h^2}{4} \tag{5}$$

Using Equation (5), we have

$$\prod_{i=0}^{n} |x - x_i| \leq \frac{h^2}{4} \prod_{i=0}^{j-1}(x - x_i) \prod_{i=j+2}^{n}(x_i - x)$$

---

*Rolle's Theorem: Let f be a function that is continuous on $[a, b]$ and differentiable on (a, b). If $f(a) = f(b) = 0$, then $f'(c) = 0$ for some point c in (a, b).

FIGURE 4.3

The sketch in Figure 4.3, showing a typical case, may be helpful. Since $x_j \leq x \leq x_{j+1}$, we have further

$$\prod_{i=0}^{n} |x - x_i| \leq \frac{h^2}{4} \prod_{i=0}^{j-1} (x_{j+1} - x_i) \prod_{i=j+2}^{n} (x_i - x_j)$$

Now use the fact that $x_i = a + ih$. Then $x_{j+1} - x_i = (j - i + 1)h$ and $x_i - x_j = (i - j)h$. Therefore,

$$\prod_{i=0}^{n} |x - x_i| \leq \frac{h^2}{4} h^j h^{n-(j+2)+1} \prod_{i=0}^{j-1} (j - i + 1) \prod_{i=j+2}^{n} (i - j)$$

$$\leq \frac{1}{4} h^{n+1} (j + 1)!(n - j)! \leq \frac{1}{4} h^{n+1} n!$$

In the last step, we use the fact that if $0 \leq j \leq n - 1$, then $(j + 1)!(n - j)! \leq n!$. This, too, is left as an exercise (Problem **2**). Hence, Inequality (4) is established.

We can now establish a bound on the interpolation error.

INTERPOLATION ERRORS THEOREM 2

Let f be a function such that $f^{(n+1)}$ is continuous on $[a, b]$ and satisfies $|f^{(n+1)}(x)| \leq M$. Let p be the polynomial of degree $\leq n$ that interpolates f at $n + 1$ equally spaced nodes in $[a, b]$, including the endpoints. Then on $[a, b]$,

$$|f(x) - p(x)| \leq \frac{1}{4(n + 1)} M \left(\frac{b - a}{n} \right)^{n+1} \tag{6}$$

Proof   Use Theorem 1 on interpolation errors and Inequality (4).   ■

EXAMPLE 1   Assess the error if $\sin x$ is replaced by an interpolation polynomial that has ten equally spaced nodes in $[0, 1.6875]$. (See the related example in Section 4.1.)

Solution   We use Theorem 2 on Interpolation Errors, taking $f(x) = \sin x$ where $n = 9$, $a = 0$, and $b = 1.6875$. Since $f^{(10)}(x) = -\sin x$, $|f^{(10)}(x)| \leq 1$. Hence in Equation (6), we can let $M = 1$. The result is

$$|\sin x - p(x)| \leq 1.34 \times 10^{-9}$$

Thus, $p(x)$ represents $\sin x$ on this interval with an error of at most two units in the ninth decimal place. Therefore, the interpolation polynomial that has ten equally spaced nodes on the interval $[0, 1.6875]$ approximates $\sin x$ to at least eight decimal digits of accuracy. In fact, a careful check on a computer would reveal that the polynomial is accurate to more decimal places. (Why?) $\square$

The error expression in polynomial interpolation can also be given in terms of divided differences:

INTERPOLATION ERRORS THEOREM 3

If p is the polynomial of degree n that interpolates the function f at nodes $x_0, x_1, \ldots, x_n$, then for any x not a node,

$$f(x) - p(x) = f[x_0, x_1, \ldots, x_n, x] \prod_{i=0}^{n} (x - x_i)$$

Proof   Let t be any point, other than a node, where $f(t)$ is defined. Let q be the polynomial of degree $\leq n + 1$ that interpolates f at $x_0, x_1, \ldots, x_n, t$. By the Newton interpolation formula [Equation (6) in Section 4.1], we have

$$q(x) = p(x) + f[x_0, x_1, \ldots, x_n, t] \prod_{i=0}^{n} (x - x_i)$$

Since $q(t) = f(t)$, this yields at once

$$f(t) = p(t) + f[x_0, x_1, \ldots, x_n, t] \prod_{i=0}^{n} (t - x_i) \qquad \blacksquare$$

The following theorem shows that there is a relationship between divided differences and derivatives.

THEOREM ON RELATING DIVIDED DIFFERENCES AND DERIVATIVES

If $f^{(n)}$ is continuous on $[a, b]$ and if $x_0, x_1, \ldots, x_n$ are any $n + 1$ distinct points in $[a, b]$, then for some ξ in (a, b),

$$f[x_0, x_1, \ldots, x_n] = \frac{1}{n!} f^{(n)}(\xi)$$

Proof   Let p be the polynomial of degree $\leq n - 1$ that interpolates f at $x_0, x_1, \ldots, x_{n-1}$. By Theorem 1 on Interpolation Errors, there is a point ξ such that

$$f(x_n) - p(x_n) = \frac{1}{n!} f^{(n)}(\xi) \prod_{i=0}^{n-1}(x_n - x_i)$$

By Theorem 3 on Interpolation Errors,

$$f(x_n) - p(x_n) = f[x_0, x_1, \ldots, x_{n-1}, x_n] \prod_{i=0}^{n-1}(x_n - x_i) \qquad ■$$

As an immediate consequence of this theorem, we observe that polynomials have only a finite number of nonzero divided differences.

> ### COROLLARY
>
> If f is a polynomial of degree n, then all divided differences $f[x_0, x_1, \ldots, x_i]$ are zero for $i \geq n + 1$.

EXAMPLE 2   Is there a cubic polynomial that takes these values?

x	1	-2	0	3	-1	7
y	-2	-56	-2	4	-16	376

Solution   If such a polynomial exists, its divided differences $f[\,,\,,\,,\,]$ would all be zero. We form a divided-difference table to check this possibility:

x	$f[\,]$	$f[\,,\,]$	$f[\,,\,,\,]$	$f[\,,\,,\,,\,]$	$f[\,,\,,\,,\,,\,]$
1	-2				
		18			
-2	-56		-9		
		27		2	
0	-2		-5		0
		2		2	
3	4		-3		0
		5		2	
-1	-16		11		
		49			
7	376				

The data can be represented by a cubic polynomial because the divided differences $f[\,,\,,\,,\,]$ are zero. From the Newton interpolation formula, this polynomial is

$$p_3(x) = -2 + 18(x - 1) - 9(x - 1)(x + 2) + 2(x - 1)(x + 2)x$$

The Maple statements

```
interp([1,-2,0,3,-1,7], [-2,-56,-2,4,-16,376], x);
convert(", horner, x);
```

give us the polynomial

$$p_3(x) = 2x^3 - 7x^2 + 5x - 2 = x(x(2x - 7) + 5) - 2 \qquad \square$$

**PROBLEMS
4.2**

1. Fill in a detail in the proof of Inequality (4) by proving Inequality (5).

2. (Continuation) Fill in another detail in the proof of Inequality (4) by showing that $(j + 1)!(n - j)! \leqq n!$ if $0 \leqq j \leqq n - 1$. Induction and a symmetry argument can be used.

3. Show directly that the maximum error associated with linear interpolation is bounded by $\frac{1}{8}(x_1 - x_0)^2 M$, where $M = \max_{x_0 \leqq x \leqq x_1} |f''(x)|$.

4. How accurately can we determine $\sin x$ by linear interpolation, given a table of $\sin x$ to ten decimal places, for x in $[0, 2]$ with $h = 0.01$?

5. (Continuation) Given the data

x	$\sin x$	$\cos x$
0.70	0.64421 76872	0.76484 21872
0.71	0.65183 37710	0.75836 18759

find approximate values of $\sin 0.705$ and $\cos 0.702$ by linear interpolation. What is the error?

6. *Linear interpolation* in a table of function values means the following: If $y_0 = f(x_0)$ and $y_1 = f(x_1)$ are tabulated values, and if $x_0 < x < x_1$, then an interpolated value of $f(x)$ is $y_0 + [(y_1 - y_0)/(x_1 - x_0)](x - x_0)$, as explained at the beginning of Section 4.1. A table of values of $\cos x$ is required so that the linear interpolation will yield five-decimal-place accuracy for any value of x in $[0, \pi]$. Assume that the tabular values are equally spaced and determine the minimum number of entries needed in this table.

7. An interpolating polynomial of degree 20 is to be used to approximate e^{-x} on the interval $[0, 2]$. How accurate will it be?

8. Let the function $f(x) = \ln x$ be approximated by an interpolation polynomial of degree 9 with ten nodes uniformly distributed in the interval $[1, 2]$. What bound can be placed on the error?

9. In Theorem 1 on Interpolation Errors, show that if $x_0 < x_1 < \cdots < x_n$ and $x_0 < x < x_n$, then $x_0 < \xi < x_n$.

10. (Continuation) In the same theorem, considering ξ as a function of x, show that $f^{(n)}[\xi(x)]$ is a continuous function of x. Note that $\xi(x)$ need not be a continuous function of x.

11. Suppose $\cos x$ is to be approximated by an interpolating polynomial of degree n, using $n + 1$ equally spaced nodes in the interval $[0, 1]$. How accurate is the approximation? (Express your answer in terms of n.) How accurate is the approximation when $n = 9$? For what values of n is the error less than 10^{-7}?

12. In interpolating with $n + 1$ equally spaced nodes on an interval, we could use $x_i = a + (2i + 1)h/2$, where $0 \leq i \leq n - 1$ and $h = (b - a)/n$. What bound can be given now for $\prod_{i=0}^{n} |x - x_i|$ when $a \leq x \leq b$? Note that we are not requiring the endpoints to be nodes.

13. Using Equation (3), show that

$$w'(t) = \sum_{\substack{i=0}}^{n} \prod_{\substack{j\neq i \\ j=0}}^{n} (t - x_j) \quad \text{and} \quad w'(x_i) = \prod_{\substack{j\neq i \\ j=0}}^{n} (x_i - x_j)$$

14. Use a divided-difference table to show that the following data can be represented by a polynomial of degree 3.

x	-2	-1	0	1	2	3
y	1	4	11	16	13	-4

15. Does every polynomial p of degree $\leq n$ obey this equation? Explain.

$$p(x) = \sum_{i=0}^{n} p[x_0, x_1, \ldots, x_i] \prod_{j=0}^{i-1} (x - x_j)$$

Hint: Use the uniqueness of the interpolating polynomial.

16. For nonuniformly distributed nodes $a = x_0 < x_1 < \cdots < x_n = b$, where $h = \max_{1 \leq i \leq n}\{(x_i - x_{i-1})\}$, show that Inequality (4) is true.

COMPUTER PROBLEMS 4.2

1. Using 21 equally spaced nodes on the interval $[-5, 5]$, find the interpolating polynomial p of degree 20 for the function $f(x) = (x^2 + 1)^{-1}$. Print the values of $f(x)$ and $p(x)$ at 41 equally spaced points, including the nodes. Observe the large discrepancy between $f(x)$ and $p(x)$.

2. (Continuation) Perform the experiment in Computer Problem **1**, using Chebyshev nodes $x_i = 5\cos(i\pi/20)$, where $0 \leq i \leq 20$, and $x_i = 5\cos[(2i + 1)\pi/42]$, where $0 \leq i \leq 20$.

3. Using procedures corresponding to the pseudocode in the text, find a polynomial of degree 13 that interpolates $f(x) = \arctan x$ on the interval $[-1, 1]$. Test numerically by taking 100 points to determine how accurate the polynomial approximation is.

4. (Continuation) Write a function procedure for $\arctan x$ that uses the polynomial of Computer Problem **3**. If x is not in the interval $[-1, 1]$, use the formula $1/\tan\theta = \cot\theta = \tan(\pi/2 - \theta)$.

5. Approximate $f(x) = \arcsin x$ on the interval $\left[-1/\sqrt{2}, 1/\sqrt{2}\right]$ by an interpolating polynomial of degree 15. Determine how accurate the approximation is by numerical tests.

6. (Continuation) Write a function procedure for $\arcsin x$, using the polynomial of Computer Problem 5. Use the equation $\sin(\pi/2 - \theta) = \cos\theta = \sqrt{1 - \sin^2\theta}$ if x is in the interval $|x| > 1/\sqrt{2}$.

7. Let $f(x) = \max\{0, 1 - x\}$. Sketch the function f. Then find interpolating polynomials p of degrees 2, 4, 8, 16, and 32 to f on the interval $[-4, 4]$, using equally spaced nodes. Print out the discrepancy $f(x) - p(x)$ at 128 equally spaced points. Then redo the problem using Chebyshev nodes.

8. Using *coef* and *eval* and an automatic plotter, fit a polynomial through the following data.

x	0.0	0.60	1.50	1.70	1.90	2.1	2.30	2.60	2.8	3.00
y	−0.8	−0.34	0.59	0.59	0.23	0.1	0.28	1.03	1.5	1.44

Does the resulting curve look like a good fit? Explain.

9. Why are the Chebyshev nodes generally better than equally spaced nodes in polynomial interpolation? The answer lies in the term $\prod_{i=0}^{n}(x - x_i)$ that occurs in the error formula. If $x_i = \cos[(2i - 1)\pi/(2n)]$, then

$$\left|\prod_{i=0}^{n}(x - x_i)\right| \leq 2^{-n}$$

for all x in $[-1, 1]$. Carry out a numerical experiment to test the given inequality for $n = 3, 7, 15$.

10. Find the polynomial p of degree ≤ 10 that interpolates $|x|$ on $[-1, 1]$ at 11 equally spaced points. Print the difference $|x| - p(x)$ at 41 equally spaced points. Then do the same with Chebyshev nodes. Compare.

4.3 Estimating Derivatives and Richardson Extrapolation

A numerical experiment outlined in Chapter 1 (at the end of Section 1.1) showed that determining the derivative of a function f at a point x is not a trivial numerical problem. Specifically, if $f(x)$ can be computed with only n digits of precision, it is difficult to calculate $f'(x)$ numerically with n digits of precision. This difficulty can be traced to the subtraction of quantities that are nearly equal. In this section, several alternatives are offered for the numerical computation of $f'(x)$ and $f''(x)$.

First-Derivative Formulas via Taylor Series

First, consider again the obvious method based on the definition of $f'(x)$. It consists of selecting one or more small values of h and writing

$$f'(x) \approx \frac{1}{h}[f(x + h) - f(x)] \tag{1}$$

What error is involved in this formula? To find out, use Taylor's Theorem from Section 1.2:

$$f(x + h) = f(x) + hf'(x) + \frac{1}{2}h^2 f''(\xi)$$

Rearranging this equation gives

$$f'(x) = \frac{1}{h}[f(x + h) - f(x)] - \frac{1}{2}hf''(\xi) \tag{2}$$

Hence, we see that approximation (1) has error term $-\frac{1}{2}hf''(\xi)$, where ξ is in the interval having endpoints x and $x + h$.

Equation (2) shows that, in general, as $h \to 0$, the difference between $f'(x)$ and the estimate $h^{-1}[f(x + h) - f(x)]$ approaches zero at the same rate that h does—that is, $\mathcal{O}(h)$. Of course, if $f''(x) = 0$, then the error term will be $\frac{1}{6}h^2 f'''(\gamma)$, which converges to zero somewhat faster at $\mathcal{O}(h^2)$. But usually $f''(x)$ is not zero.

Equation (2) gives the **truncation** error for this numerical procedure—namely, $-\frac{1}{2}hf''(\xi)$. It is an error that is present even if the calculations are performed with *infinite* precision; it is due to our imitating the mathematical limit process by means of an approximation formula. Additional (and worse) errors must be expected when calculations are performed on a computer with finite word length.

As we saw in Newton's method (Chapter 3) and will see in the Romberg method (Chapter 5), it is advantageous to have the convergence of numerical processes occur with higher powers of some quantity approaching zero. In the present situation, we want an approximation to $f'(x)$ in which the error behaves like $\mathcal{O}(h^2)$. One such method is easily obtained with the aid of the following two Taylor series:

$$\begin{cases} f(x + h) = f(x) + hf'(x) + \dfrac{1}{2!}h^2 f''(x) + \dfrac{1}{3!}h^3 f'''(x) + \dfrac{1}{4!}h^4 f^{(4)}(x) + \cdots \\[2mm] f(x - h) = f(x) - hf'(x) + \dfrac{1}{2!}h^2 f''(x) - \dfrac{1}{3!}h^3 f'''(x) + \dfrac{1}{4!}h^4 f^{(4)}(x) - \cdots \end{cases} \tag{3}$$

By subtraction, we obtain

$$f(x + h) - f(x - h) = 2hf'(x) + \frac{2}{3!}h^3 f'''(x) + \frac{2}{5!}h^5 f^{(5)}(x) + \cdots$$

This leads to a very important formula for approximating $f'(x)$:

$$f'(x) = \frac{1}{2h}[f(x+h) - f(x-h)] - \frac{h^2}{3!}f'''(x) - \frac{h^4}{5!}f^{(5)}(x) - \cdots \tag{4}$$

Expressed otherwise,

$$f'(x) \approx \frac{1}{2h}[f(x+h) - f(x-h)] \tag{5}$$

with an error whose leading term is $-\frac{1}{6}h^2 f'''(x)$, which makes it $\mathcal{O}(h^2)$.

By using Taylor's Theorem with its error term, we could have obtained the following two expressions:

$$f(x+h) = f(x) + hf'(x) + \frac{1}{2}h^2 f''(x) + \frac{1}{6}h^3 f'''(\xi_1)$$

$$f(x-h) = f(x) - hf'(x) + \frac{1}{2}h^2 f''(x) - \frac{1}{6}h^3 f'''(\xi_2)$$

Then the subtraction would lead to

$$f'(x) = \frac{1}{2h}[f(x+h) - f(x-h)] - \frac{1}{6}h^2 \left[\frac{f'''(\xi_1) + f'''(\xi_2)}{2} \right]$$

The error term here can be simplified by the following reasoning: The expression $\frac{1}{2}[f'''(\xi_1) + f'''(\xi_2)]$ is the average of two values of f''' on the interval $[x-h, x+h]$. It therefore lies between the least and greatest values of f''' on this interval. If f''' is continuous on this interval, then this average value is assumed at some point ξ. Hence, the formula with its error term can be written as

$$f'(x) = \frac{1}{2h}[f(x+h) - f(x-h)] - \frac{1}{6}h^2 f'''(\xi)$$

This is based on the sole assumption that f''' is continuous on $[x-h, x+h]$. This formula for numerical differentiation turns out to be very useful in the numerical solution of certain differential equations, as we shall see in Chapter 12 (on boundary value problems) and Chapter 13 (on partial differential equations).

Returning now to Equation (4), we write it in a simpler form:

$$f'(x) = \frac{1}{2h}[f(x+h) - f(x-h)] + a_2 h^2 + a_4 h^4 + a_6 h^6 + \cdots \tag{6}$$

in which the constants $a_2, a_4, \ldots$ depend on f and x. When such information is available about a numerical process, it is possible to use a powerful technique known as Richardson extrapolation to wring more accuracy out of the method. This procedure is now explained, using Equation (6) as our model.

Richardson Extrapolation

Holding f and x fixed, we define a function of h by the formula

$$\phi(h) = \frac{1}{2h}[f(x+h) - f(x-h)] \qquad (7)$$

From Equation (6), we see that $\phi(h)$ is an approximation to $f'(x)$ with error of order $\mathcal{O}(h^2)$. Our objective is to compute $\lim_{h\to 0} \phi(h)$ because this is the quantity $f'(x)$ that we wanted in the first place. Obviously, we can take any convenient sequence h_n that converges to zero, calculate $\phi(h_n)$ from Equation (7), and use these as approximations to $f'(x)$.

But something much more clever can be done. Suppose we compute $\phi(h)$ for some h and then compute $\phi(h/2)$. By Equation (6), we have

$$\phi(h) = f'(x) - a_2h^2 - a_4h^4 - a_6h^6 - \cdots$$

$$\phi\left(\frac{h}{2}\right) = f'(x) - a_2\left(\frac{h}{2}\right)^2 - a_4\left(\frac{h}{2}\right)^4 - a_6\left(\frac{h}{2}\right)^6 - \cdots$$

We can eliminate the dominant term in the error series by simple algebra. To do so, multiply the bottom equation by 4 and subtract the result from the top equation. The result is

$$\phi(h) - 4\phi\left(\frac{h}{2}\right) = -3f'(x) - \frac{3}{4}a_4h^4 - \frac{15}{16}a_6h^6 - \cdots$$

We divide by -3 and rearrange this to get

$$\phi\left(\frac{h}{2}\right) + \frac{1}{3}\left[\phi\left(\frac{h}{2}\right) - \phi(h)\right] = f'(x) + \frac{1}{4}a_4h^4 + \frac{5}{16}a_6h^6 + \cdots$$

This is a miraculous discovery. Simply by adding $\frac{1}{3}[\phi(h/2) - \phi(h)]$ to $\phi(h/2)$, we have apparently improved the precision to $\mathcal{O}(h^4)$ because the error series that accompanies this new combination begins with $\frac{1}{4}a_4h^4$. Since h will be small, this is a dramatic improvement. And now, to top it off, note that the same procedure can be repeated over and over again to "kill" higher and higher terms in the error. This is **Richardson extrapolation**.

Essentially the same situation arises in the derivation of Romberg's algorithm in Chapter 5. Therefore, it is desirable to have a general discussion of the procedure here. We start with an equation that includes both situations. Let ϕ be a function such that

$$\phi(h) = L - \sum_{k=1}^{\infty} a_{2k}h^{2k} \qquad (8)$$

where the coefficients a_{2k} are not known. Equation (8) is not interpreted as the *definition* of ϕ but rather as a *property* that ϕ possesses. It is assumed that $\phi(h)$ can be computed for any $h > 0$ and that our objective is to approximate L accurately using ϕ.

Select a convenient h, and compute the numbers

$$D(n, 0) = \phi\left(\frac{h}{2^n}\right) \qquad (n \geq 0) \tag{9}$$

Because of Equation (8), we have

$$D(n, 0) = L + \sum_{k=1}^{\infty} A(k, 0)\left(\frac{h}{2^n}\right)^{2k}$$

where $A(k, 0) = -a_{2k}$. These quantities $D(n, 0)$ give a crude estimate of the unknown number $L = \lim_{x \to 0} \phi(x)$. More accurate estimates are obtained via Richardson extrapolation. The extrapolation formula is

$$D(n, m) = \frac{4^m}{4^m - 1}D(n, m - 1) - \frac{1}{4^m - 1}D(n - 1, m - 1) \qquad (1 \leq m \leq n) \tag{10}$$

RICHARDSON EXTRAPOLATION THEOREM

The quantities $D(n, m)$ defined in the Richardson extrapolation process (10) obey the equation

$$D(n, m) = L + \sum_{k=m+1}^{\infty} A(k, m)\left(\frac{h}{2^n}\right)^{2k} \qquad (0 \leq m \leq n) \tag{11}$$

Proof   Equation (11) is true by hypothesis if $m = 0$. For the purpose of an inductive proof, we *assume* that Equation (11) is valid for an arbitrary value of $m - 1$, and we prove that Equation (11) is then valid for m. Now from Equations (10) and (11) for a fixed value m, we have

$$D(n, m) = \frac{4^m}{4^m - 1}\left[L + \sum_{k=m}^{\infty} A(k, m - 1)\left(\frac{h}{2^n}\right)^{2k}\right]$$

$$- \frac{1}{4^m - 1}\left[L + \sum_{k=m}^{\infty} A(k, m - 1)\left(\frac{h}{2^{n-1}}\right)^{2k}\right]$$

After simplification, this becomes

$$D(n, m) = L + \sum_{k=m}^{\infty} A(k, m - 1)\left(\frac{4^m - 4^k}{4^m - 1}\right)\left(\frac{h}{2^n}\right)^{2k} \tag{12}$$

Thus, we are led to define

$$A(k,m) = A(k,m-1)\left(\frac{4^m - 4^k}{4^m - 1}\right)$$

At the same time, we notice that $A(m,m) = 0$. Hence, Equation (12) can be written as

$$D(n,m) = L + \sum_{k=m+1}^{\infty} A(k,m)\left(\frac{h}{2^n}\right)^{2k}$$

Equation (11) is true for m, and the induction is complete.   ■

The significance of Equation (11) is that the summation *begins* with the term $(h/2^n)^{2m}$. Since $h/2^n$ is small, this indicates that the numbers $D(n,m)$ are approaching L very rapidly.

In practice, one can arrange the quantities in a two-dimensional triangular array as follows:

$$
\begin{array}{lllll}
D(0,0) & & & & \\
D(1,0) & D(1,1) & & & \\
D(2,0) & D(2,1) & D(2,2) & & \quad\quad\text{(13)}\\
\vdots & \vdots & \vdots & \ddots & \\
D(N,0) & D(N,1) & D(N,2) & \cdots & D(N,N)
\end{array}
$$

The main features of a code to generate such an array are as follows:

1. Write a procedure function for ϕ.

2. Decide on suitable values for N and h.

3. For $i = 0, 1, \ldots, N$, compute $D(i,0) = \phi(h/2^i)$.

4. For $0 \leq i \leq j \leq N$, compute

$$D(i,j) = D(i,j-1) + (4^j - 1)^{-1}[D(i,j-1) - D(i-1,j-1)]$$

Notice that in this algorithm, the computation of $D(i,j)$ follows Equation (10) but has been rearranged slightly to improve its numerical properties.

EXAMPLE 1   Write a procedure to compute the derivative of a function at a point by using Equation (5) and Richardson extrapolation.

Solution   The input to the procedure will be a function f, a specific point x, a value of h, and a number n signifying how many rows in the array (13) are to be computed. The output will be the array (13). Here is a suitable pseudocode:

procedure *deriv*$(f, x, n, h, (d_{ij}))$
real array $(d_{ij})_{0:n \times 0:n}$

```
integer i, j, n
real h, x
interface external function f
for i = 0 to n do
    d_{i0} ← [f(x + h) − f(x − h)]/(2h)
    for j = 0 to i − 1 do
        d_{i,j+1} ← d_{ij} + (d_{ij} − d_{i−1,j})/(4^{j+1} − 1)
    end do
    h ← h/2
end do
end procedure deriv
```

To test the procedure, we choose $f(x) = \sin x$, where $h = 1$ and $x_0 = \pi/3$. Then $f'(x) = \cos x$ and $f'(x_0) = \frac{1}{2}$. A pseudocode is written as follows:

```
program main
real array (d_{ij})_{0:n×0:n}
integer parameter n ← 10
real parameter h ← 1
interface external function f
call deriv(f, π/3, n, h, (d_{ij}))
output (d_{ij})
end program main

real function f(x)
real x
f ← sin(x)
end function f
```

The computer output, not reproduced here, is the triangular array (d_{ij}) with indices $0 \leq j \leq i \leq 10$. The most accurate value, 0.5000029802, is $(d_{4,4})$.

The values d_{i0}, which are obtained solely by Equations (9) and (7) without any extrapolation, are not as accurate, having no more than nine correct digits. ☐

For such a simple function, we can use Maple to find the derivative exactly and then evaluate the numerical answer as follows.

```
diff(sin(x), x);
evalf(subs(x = Pi/3, "));
```

The answer returned is $\cos(\frac{1}{3}\pi) \approx 0.5000000002$.

First-Derivative Formulas via Interpolation Polynomials

An important general strategem can be used to approximate derivatives (as well as integrals and other quantities). The function f is first approximated by a polynomial p so that $f \approx p$. Then we simply proceed to the approximate equation $f'(x) \approx p'(x)$

as a consequence. Of course, this strategy should be used very cautiously because the behavior of the interpolating polynomial can be oscillatory.

In practice, the approximating polynomial p is often determined by interpolation at a few points. For example, suppose that p is the polynomial of degree ≤ 1 that interpolates f at two nodes, x_0 and x_1. Then from Equation (6) in Section 4.1 with $n = 1$, we have

$$p_1(x) = f(x_0) + f[x_0, x_1](x - x_0)$$

Consequently,

$$f'(x) \approx p_1'(x) = f[x_0, x_1] = \frac{f(x_1) - f(x_0)}{x_1 - x_0} \tag{14}$$

If $x_0 = x$ and $x_1 = x + h$ (see Figure 4.4), this formula is one previously considered—namely, Equation (1):

$$f'(x) \approx \frac{1}{h}[f(x + h) - f(x)] \tag{15}$$

If $x_0 = x - h$ and $x_1 = x + h$ (see Figure 4.5), the resulting formula is Equation (5):

$$f'(x) \approx \frac{1}{2h}[f(x + h) - f(x - h)] \tag{16}$$

Now consider interpolation with three nodes, x_0, x_1, and x_2. The interpolating polynomial is obtained from Equation (6) in Section 4.1:

$$p_2(x) = f(x_0) + f[x_0, x_1](x - x_0) + f[x_0, x_1, x_2](x - x_0)(x - x_1)$$

and its derivative is

$$p_2'(x) = f[x_0, x_1] + f[x_0, x_1, x_2](2x - x_0 - x_1) \tag{17}$$

Here the right-hand side consists of two terms: The first is the previous estimate in Equation (14), and the second is a refinement or correction. If Equation (17) is used

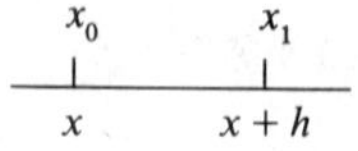

FIGURE 4.4

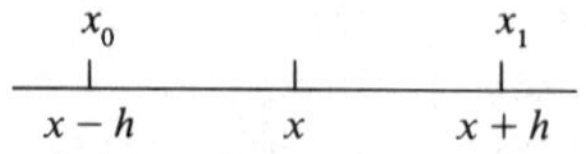

$$\begin{array}{cc} x_0 & x_1 \\ | & | \\ x & x+h \end{array}$$

FIGURE 4.5

$$\begin{array}{ccc} x_0 & & x_1 \\ | & | & | \\ x-h & x & x+h \end{array}$$

to evaluate $f'(x)$ when $x = \frac{1}{2}(x_0 + x_1)$, as in Equation (16), then the correction term in Equation (17) is zero. Thus, the first term in this case must be more accurate than in other cases because the correction term adds nothing. This is why Equation (16) is more accurate than (15).

   An analysis of the errors in this general procedure goes as follows: Suppose that p_n is the polynomial of least degree that interpolates f at the nodes $x_0, x_1, \ldots, x_n$. Then according to Theorem 1 on Interpolating Errors in Section 4.2,

$$f(x) - p_n(x) = \frac{1}{(n+1)!} f^{(n+1)}(\xi) w(x)$$

with ξ dependent on x and $w(x) = (x - x_0)(x - x_1) \cdots (x - x_n)$. Differentiating gives

$$f'(x) - p'_n(x) = \frac{1}{(n+1)!} w(x) \frac{d}{dx} f^{(n+1)}(\xi) + \frac{1}{(n+1)!} f^{(n+1)}(\xi) w'(x) \qquad \textbf{(18)}$$

Here we had to assume that $f^{(n+1)}(\xi)$ is differentiable as a function of x, a fact that is known if $f^{(n+2)}$ exists and is continuous.

   The first observation to make about the error formula in Equation (18) is that $w(x)$ vanishes at each node, so if the evaluation is at a node x_i, the resulting equation is simpler:

$$f'(x_i) = p'_n(x_i) + \frac{1}{(n+1)!} f^{(n+1)}(\xi) w'(x_i)$$

For example, taking just two points x_0 and x_1, we obtain with $n = 1$ and $i = 0$,

$$f'(x_0) = f[x_0, x_1] + \frac{1}{2} f''(\xi) \frac{d}{dx} [(x - x_0)(x - x_1)] \Big|_{x = x_0}$$

$$= f[x_0, x_1] + \frac{1}{2} f''(\xi)(x_0 - x_1)$$

This is Equation (2) in disguise when $x_0 = x$ and $x_1 = x + h$. Similar results follow with $n = 1$ and $i = 1$.

   The second observation to make about Equation (18) is that it becomes simpler if x is chosen as a point where $w'(x) = 0$. For instance, if $n = 1$, then w is a quadratic function that vanishes at the two nodes x_0 and x_1. Because a parabola is symmetric about its axis, $w'[(x_0 + x_1)/2] = 0$. The resulting formula is

$$f'\left(\frac{x_0 + x_1}{2}\right) = f[x_0, x_1] - \frac{1}{8}(x_1 - x_0)^2 \frac{d}{dx} f''(\xi)$$

   As a final example, consider four interpolation points x_0, x_1, x_2, and x_3. The interpolating polynomial from Equation (6) in Section 4.1 with $n = 3$ is

$$p_3(x) = f(x_0) + f[x_0, x_1](x - x_0) + f[x_0, x_1, x_2](x - x_0)(x - x_1)$$
$$+ f[x_0, x_1, x_2, x_3](x - x_0)(x - x_1)(x - x_2)$$

Its derivative is

$$p_3'(x) = f[x_0, x_1] + f[x_0, x_1, x_2](2x - x_0 - x_1)$$
$$+ f[x_0, x_1, x_2, x_3]((x - x_1)(x - x_2)$$
$$+ (x - x_0)(x - x_2) + (x - x_0)(x - x_1))$$

A useful special case occurs if $x_0 = x - h$, $x_1 = x + h$, $x_2 = x - 2h$, and $x_3 = x + 2h$ (see Figure 4.6). The resulting formula is

$$f'(x) \approx \frac{1}{2h}[f(x + h) - f(x - h)] \tag{19}$$

$$- \frac{1}{12h}[f(x + 2h) - 2f(x + h) + 2f(x - h) - f(x - 2h)]$$

which has been arranged in the form in which it probably should be computed: a principal term plus a correction or refining term.

Second-Derivative Formulas via Taylor Series

In the numerical solution of differential equations, it is often necessary to approximate second derivatives. We shall derive the most important formula for accomplishing this. Simply *add* the Taylor series (3) for $f(x + h)$ and $f(x - h)$. The result is

$$f(x + h) + f(x - h) = 2f(x) + h^2 f''(x) + 2\left[\frac{1}{4!}h^4 f^{(4)}(x) + \cdots\right]$$

When this is rearranged, we get

$$f''(x) = \frac{1}{h^2}[f(x + h) - 2f(x) + f(x - h)] + E$$

where the error series is

$$E = -2\left[\frac{1}{4!}h^2 f^{(4)}(x) + \frac{1}{6!}h^4 f^{(6)}(x) + \cdots\right]$$

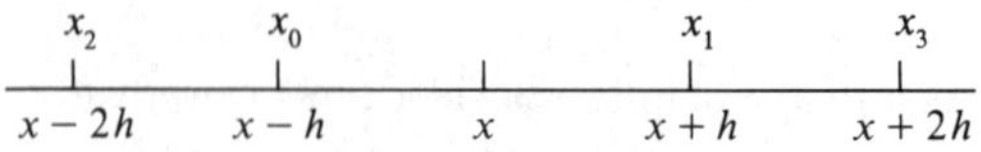

FIGURE 4.6

x_2	x_0		x_1	x_3
$x - 2h$	$x - h$	x	$x + h$	$x + 2h$

By carrying out the same process using Taylor's formula with a remainder, one can show that E is also given by

$$E = -\frac{1}{12}h^2 f^{(4)}(\xi)$$

for some ξ in the interval $(x - h, x + h)$. Hence, we have the approximation

$$f''(x) \approx \frac{1}{h^2}[f(x + h) - 2f(x) + f(x - h)] \qquad (20)$$

with error $\mathcal{O}(h^2)$.

PROBLEMS 4.3

1. Determine the error term for the formula

$$f'(x) \approx \frac{1}{4h}[f(x + 3h) - f(x - h)]$$

2. Using Taylor series, establish the error term for the formula

$$f'(0) \approx \frac{1}{2h}[f(2h) - f(0)]$$

3. Derive the approximation formula

$$f'(x) \approx \frac{1}{2h}[4f(x + h) - 3f(x) - f(x + 2h)]$$

and show that its error term is of the form $\frac{1}{3}h^2 f'''(\xi)$.

4. Can you find an approximation formula for $f'(x)$ that has error term $\mathcal{O}(h^3)$ and involves only two evaluations of the function f? Prove or disprove.

5. Averaging the forward-difference formula $f'(x) \approx [f(x + h) - f(x)]/h$ and the backward-difference formula $f'(x) \approx [f(x) - f(x - h)]/h$, each with error term $\mathcal{O}(h)$, results in the central-difference formula $f'(x) \approx [f(x + h) - f(x - h)]/(2h)$ with error $\mathcal{O}(h^2)$. Show why. *Hint:* Determine at least the first term in the error series for each formula.

6. Derive the formula and its error term for:

   a. Equation (19)

   b. Equation (20)

7. Criticize the following analysis. By Taylor's formula, we have

$$f(x + h) - f(x) = hf'(x) + \frac{h^2}{2}f''(x) + \frac{h^3}{6}f'''(\xi)$$

$$f(x - h) - f(x) = -hf'(x) + \frac{h^2}{2}f''(x) - \frac{h^3}{6}f'''(\xi)$$

So by adding, we obtain an *exact* expression for $f''(x)$:

$$f(x + h) + f(x - h) - 2f(x) = h^2 f''(x)$$

8. Criticize the following analysis. By Taylor's formula, we have

$$f(x + h) - f(x) = hf'(x) + \frac{h^2}{2}f''(x) + \frac{h^3}{6}f'''(\xi_1)$$

$$f(x - h) - f(x) = -hf'(x) + \frac{h^2}{2}f''(x) - \frac{h^3}{6}f'''(\xi_2)$$

Therefore,

$$\frac{1}{h^2}\left[f(x + h) - 2f(x) + f(x - h)\right] = f''(x) + \frac{h}{6}[f'''(\xi_1) - f'''(\xi_2)]$$

The error in the approximation formula for f'' is thus $\mathcal{O}(h)$.

9. A certain calculation requires an approximation formula for $f'(x) + f''(x)$. How well does the expression

$$\left(\frac{2 + h}{2h^2}\right)f(x + h) - \left(\frac{2}{h^2}\right)f(x) + \left(\frac{2 - h}{2h^2}\right)f(x - h)$$

serve? Derive this approximation and its error term.

10. Derive the two formulas

a. $f'(x) \approx \dfrac{1}{4h}[f(x + 2h) - f(x - 2h)]$

b. $f''(x) \approx \dfrac{1}{4h^2}[f(x + 2h) - 2f(x) + f(x - 2h)]$

and establish formulas for the errors in using them.

11. Derive the following rules for estimating derivatives

a. $f'''(x) \approx \dfrac{1}{2h^3}[f(x + 2h) - 2f(x + h) + 2f(x - h) - f(x - 2h)]$

b. $f^{(4)}(x) \approx \dfrac{1}{h^4}[f(x + 2h) - 4f(x + h) + 6f(x) - 4f(x - h) + f(x + 2h)]$

and their error terms. Which is more accurate? *Hint*: Consider the Taylor series for $D(h) \equiv f(x + h) - f(x - h)$ and $S(h) \equiv f(x + h) + f(x - h)$.

12. The values of a function f are given at three points x_0, x_1, and x_2. If a quadratic interpolating polynomial is used to estimate $f'(x)$ at $x = \frac{1}{2}(x_0 + x_1)$, what formula will result?

13. Establish the formula

$$f''(x) \approx \frac{2}{h^2} \left[\frac{f(x_0)}{(1 + \alpha)} - \frac{f(x_1)}{\alpha} + \frac{f(x_2)}{\alpha(\alpha + 1)} \right]$$

in the following two ways using the unevenly spaced points $x_0 < x_1 < x_2$, where $x_1 - x_0 = h$ and $x_2 - x_1 = \alpha h$. Notice that this formula reduces to the standard central-difference formula (20) when $\alpha = 1$.

a. Approximate $f(x)$ by the Newton interpolating polynomial of degree 2.

b. Calculate the undetermined coefficients A, B, and C in the expression $f''(x) \approx Af(x_0) + Bf(x_1) + Cf(x_2)$ by making it exact for the three polynomials 1, $x - x_1$, and $(x - x_1)^2$ and thus exact for all polynomials of degree ≤ 2.

14. (Continuation) Using Taylor series, show that

$$f'(x_1) = \frac{f(x_2) - f(x_0)}{x_2 - x_0} + (\alpha - 1)\frac{h}{2}f''(x_1) + \mathcal{O}(h^2)$$

Establish that the error for approximating $f'(x_1)$ by $[f(x_2) - f(x_0)]/(x_2 - x_0)$ is $\mathcal{O}(h^2)$ when x_1 is midway between x_0 and x_2 but only $\mathcal{O}(h)$ otherwise.

15. Show how Richardson extrapolation would work on Formula (20).

16. If $\phi(h) = L - c_1h - c_2h^2 - c_3h^3 - \cdots$, then what combination of $\phi(h)$ and $\phi(h/2)$ should give an accurate estimate of L?

17. (Continuation) State and prove a theorem analogous to the Theorem on Richardson Extrapolation for the situation of Problem **16**.

18. If $\phi(h) = L - c_1h^{1/2} - c_2h^{2/2} - c_3h^{3/2} - \cdots$, then what combination of $\phi(h)$ and $\phi(h/2)$ should give an accurate estimate of L?

19. Show that Richardson extrapolation can be carried out for any two values of h. Thus, if $\phi(h) = L - \mathcal{O}(h^p)$, then from $\phi(h_1)$ and $\phi(h_2)$, a more accurate estimate of L is given by

$$\phi(h_2) + \frac{h_2^p}{h_1^p - h_2^p}[\phi(h_2) - \phi(h_1)]$$

20. Consider a function ϕ such that $\lim_{h \to p} \phi(h) = L$ and $L - \phi(h) \approx ce^{-1/h}$ for some constant c. By combining $\phi(h)$, $\phi(h/2)$, and $\phi(h/3)$, find an accurate estimate of L.

COMPUTER PROBLEMS 4.3

1. Find $f'(0.25)$ as accurately as possible, using only the function corresponding to the pseudocode below and a method for numerical differentiation.

real function $f(x)$
integer i
real a, b, c, x

$$a \leftarrow 1$$
$$b \leftarrow \cos(x)$$
for $i = 1$ **to** 5 **do**
$\quad c \leftarrow b$
$\quad b \leftarrow \sqrt{ab}$
$\quad a \leftarrow (a + c)/2$
end do
$f \leftarrow 2 \arctan(1)/a$
end function f

2. Carry out a numerical experiment to compare the accuracy of Formulas (5) and (19) on a function f whose derivative can be computed precisely. Take a sequence of values for h, such as 4^{-n} with $0 \leq n \leq 24$.

3. Test procedure *deriv* on the following functions at the points indicated in a single computer run. Interpret the results.

   a. $f(x) = \cos x$ at $x = 0$

   b. $f(x) = \arctan x$ at $x = 1$

   c. $f(x) = |x|$ at $x = 0$

4. Write and test a procedure similar to *deriv* that computes $f''(x)$ with repeated Richardson extrapolation.

5

NUMERICAL INTEGRATION

In electrical field theory, it is proved that the magnetic field induced by a current flowing in a circular loop of wire has intensity

$$H(x) = \frac{4Ir}{r^2 - x^2} \int_0^{\pi/2} \left[1 - \left(\frac{x}{r}\right)^2 \sin^2 \theta \right]^{1/2} d\theta$$

where I is the current, r the radius of the loop, and x the distance from the center to the point where the magnetic intensity is being computed ($0 \leq x \leq r$). If I, r, and x are given, we have a nasty integral to evaluate. It is an **elliptic** integral and not expressible in terms of familiar functions. But H *can* be computed precisely by the methods of this chapter. For example, if $I = 15.3$, $r = 120$, and $x = 84$, we find $H = 1.35566\,1135$ accurate to nine decimals.

5.1 Definite Integral

Elementary calculus focuses largely on two important processes of mathematics: differentiation and integration. In Section 1.1, numerical differentiation was considered briefly, and then it was taken up again in Section 4.3. In this chapter, the process of integration is examined from the standpoint of numerical mathematics.

Definite and Indefinite Integrals

It is customary to distinguish two types of integrals, the definite and the indefinite integral. The indefinite integral of a function is another *function* or a class of functions, whereas the definite integral of a function over a fixed interval is a *number*. For example,

$$\text{Indefinite integral}: \qquad \int x^2 \, dx = \frac{1}{3}x^3 + C$$

$$\text{Definite integral}: \qquad \int_0^2 x^2 \, dx = \frac{8}{3}$$

Actually, a function has not just one but many indefinite integrals. These differ from each other by constants. Thus, in the preceding example, any constant value may be assigned to C and the result is still an indefinite integral. In elementary calculus, the concept of an indefinite integral is identical with the concept of an antiderivative. An antiderivative of a function f is any function F with the property that $F' = f$.

The definite and indefinite integrals are related by the **Fundamental Theorem of Calculus**,* which states that $\int_a^b f(x) \, dx$ can be computed by first finding an antiderivative F of f and then evaluating $F(b) - F(a)$. Thus, using traditional notation, we have

$$\int_1^3 (x^2 - 2) \, dx = \frac{x^3}{3} - 2x \Big|_1^3 = \left(\frac{27}{3} - 6\right) - \left(\frac{1}{3} - 2\right) = \frac{14}{3}$$

As another example of the Fundamental Theorem of Calculus, we can write

$$\int_a^b F'(x) \, dx = F(b) - F(a)$$

$$\int_a^x F'(t) \, dt = F(x) - F(a)$$

If this second equation is differentiated with respect to x, the result is (and here we have put $f = F'$)

$$\frac{d}{dx} \int_a^x f(t) \, dt = f(x)$$

This last equation shows that $\int_a^x f(t) \, dt$ must be an antiderivative (indefinite integral) of f.

The foregoing technique for computing definite integrals is virtually the only one emphasized in elementary calculus. The definite integral of a function, however,

---

****Fundamental Theorem of Calculus:** If f is continuous on the interval $[a, b]$ and F is an antiderivative of f, then

$$\int_a^b f(x) \, dx = F(b) - F(a)$$

has an interpretation as the area under a curve, and so the existence of a numerical value for $\int_a^b f(x)\,dx$ should not depend logically on our limited ability to find antiderivatives. Thus, for instance,

$$\int_0^1 e^{x^2}\,dx$$

has a precise numerical value despite the fact that there is no elementary function F such that $F'(x) = e^{x^2}$. By the preceding remarks, e^{x^2} does have antiderivatives, one of which is

$$F(x) = \int_0^x e^{t^2}\,dt$$

However, this form of the function F is of no help in obtaining the numerical value sought.

Lower and Upper Sums

The existence of the definite integral of a nonnegative function f on a closed interval $[a, b]$ is based on an interpretation of that integral as the area under the graph of f. The definite integral is defined by means of two concepts, the *lower sums of f* and the *upper sums of f*; these are approximations to the area under the graph.

Let P be a **partition** of the interval $[a, b]$ given by

$$P = \{a = x_0 < x_1 < x_2 < \cdots < x_{n-1} < x_n = b\}$$

with partition points $x_0, x_1, x_2, \ldots, x_n$ that divide the interval $[a, b]$ into n subintervals $[x_i, x_{i+1}]$. Now denote by m_i the greatest lower bound (*infimum*) of $f(x)$ on the subinterval $[x_i, x_{i+1}]$. In symbols,

$$m_i = \inf\{f(x) : x_i \leqq x \leqq x_{i+1}\}$$

Likewise, we denote by M_i the least upper bound (*supremum*) of $f(x)$ on $[x_i, x_{i+1}]$. Thus,

$$M_i = \sup\{f(x) : x_i \leqq x \leqq x_{i+1}\}$$

The **lower sums** and **upper sums** of f corresponding to the given partition P are defined to be

$$L(f; P) = \sum_{i=0}^{n-1} m_i(x_{i+1} - x_i)$$

$$U(f; P) = \sum_{i=0}^{n-1} M_i(x_{i+1} - x_i)$$

If f is a positive function, these two quantities can be interpreted as estimates of the area under the curve for f. These sums are shown in Figure 5.1.

EXAMPLE 1 What are the numerical values of the upper and lower sums for $f(x) = x^2$ on the interval $[0, 1]$ if the partition is $P = \{0, \frac{1}{4}, \frac{1}{2}, \frac{3}{4}, 1\}$?

Solution We want the value of

$$U(f; P) = M_0(x_1 - x_0) + M_1(x_2 - x_1) + M_2(x_3 - x_2) + M_3(x_4 - x_3)$$

Since f is increasing on $[0, 1]$, $M_0 = f(x_1) = \frac{1}{16}$. Similarly, $M_1 = f(x_2) = \frac{1}{4}$, $M_2 = f(x_3) = \frac{9}{16}$, and $M_3 = f(x_4) = 1$. The widths of the subintervals are all equal to $\frac{1}{4}$. Hence,

$$U(f; P) = \frac{1}{4}\left(\frac{1}{16} + \frac{1}{4} + \frac{9}{16} + 1\right) = \frac{15}{32}$$

In the same way, we find that $m_0 = f(x_0) = 0$, $m_1 = \frac{1}{16}$, $m_2 = \frac{1}{4}$, and $m_3 = \frac{9}{16}$. Hence,

$$L(f; P) = \frac{1}{4}\left(0 + \frac{1}{16} + \frac{1}{4} + \frac{9}{16}\right) = \frac{7}{32}$$

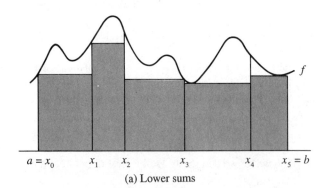

(a) Lower sums

(b) Upper sums

FIGURE 5.1

If we had no other way of calculating $\int_0^1 x^2\, dx$, we would take a value halfway between $U(f;P)$ and $L(f;P)$ as the best estimate. This number is $\frac{11}{32}$. The correct value is $\frac{1}{3}$, and the error is $\frac{11}{32} - \frac{1}{3} = \frac{1}{96}$. □

It is intuitively clear that the upper sum *overestimates* the area under the curve, and the lower sum *underestimates* it. Therefore, the expression $\int_a^b f(x)\, dx$, which we are trying to define, is *required* to satisfy the basic inequality

$$L(f;P) \leqq \int_a^b f(x)\, dx \leqq U(f;P) \tag{1}$$

for all partitions P. It turns out that if f is a *continuous* function defined on $[a,b]$, then Inequality (1) does indeed define the integral. That is, there is one and only one real number that is greater than or equal to all lower sums of f and less than or equal to all upper sums of f. This unique number (depending on f, a, and b) is defined to be $\int_a^b f(x)\, dx$.

Riemann-Integrable Functions

We consider the least upper bound (supremum) of the set of all numbers $L(f;P)$ obtained when P is allowed to range over all partitions of the interval $[a,b]$. This is abbreviated $\sup_P L(f;P)$. Similarly, we consider the greatest lower bound (infimum) of $U(f;P)$ when P ranges over all partitions of $[a,b]$. This is denoted by $\inf_P U(f;P)$. Now if these two numbers are the same—that is, if

$$\inf_P U(f;P) = \sup_P L(f;P) \tag{2}$$

then we say that f is *Riemann-integrable* on $[a,b]$ and define $\int_a^b f(x)\, dx$ to be the common value obtained in Equation (2). The important result mentioned above can be stated formally as follows.

RIEMANN'S THEOREM

Every continuous function defined on a closed and bounded interval of the real line is Riemann-integrable.

There are plenty of functions that are *not* Riemann-integrable. The simplest is known as the **Dirichlet function**:

$$d(x) = \begin{cases} 0 & \text{if } x \text{ is rational} \\ 1 & \text{if } x \text{ is irrational} \end{cases}$$

For any interval $[a,b]$ and for any partition P of $[a,b]$, we have $L(d;P) = 0$ and $U(d;P) = b - a$. Hence,

$$0 = \sup_P L(d; P) < \inf_P U(d; P) = b - a$$

In calculus, it is proved not only that the Riemann integral of a continuous function on $[a, b]$ exists but also that it can be obtained by two limits:

$$\lim_{n \to \infty} L(f; P_n) = \int_a^b f(x)\, dx = \lim_{n \to \infty} U(f; P_n)$$

in which $P_0, P_1, \ldots$ is any sequence of partitions with the property that the length of the largest subinterval in P_n converges to zero as $n \to \infty$. Furthermore, if it is so arranged that P_{n+1} is obtained from P_n by adding new points (and not deleting points), then the lower sums converge *upward* to the integral and the upper sums converge *downward* to the integral. From the numerical standpoint, this is a desirable feature of the process because at each step an interval that contains the unknown number $\int_a^b f(x)\, dx$ will be available. Moreover, these intervals shrink in width at each succeeding step.

Examples and Pseudocode

The process just described can easily be carried out on a computer. To illustrate, we select the function $f(x) = e^{-x^2}$ and the interval $[0, 1]$; that is, we consider

$$\int_0^1 e^{-x^2}\, dx$$

This function is of great importance in statistics, but its indefinite integral cannot be obtained by the elementary techniques of calculus. For partitions, we take equally spaced points in $[0, 1]$. Thus, if there are to be n subintervals in P_n, then we define $P_n = \{x_0, x_1, \ldots, x_n\}$, where $x_i = ih$ for $0 \leq i \leq n$ and $h = 1/n$. Since e^{-x^2} is *decreasing* on $[0, 1]$, the least value of f on the subinterval $[x_i, x_{i+1}]$ occurs at x_{i+1}. Similarly, the greatest value occurs at x_i. Hence, $m_i = f(x_{i+1})$ and $M_i = f(x_i)$. Putting this into the formulas for the upper and lower sums, we obtain for this function

$$L(f; P_n) = \sum_{i=0}^{n-1} h f(x_{i+1}) = h \sum_{i=0}^{n-1} e^{-x_{i+1}^2}$$

$$U(f; P_n) = \sum_{i=0}^{n-1} h f(x_i) = h \sum_{i=0}^{n-1} e^{-x_i^2}$$

Since these sums are almost the same, it is more economical to compute $L(f; P_n)$ by the given formula and to obtain $U(f; P_n)$ by observing that

$$U(f; P_n) = h f(x_0) + L(f; P_n) - h f(x_n) = L(f; P_n) + h(1 - e^{-1})$$

The last equation also shows that the interval defined by Inequality (1) is of width $h(1 - e^{-1})$ for this problem.

Here is pseudocode to carry out this experiment with $n = 1000$:

```
program sums
integer parameter n ← 1000
real parameter a ← 0, b ← 1
integer i
real h, sum, sumL, sumU, x
h ← (b − a)/n
sum ← 0
for i = n to 1 step −1 do
    x ← a + ih
    sum ← sum + f(x)
end do
sumL ← (sum)h
sumU ← sumL + h[f(a) − f(b)]
output sumL, sumU
end program sums

real function f(x)
real x
f ← 1/e^(x²)
end function f
```

A few comments about this pseudocode may be helpful. First, a subscripted variable is not needed in the program for the points x_i. Each point is labeled x. After it is defined and used, it need not be saved. Next, observe that the program has been written so that only one line of code needs to be changed if another value of n is required. Finally, the numbers $e^{-x_i^2}$ are added in order of *ascending* magnitude to reduce roundoff error. However, roundoff errors in the computer are negligible compared with the error in our final estimation of the integral. This code can be used with any function that is decreasing on $[a, b]$ because with that assumption, $U(f; P)$ can be easily obtained from $L(f; P)$ (see Problem **8**).

The computer program corresponding to the pseudocode produces as output the following values of the lower and upper sums:

$$sum_L = 0.74651 \qquad sum_U = 0.74714$$

A good approximate value for the integral can be computed using the Maple statements

```
int(exp(-x*x), x=0..1);
evalf(");
```

and we obtain

$$\int_0^1 e^{-x^2}\, dx = \frac{1}{2}\sqrt{\pi}\,\text{erf}(1) \approx 0.74682\,41330$$

The function

$$\frac{2}{\sqrt{\pi}} \int_0^x e^{-t^2}\, dt$$

is known as the **error function** and is usually denoted by erf(x). A tabulation of this function (and many others) can be found in mathematical handbooks, such as *CRC Standard Mathematical Tables* or Abramowitz and Stegun [1964].

At this juncture the reader is urged to program this experiment or one like it. The experiment shows how the computer can mimic the abstract definition of the Riemann integral, at least in cases where the numbers m_i and M_i can be obtained easily. Another conclusion that can be drawn from the experiment is that the direct translation of a definition into a computer algorithm may leave much to be desired in *precision*. With 999 evaluations of the function, the error is still about 0.0003 (absolute). We shall soon see that more sophisticated algorithms (such as Romberg's) improve this situation dramatically.

EXAMPLE 2   If the integral $\int_0^\pi e^{\cos x}\, dx$ is to be computed with absolute error less than $\frac{1}{2} \times 10^{-3}$, and if we are going to use upper and lower sums computed with a uniform partition, how many subintervals are needed?

Solution   The integrand, $f(x) = e^{\cos x}$, is a decreasing function on the interval $[0, \pi]$. Hence, in the formulas for $U(f; P)$ and $L(f; P)$, we have

$$m_i = f(x_{i+1}) \quad \text{and} \quad M_i = f(x_i)$$

Let P be a partition of $[0, \pi]$ by $n + 1$ equally spaced points, $0 = x_0 < \cdots < x_n = \pi$. Then there will be n subintervals, all of width π/n. Hence,

$$L(f; P) = \frac{\pi}{n} \sum_{i=0}^{n-1} m_i = \frac{\pi}{n} \sum_{i=0}^{n-1} f(x_{i+1}) \tag{3}$$

$$U(f; P) = \frac{\pi}{n} \sum_{i=0}^{n-1} M_i = \frac{\pi}{n} \sum_{i=0}^{n-1} f(x_i) \tag{4}$$

The correct value of the integral lies in the interval between $L(f; P)$ and $U(f; P)$. We take the *midpoint* of the interval as the best estimate, thus obtaining an error of at most $\frac{1}{2}[U(f; P) - L(f; P)]$—that is, the length of half the interval. To meet the error criterion imposed in the problem, we must have

$$\frac{1}{2}[U(f; P) - L(f; P)] < \frac{1}{2} \times 10^{-3}$$

From Formulas (3) and (4), we can calculate the difference between the upper and lower sums. This leads to $(\pi/n)(e^1 - e^{-1}) < 10^{-3}$. With the aid of a calculator, we determine that $n \geq 7386$. $\square$

**PROBLEMS
5.1**

1. If we estimate $\int_0^1 (x^2 + 2)^{-1} \, dx$ by means of a lower sum using the partition $P = \{0, \frac{1}{2}, 1\}$, what is the result?

2. What is the result if we estimate $\int_1^2 x^{-1} \, dx$ by means of the upper sum using the partition $P = \{1, \frac{3}{2}, 2\}$?

3. Show that if $\theta_i \geq 0$ and $\sum_{i=0}^n \theta_i = 1$, then $\sum_{i=0}^n \theta_i a_i$ lies between the least and the greatest of the numbers a_i.

4. Establish the **midpoint rule** for estimating an integral:

$$\int_a^b f(x) \, dx \approx \sum_{i=0}^{n-1} (x_{i+1} - x_i) f\left(\frac{x_{i+1} + x_i}{2}\right)$$

5. (Continuation) Show the relationship between the midpoint rule and the upper and lower sums.

6. (Continuation) Show that the midpoint rule for equal subintervals is given by

$$\int_a^b f(x) \, dx \approx h \sum_{i=0}^{n-1} f\left(x_i + \frac{1}{2}h\right)$$

where $h = (b - a)/n$, $x_i = a + ih$, and $0 \leq i \leq n$.

7. Calculate an approximate value of $\int_0^\alpha [(e^x - 1)/x] \, dx$ for $\alpha = 10^{-4}$ correct to 14 decimal places. *Hint:* Use Taylor series.

8. For a decreasing function $f(x)$ over an interval $[a, b]$ with n uniform subintervals, show that the difference between the upper sum and the lower sum is given by $[(b - a)/n][f(a) - f(b)]$.

9. (Continuation) Repeat Problem **8** for an increasing function.

10. If upper and lower sums are used with regularly spaced points to compute $\int_2^5 (dx/\log x)$, how many points are needed to achieve an accuracy of $\frac{1}{2} \times 10^{-4}$?

11. Let f be an increasing function. If $\int_0^1 f(x) \, dx$ is to be estimated by using the method of upper and lower sums, taking n equally spaced points, what is the worst possible error?

12. If f is a (strictly) increasing function on $[a, b]$, and if $\alpha = f(a)$ and $\beta = f(b)$, then $f^{-1}(x)$ is well defined for $\alpha \leq x \leq \beta$. Discover the relationship between $\int_a^b f(x) \, dx$ and $\int_\alpha^\beta f^{-1}(x) \, dx$.

**COMPUTER
PROBLEMS
5.1**

1. Estimate the definite integral $\int_0^1 x^{-1} \sin x \, dx$ by computing the upper and lower sums, using 800 points in the interval. The integrand is defined to be 1 at $x = 0$. The function is decreasing, and this fact should be shown by calculus. (For a decreasing function f, $f' < 0$.) *Note:* The function

$$\text{Si}(x) = \int_0^x t^{-1} \sin t \, dt$$

is an important special function known as the **sine integral**. It is represented by a Taylor series that converges for all real or complex values of x. The easiest way to obtain this series is to start with the series for $\sin t$, divide by t, and integrate term by term:

$$\text{Si}(x) = \int_0^x t^{-1} \sin t \, dt = \int_0^x \sum_{n=0}^{\infty} (-1)^n \frac{t^{2n}}{(2n+1)!} \, dt$$

$$= \sum_{n=0}^{\infty} (-1)^n \frac{x^{2n+1}}{(2n+1)!(2n+1)} = x - \frac{x^3}{18} + \frac{x^5}{600} - \frac{x^7}{35280} + \cdots$$

This series is rapidly convergent. For example, from only the terms shown, Si(1) is computed to be 0.94608 27 with an error of at most four units in the last digit shown.

2. Write a general-purpose procedure to estimate integrals of decreasing functions by the method of upper and lower sums with a uniform partition. Give the procedure the calling sequence

   real function *integral*($f, a, b, \varepsilon, n, sum_L, sum_U$)

   where f is the function name, a and b are the endpoints of the interval, and ε is the tolerance. The procedure determines n so that $sum_U - sum_L < 2\varepsilon$. The procedure returns the average of sum_U and sum_L. Test it on the sine integral of Computer Problem 1 using $\varepsilon = \frac{1}{2} \times 10^{-5}$.

3. From calculus, the length of a curve is $\int_a^b \sqrt{1 + [f'(x)]^2} \, dx$, where f is a function whose graph is the curve on the interval $a \leq x \leq b$. Find the length of the ellipse $y^2 + 4x^2 = 1$.

4. The logarithmic integral is a special mathematical function defined by the equation

$$\text{li}(x) = \int_2^x \frac{dt}{\ln t}$$

   For large x, the number of prime integers less than or equal to x is closely approximated by li(x). For example, there are 46 primes less than 200, and li(200) is around 50. Find li(200) with three significant figures by means of upper and lower sums. Determine the number of partition points needed *prior* to writing the program.

5.2 Trapezoid Rule

Description

The next method we consider is an improvement over the coarse method of the preceding section. Moreover, it is an important ingredient of the Romberg algorithm of the next section. This method is called the **trapezoid rule** and is based on an

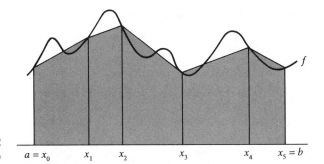

FIGURE 5.2
Trapezoid rule

$a = x_0$ x_1 x_2 x_3 x_4 $x_5 = b$

estimation of the area beneath a curve using trapezoids. Again, the estimation of $\int_a^b f(x)\,dx$ is approached by first dividing the interval $[a, b]$ into subintervals according to the **partition**: $P = \{a = x_0 < x_1 < x_2 < \cdots < x_n = b\}$. For each such partition of the interval (the partition points x_i need not be uniformly spaced), an estimation of the integral by the trapezoid rule is obtained. We denote it by $T(f; P)$. Figure 5.2 shows what the trapezoids are. A typical trapezoid has the subinterval $[x_i, x_{i+1}]$ as its base, and the two vertical sides are $f(x_i)$ and $f(x_{i+1})$ (see Figure 5.3). The area is equal to the base times the average height:

$$A_i = \frac{1}{2}(x_{i+1} - x_i)[f(x_i) + f(x_{i+1})]$$

Hence, the total area of the trapezoids is

$$T(f; P) = \sum_{i=0}^{n-1} A_i = \frac{1}{2} \sum_{i=0}^{n-1} (x_{i+1} - x_i)[f(x_i) + f(x_{i+1})]$$

Alternative terminology is **compound** or **composite trapezoid rule**.

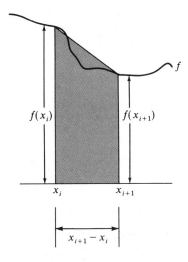

$f(x_i)$ $f(x_{i+1})$ f

x_i x_{i+1}

$x_{i+1} - x_i$

FIGURE 5.3

Uniform Spacing

In practice and in the Romberg algorithm (discussed in the next section), the trapezoid rule is used with a *uniform* partition of the interval. This means that the division points x_i are equally spaced: $x_i = a + ih$, where $h = (b - a)/n$ and $0 \leq i \leq n$. In this case, the formula for $T(f; P)$ can be given in simpler form because $x_{i+1} - x_i = h$. Thus,

$$T(f; P) = \frac{h}{2} \sum_{i=0}^{n-1} [f(x_i) + f(x_{i+1})]$$

It should be emphasized that in order to economize the amount of arithmetic, the computationally preferable formula for the trapezoid rule is

$$T(f; P) = h \left\{ \sum_{i=1}^{n-1} f(x_i) + \frac{1}{2}[f(x_0) + f(x_n)] \right\} \qquad (1)$$

To illustrate, we consider the integral $\int_0^1 e^{-x^2} \, dx$, which was approximated in the previous section by lower and upper sums. Here is pseudocode for Equation (1) with $n = 60$ and $f(x) = e^{-x^2}$:

```
program trapezoid
integer parameter n ← 60
real parameter a ← 0, b ← 1
integer i
real h, sum, x
h ← (b − a)/n
sum ← ½[f(a) + f(b)]
for i = 1 to n − 1 do
    x ← a + ih
    sum ← sum + f(x)
end do
sum ← (sum)h
output sum
end trapezoid

real function f(x)
real x
f ← 1/e^(x²)
end function f
```

The computer output 0.74681 is closer to the exact value but still not accurate to all digits shown. (Why?)

EXAMPLE 1   Compute $\int_0^1 (\sin x/x) \, dx$ by using the trapezoid rule with six uniform points.

Solution   The function values are arranged in a table as follows:

x_i	$f(x_i)$
0.0	1.00000
0.2	0.99335
0.4	0.97355
0.6	0.94107
0.8	0.89670
1.0	0.84147

Notice that we have assigned the value $\sin x / x = 1$ at $x = 0$. Then

$$T(f; P) = 0.2 \sum_{i=1}^{4} f(x_i) + (0.1)[f(x_0) + f(x_5)]$$

$$= (0.2)(3.80467) + (0.1)(1.84147)$$

$$= 0.94508$$

This result is not accurate to all the digits shown because only five subintervals were used. Using the Maple command

```
evalf(Si(1));
```

we obtain $\text{Si}(1) \approx 0.9460830704$. (Refer to Computer Problem **1** of Section 5.1.) We shall see later how to determine a suitable value for n to obtain a desired accuracy using the Trapezoid rule. $\square$

Error Analysis

We now analyze the error incurred in using the trapezoid rule to estimate $\int_a^b f(x) \, dx$. We shall establish the following result:

THEOREM ON PRECISION OF TRAPEZOID RULE

If f'' exists and is continuous on the interval $[a, b]$, and if the trapezoid rule T with uniform spacing h is used to estimate the integral $I = \int_a^b f(x) \, dx$, then for some ζ in (a, b),

$$I - T = -\frac{1}{12}(b - a)h^2 f''(\zeta) = \mathcal{O}(h^2)$$

Proof The first step in the analysis is to prove the above result when $a = 0$, $b = 1$, and $h = 1$. In this case, we have to show

$$\int_0^1 f(x) \, dx - \frac{1}{2}[f(0) + f(1)] = -\frac{1}{12} f''(\zeta) \tag{2}$$

This is easily established with the aid of the error formula for polynomial interpolation (see Section 4.2). To do so, let p be the polynomial of degree 1 that interpolates f at 0 and 1. Then p is given by

$$p(x) = f(0) + [f(1) - f(0)]x$$

Hence,

$$\int_0^1 p(x)\,dx = f(0) + \frac{1}{2}[f(1) - f(0)]$$

$$= \frac{1}{2}[f(0) + f(1)]$$

By the error formula that governs polynomial interpolation [Equation (2) in Section 4.2], we have (here, of course, $n = 1$, $x_0 = 0$, and $x_1 = 1$)

$$f(x) - p(x) = \frac{1}{2}f''[\xi(x)]x(x - 1) \tag{3}$$

with $\xi(x)$ depending on x in $(0, 1)$. From Equation (3), it follows that

$$\int_0^1 f(x)\,dx - \int_0^1 p(x)\,dx = \frac{1}{2}\int_0^1 f''[\xi(x)]x(x - 1)\,dx$$

That $f''[\xi(x)]$ is continuous can be proved by solving Equation (3) for $f''[\xi(x)]$ and verifying the continuity. (See Problem **10** in Section 4.2.) Notice that $x(x - 1)$ does not change sign in the interval $[0, 1]$. Hence, by the **Mean-Value Theorem for Integrals**,[*] there is a point $x = s$ for which

$$\int_0^1 f''[\xi(x)]x(x - 1)\,dx = f''[\xi(s)]\int_0^1 x(x - 1)\,dx$$

$$= f''(\zeta)\left(-\frac{1}{6}\right)$$

By putting all these equations together, we obtain Equation (2).

From Equation (2), by making a change of variable, we obtain

$$\int_a^b f(x)\,dx = \frac{b - a}{2}[f(a) + f(b)] - \frac{1}{12}(b - a)^3 f''(\xi) \tag{4}$$

---

[*]**Mean-Value Theorem for Integrals**: If f is continuous and if g is Riemann-integrable with $g(x) \geq 0$ [or $g(x) \leq 0$] on $[a, b]$, then there exists a point ξ such that $a \leq \xi \leq b$ and $\int_a^b f(x)g(x)\,dx = f(\xi)\int_a^b g(x)\,dx$.

The details of this are as follows: Let $g(t) = f(a + t(b - a))$ and $x = a + (b - a)t$. Thus, as t traverses the interval $[0, 1]$, x traverses the interval $[a, b]$. Also, $dx = (b - a)\,dt$, $g'(t) = f'[a + t(b - a)](b - a)$ and $g''(t) = f''[a + t(b - a)](b - a)^2$. Hence, by Equation (2),

$$\int_a^b f(x)\,dx = (b - a) \int_0^1 f[a + t(b - a)]\,dt$$

$$= (b - a) \int_0^1 g(t)\,dt$$

$$= (b - a) \left\{ \frac{1}{2}[g(0) + g(1)] - \frac{1}{12}g''(\zeta) \right\}$$

$$= \frac{b - a}{2}[f(a) + f(b)] - \frac{(b - a)^3}{12}f''(\xi)$$

This is the trapezoid rule and error term for the interval $[a, b]$ with only one subinterval, which is the entire interval. Thus, the error term is $\mathcal{O}(h^3)$ with $h = b - a$. Here, ξ is in (a, b).

Now let the interval $[a, b]$ be divided into n equal subintervals by points $x_0, x_1, \ldots, x_n$ with spacing h. Applying Formula (4) to each subinterval, we have

$$\int_{x_i}^{x_{i+1}} f(x)\,dx = \frac{h}{2}[f(x_i) + f(x_{i+1})] - \frac{1}{12}h^3 f''(\xi_i) \tag{5}$$

where $x_i < \xi_i < x_{i+1}$. We use this result over the interval $[a, b]$, obtaining

$$\int_a^b f(x)\,dx = \sum_{i=0}^{n-1} \int_{x_i}^{x_{i+1}} f(x)\,dx = \frac{h}{2} \sum_{i=0}^{n-1} [f(x_i) + f(x_{i+1})] - \frac{h^3}{12} \sum_{i=0}^{n-1} f''(\xi_i) \tag{6}$$

The final term in Equation (6) is the error term, and we can simplify it in the following way: Since $h = (b - a)/n$, we have

$$-\frac{h^3}{12} \sum_{i=0}^{n-1} f''(\xi_i) = -\frac{b - a}{12}h^2 \left[\frac{1}{n} \sum_{i=0}^{n-1} f''(\xi_i) \right] = -\frac{b - a}{12}h^2 f''(\zeta)$$

Here we have reasoned that the average $[1/n]\sum_{i=0}^{n-1} f''(\xi_i)$ lies between the least and greatest values of f'' on the interval (a, b). Hence, by the **Intermediate-Value Theorem of Continuous Functions**,[*] it is $f''(\zeta)$ for some point ζ in (a, b). This completes our proof of the error formula. ∎

---

[*]**Intermediate-Value Theorem of Continuous Functions**: If the function g is continuous on an interval $[a, b]$, then for each c between $g(a)$ and $g(b)$, there is a point ξ in $[a, b]$ for which $g(\xi) = c$.

EXAMPLE 2   Use Taylor series to represent the error in the trapezoid rule by an infinite series.

Solution   Equation (4) is equivalent to

$$\int_a^{a+h} f(x)\,dx = \frac{h}{2}[f(a) + f(a+h)] - \frac{1}{12}h^3 f''(\xi)$$

Let

$$F(t) = \int_a^t f(x)\,dx$$

The Taylor series for F is

$$F(a+h) = F(a) + hF'(a) + \frac{h^2}{2}F''(a) + \frac{h^3}{3!}F'''(a) + \cdots$$

By the Fundamental Theorem of Calculus, $F' = f$, and we observe that $F(a) = 0$, $F'' = f'$, $F''' = f''$, and so on. Hence, we have

$$\int_a^{a+h} f(x)\,dx = hf(a) + \frac{h^2}{2}f'(a) + \frac{h^3}{3!}f''(a) + \cdots$$

The Taylor series for f is

$$f(a+h) = f(a) + hf'(a) + \frac{h^2}{2}f''(a) + \frac{h^3}{3!}f'''(a) + \cdots$$

Adding $f(a)$ to both sides of this equation and then multiplying by $\frac{1}{2}h$, we get

$$\frac{h}{2}[f(a) + f(a+h)] = hf(a) + \frac{h^2}{2}f'(a) + \frac{h^3}{4}f''(a) + \cdots$$

Subtracting, we have

$$\int_a^{a+h} f(x)\,dx - \frac{h}{2}[f(a) + f(a+h)] = -\frac{1}{12}h^3 f''(a) + \cdots$$ ☐

Applying the Error Formula

How can an error formula like the one just derived be used? Our first application is in predicting how small h must be in order to attain a specified precision in the trapezoid rule.

EXAMPLE 3   If the trapezoid rule is to be used to compute

$$\int_0^1 e^{-x^2}\, dx$$

with an error of at most $\frac{1}{2} \times 10^{-4}$, how many points should be used?

Solution The error formula is

$$-\frac{b-a}{12}h^2 f''(\zeta)$$

In this example, $f(x) = e^{-x^2}$, $f'(x) = -2xe^{-x^2}$, and $f''(x) = (4x^2 - 2)e^{-x^2}$. Thus, $|f''(x)| \leq 2$ on the interval $[0, 1]$, and the error in absolute value will be no greater than $\frac{1}{6}h^2$. To have an error of at most $\frac{1}{2} \times 10^{-4}$, we require

$$\frac{1}{6}h^2 \leq \frac{1}{2} \times 10^{-4} \quad \text{or} \quad h \leq 0.01732$$

In this example, $h = 1/n$ so we require $n \geq 58$. Hence, at least 59 points are needed for the desired accuracy. Using the Maple commands

```
with(student);
trapezoid(exp(-x*x), x=0..1, 58);
evalf(");
```

we obtain 0.74680 59064. ☐

EXAMPLE 4 How many subintervals are needed to approximate

$$\int_0^1 \frac{\sin x}{x}\, dx$$

with error not to exceed $\frac{1}{2} \times 10^{-5}$ using the trapezoid rule? Here the integrand, $f(x) = x^{-1}\sin x$, is defined to be 1 when x is 0.

Solution We wish to establish a bound on $f''(x)$ for x in the range $[0, 1]$. Taking derivatives in the usual way is not satisfactory because each term contains x with a negative power, and it is difficult to find an upper bound on $|f''(x)|$. However, using Taylor series, we have

$$f(x) = 1 - \frac{x^2}{3!} + \frac{x^4}{5!} - \frac{x^6}{7!} + \frac{x^8}{9!} - \cdots$$

$$f'(x) = -\frac{2x}{3!} + \frac{4x^3}{5!} - \frac{6x^5}{7!} + \frac{8x^7}{9!} - \cdots$$

$$f''(x) = -\frac{2}{3!} + \frac{3 \times 4x^2}{5!} - \frac{5 \times 6x^4}{7!} + \frac{7 \times 8x^6}{9!} - \cdots$$

Thus, on the interval $[0, 1]$, $|f''(x)|$ cannot exceed $\frac{1}{2}$ because

$$\frac{2}{3!} + \frac{3 \times 4}{5!} + \frac{5 \times 6}{7!} + \frac{7 \times 8}{9!} + \cdots$$

$$< \frac{1}{3} + \frac{1}{10} + \frac{1}{24}\left(\frac{1}{2} + \frac{1}{4} + \frac{1}{8} + \cdots\right) < \frac{1}{2}$$

Therefore, the error term $|(b - a)h^2 f''(\zeta)/12|$ cannot exceed $h^2/24$. For this to be less than $\frac{1}{2} \times 10^{-5}$, it suffices to take $h < \sqrt{1.2 \times 10^{-2}}$ or $n > (1/\sqrt{1.2})10^2 = 91.3$. So for this example, we need at least 92 subintervals. $\qquad\square$

Recursive Trapezoid Formula for 2^n Equal Subintervals

In the next section, we require a formula for the trapezoid rule when the interval $[a, b]$ is subdivided into 2^n equal parts. By Formula (1), we have

$$T(f; P) = h \sum_{i=1}^{n-1} f(x_i) + \frac{h}{2}[f(x_0) + f(x_n)]$$

$$= h \sum_{i=1}^{n-1} f(a + ih) + \frac{h}{2}[f(a) + f(b)]$$

If we now replace n by 2^n and use $h = (b - a)/2^n$, the preceding formula becomes

$$R(n, 0) = h \sum_{i=1}^{2^n-1} f(a + ih) + \frac{h}{2}[f(a) + f(b)] \qquad (7)$$

Here we have introduced the notation that will be used in the next section on the Romberg algorithm—namely, $R(n, 0)$. It denotes the result of applying the trapezoid rule with 2^n equal subintervals.

In the Romberg algorithm, it will also be necessary to have a means of computing $R(n, 0)$ from $R(n - 1, 0)$ without involving unneeded evaluations of f. For example, the computation of $R(2, 0)$ utilizes the values of f at the five points a, $a + (b - a)/4$, $a + 2(b - a)/4$, $a + 3(b - a)/4$, and b. In computing $R(3, 0)$, we need values of f at these five points, as well as at four new points: $a + (b - a)/8$, $a + 3(b - a)/8$, $a + 5(b - a)/8$, and $a + 7(b - a)/8$ (see Figure 5.4). The computation should take advantage of the previously computed result. The manner of doing so is now explained.

If $R(n - 1, 0)$ has been computed and $R(n, 0)$ is to be computed, we use the identity

$$R(n, 0) = \frac{1}{2}R(n - 1, 0) + \left[R(n, 0) - \frac{1}{2}R(n - 1, 0)\right]$$

It is desirable to compute the bracketed expression with as little additional work as possible. Fixing $h = (b - a)/2^n$ for the analysis and putting

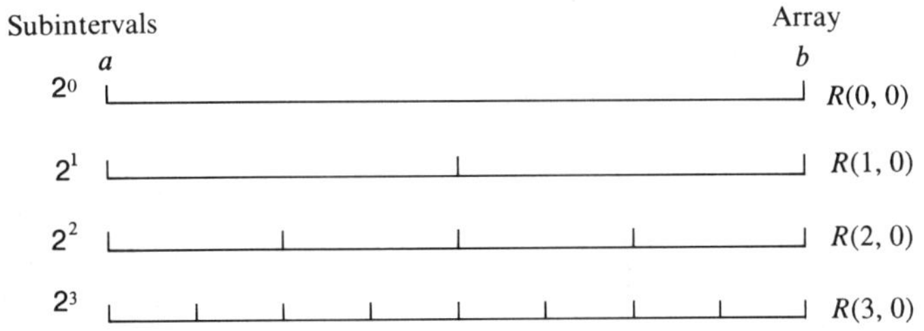

Subintervals   Array

FIGURE 5.4
2^n equal subintervals

$$C = \frac{h}{2}[f(a) + f(b)]$$

we have from Equation (7)

$$R(n,0) = h \sum_{i=1}^{2^n-1} f(a + ih) + C \qquad \qquad \textbf{(8)}$$

$$R(n-1,0) = 2h \sum_{j=1}^{2^{n-1}-1} f(a + 2jh) + 2C \qquad \qquad \textbf{(9)}$$

Notice that the subintervals for $R(n-1,0)$ are *twice* the size of those for $R(n,0)$. Now from Equations (8) and (9), we have

$$R(n,0) - \frac{1}{2}R(n-1,0) = h \sum_{i=1}^{2^n-1} f(a + ih) - h \sum_{j=1}^{2^{n-1}-1} f(a + 2jh)$$

$$= h \sum_{k=1}^{2^{n-1}} f[a + (2k-1)h]$$

Here we have taken account of the fact that each term in the first sum that corresponds to an *even* value of i is *cancelled* by a term in the second sum. This leaves only terms that correspond to *odd* values of i.

To summarize:

RECURSIVE TRAPEZOID FORMULA

If $R(n-1,0)$ is available, then $R(n,0)$ can be computed by the formula

$$R(n,0) = \frac{1}{2}R(n-1,0) + h \sum_{k=1}^{2^{n-1}} f[a + (2k-1)h] \qquad (n \geq 1) \qquad \textbf{(10)}$$

using $h = (b-a)/2^n$. Here $R(0,0) = \frac{1}{2}(b-a)[f(a) + f(b)]$.

This formula allows us to compute a sequence of approximations to a definite integral using the trapezoid rule without reevaluating the integrand at points where it has already been evaluated.

PROBLEMS 5.2

1. Obtain an upper bound on the absolute error when we compute $\int_0^6 \sin x^2\, dx$ by means of the trapezoid rule using 101 equally spaced points.

2. What is the numerical value of the trapezoid rule applied to $f(x) = x^{-1}$ using the points 1, $\frac{4}{3}$, and 2?

3. Compute an approximate value of $\int_0^1 (x^2 + 1)^{-1}\, dx$ by using the trapezoid rule with three points. Then compare with the actual value of the integral. Next, determine the error formula and numerically verify an upper bound on it.

4. (Continuation) Having computed $R(1,0)$ in Problem **3**, compute $R(2,0)$ by using Formula (10).

5. If the trapezoid rule is used to compute $\int_2^5 \sin x\, dx$ with $h = 0.01$, give a realistic bound on the error.

6. Consider the function $f(x) = |x|$ on the interval $[-1, 1]$. Calculate the results of applying the following rules to approximate $\int_{-1}^{1} f(x)\, dx$. Account for the differences in the results and compare with the true solution. Use the

   a. lower sums

   b. upper sums

   c. trapezoid rule

   with uniform spacing $h = 2, 1, \frac{1}{2}, \frac{1}{4}$.

7. Let f be a decreasing function on $[a, b]$. Let P be a partition of the interval. Show that

$$T(f;P) = \frac{1}{2}[L(f;P) + U(f;P)]$$

   where T, L, and U are the trapezoid rule, the lower sums, and the upper sums, respectively.

8. Show that for any function f and any partition P,

$$L(f;P) \le T(f;P) \le U(f;P)$$

9. Give an example of a function f and a partition P for which $L(f;P)$ is a better estimate of $\int_a^b f(x)\, dx$ than is $T(f;P)$.

10. Let f be a continuous function and let P_n, for $n = 0, 1, \ldots$, be partitions of $[a, b]$ such that the width of the largest subinterval in P_n converges to zero as $n \to \infty$. Show that $T(f;P_n)$ converges to $\int_a^b f(x)\, dx$ as $n \to \infty$. *Hint*: Use Problem **8** and known facts about upper and lower sums.

11. A function is said to be **convex** if its graph lies beneath every chord drawn between two points of the graph. What is the relationship of $L(f;P)$, $U(f;P)$, $T(f;P)$, and $\int_a^b f(x)\,dx$ for such a function?

12. How large must n be if the trapezoid rule with equal subintervals is to estimate $\int_0^2 e^{-x^2}\,dx$ with an error not exceeding 10^{-6}?

13. Show that

$$\int_a^b f(x)\,dx - \frac{b-a}{2}[f(a) + f(b)] = -\sum_{k=3}^{\infty} \frac{k-2}{2 \times k!}(b-a)^k f^{(k-1)}(a)$$

14. The (left) **rectangle rule** for numerical integration is like the upper and lower sums but simpler:

$$\int_a^b f(x)\,dx \approx \sum_{i=0}^{n-1} (x_{i+1} - x_i)f(x_i)$$

Here the partition is $P = \{a = x_0 < x_1 < x_2 < \cdots < x_n = b\}$. As in Problem 10, show that the rectangle rule converges to the integral as $n \to \infty$.

15. (Continuation) For uniform spacing, the rectangle rule reads as follows:

$$\int_a^b f(x)\,dx \approx h \sum_{i=0}^{n-1} f(x_i)$$

where $h = (b-a)/n$ and $x_i = a + ih$ for $0 \leq i \leq n$. Find an expression for the error involved in this latter formula.

16. How large must n be if the trapezoid rule in Equation (1) is to estimate $\int_0^\pi \sin x\,dx$ with error $\leq 10^{-12}$? Will the estimate be too big or too small?

17. In the trapezoid rule, the spacing need not be uniform. Establish the formula

$$\int_a^b f(x)\,dx \approx \frac{1}{2}\sum_{i=1}^{n-1}(h_{i-1} + h_i)f(x_i) + \frac{1}{2}[h_0 f(x_0) + h_{n-1}f(x_n)]$$

where $h_i = x_{i+1} - x_i$ and $a = x_0 < x_1 < x_2 < \cdots < x_n = b$.

18. What formula results from using the trapezoid rule on $f(x) = x^2$, with interval $[0, 1]$ and $n + 1$ equally spaced points? Simplify your result by using the fact that $1^2 + 2^2 + 3^2 + \cdots + n^2 = \frac{1}{6}n(2n + 1)(n + 1)$. Show that as $n \to \infty$, the trapezoidal estimate converges to the correct value, $\frac{1}{3}$.

19. Prove that if a function is concave downward, then the trapezoid rule underestimates the integral.

20. Compute two approximate values for $\int_1^2 dx/x^2$ using $h = \frac{1}{2}$ with lower sums and the trapezoid rule.

21. From Problems **14** and **15**, the rectangle rule for a single interval is given by

$$\int_a^b f(x)\,dx = (b-a)f(a) + \frac{1}{2}(b-a)^2 f'(\zeta)$$

Establish the rectangle rule and its error term when the interval $[a, b]$ is partitioned into 2^n uniform subintervals, each of size h. Simplify the results.

22. Consider $\int_1^2 dx/x^3$. What is the result of using the trapezoid rule with the partition points 1, $\frac{3}{2}$, and 2?

23. If the trapezoid rule is used with $h = 0.01$ to compute $\int_2^5 \sin x\,dx$, what numerical value will the error not exceed? (Use the absolute value of error.) Give the best answer based on the error formula.

24. Approximate $\int_0^2 2^x\,dx$ using the trapezoid rule with $h = \frac{1}{2}$.

25. Consider $\int_0^1 dx/(x^2 + 2)$. What is the result of using the trapezoid rule with 0, $\frac{1}{2}$, and 1 as partition points?

26. What is a reasonable bound on the error when we use the trapezoid rule on $\int_0^4 \cos x^3\,dx$ taking 201 equally spaced points (including endpoints)?

27. We want to approximate $\int_1^2 f(x)\,dx$ given the table of values

x	1	5/4	3/2	7/4	2
$f(x)$	10	8	7	6	5

Compute an estimate by the trapezoid rule. Can upper and lower sums be computed from the given data?

28. Consider the integral $I(h) \equiv \int_a^{a+h} f(x)\,dx$. Establish an expression for the error term for each of the following rules.

   a. $I(h) \approx hf(a + h)$
   b. $I(h) \approx hf(a + h) - \frac{1}{2}h^2 f'(a)$
   c. $I(h) \approx hf(a)$
   d. $I(h) \approx hf(a) - \frac{1}{2}h^2 f'(a)$

   For each, determine the corresponding general rule and error terms for the integral $\int_a^b f(x)\,dx$, where the partition is uniform; that is, $x_i = a + ih$ and $h = (b-a)/n$ for $0 \leq i \leq n$.

29. Obtain an expression for the error term of the midpoint rule for **(a)** one subinterval, **(b)** n unequal subintervals, and **(c)** n uniform subintervals; that is, for

   a. $\displaystyle\int_a^{a+h} f(x)\,dx \approx hf\left(a + \frac{1}{2}h\right)$

   b. $\displaystyle\int_a^b f(x)\,dx \approx \sum_{i=0}^{n-1} h_i f\left(x_i + \frac{1}{2}h_i\right)$

   c. $\displaystyle\int_a^b f(x)\,dx \approx h\sum_{i=0}^{n-1} f\left(a + \left(i + \frac{1}{2}\right)h\right)$

where $h_i = x_{i+1} - x_i$ and $h = (b - a)/n$. (The midpoint rule was introduced in Problems **4–6** of Section 5.1.)

30. Show that there exist coefficients $w_0, w_1, \ldots, w_n$ depending on $x_0, x_1, \ldots, x_n$ and on a, b such that

$$\int_a^b p(x)\, dx = \sum_{i=0}^n w_i p(x_i)$$

for all polynomials p of degree $\leqq n$. *Hint*: Use the Lagrange form of the interpolating polynomials from Section 4.1.

31. Show that when the trapezoid rule is applied to $\int_a^b e^x\, dx$ using equally spaced points, the relative error is exactly $1 - (h/2) - [h/(e^h - 1)]$.

32. Establish the following error formula for the trapezoid rule with unequal spacing of points:

$$\int_a^b f(x)\, dx = \sum_{i=0}^{n-1} \frac{h_i}{2}[f(x_i) + f(x_{i+1})] - \frac{1}{12}(b - a)h^2 f''(\xi)$$

where $\xi \in (a, b)$, $h_i = x_{i+1} - x_i$, and $\min_i h_i \leqq h \leqq \max_i h_i$. (The trapezoid rule with nonuniform spacing was introduced in Problem **17.**)

COMPUTER PROBLEMS 5.2

1. Write

  real function *trapezoid_ uniform* (f, a, b, n)

to calculate $\int_a^b f(x)\, dx$ using the trapezoid rule with n equal subintervals.

2. (Continuation) Test the code written in Computer Problem **1** on the following functions. In each case, compare with the correct answer.

  a. $\displaystyle\int_0^\pi \sin x\, dx$

  b. $\displaystyle\int_0^1 e^x\, dx$

  c. $\displaystyle\int_0^1 \arctan x\, dx$

3. Compute π from an integral of the form $c \int_a^b dx/(1 + x^2)$.

4. Compute an approximate value for the integral $\int_0^{0.8}(\sin x/x)\, dx$.

5. Compute these integrals by using a large value for the upper limit and applying a numerical method. Then compute them by first making the indicated change of variable.

  a. $\displaystyle\int_0^\infty e^{-x^2}\, dx$, using $x = -\ln t$

b. $\int_1^\infty x^{-1} \sin x \, dx$, using $x = t^{-1}$

c. $\int_0^\infty \sin x^2 \, dx$, using $x = \tan t$

Note: Part **a** is a multiple of a *probability integral*, **b** is a *sine integral*, and **c** is a *Fresnel integral*.

5.3 Romberg Algorithm

Description

The Romberg algorithm produces a triangular array of numbers, all of which are numerical estimates of the definite integral $\int_a^b f(x) \, dx$. The array is denoted here by the notation

$$R(0,0)$$

$$R(1,0) \quad R(1,1)$$

$$R(2,0) \quad R(2,1) \quad R(2,2)$$

$$R(3,0) \quad R(3,1) \quad R(3,2) \quad R(3,3)$$

$$\vdots \qquad \vdots \qquad \vdots \qquad \vdots \qquad \ddots$$

$$R(n,0) \quad R(n,1) \quad R(n,2) \quad R(n,3) \quad \cdots \quad R(n,n)$$

The first column of this table contains estimates of the integral obtained by the recursive trapezoid formula with decreasing values of the step size. Explicitly, $R(n,0)$ is the result of applying the trapezoid rule with 2^n equal subintervals. The first of them, $R(0,0)$, is obtained with just one trapezoid:

$$R(0,0) = \frac{1}{2}(b - a)[f(a) + f(b)]$$

Similarly, $R(1,0)$ is obtained with two trapezoids:

$$R(1,0) = \frac{1}{4}(b - a)\left[f(a) + f\left(\frac{a+b}{2}\right)\right] + \frac{1}{4}(b - a)\left[f\left(\frac{a+b}{2}\right) + f(b)\right]$$

$$= \frac{1}{4}(b - a)[f(a) + f(b)] + \frac{1}{2}(b - a) f\left(\frac{a+b}{2}\right)$$

$$= \frac{1}{2}R(0,0) + \frac{1}{2}(b - a) f\left(\frac{a+b}{2}\right)$$

These formulas agree with those developed in the preceding section. In particular, note that $R(n, 0)$ is obtained easily from $R(n - 1, 0)$ if Equation (10) in Section 5.2 is used; that is,

$$R(n, 0) = \frac{1}{2}R(n - 1, 0) + h \sum_{k=1}^{2^{n-1}} f[a + (2k - 1)h] \tag{1}$$

where $h = (b - a)/2^n$ and $n \geq 1$.

The second and successive columns in the Romberg array are generated by the extrapolation formula

$$R(n, m) = R(n, m - 1) + \frac{1}{4^m - 1}[R(n, m - 1) - R(n - 1, m - 1)] \tag{2}$$

with $n \geq 1$ and $m \geq 1$. This formula will be derived later using the theory of Richardson extrapolation from Section 4.3.

EXAMPLE 1   If $R(4, 2) = 8$ and $R(3, 2) = 1$, what is $R(4, 3)$?

Solution   From Equation (2), we have

$$R(4, 3) = R(4, 2) + \frac{1}{63}[R(4, 2) - R(3, 2)]$$

$$= 8 + \frac{1}{63}(8 - 1) = \frac{73}{9} \qquad \square$$

Pseudocode

We now develop computational formulas for the Romberg algorithm. By replacing n with i and m with j in Equation (2), we obtain, for $i \geq 1$ and $j \geq 1$,

$$R(i, j) = R(i, j - 1) + \frac{1}{4^j - 1}[R(i, j - 1) - R(i - 1, j - 1)]$$

and

$$R(i, 0) = \frac{1}{2}R(i - 1, 0) + h \sum_{k=1}^{2^{i-1}} f[a + (2k - 1)h]$$

The range of the summation is $1 \leq k \leq 2^{i-1}$, so that $1 \leq 2k - 1 \leq 2^i - 1$.

One way to generate the Romberg array is to compute a reasonable number of terms in the first column, $R(0, 0)$ up to $R(n, 0)$, and then use the extrapolation Formula (2) to construct columns $1, 2, \ldots, n$ in order. Another way is to compute the array row by row. Observe, for example, that $R(1, 1)$ can be computed by the extrapolation formula as soon as $R(1, 0)$ and $R(0, 0)$ are available. The procedure *romberg* computes, row by row, n rows and columns of the Romberg array for a function f and a specified interval $[a, b]$.

procedure *romberg* $(f, a, b, n, (r_{ij}))$
real array $(r_{ij})_{0:n \times 0:n}$
integer i, j, k, n
real a, b, h, sum
interface external function f
$h \leftarrow b - a$
$r_{00} \leftarrow (h/2)[f(a) + f(b)]$
for $i = 1$ **to** n **do**
  $h \leftarrow h/2$
  $sum \leftarrow 0$
  for $k = 1$ **to** $2^i - 1$ **step** 2 **do**
    $sum \leftarrow sum + f(a + kh)$
  end do
  $r_{i0} \leftarrow \frac{1}{2} r_{i-1,0} + (sum)h$
  for $j = 1$ **to** i **do**
    $r_{ij} \leftarrow r_{i,j-1} + (r_{i,j-1} - r_{i-1,j-1})/(4^j - 1)$
  end do
end do
end procedure *romberg*

This procedure is used with a main program and a function procedure (for computing values of the function f). In the main program and perhaps in the procedure *romberg*, some language-specific interface must be included to indicate that the first argument is an external function. Remember that in the Romberg algorithm as described, the number of subintervals is 2^n. Thus, a modest value of n should be chosen—for example, $n = 5$. A more sophisticated program would include automatic tests to terminate the calculation as soon as the error reaches a preassigned tolerance.

Derivation

Now we explain the source of Equation (2), which is used for constructing the successive columns of the Romberg array. We begin with a formula that expresses the error in the trapezoid rule over 2^{n-1} subintervals:

$$\int_a^b f(x)\, dx = R(n - 1, 0) + a_2 h^2 + a_4 h^4 + a_6 h^6 + \cdots \tag{3}$$

Here $h = (b - a)/2^{n-1}$ and the coefficients a_i depend on f but not on h. This equation is one form of the *Euler-Maclaurin formula* and is given here without proof. (See Young and Gregory [1972].) In this equation, $R(n - 1, 0)$ denotes a typical element of the first column in the Romberg array; hence, it is one of the trapezoidal estimates of the integral. Notice particularly that the error is expressed in powers of h^2, and the error series is $\mathcal{O}(h^2)$. For our purposes, it is not necessary to know the coefficients, but, in fact, they have definite expressions in terms of f and its derivatives. For the theory to work smoothly, it is assumed that f possesses derivatives of all orders on the interval $[a, b]$.

The reader should now recall the theory of Richardson extrapolation as outlined in Section 4.3. That theory is applicable because of Equation (3). In Equation (8) of Section 4.3, $L = \phi(h) + \sum_{k=1}^{\infty} a_{2k}h^{2k}$. Here L is the value of the integral and $\phi(h)$ is $R(n-1, 0)$, the trapezoidal estimate of L using subintervals of size h. Equation (10) of Section 4.3 gives the approximate extrapolation formula, which in this situation is Equation (2) in Section 5.3.

We briefly review this procedure. Replacing n with $n + 1$ and h with $h/2$ in Equation (3), we have

$$\int_a^b f(x)\,dx = R(n, 0) + \frac{1}{4}a_2h^2 + \frac{1}{16}a_4h^4 + \frac{1}{64}a_6h^6 + \cdots \tag{4}$$

Subtract Equation (3) from 4 times Equation (4) to obtain

$$\int_a^b f(x)\,dx = R(n, 1) - \frac{1}{4}a_4h^4 - \frac{5}{16}a_6h^6 - \cdots \tag{5}$$

where

$$R(n, 1) = R(n, 0) + \frac{1}{3}[R(n, 0) - R(n-1, 0)] \qquad (n \geq 1)$$

Note that this is the first case ($m = 1$) of the extrapolation Formula (2). Now $R(n, 1)$ should be considerably more accurate than $R(n, 0)$ or $R(n-1, 0)$ because its error formula begins with an h^4 term. Hence, the error series is now $\mathcal{O}(h^4)$. This process can be repeated using Equation (5) slightly modified as the starting point—that is, with n replaced by $n - 1$ and with h replaced by $2h$. Then combine the two equations appropriately to eliminate the h^4 term. The result is a new combination of elements from column 2 in the Romberg array:

$$\int_a^b f(x)\,dx = R(n, 2) + \frac{1}{4^3}a_6h^6 + \frac{21}{4^5}a_8h^8 + \cdots \tag{6}$$

where

$$R(n, 2) = R(n, 1) + \frac{1}{15}[R(n, 1) - R(n-1, 1)] \qquad (n \geq 2)$$

which agrees with Equation (2) again. Thus, $R(n, 2)$ is an even more accurate approximation to the integral because its error series is $\mathcal{O}(h^6)$.

The basic assumption on which of all this analysis depends is that Equation (3) is valid for the function f being integrated. Of course, in practice, we will use a modest number of rows in the Romberg algorithm, and only this number of terms in Equation (3) is needed. Here is the theorem that governs the situation:

THEOREM (EULER-MACLAURIN FORMULA AND ERROR TERM)

If $f^{(2m)}$ exists and is continuous on the interval $[a, b]$, then

$$\int_a^b f(x)\, dx = \frac{h}{2} \sum_{i=0}^{n-1} [f(x_i) + f(x_{i+1})] + E$$

where $h = (b - a)/n$, $x_i = a + ih$ for $0 \leq i \leq n$, and

$$E = \sum_{k=1}^{m-1} A_{2k} h^{2k} [f^{(2k-1)}(a) - f^{(2k-1)}(b)] - A_{2m}(b - a)h^{2m} f^{(2m)}(\xi)$$

for some ξ in the interval (a, b).

In this theorem, the A_k's are constants (related to the *Bernoulli numbers*) and ξ is some point in the interval (a, b). The interested reader should refer to Young and Gregory [1972, vol. 1, p. 374]. It turns out that the A_k's can be defined by the equation

$$\frac{x}{e^x - 1} = \sum_{k=0}^{\infty} A_k x^k \tag{7}$$

Observe that in the Euler-Maclaurin formula, the right-hand side contains the trapezoid rule and an error term, E. Furthermore, E can be expressed as a finite sum in ascending powers of h^2. This theorem gives the formal justification (and the details) of Equation (3).

If the integrand f does not possess a large number of derivatives but is at least Riemann-integrable, then the Romberg algorithm still converges in the following sense: The limit of each *column* in the array equals the integral:

$$\lim_{n \to \infty} R(n, m) = \int_a^b f(x)\, dx \qquad (m \geq 0)$$

The convergence of the first column is easily justified by referring to the upper and lower sums. (See Problem **10** in Section 5.2.) After the convergence of the first column has been established, the convergence of the remaining columns can be proved using Equation (2). (See Problems **21** and **22**.)

In practice, we may not know whether the function f whose integral we seek satisfies the smoothness criterion upon which the theory depends. Then it would not be known whether Equation (3) is valid for f. One way of testing this in the course of the Romberg algorithm is to compute the ratios

$$\frac{R(n,m) - R(n-1,m)}{R(n+1,m) - R(n,m)}$$

and to note whether they are close to 4^m. Let us verify, at least for the case $m = 0$, that this ratio is near 4 for a function that obeys Equation (3).

If we subtract Equation (4) from (3), the result is

$$R(n,0) - R(n-1,0) = \frac{3}{4}a_2h^2 + \frac{15}{16}a_4h^4 + \frac{63}{64}a_6h^6 + \cdots \tag{8}$$

If we write down the same equation for the *next* value of n, then the h of that equation is half the value of h used in Equation (8). Hence,

$$R(n+1,0) - R(n,0) = \frac{3}{4^2}a_2h^2 + \frac{15}{16^2}a_4h^4 + \frac{63}{64^2}a_6h^6 + \cdots \tag{9}$$

Equations (8) and (9) are now used to express the ratio mentioned previously:

$$\frac{R(n,0) - R(n-1,0)}{R(n+1,0) - R(n,0)} = 4 \left[\frac{1 + \frac{5}{4}\left(\frac{a_4}{a_2}\right)h^2 + \frac{21}{16}\left(\frac{a_6}{a_2}\right)h^4 + \cdots}{1 + \frac{5}{4^2}\left(\frac{a_4}{a_2}\right)h^2 + \frac{21}{16^2}\left(\frac{a_6}{a_2}\right)h^4 + \cdots} \right]$$

$$= 4\left[1 + \frac{15}{4^2}\left(\frac{a_4}{a_2}\right)h^2 + \cdots \right]$$

For small values of h, this expression is close to 4.

General Extrapolation

In closing, we return to the extrapolation process that is the heart of the Romberg algorithm. The process is Richardson extrapolation, which was discussed in Section 4.3. It is an example of a general dictum in numerical mathematics that if anything at all is known about the errors in a process, then that knowledge can be exploited to improve the process.

The only type of extrapolation illustrated so far (in this section and Section 4.3) has been the so-called h^2 extrapolation. It applies to a numerical process in which the error series is of the form

$$E = a_2h^2 + a_4h^4 + a_6h^6 + \cdots$$

In this case, the errors behave like $\mathcal{O}(h^2)$ as $h \to 0$, but the basic idea of Richardson extrapolation has much wider applicability. We could apply extrapolation if we knew, for example, that

$$E = ah^\alpha + bh^\beta + ch^\gamma + \cdots$$

provided that $0 < \alpha < \beta < \gamma < \cdots$. It is sufficient to see how to annihilate the first term of the error expansion because the succeeding steps would be similar.

Suppose therefore that

$$L = \phi(h) + ah^\alpha + bh^\beta + ch^\gamma + \cdots \qquad (10)$$

Here L is a mathematical entity that is approximated by a formula $\phi(h)$ depending on h with the error series $ah^\alpha + bh^\beta + \cdots$. It follows that

$$L = \phi\left(\frac{h}{2}\right) + a\left(\frac{h}{2}\right)^\alpha + b\left(\frac{h}{2}\right)^\beta + c\left(\frac{h}{2}\right)^\gamma + \cdots$$

Hence, if we multiply this by 2^α, we get

$$2^\alpha L = 2^\alpha \phi\left(\frac{h}{2}\right) + ah^\alpha + 2^\alpha b\left(\frac{h}{2}\right)^\beta + 2^\alpha c\left(\frac{h}{2}\right)^\gamma + \cdots$$

By subtracting Equation (10) from this equation, we rid ourselves of the h^α term:

$$(2^\alpha - 1)L = 2^\alpha \phi\left(\frac{h}{2}\right) - \phi(h) + (2^{\alpha-\beta} - 1)bh^\beta + (2^{\alpha-\gamma} - 1)ch^\gamma + \cdots$$

This we rewrite as

$$L = \frac{2^\alpha}{2^\alpha - 1}\phi\left(\frac{h}{2}\right) - \frac{1}{2^\alpha - 1}\phi(h) + \tilde{b}h^\beta + \tilde{c}h^\gamma + \cdots \qquad (11)$$

Thus, the special linear combination

$$\frac{2^\alpha}{2^\alpha - 1}\phi\left(\frac{h}{2}\right) - \frac{1}{2^\alpha - 1}\phi(h) = \phi\left(\frac{h}{2}\right) + \frac{1}{2^\alpha - 1}\left[\phi\left(\frac{h}{2}\right) - \phi(h)\right] \qquad (12)$$

should be a more accurate approximation to L than either $\phi(h)$ or $\phi(h/2)$ because their error series, in Equations (10) and (11), improve from $\mathcal{O}(h^\alpha)$ to $\mathcal{O}(h^\beta)$ as $h \to 0$ with $\beta > \alpha > 0$. Notice that when $\alpha = 2$, the combination in Equation (12) is the one we have already used for the second column in the Romberg array.

Extrapolation of the same type can be used in still more general situations, as illustrated next (and in the problems).

EXAMPLE 2   If ϕ is a function with the property

$$\phi(x) = L + a_1 x^{-1} + a_2 x^{-2} + a_3 x^{-3} + \cdots$$

how can L be estimated using Richardson extrapolation?

Solution   Obviously $L = \lim_{x\to\infty} \phi(x)$, and thus L can be estimated by evaluating $\phi(x)$ for a succession of ever-larger values of x. In order to use extrapolation, write

$$\phi(x) = L + a_1 x^{-1} + a_2 x^{-2} + a_3 x^{-3} + \cdots$$

$$\phi(2x) = L + 2^{-1} a_1 x^{-1} + 2^{-2} a_2 x^{-2} + 2^{-3} a_3 x^{-3} + \cdots$$

$$2\phi(2x) = 2L + a_1 x^{-1} + 2^{-1} a_2 x^{-2} + 2^{-2} a_3 x^{-3} + \cdots$$

$$2\phi(2x) - \phi(x) = L - 2^{-1} a_2 x^{-2} - 3 \cdot 2^{-2} a_3 x^{-3} - \cdots$$

Thus, having computed $\phi(x)$ and $\phi(2x)$, we should compute $\psi(x) = 2\phi(2x) - \phi(x)$. It should be a better approximation to L because its error series begins with x^{-2} and is $\mathcal{O}(x^{-2})$ as $x \to \infty$. This process can be repeated, as in the Romberg algorithm.   □

This is a concrete illustration of the preceding example. We want to estimate $\lim_{x\to\infty} \phi(x)$ from the following table of numerical values:

x	1	2	4	8	16	32	64	128
$\phi(x)$	21.1100	16.4425	14.3394	13.3455	12.8629	12.6253	12.5073	12.4486

A tentative hypothesis is that ϕ has the form in the preceding example. When we compute the values of the function $\psi(x) = 2\phi(2x) - \phi(x)$, we get a new table of values:

x	1	2	4	8	16	32	64
$\psi(x)$	11.7750	12.2363	12.3516	12.3803	12.3877	12.3893	12.3899

It seems reasonable, therefore, to believe that the value of $\lim_{x\to\infty} \phi(x)$ is approximately 12.3899. Then if we do another extrapolation, we should compute $\theta(x) = [4\psi(2x) - \psi(x)]/3$; values for this table are:

x	1	2	4	8	16	32
$\theta(x)$	12.3901	12.3900	12.3899	12.3902	12.3898	12.3901

For the precision of the given data, we would conclude that $\lim_{x\to\infty} \phi(x) = 12.3900$ to within roundoff error.

PROBLEMS 5.3

1. In the Romberg algorithm, what is $R(5, 3)$ if $R(5, 2) = 12$ and $R(4, 2) = -51$?

2. If $R(3, 2) = -54$ and $R(4, 2) = 72$, what is $R(4, 3)$?

3. Compute $R(5, 2)$ from $R(3, 0) = R(4, 0) = 8$ and $R(5, 0) = -4$.

4. Let $f(x) = 2^x$. Approximate $\int_0^4 f(x)\, dx$ by the trapezoid rule using partition points $0, 2$, and 4. Repeat by using partition points $0, 1, 2, 3$, and 4. Now apply Romberg extrapolation to obtain a better approximation.

5. By the Romberg algorithm, approximate $\int_0^2 4\,dx/(1+x^2)$ by evaluating $R(1,1)$.

6. Using the Romberg scheme, establish a numerical value for the approximation

$$\int_0^1 e^{-(10x)^2}\,dx \approx R(1,1)$$

Compute the approximation to only three decimal places of accuracy.

7. We are going to use the Romberg method to estimate $\int_0^1 \sqrt{x}\cos x\,dx$. Will the method work? Will it work well? Explain.

8. By combining $R(0,0)$ and $R(1,0)$ for the partition $P = \{-h < 0 < h\}$, determine $R(1,1)$.

9. In calculus, a technique of integration by substitution is developed. For example, if the substitution $x = z^2$ is made in the integral $\int_0^1 (e^x/\sqrt{x})\,dx$, the result is $2\int_0^1 e^{z^2}\,dz$. Verify this and discuss the numerical aspects of this example. Which form is likely to produce a more accurate answer by the Romberg method?

10. How many evaluations of the function (integrand) are needed if the Romberg array with n rows and n columns is to be constructed?

11. Using Equation (2), fill in the following diagram with coefficients used in the Romberg algorithm:

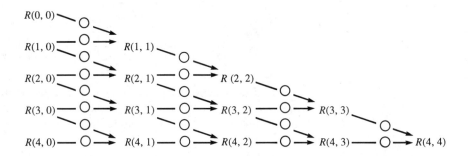

12. Derive the quadrature rule for $R(1,1)$ in terms of the function f evaluated at partition points a, $a + h$, and $a + 2h$, where $h = (b - a)/2$. Do the same for $R(n,1)$ with $h = (b - a)/2^n$.

13. (Continuation) Derive the quadrature rule $R(2,2)$ in terms of the function f evaluated at a, $a + h$, $a + 2h$, $a + 3h$, and b, where $h = (b - a)/4$.

14. We want to compute $X = \lim_{n\to\infty} S_n$, and we have already computed the two numbers $u = S_{10}$ and $v = S_{30}$. It is known that $X = S_n + Cn^{-3}$. What is X in terms of u and v?

15. If the Romberg algorithm is operating on a function that possesses continuous derivatives of all orders on the interval of integration, then what is a bound on the quantity $|\int_a^b f(x)\,dx - R(n,m)|$ in terms of h?

16. Show that the precise form of Equation (5) is

$$\int_a^b f(x)\,dx = R(n, 1) - \sum_{j=1}^{\infty} \left(\frac{4^j - 1}{3 \times 4^j}\right) a_{2j+2} h^{2j+2}$$

17. Derive Equation (6) and show that its precise form is

$$\int_a^b f(x)\,dx = R(n, 2) + \sum_{j=2}^{\infty} \left(\frac{4^j - 1}{3 \times 4^j}\right)\left(\frac{4^{j-1} - 1}{15 \times 4^{j-1}}\right) a_{2j+2} h^{2j+2}$$

18. Use the fact that the coefficients in Equation (3) have the form

$$a_k = c_k[f^{(k-1)}(b) - f^{(k-1)}(a)]$$

to prove that $\int_a^b f(x)\,dx = R(n, m)$ if f is a polynomial of degree $\leq 2m - 2$.

19. In the Romberg algorithm, $R(n, 0)$ denotes an estimate of $\int_a^b f(x)\,dx$ with sub-intervals of size $h = (b - a)/2^n$. If it were known that

$$\int_a^b f(x)\,dx = R(n, 0) + a_3 h^3 + a_6 h^6 + \cdots$$

how would we have to modify the Romberg algorithm?

20. Show that if f'' is continuous, then the first column in the Romberg array converges to the integral in such a way that the error at the nth step is bounded in magnitude by a constant times 4^{-n}.

21. Assuming that the first column of the Romberg array converges to $\int_a^b f(x)\,dx$, show that the second column does also.

22. (Continuation) In Problem **21**, we established the elementary property that if $\lim_{n \to \infty} R(n, 0) = \int_a^b f(x)\,dx$, then $\lim_{n \to \infty} R(n, 1) = \int_a^b f(x)\,dx$. Show that

$$\lim_{n \to \infty} R(n, 2) = \lim_{n \to \infty} R(n, 3) = \cdots = \lim_{n \to \infty} R(n, n) = \int_a^b f(x)\,dx$$

23. a. Prove that the coefficients $A_0, A_1, \ldots$ that occur in the Euler-Maclaurin formula can be generated recursively by the algorithm

$$\begin{cases} A_0 = 1 \\ A_k = -\sum_{j=1}^{k} \frac{A_{k-j}}{(j + 1)!} \end{cases}$$

Hint: Use Formula (7).

b. Determine A_k for $1 \leq k \leq 6$.

24. Evaluate E in the theorem on the Euler-Maclaurin formula for this special case: $a = 0, b = 2\pi, f(x) = 1 + \cos 4x, n = 4$, and m arbitrary.

25. Suppose that we want to estimate $Z = \lim_{h \to 0} f(h)$ and that we calculate $f(1)$, $f(2^{-1}), f(2^{-2}), f(2^{-3}), \ldots, f(2^{-10})$. Then suppose also that it is known that $Z = f(h) + ah^2 + bh^4 + ch^6$. Show how to obtain an improved estimate of Z from the 11 numbers already computed. Show how Z can be determined exactly from any 4 of the 11 computed numbers.

26. Show how Richardson extrapolation works on a sequence $x_1, x_2, x_3, \ldots$ that converges to L as $n \to \infty$ in such a way that $L - x_n = a_2 n^{-2} + a_3 n^{-3} + a_4 n^{-4} + \cdots$.

27. Let x_n be a sequence that converges to L as $n \to \infty$. If $L - x_n$ is known to be of the form $a_3 n^{-3} + a_4 n^{-4} + \cdots$ (in which the coefficients are unknown), how can the convergence of the sequence be accelerated by taking combinations of x_n and x_{n+1}?

COMPUTER PROBLEMS 5.3

1. Design and carry out an experiment using the Romberg algorithm. *Suggestions*: For a function that possesses many continuous derivatives on the interval, the method should work well. Try such a function first. If you choose one whose integral you can compute by other means, you will acquire a better understanding of the accuracy in the Romberg algorithm. For example, definite integrals for $\int (1 + x)^{-1} dx = \ln(1 + x)$, $\int e^x dx = e^x$, and $\int (1 + x^2)^{-1} dx = \arctan x$ are good to try.

2. Test the Romberg algorithm on a *bad* function, such as $\sqrt{x}$ on $[0, 1]$. Why is it bad?

3. Compute eight rows and columns in the Romberg array for $\int_{1.3}^{2.19} x^{-1} \sin x \, dx$.

4. The transcendental number π is the area of a circle whose radius is 1. Show that $\pi = 8 \int_0^{1/\sqrt{2}} (\sqrt{1 - x^2} - x) dx$ with the help of a diagram, and use this integral to approximate π by the Romberg method.

5. The **Bessel function** of order 0 is defined by the equation

$$J_0(x) = \frac{1}{\pi} \int_0^\pi \cos(x \sin \theta) \, d\theta$$

Calculate $J_0(1)$ by applying the Romberg algorithm to the integral.

6. Recode the Romberg procedure so that *all* the trapezoid rule results are computed *first* and stored in the first column. Then in a separate procedure

extrapolate$(n, (r_i))$

carry out Richardson extrapolation and store the results in the lower triangular part of the (r_i) array. What are the advantages and disadvantages of this procedure over the routine given in the text? Test on the two integrals $\int_0^4 dx/(1 + x)$ and $\int_{-1}^1 e^x dx$ using only one computer run.

7. Apply the Romberg method to estimate $\int_0^\pi (2 + \sin 2x)^{-1}\, dx$. Observe the high precision obtained in the first column of the array—that is, by the simple trapezoidal estimates.

8. Compute $\int_0^\pi x \cos 3x\, dx$ by the Romberg algorithm using $n = 6$. What is the correct answer?

9. An integral of the form $\int_0^\infty f(x)\, dx$ can be transformed into an integral on a finite interval by making a change of variable. Verify, for instance, that the substitution $x = -\ln y$ changes the integral $\int_0^\infty f(x)\, dx$ into $\int_0^1 y^{-1} f(-\ln y)\, dy$. Use this idea to compute $\int_0^\infty [e^{-x}/(1 + x^2)]\, dx$ by means of the Romberg algorithm, using 128 evaluations of the transformed function.

10. Calculate

$$\int_0^1 \frac{\sin x}{\sqrt{x}}\, dx$$

by the Romberg algorithm. *Hint*: Consider making a change of variable.

11. Calculate $\int_0^\infty e^{-x}\sqrt{1 - \sin x}\, dx$ by the Romberg algorithm.

12. Compute $\log 2$ by using the Romberg algorithm on a suitable integral.

5.4   An Adaptive Simpson's Scheme

In this section, we first derive a method known as Simpson's rule for numerical integration and then develop an adaptive scheme based on it for obtaining a numerical approximation for the integral

$$\int_a^b f(x)\, dx$$

In this adaptive algorithm, the partitioning of the interval $[a, b]$ is not selected beforehand but is automatically determined. The partition is generated adaptively so that more and smaller subintervals are used in some parts of the interval and fewer and larger subintervals are used in other parts.

Simpson's Rule

A numerical integration rule over two equal subintervals with partition points a, $a + h$, and $a + 2h = b$ is the widely used **Simpson's rule**:

$$\int_a^{a+2h} f(x)\, dx \approx \frac{h}{3}[f(a) + 4f(a + h) + f(a + 2h)] \tag{1}$$

Its error can be established by using the Taylor series from Section 1.2:

$$f(a + h) = f + hf' + \frac{1}{2!}h^2 f'' + \frac{1}{3!}h^3 f''' + \frac{1}{4!}h^4 f^{(4)} + \cdots$$

where the functions $f, f', f'', \ldots$ on the right-hand side are evaluated at a. Now replacing h by $2h$, we have

$$f(a + 2h) = f + 2hf' + 2h^2 f'' + \frac{4}{3}h^3 f''' + \frac{2^4}{4!}h^4 f^{(4)} + \cdots$$

Using these two series, we obtain

$$f(a) + 4f(a + h) + f(a + 2h) = 6f + 6hf' + 4h^2 f'' + 2h^3 f''' + \frac{20}{4!}h^4 f^{(4)} + \cdots$$

and, thereby,

$$\frac{h}{3}[f(a) + 4f(a + h) + f(a + 2h)] = 2hf + 2h^2 f' + \frac{4}{3}h^3 f'' \tag{2}$$

$$+ \frac{2}{3}h^4 f''' + \frac{20}{3 \cdot 4!}h^5 f^{(4)} + \cdots$$

Hence, we have a series for the right side of Equation (1). Now let's find one for the left side. The Taylor series for $F(a + 2h)$ is

$$F(a + 2h) = F(a) + 2hF'(a) + 2h^2 F''(a) + \frac{4}{3}h^3 F'''(a)$$

$$+ \frac{2}{3}h^4 F^{(4)}(a) + \frac{2^5}{5!}h^5 F^{(5)}(a) + \cdots$$

Let

$$F(x) = \int_a^x f(t)\, dt$$

By the Fundamental Theorem of Calculus, $F' = f$. Also, we observe that $F(a) = 0$ and $F(a + 2h)$ is the integral on the left side of Equation (1). Since $F'' = f'$, $F''' = f''$, and so on, we have

$$\int_a^{a+2h} f(x)\, dx = 2hf + 2h^2 f' + \frac{4}{3}h^3 f'' + \frac{2}{3}h^4 f''' + \frac{2^5}{5 \cdot 4!}h^5 f^{(4)} + \cdots \tag{3}$$

Subtracting (2) from (3), we obtain

$$\int_a^{a+2h} f(x)\, dx = \frac{h}{3}[f(a) + 4f(a + h) + f(a + 2h)] - \frac{h^5}{90}f^{(4)} - \cdots$$

A more detailed analysis will show that the error term for Simpson's rule (1) is $-(h^5/90)f^{(4)}(\xi) = \mathcal{O}(h^5)$ as $h \to 0$, for some ξ between a and $a + 2h$. We can rewrite Simpson's rule over the interval $[a, b]$ as

$$\int_a^b f(x)\, dx \approx \frac{(b-a)}{6}\left[f(a) + 4f\left(\frac{a+b}{2}\right) + f(b)\right]$$

with error term

$$-\frac{1}{90}\left(\frac{b-a}{2}\right)^5 f^{(4)}(\xi)$$

for some ξ in (a, b).

Adaptive Simpson's Algorithm

In the adaptive process, we divide the interval $[a, b]$ into two subintervals and then decide whether each of them is to be divided into more subintervals. This procedure is continued until some specified accuracy is obtained throughout the entire interval $[a, b]$. Since the integrand f varies in its behavior on the interval $[a, b]$, we expect the final partitioning not to be uniform but to vary in the spacing and density of the partition points.

We now develop the test for deciding whether subintervals should continue to be divided. One application of Simpson's rule over the interval $[a, b]$ can be written as

$$I \equiv \int_a^b f(x)\, dx = S(a, b) + E(a, b)$$

where

$$S(a, b) = \frac{(b-a)}{6}\left[f(a) + 4f\left(\frac{a+b}{2}\right) + f(b)\right]$$

and

$$E(a, b) = -\frac{1}{90}\left(\frac{b-a}{2}\right)^5 f^{(4)}$$

Here we assume that the point at which $f^{(4)}$ is evaluated remains constant. Letting $h = b - a$, we have

$$I = S^{(1)} + E^{(1)} \tag{4}$$

where

$$S^{(1)} = S(a, b)$$

and

$$E^{(1)} = -\frac{1}{90}\left(\frac{h}{2}\right)^5 f^{(4)}$$

Now two applications of Simpson's rule over the interval $[a, b]$ gives

$$I = S^{(2)} + E^{(2)} \tag{5}$$

where

$$S^{(2)} = S(a, c) + S(c, b)$$

and

$$E^{(2)} = -\frac{1}{90}\left(\frac{h/2}{2}\right)^5 f^{(4)} - \frac{1}{90}\left(\frac{h/2}{2}\right)^5 f^{(4)}$$

$$= \frac{1}{16}E^{(1)}$$

Here $c = (a + b)/2$ as in Figure 5.5. Subtracting Equation (5) from (4), we have

$$S^{(2)} - S^{(1)} = E^{(1)} - E^{(2)} = 15E^{(2)}$$

From this equation and Equation (4), we have

$$|I - S^{(2)}| = \frac{1}{15}|S^{(2)} - S^{(1)}| \tag{6}$$

Hence, we utilize the following criterion to guide the adaptive process:

$$\frac{1}{15}|S^{(2)} - S^{(1)}| < \varepsilon \tag{7}$$

If test (7) is not satisfied, the interval $[a, b]$ is split into two subintervals, $[a, c]$ and $[c, b]$, where c is the midpoint $c = (a + b)/2$. On each of these subintervals we again use test (7) with ε replaced by $\varepsilon/2$ so that the resulting tolerance will be ε over the entire interval $[a, b]$. A recursive procedure handles this quite nicely.

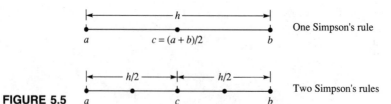

FIGURE 5.5

To see why we take $\varepsilon/2$ on each subinterval, recall that

$$I = \int_a^b f(x)\,dx = \int_a^c f(x)\,dx + \int_c^b f(x)\,dx = I_L + I_R$$

If S is the sum of approximations $S_L^{(2)}$ over $[a, c]$ and $S_R^{(2)}$ over $[c, b]$, we have

$$|I - S| = |I_L + I_R - S_L^{(2)} - S_R^{(2)}| \leqq |I_L - S_L^{(2)}| + |I_R - S_R^{(2)}|$$

$$= \frac{1}{15}|S_L^{(2)} - S_L^{(1)}| + \frac{1}{15}|S_R^{(2)} - S_R^{(1)}|$$

using Equation (6). Hence, if we require

$$\frac{1}{15}|S_L^{(2)} - S_L^{(1)}| \leqq \frac{\varepsilon}{2} \quad \text{and} \quad \frac{1}{15}|S_R^{(2)} - S_R^{(1)}| \leqq \frac{\varepsilon}{2}$$

then $|I - S| \leqq \varepsilon$ over the entire interval $[a, b]$.

Pseudocode

We now describe an adaptive Simpson recursive procedure. The interval $[a, b]$ is partitioned into four subintervals of width $(b - a)/4$. Two Simpson approximations are computed using two double-width subintervals and four single-width subintervals; that is,

$$one\_simpson \leftarrow \frac{h}{6}\left[f(a) + 4f\left(\frac{a + b}{2}\right) + f(b) \right]$$

$$two\_simpson \leftarrow \frac{h}{12}\left[f(a) + 4f\left(\frac{a + c}{2}\right) + 2f(c) \right.$$

$$\left. + 4f\left(\frac{c + b}{2}\right) + f(b) \right]$$

where $h = b - a$ and $c = (a + b)/2$.

According to Inequality (6), if one_simpson and two_simpson agree to within 15ε, then the interval $[a, b]$ does not need to be subdivided further in order to obtain an accurate approximation to the integral $\int_a^b f(x)\,dx$. In this case, the value of two_simpson is used as the approximate value of the integral over the interval $[a, b]$. If the desired accuracy for the integral has not been obtained, then the interval $[a, b]$ is divided in half. The subintervals $[a, c]$ and $[c, b]$, where $c = (a + b)/2$, are used in a recursive call to the adaptive Simpson procedure with tolerance $\varepsilon/2$ on each. This procedure terminates whenever all subintervals satisfy Inequality (6). Alternatively, a maximum number of allowable levels of subdividing intervals is used as well to terminate the procedure prematurely. The recursive procedure provides an elegant and simple way to keep track of which subintervals satisfy the tolerance test and which need to be divided further.

Example Using Adaptive Simpson Procedure

The main program for calling the adaptive Simpson procedure can best be presented in terms of a concrete example. An approximate value for the integral

$$\int_0^{2\pi} \left[\frac{\cos(2x)}{e^x} \right] dx$$

is desired with accuracy $\frac{1}{2} \times 10^{-4}$. A function procedure f is written for the integrand. Its name is the first argument in the procedure, and necessary interface statements are needed here and in the main program. Other arguments are the values of the upper and lower limits a and b of the integral, the desired accuracy ε, the level of the current subinterval, and the maximum level depth. Here is the pseudocode:

recursive real function $simpson(f, a, b, \varepsilon, level, level_{max})$ **&** $result(simpson\_result)$
integer $level, level_{max}$
real a, b, c, d, e, h
interface external function f
$level \leftarrow level + 1$
$h \leftarrow b - a$
$c \leftarrow (a + b)/2$
$one\_simpson \leftarrow h[f(a) + 4f(c) + f(b)]/6$
$d \leftarrow (a + c)/2$
$e \leftarrow (c + b)/2$
$two\_simpson \leftarrow h[f(a) + 4f(d) + 2f(c) + 4f(e) + f(b)]/12$
if $level > level_{max}$ **then**
   $simpson\_result \leftarrow two\_simpson$
   output "an error message"
else
   if $|two\_simpson - one\_simpson| < 15\varepsilon$ **then**
      $simpson\_result \leftarrow two\_simpson$
   else
      $left\_simpson \leftarrow simpson(f, a, c, \varepsilon/2, level, level_{max})$
      $right\_simpson \leftarrow simpson(f, c, b, \varepsilon/2, level, level_{max})$
      $simpson\_result \leftarrow left\_simpson + right\_simpson$
   end if
end if
end function $simpson$

By coding and running this sample pseudocode on a computer, we obtain an approximate value of 0.1996223 as the computed value for the integral.

Using the Maple commands

```
Digits := 20;
int(cos(2*x)/exp(x), x=0..2*Pi);
evalf(");
```

we obtain $0.19962\,65114\,53658\,40224$. Maple determines the exact answer to be $\frac{1}{5}(1 - e^{-2\pi})$ and then evaluates it.

PROBLEMS 5.4

1. Establish the general Simpson's rule over n (even) subintervals

$$\int_a^b f(x)\,dx \approx \frac{h}{3}\left\{[f(a) + f(b)] + 4\sum_{i=1}^{n/2} f[a + (2i - 1)h] + 2\sum_{i=1}^{(n-2)/2} f(a + 2ih)\right\}$$

where $x_i = a + ih$ for $0 \leq i \leq n$ and $h = (b - a)/n$. Derive the general error term

$$-\frac{1}{180}(b - a)h^4 f^{(4)}(\xi)$$

for some $\xi \in (a, b)$.

2. (Continuation) Simpson's rule for calculating $\int_a^b f(x)\,dx$ can be written as

$$S_{n-1} = \frac{h}{3}[f(x_0) + 4f(x_1) + 2f(x_2) + \cdots + 4f(x_{n-1}) + f(x_n)]$$

with $h = (b - a)/n$ when n is even. Its error is of the form Ch^4. Show how two values of S_k can be combined to obtain a more accurate estimate of the integral.

3. A numerical integration scheme that is not as well known is **Simpson's $\frac{3}{8}$ rule** over three subintervals:

$$\int_a^{a+3h} f(x)\,dx \approx \frac{3h}{8}[f(a) + 3f(a + h) + 3f(a + 2h) + f(a + 3h)]$$

Establish the error term for this rule and explain why this rule is overshadowed by Simpson's rule.

4. (Continuation) Using Problem **3**, establish the general Simpson's $\frac{3}{8}$ rule over n (divisible by 3) subintervals. Derive the general error term.

5. Compute $\int_0^1 (1 + x^2)^{-1}\,dx$ by Simpson's rule, using the three partition points $x = 0, 0.5$, and 1. Compare with the true solution.

6. Consider the integral $\int_0^1 \sin(\pi x^2/2)\,dx$. Suppose that we wish to integrate numerically, with an error of magnitude less than 10^{-3}.

   a. What interval width h is needed if we wish to use the trapezoid rule?

   b. Simpson's rule?

   c. Simpson's $\frac{3}{8}$ rule?

7. A function f has the values shown.

x	1	1.25	1.5	1.75	2
$f(x)$	10	8	7	6	5

a. Use Simpson's rule and the function values at $x = 1, 1.5$, and 2 to approximate $\int_1^2 f(x)\,dx$.

b. Repeat the preceding part using $x = 1, 1.25, 1.5, 1.75$, and 2.

c. Use the results from parts **a** and **b** along with the error terms to establish an improved approximation.

8. Find an approximate value of $\int_1^2 x^{-1}\,dx$ using Simpson's rule with $h = 0.25$. Give a bound on the error.

COMPUTER PROBLEMS 5.4

1. Find approximate values for the two integrals

$$4 \int_0^1 \frac{dx}{1 + x^2} \qquad 8 \int_0^{1/\sqrt{2}} (\sqrt{1 - x^2} - x)\,dx$$

Use *simpson* with $\varepsilon = \frac{1}{2} \times 10^{-5}$ and *level*$_{max}$ = 4. Sketch the curves of the integrand $f(x)$ in each case and show how *simpson* partitions the intervals. You may want to print the intervals at which new values are added to *simpson_result* in procedure *simpson* and also to print values of $f(x)$ over the entire interval $[a, b]$ in order to sketch the curves.

2. Figure out how to save function evaluations in procedure *simpson* so that the integrand $f(x)$ is evaluated only once at each partition point. Test the modified code using the example in the text; that is,

$$\int_0^{2\pi} \cos(2x)e^{-x}\,dx$$

with $\varepsilon = 5.0 \times 10^{-5}$ and *level*$_{max}$ = 4.

3. Modify and test the code in this section so that it stores the partition points and function values. Using an automatic plotter and the modified code, repeat Computer Problem **1**, and plot the resulting partition points and function values.

4. Write and test code similar to that in this section but based on a different quadrature rule.

5.5 Gaussian Quadrature Formulas

Description

Most numerical integration formulas conform to the following pattern:

$$\int_a^b f(x)\,dx \approx A_0 f(x_0) + A_1 f(x_1) + \cdots + A_n f(x_n) \tag{1}$$

To use such a formula, it is only necessary to know the *nodes* $x_0, x_1, \ldots, x_n$ and the *weights* $A_0, A_1, \ldots, A_n$. There are tables that list the numerical values of the nodes and weights.

Where do formulas such as (1) come from? One major source is the theory of polynomial interpolation as presented in Chapter 4. If the nodes have been fixed, then there is a corresponding Lagrange interpolation formula:

$$p(x) = \sum_{i=0}^{n} f(x_i)\ell_i(x) \quad \text{where} \quad \ell_i(x) = \prod_{\substack{j \neq i \\ j=0}}^{n} \left(\frac{x - x_j}{x_i - x_j} \right)$$

This formula [Equation (12) from Section 4.1] provides a polynomial p of degree at most $\leq n$ that interpolates f at the nodes; that is, $p(x_i) = f(x_i)$ for $0 \leq i \leq n$. If the circumstances are favorable, p will be a good approximation to f, and $\int_a^b p(x)\,dx$ will be a good approximation to $\int_a^b f(x)\,dx$. Therefore,

$$\int_a^b f(x)\,dx \approx \int_a^b p(x)\,dx = \sum_{i=0}^{n} f(x_i) \int_a^b \ell_i(x)\,dx = \sum_{i=0}^{n} A_i f(x_i) \qquad (2)$$

where we have put

$$A_i = \int_a^b \ell_i(x)\,dx$$

From the way in which Formula (2) has been derived, we know that it will give correct values for the integral of every polynomial of degree at most n.

EXAMPLE 1 Determine the quadrature formula of the form (1) when the interval is $[-2, 2]$ and the nodes are $-1, 0$, and 1.

Solution The functions ℓ_i are given above. Thus, we have

$$\ell_0(x) = \prod_{\substack{j \neq i \\ j=0}}^{2} \left(\frac{x - x_j}{x_0 - x_j} \right) = \frac{1}{2}x(x - 1)$$

Similarly, $\ell_1(x) = -(x+1)(x-1)$ and $\ell_2(x) = \frac{1}{2}x(x+1)$. The weights are obtained by integrating these functions. For example,

$$A_0 = \int_{-2}^{2} \ell_0(x)\,dx = \frac{1}{2}\int_{-2}^{2} (x^2 - x)\,dx = \frac{8}{3}$$

Similarly, $A_1 = -\frac{4}{3}$ and $A_2 = \frac{8}{3}$. Therefore, the quadrature formula is

$$\int_{-2}^{2} f(x)\,dx \approx \frac{8}{3}f(-1) - \frac{4}{3}f(0) + \frac{8}{3}f(1)$$

As a check on the work, one can verify that the formula gives correct values for the three functions $f(x) = 1$, x, and x^2. $\square$

Gaussian Nodes and Weights

In the preceding discussion, the nodes have been arbitrary, although for practical reasons they should belong to the interval in which the integration is to be carried out. The great mathematician Karl Friedrich Gauss (1777–1855) discovered that by a special placement of the nodes the accuracy of the numerical integration process could be greatly increased. Here is Gauss's remarkable result.

THEOREM ON GAUSSIAN QUADRATURE

Let q be a polynomial of degree $n + 1$ such that

$$\int_a^b x^k q(x)\, dx = 0 \qquad (0 \leqq k \leqq n)$$

Let $x_0, x_1, \ldots, x_n$ be the zeros of q. Then the formula

$$\int_a^b f(x)\, dx \approx \sum_{i=0}^n A_i f(x_i) \quad \text{where} \quad A_i = \int_a^b \ell_i(x)\, dx \qquad \textbf{(3)}$$

with these x_i's as nodes will be exact for all polynomials of degree at most $2n + 1$.

Proof   Let f be any polynomial of degree $\leqq 2n + 1$. Dividing f by q, we obtain a quotient p and a remainder r, both of which have degree at most n. So

$$f = pq + r$$

By our hypothesis, $\int_a^b q(x)p(x)\, dx = 0$. Furthermore, because each x_i is a root of q, we have $f(x_i) = p(x_i)q(x_i) + r(x_i) = r(x_i)$. Finally, since r has degree at most n, Formula (3) will give $\int_a^b r(x)\, dx$ precisely. Hence,

$$\int_a^b f(x)\, dx = \int_a^b p(x)q(x)\, dx + \int_a^b r(x)\, dx = \int_a^b r(x)\, dx$$

$$= \sum_{i=0}^n A_i r(x_i) = \sum_{i=0}^n A_i f(x_i) \qquad \blacksquare$$

To summarize: With arbitrary nodes, Formula (3) will be exact for all polynomials of degree $\leqq n$. With the Gaussian nodes, Formula (3) will be exact for all polynomials of degree $\leqq 2n + 1$.

The quadrature formulas that arise as applications of this theorem are called **Gaussian** or **Gauss-Legendre quadrature formulas**. There is a different formula for each interval $[a, b]$ and each value of n. There are also more general Gaussian formulas to give approximate values of integrals such as

$$\int_0^\infty f(x)e^{-x}\, dx \qquad \int_{-1}^1 f(x)(1 - x^2)^{1/2}\, dx \qquad \int_{-\infty}^\infty f(x)e^{-x^2}\, dx \qquad \text{etc.}$$

We now derive a Gaussian formula that is not too complicated.

EXAMPLE 2   Determine the Gaussian quadrature formula with three Gaussian nodes and three weights for the integral $\int_{-1}^1 f(x)\, dx$.

Solution   We must find the polynomial q referred to in the theorem and then compute its roots. The degree of q is 3, and so q has the form

$$q(x) = c_0 + c_1 x + c_2 x^2 + c_3 x^3$$

The conditions that q must satisfy are

$$\int_{-1}^1 q(x)\, dx = \int_{-1}^1 x q(x)\, dx = \int_{-1}^1 x^2 q(x)\, dx = 0$$

If we let $c_0 = c_2 = 0$, then $q(x) = c_1 x + c_3 x^3$ and so

$$\int_{-1}^1 q(x)\, dx = \int_{-1}^1 x^2 q(x)\, dx = 0$$

because the integral of an odd function over a symmetric interval is 0. To obtain c_1 and c_3, we impose the condition

$$\int_{-1}^1 x(c_1 x + c_3 x^3)\, dx = 0$$

A convenient solution of this is $c_1 = -3$ and $c_3 = 5$. Hence,

$$q(x) = 5x^3 - 3x$$

The roots of q are $-\sqrt{3/5}$, 0, and $\sqrt{3/5}$. These, then, are the Gaussian nodes for the quadrature formula.

To obtain the weights A_0, A_1, and A_2, we use a procedure known as the **method of undetermined coefficients**. We want to select A_0, A_1, and A_2 in the formula

$$\int_{-1}^1 f(x)\, dx \approx A_0 f\left(-\sqrt{\frac{3}{5}}\right) + A_1 f(0) + A_2 f\left(\sqrt{\frac{3}{5}}\right) \tag{4}$$

so that the approximate equality ($\approx$) is an exact equality ($=$) whenever f is of the form $ax^2 + bx + c$. Since integration is a linear process, Formula (4) will be exact for all polynomials of degree ≤ 2 if it is exact for these three: 1, x, and x^2. We arrange the calculations in a tabular form.

f	Left-hand side	Right-hand side
1	$\int_{-1}^{1} dx = 2$	$A_0 + A_1 + A_2$
x	$\int_{-1}^{1} x\,dx = 0$	$-\sqrt{\frac{3}{5}}A_0 + \sqrt{\frac{3}{5}}A_2$
x^2	$\int_{-1}^{1} x^2\,dx = \frac{2}{3}$	$\frac{3}{5}A_0 + \frac{3}{5}A_2$

The left-hand side of Equation (4) will equal the right-hand side for all quadratic polynomials when A_0, A_1, and A_2 satisfy the equations

$$\begin{cases} A_0 + A_1 + A_2 = 2 \\ A_0 \quad\quad - A_2 = 0 \\ A_0 \quad\quad + A_2 = \frac{10}{9} \end{cases}$$

Clearly, the weights are $A_0 = A_2 = \frac{5}{9}$ and $A_1 = \frac{8}{9}$. Therefore, the final formula is

$$\int_{-1}^{1} f(x)\,dx \approx \frac{5}{9} f\left(-\sqrt{\frac{3}{5}}\right) + \frac{8}{9} f(0) + \frac{5}{9} f\left(\sqrt{\frac{3}{5}}\right) \tag{5}$$

It will integrate all quintic polynomials correctly. For example, $\int_{-1}^{1} x^4\,dx = \frac{2}{5}$, and the formula also yields the value $\frac{2}{5}$ for this function. $\square$

Legendre Polynomials

Much more could be said about Gaussian quadrature formulas. In particular, there are efficient methods for generating the special polynomials whose roots are used as nodes in the quadrature formula. If we specialize to the integral $\int_{-1}^{1} f(x)\,dx$ and standardize q_n so that $q_n(1) = 1$, then these polynomials are called **Legendre** polynomials. Thus, the roots of the Legendre polynomials are the nodes for Gaussian quadrature on the interval $[-1, 1]$. The first few Legendre polynomials are

$$q_0(x) = 1$$

$$q_1(x) = x$$

$$q_2(x) = \frac{3}{2}x^2 - \frac{1}{2}$$

$$q_3(x) = \frac{5}{2}x^3 - \frac{3}{2}x$$

They can be generated by a three-term recurrence relation

$$q_n(x) = \left(\frac{2n-1}{n}\right)xq_{n-1}(x) - \left(\frac{n-1}{n}\right)q_{n-2}(x) \qquad (n \geq 2) \qquad \textbf{(6)}$$

With no new ideas, we can treat integrals of the form $\int_a^b f(x)w(x)\,dx$. Here, $w(x)$ should be a fixed positive function on (a, b) for which the integrals $\int_a^b x^n w(x)\,dx$ all exist, for $n = 0, 1, 2, \ldots$. Important examples for the interval $[-1, 1]$ are given by $w(x) = (1 - x^2)^{-1/2}$ and $w(x) = (1 - x^2)^{1/2}$. The corresponding theorem is as follows:

THEOREM

Let q be a polynomial of degree $n + 1$ such that

$$\int_a^b x^k q(x)w(x)\,dx = 0 \qquad (0 \leq k \leq n)$$

Let $x_0, x_1, \ldots, x_n$ be the roots of q. Then the formula

$$\int_a^b f(x)w(x)\,dx \approx \sum_{i=0}^n A_i f(x_i)$$

where

$$A_i = \int_a^b \ell_i(x)w(x)\,dx$$

with these x_i as nodes will be exact whenever f is a polynomial of degree at most $2n + 1$.

An important case where Gaussian formulas have an advantage occurs in integrating a function that is infinite at one end of the interval. The reason for this advantage is that the nodes in Gaussian quadrature are always *interior* points of the interval. Thus, for example, in computing $\int_0^1 x^{-1} \sin x\,dx$, we can safely use the statement $y \leftarrow \sin(x)/x$ with a Gaussian formula because the value at $x = 0$ will not be required. More difficult integrals such as $\int_0^1 (x^2 - 1)^{1/3}[\sin(e^x - 1)]^{-1/2}\,dx$ can be computed directly with a Gaussian formula in spite of the singularity at 0. Of course, we are referring to integrals that are well-defined and finite in spite of a singularity. A typical case is $\int_0^1 x^{-1/2}\,dx$.

The nodes and weights for several values of n in the Gaussian quadrature formula

$$\int_{-1}^1 f(x)\,dx \approx \sum_{i=0}^n A_i f(x_i)$$

are given in Table 5.1. The numerical values of nodes and weights for various values of n up to 95 can be found in Abramowitz and Stegun [1964]. See also Stroud and Secrest [1966]. Since these nodes and weights are mostly irrational numbers, they are not used in computations by hand as much as are simpler rules that involve integer and rational values. However, in programs for automatic computation, it does not matter whether a formula looks elegant, and the Gaussian quadrature formulas usually give greater accuracy with fewer function evaluations. The choice of quadrature formulas depends on the specific application being considered, and the reader should consult more advanced references for guidelines. See, for example, Davis and Rabinowitz [1984], Ghizetti and Ossiccini [1970], or Krylov [1962].

**PROBLEMS
5.5**

1. A Gaussian quadrature rule for the interval $[-1, 1]$ can be used on the interval $[a, b]$ by applying a suitable linear transformation. Approximate

$$\int_0^2 e^{-x^2}\, dx$$

using the transformed rule from Table 5.1 with $n = 1$.

2. Using Table 5.1, show directly that the Gaussian quadrature rule is exact for the polynomials $1, x, x^2, \ldots, x^{2n+1}$ when

   a. $n = 1$

   b. $n = 3$

   c. $n = 4$

3. For how high a degree polynomial is Formula (5) exact? Verify your answer by continuing the method of undetermined coefficients until an equation is not satisfied.

4. Verify parts of Table 5.1 by finding the roots of q_n and using the method of undetermined coefficients to establish the Gaussian quadrature formula on the interval $[-1, 1]$ for the following.

   a. $n = 1$

   b. $n = 3$

   c. $n = 4$

5. Construct a rule of the form

$$\int_{-1}^1 f(x)\, dx \approx \alpha f\left(-\frac{1}{2}\right) + \beta f(0) + \gamma f\left(\frac{1}{2}\right)$$

that is exact for all polynomials of degree ≤ 2; that is, determine values for α, β, and γ. *Hint*: Make the relation exact for 1, x, and x^2 and find a solution of the resulting equations. If it is exact for these polynomials, it is exact for all polynomials of degree ≤ 2.

TABLE 5.1 Gaussian quadrature nodes and weights

n	Nodes x_i	Weights A_i
1	$-\sqrt{\dfrac{1}{3}}$	1
	$+\sqrt{\dfrac{1}{3}}$	1
2	$-\sqrt{\dfrac{3}{5}}$	$\dfrac{5}{9}$
	0	$\dfrac{8}{9}$
	$+\sqrt{\dfrac{3}{5}}$	$\dfrac{5}{9}$
3	$-\sqrt{\dfrac{1}{7}\left(3-4\sqrt{0.3}\right)}$	$\dfrac{1}{2}+\dfrac{1}{12}\sqrt{\dfrac{10}{3}}$
	$-\sqrt{\dfrac{1}{7}\left(3+4\sqrt{0.3}\right)}$	$\dfrac{1}{2}-\dfrac{1}{12}\sqrt{\dfrac{10}{3}}$
	$+\sqrt{\dfrac{1}{7}\left(3-4\sqrt{0.3}\right)}$	$\dfrac{1}{2}+\dfrac{1}{12}\sqrt{\dfrac{10}{3}}$
	$+\sqrt{\dfrac{1}{7}\left(3+4\sqrt{0.3}\right)}$	$\dfrac{1}{2}-\dfrac{1}{12}\sqrt{\dfrac{10}{3}}$
4	$-\sqrt{\dfrac{1}{9}\left(5-2\sqrt{\dfrac{10}{7}}\right)}$	$0.3\left(\dfrac{-0.7+5\sqrt{0.7}}{-2+5\sqrt{0.7}}\right)$
	$-\sqrt{\dfrac{1}{9}\left(5+2\sqrt{\dfrac{10}{7}}\right)}$	$0.3\left(\dfrac{0.7+5\sqrt{0.7})}{2+5\sqrt{0.7}}\right)$
	0	$\dfrac{128}{225}$
	$+\sqrt{\dfrac{1}{9}\left(5-2\sqrt{\dfrac{10}{7}}\right)}$	$0.3\left(\dfrac{-0.7+5\sqrt{0.7}}{-2+5\sqrt{0.7}}\right)$
	$+\sqrt{\dfrac{1}{9}\left(5+2\sqrt{\dfrac{10}{7}}\right)}$	$0.3\left(\dfrac{0.7+5\sqrt{0.7}}{2+5\sqrt{0.7}}\right)$

6. Establish a numerical integration formula of the form

$$\int_a^b f(x)\,dx \approx Af(a) + Bf'(b)$$

that is accurate for polynomials of as high a degree as possible.

7. Derive a formula for $\int_a^{a+h} f(x)\,dx$ in terms of $f(a)$, $f(a+h)$, and $f(a+2h)$ that is correct for polynomials of as high degree as possible. *Hint*: Use polynomials 1, $x - a$, $(x - a)^2$, and so on.

8. Derive a formula of the form

$$\int_a^b f(x)\,dx \approx w_0 f(a) + w_1 f(b) + w_2 f'(a) + w_3 f'(b)$$

that is exact for polynomials of the highest degree possible.

9. Derive the Gaussian quadrature rule of the form

$$\int_{-1}^1 f(x)x^2\,dx \approx af(-\alpha) + bf(0) + cf(\alpha)$$

that is exact for all polynomials of as high a degree as possible; that is, determine α, a, b, and c.

10. Determine a formula of the form

$$\int_0^h f(x)\,dx \approx w_0 f(0) + w_1 f(h) + w_2 f''(0) + w_3 f''(h)$$

that is exact for polynomials of as high a degree as possible.

11. Derive a numerical integration formula of the form

$$\int_{x_{n-1}}^{x_{n+1}} f(x)\,dx \approx Af(x_n) + Bf'(x_{n-1}) + Cf''(x_{n+1})$$

for uniformly spaced points x_{n-1}, x_n, and x_{n+1} with spacing h. The formula should be exact for polynomials of as high a degree as possible. *Hint*: Consider

$$\int_{-h}^h f(x)\,dx \approx Af(0) + Bf'(-h) + Cf''(h)$$

12. By the method of undetermined coefficients, derive a numerical integration formula of the form

$$\int_{-2}^{+2} f(x)\,dx \approx Af(-1) + Bf(0) + Cf(+1)$$

that is exact for polynomials of degree ≤ 2. Is it exact for polynomials of degree greater than 2?

13. Determine A, B, C, and D for a formula of the form

$$Af(-h) + Bf(0) + Cf(h) = hDf'(h) + \int_{-h}^{h} f(x)\,dt$$

that is accurate for polynomials of as high a degree as possible.

14. The numerical integration rule

$$\int_{0}^{3h} f(x)\,dx \approx \frac{3h}{8}[f(0) + 3f(h) + 3f(2h) + f(3h)]$$

is exact for polynomials of degree $\leq n$. Determine the largest value of n for which this assertion is true.

15. (Adams-Bashforth-Moulton formulas) Verify that the numerical integration formulas

a. $\int_{t}^{t+h} g(s)\,ds \approx \frac{h}{24}[55g(t) - 59g(t - h) + 37g(t - 2h) - 9g(t - 3h)]$

b. $\int_{t}^{t+h} g(s)\,ds \approx \frac{h}{24}[9g(t + h) + 19g(t) - 5g(t - h) + g(t - 2h)]$

are exact for polynomials of third degree. *Note:* These two formulas can also be derived by replacing the two integrands g with two interpolating polynomials from Chapter 4 using nodes $(t, t - h, t - 2h, t - 3h)$ or nodes $(t + h, t, t - h, t - 2h)$, respectively.

16. Prove the second theorem in this section.

<hr>

**COMPUTER
PROBLEMS
5.5**

1. With the transformation $t = [2x - (b + a)]/(b - a)$, a Gaussian quadrature rule of the form

$$\int_{-1}^{1} f(t)\,dt \approx \sum_{i=0}^{n} A_i f(t_i)$$

can be used over the interval $[a, b]$; that is,

$$\int_{a}^{b} f(x)\,dx = \frac{1}{2}(b - a) \int_{-1}^{1} f\left[\frac{1}{2}(b - a)t + \frac{1}{2}(b + a)\right] dt$$

Write a program to evaluate the integral $\int_{0}^{2} e^{-\cos^2 x}\,dx$ using Formula (5).

2. (Continuation) By use of Computer Problem 1, compute an approximate value of the integral $\int_{0}^{1} dx/\sqrt{x}$.

3. Compute $\int_{0}^{1} x^{-1} \sin x\,dx$ by the Gaussian Formula (5) suitably modified as in Computer Problem 1.

4. Write a procedure for evaluating $\int_a^b f(x)\,dx$ by first subdividing the interval into n equal subintervals and then using the three-point Gaussian Formula (5) modified to apply to the n different subintervals. The function f and the integer n will be furnished to the procedure.

5. (Continuation) Test the procedure written in Computer Problem **4** on these examples.

   a. $\int_0^1 x^5\,dx$, using $n = 1, 2, 10$

   b. $\int_0^1 x^{-1} \sin x\,dx$, using $n = 1, 2, 3, 4$

6 SYSTEMS OF LINEAR EQUATIONS

A simple electrical network contains a number of resistances and a single source of electromotive force (a battery) as shown in Figure 6.1. Using Kirchhoff's laws and Ohm's law, we can write a system of equations that govern this circuit. If x_1, x_2, x_3, and x_4 are the loop currents as shown, then the equations are

$$\begin{cases} 15x_1 - 2x_2 - 6x_3 = 300 \\ -2x_1 + 12x_2 - 4x_3 - x_4 = 0 \\ -6x_1 - 4x_2 + 19x_3 - 9x_4 = 0 \\ - x_2 - 9x_3 + 21x_4 = 0 \end{cases}$$

Systems of equations like this, even those that contain hundreds of unknowns, can be solved by using the methods developed in this chapter. The solution to the preceding system, for instance, is $x_1 = 26.5$, $x_2 = 9.35$, $x_3 = 13.3$, and $x_4 = 6.13$.

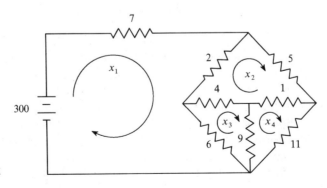

FIGURE 6.1
Electrical network

6.1 Naive Gaussian Elimination

Our objective in this chapter is to develop a good program for solving a system of n linear equations in n unknowns:

$$
\begin{cases}
a_{11}x_1 + a_{12}x_2 + a_{13}x_3 + \cdots + a_{1n}x_n = b_1 \\
a_{21}x_1 + a_{22}x_2 + a_{23}x_3 + \cdots + a_{2n}x_n = b_2 \\
a_{31}x_1 + a_{32}x_2 + a_{33}x_3 + \cdots + a_{3n}x_n = b_3 \\
\quad \vdots \qquad\quad \vdots \qquad\quad \vdots \qquad\qquad\quad \vdots \qquad \vdots \\
a_{i1}x_1 + a_{i2}x_2 + a_{i3}x_3 + \cdots + a_{in}x_n = b_i \\
\quad \vdots \qquad\quad \vdots \qquad\quad \vdots \qquad\qquad\quad \vdots \qquad \vdots \\
a_{n1}x_1 + a_{n2}x_2 + a_{n3}x_3 + \cdots + a_{nn}x_n = b_n
\end{cases}
\tag{1}
$$

In compact form, this system can be written as

$$
\sum_{j=1}^{n} a_{ij}x_j = b_i \qquad (1 \leq i \leq n)
$$

In these equations, a_{ij} and b_i are prescribed real numbers (data) and the unknowns x_j are to be determined. Subscripts on the letter a are separated by a comma only if necessary for clarity—for example, in $a_{32,75}$ but not in a_{ij}.

Numerical Example

In this section, the simplest form of Gaussian elimination is explained. The adjective *naive* applies because this form is not suitable for automatic computation unless essential modifications are made. We illustrate with a specific example that has four equations and four unknowns:

$$
\begin{cases}
6x_1 - 2x_2 + 2x_3 + 4x_4 = 16 \\
12x_1 - 8x_2 + 6x_3 + 10x_4 = 26 \\
3x_1 - 13x_2 + 9x_3 + 3x_4 = -19 \\
-6x_1 + 4x_2 + x_3 - 18x_4 = -34
\end{cases}
\tag{2}
$$

In the first step of the elimination procedure, certain multiples of the first equation are subtracted from the second, third, and fourth equations so as to eliminate x_1 from these equations. Thus, we want to create 0's as coefficients for each x_1 below the first (where 12, 3, and -6 now stand). It is clear that we should subtract 2 times the first equation from the second. (This multiplier is simply the quotient $\frac{12}{6}$.) Likewise, we should subtract $\frac{1}{2}$ times the first equation from the third. (Again, this multiplier is just $\frac{3}{6}$.) Finally, we should subtract -1 times the first equation from the fourth. When all of this has been done, the result is

$$\begin{cases} 6x_1 - 2x_2 + 2x_3 + 4x_4 = 16 \\ \quad - 4x_2 + 2x_3 + 2x_4 = -6 \\ \quad - 12x_2 + 8x_3 + \quad x_4 = -27 \\ \quad 2x_2 + 3x_3 - 14x_4 = -18 \end{cases} \tag{3}$$

Note that the first equation was not altered in this process, although it was used to produce the 0 coefficients in the other equations. In this context, it is called the **pivot** equation.

Notice, also, that Systems (2) and (3) are *equivalent* in the following technical sense: Any solution of (2) is also a solution of (3), and vice versa. This follows at once from the fact that if equal quantities are added to equal quantities, the resulting quantities are equal. One can get System (2) from System (3) by adding 2 times the first equation to the second, and so on.

In the second step of the process, we mentally ignore the first equation and the first column of coefficients. This leaves a system of three equations with three unknowns. The same process is now repeated using the top equation in the smaller system as the current pivot equation. Thus, we begin by subtracting 3 times the second equation from the third. (The multiplier is just the quotient $\frac{-12}{-4}$.) Then we subtract $-\frac{1}{2}$ times the second equation from the fourth. After doing the arithmetic, we arrive at

$$\begin{cases} 6x_1 - 2x_2 + 2x_3 + \quad 4x_4 = 16 \\ \quad - 4x_2 + 2x_3 + \quad 2x_4 = -6 \\ \quad 2x_3 - \quad 5x_4 = -9 \\ \quad 4x_3 - 13x_4 = -21 \end{cases} \tag{4}$$

The final step consists in subtracting 2 times the third equation from the fourth. The result is

$$\begin{cases} 6x_1 - 2x_2 + 2x_3 + 4x_4 = 16 \\ \quad - 4x_2 + 2x_3 + 2x_4 = -6 \\ \quad 2x_3 - 5x_4 = -9 \\ \quad - 3x_4 = -3 \end{cases} \tag{5}$$

This system is said to be in **upper triangular** form. It is equivalent to System (2).

This completes the first phase (**forward elimination**) in the Gauss algorithm. The second phase (**back substitution**) is solving System (5) for the unknowns *starting at the bottom*. Thus, from the fourth equation,

$$x_4 = \frac{-3}{-3} = 1$$

Putting $x_4 = 1$ in the third equation gives us

$$2x_3 - 5 = -9$$

whence

$$x_3 = \frac{-4}{2} = -2$$

and so on. The solution is

$$x_1 = 3 \qquad x_2 = 1 \qquad x_3 = -2 \qquad x_4 = 1$$

This example can be solved directly using Maple as follows.

```
with(linalg):
A := array([[ 6,  -2, 2,   4],
            [12,  -8, 6,  10],
            [ 3, -13, 9,   3],
            [-6,   4, 1, -18]]);
b := array( [16, 26, -19, -34]);
linsolve(A, b);
```

Algorithm

To simplify the discussion, we write System (1) in matrix-vector form. The coefficient elements a_{ij} form an $n \times n$ square array, or matrix. The unknowns x_i and the right-hand-side elements b_i form $n \times 1$ arrays, or vectors.* Hence, we have

$$\begin{bmatrix} a_{11} & a_{12} & a_{13} & \cdots & a_{1n} \\ a_{21} & a_{22} & a_{23} & \cdots & a_{2n} \\ a_{31} & a_{32} & a_{33} & \cdots & a_{3n} \\ \vdots & \vdots & \vdots & & \vdots \\ a_{i1} & a_{i2} & a_{i3} & \cdots & a_{in} \\ \vdots & \vdots & \vdots & & \vdots \\ a_{n1} & a_{n2} & a_{n3} & \cdots & a_{nn} \end{bmatrix} \begin{bmatrix} x_1 \\ x_2 \\ x_3 \\ \vdots \\ x_i \\ \vdots \\ x_n \end{bmatrix} = \begin{bmatrix} b_1 \\ b_2 \\ b_3 \\ \vdots \\ b_i \\ \vdots \\ b_n \end{bmatrix} \qquad (6)$$

or

$$Ax = b$$

Operations between equations correspond to operations between rows in this notation. We shall use these two words interchangeably.

---

*To save space, we occasionally write a vector as $[x_1, x_2, \ldots, x_n]^T$, with the T standing for the transpose. It tells us that this is an $n \times 1$ array or vector and *not* $1 \times n$, as would be indicated without the transpose. (See Appendix A for linear algebra notation and concepts.)

Now let us organize the naive Gaussian elimination algorithm for the general system, which contains n equations and n unknowns. In the forward elimination phase of the process, there are $n - 1$ principal steps. The first of these steps uses the first equation to produce $n - 1$ zeros as coefficients for each x_1 in all but the first equation. This is done by subtracting appropriate multiples of the first equation from the others. In this process, we refer to the first equation as the first **pivot equation**. For each remaining equation ($2 \leq i \leq n$), we compute

$$\begin{cases} a_{ij} \leftarrow a_{ij} - \left(\dfrac{a_{i1}}{a_{11}}\right) a_{1j} & (1 \leq j \leq n) \\[2ex] b_i \leftarrow b_i - \left(\dfrac{a_{i1}}{a_{11}}\right) b_1 \end{cases}$$

The symbol $\leftarrow$ indicates a *replacement*. Thus, the content of the memory location allocated to a_{ij} is replaced by $a_{ij} - (a_{i1}/a_{11})a_{1j}$, and so on. This is accomplished by the line of pseudocode

$$a_{ij} \leftarrow a_{ij} - (a_{i1}/a_{11})a_{1j}$$

Note that the quantities (a_{i1}/a_{11}) are the *multipliers*. The new coefficient of x_1 in the ith equation will be 0 because $a_{i1} - (a_{i1}/a_{11})a_{11} = 0$.

After the first step, the system will be of the form

$$\begin{bmatrix} a_{11} & a_{12} & a_{13} & \cdots & a_{1n} \\ 0 & a_{22} & a_{23} & \cdots & a_{2n} \\ 0 & a_{23} & a_{33} & \cdots & a_{3n} \\ \vdots & \vdots & \vdots & & \vdots \\ 0 & a_{i2} & a_{i3} & \cdots & a_{in} \\ \vdots & \vdots & \vdots & & \vdots \\ 0 & a_{n2} & a_{n3} & \cdots & a_{nn} \end{bmatrix} \begin{bmatrix} x_1 \\ x_2 \\ x_3 \\ \vdots \\ x_i \\ \vdots \\ x_n \end{bmatrix} = \begin{bmatrix} b_1 \\ b_2 \\ b_3 \\ \vdots \\ b_i \\ \vdots \\ b_n \end{bmatrix}$$

From here on, we will not alter the first equation, nor will we alter any of the coefficients for x_1 (since a multiplier times 0 subtracted from 0 is still 0). Thus, we can mentally ignore the first row and the first column and repeat the process on the smaller system. With the second equation as the pivot equation, we compute for each remaining equation ($3 \leq i \leq n$)

$$\begin{cases} a_{ij} \leftarrow a_{ij} - (a_{i2}/a_{22})a_{2j} & (2 \leq j \leq n) \\[1ex] b_i \leftarrow b_i - (a_{i2}/a_{22})b_2 \end{cases}$$

Just prior to the kth step in the forward elimination, the system will appear as follows:

$$
\begin{bmatrix}
a_{11} & a_{12} & a_{13} & \cdots & & \cdots & & \cdots & a_{1n} \\
 & a_{22} & a_{23} & \cdots & & \cdots & & \cdots & a_{2n} \\
 & & a_{33} & \cdots & & \cdots & & \cdots & a_{3n} \\
 & & & \ddots & & & & & \vdots \\
 & & & & a_{kk} & \cdots & a_{kj} & \cdots & a_{kn} \\
 & & & & \vdots & & \vdots & & \vdots \\
 & & & & a_{ik} & \cdots & a_{ij} & \cdots & a_{in} \\
 & & & & \vdots & & \vdots & & \vdots \\
 & & & & a_{nk} & \cdots & a_{nj} & \cdots & a_{nn}
\end{bmatrix}
\begin{bmatrix}
x_1 \\ x_2 \\ x_3 \\ \vdots \\ x_k \\ \vdots \\ x_i \\ \vdots \\ x_n
\end{bmatrix}
=
\begin{bmatrix}
b_1 \\ b_2 \\ b_3 \\ \vdots \\ b_k \\ \vdots \\ b_i \\ \vdots \\ b_n
\end{bmatrix}
$$

Here a wedge of 0 coefficients has been created, and the first k equations have been processed and are now fixed. Using the kth equation as the pivot equation, we select multipliers in order to create 0's as coefficients for each x_k below the a_{kk} coefficient. Hence, we compute for each remaining equation ($k + 1 \leq i \leq n$)

$$
\begin{cases}
a_{ij} \leftarrow a_{ij} - (a_{ik}/a_{kk})a_{kj} & (k \leq j \leq n) \\
b_i \leftarrow b_i - (a_{ik}/a_{kk})b_k
\end{cases}
$$

Obviously, we must assume that all the divisors in this algorithm are nonzero.

Pseudocode

We now consider the pseudocode for forward elimination. The coefficient array is stored as a double-subscripted array (a_{ij}); the right side of the system of equations is stored as a single-subscripted array (b_i); the solution is computed and stored in a single-subscripted array (x_i). It is easy to see that the following lines of pseudocode carry out the forward elimination phase of naive Gaussian elimination:

real array $(a_{ij})_{n \times n}, (b_i)_n$
integer i, j, k
$\quad \vdots$
for $k = 1$ **to** $n - 1$ **do**
$\quad$ **for** $i = k + 1$ **to** n **do**
$\quad \quad$ **for** $j = k$ **to** n **do**
$\quad \quad \quad a_{ij} \leftarrow a_{ij} - (a_{ik}/a_{kk})a_{kj}$
$\quad \quad$ **end do**
$\quad \quad b_i \leftarrow b_i - (a_{ik}/a_{kk})b_k$
$\quad$ **end do**
end do

Since the multiplier a_{ik}/a_{kk} does not depend on j, it should be moved outside the j loop. Notice also that the new values in column k will be 0, at least theoretically, because when $j = k$, we have

$$a_{ij} \leftarrow a_{ik} - (a_{ik}/a_{kk})a_{kk}$$

Since we expect this to be 0, no purpose is served in computing it. The location where the 0 is being created is a good place to store the multiplier. If these remarks are put into practice, the pseudocode will look like this:

real array $(a_{ij})_{n \times n}, (b_i)_n$
integer i, j, k
real *xmult*
   $\vdots$
for $k = 1$ **to** $n - 1$ **do**
   for $i = k + 1$ **to** n **do**
      xmult $\leftarrow a_{ik}/a_{kk}$
      $a_{ik} \leftarrow$ *xmult*
      for $j = k + 1$ **to** n **do**
         $a_{ij} \leftarrow a_{ij} - ($*xmult*$)a_{kj}$
      end do
      $b_i \leftarrow b_i - ($*xmult*$)b_k$
   end do
end do

At the beginning of the back substitution phase, the linear system is of the form

$$\begin{cases} a_{11}x_1 + a_{12}x_2 + a_{13}x_3 + \cdots & \cdots + & a_{1n}x_n = b_1 \\ a_{22}x_2 + a_{23}x_3 + \cdots & \cdots + & a_{2n}x_n = b_2 \\ a_{33}x_3 + \cdots & + & a_{3n}x_n = b_3 \\ \ddots & \vdots & \vdots \\ a_{ii}x_i + a_{i,i+1}x_{i+1} + & \cdots + & a_{in}x_n = b_i \\ \ddots & \vdots & \vdots \\ a_{n-1,n-1}x_{n-1} + & a_{n-1,n}x_n = b_{n-1} \\ & a_{nn}x_n = b_n \end{cases}$$

where the a_{ij}'s and b_i's are *not* the original ones from System (6) but instead are the ones that have been altered by the elimination process.

The back substitution starts by solving the nth equation for x_n:

$$x_n = \frac{b_n}{a_{nn}}$$

Then, using the $(n - 1)$th equation, we solve for x_{n-1}:

$$x_{n-1} = \frac{b_{n-1} - a_{n-1,n}x_n}{a_{n-1,n-1}}$$

We continue working upward, recovering each x_i by the formula

$$x_i = \frac{1}{a_{ii}} \left(b_i - \sum_{j=i+1}^{n} a_{ij}x_j \right) \qquad (i = n-1, n-2, \ldots, 1)$$

Here is pseudocode to do this:

real array $(a_{ij})_{n \times n}, (x_i)_n$
integer i, j, n
real *sum*
$\vdots$
$x_n \leftarrow b_n/a_{nn}$
for $i = n - 1$ **to** 1 **step** $- 1$ **do**
   sum $\leftarrow b_i$
   for $j = i + 1$ **to** n **do**
      sum $\leftarrow$ *sum* $- a_{ij}x_j$
   end do
   $x_i \leftarrow$ *sum*$/a_{ii}$
end do

    Now we put these segments of pseudocode together to form a procedure, called *ngauss*, which is intended to solve a system of n linear equations in n unknowns by the method of naive Gaussian elimination. This pseudocode serves a didactic purpose only; a more robust pseudocode will be developed in the next section.

procedure *ngauss*$(n, (a_{ij}), (b_i), (x_i))$
real array $(a_{ij})_{n \times n}, (b_i)_n, (x_i)_n$
integer i, j, k, n
real *sum*, *xmult*
for $k = 1$ **to** $n - 1$ **do**
  for $i = k + 1$ **to** n **do**
    xmult $\leftarrow a_{ik}/a_{kk}$
    $a_{ik} \leftarrow$ *xmult*
    for $j = k + 1$ **to** n **do**
      $a_{ij} \leftarrow a_{ij} - ($*xmult*$)a_{kj}$
    end do
    $b_i \leftarrow b_i - ($*xmult*$)b_k$
  end do
end do
$x_n \leftarrow b_n/a_{n,n}$
for $i = n - 1$ **to** 1 **step** $- 1$ **do**
   sum $\leftarrow b_i$
   for $j = i + 1$ **to** n **do**
      sum $\leftarrow$ *sum* $- a_{ij}x_j$

end do
$$x_i \leftarrow sum/a_{ii}$$
end do
end procedure *ngauss*

Testing the Pseudocode

One good way to test a procedure is to set up an artificial problem whose solution is known beforehand. Sometimes the test problem will include a parameter that can be changed to vary the difficulty. The next example illustrates this.

Fixing a value of n, define the polynomial

$$p(t) = 1 + t + t^2 + \cdots + t^{n-1} = \sum_{j=1}^{n} t^{j-1}$$

The coefficients in this polynomial are all equal to 1. We shall try to recover these known coefficients from n values of the polynomial. We use the values of $p(t)$ at the integers $t = 1 + i$ for $i = 1, 2, \ldots, n$. Denoting the coefficients in the polynomial $x_1, x_2, \ldots, x_n$, we should have

$$\sum_{j=1}^{n} (1 + i)^{j-1} x_j = \frac{1}{i} [(1 + i)^n - 1] \qquad (1 \leq i \leq n) \tag{7}$$

Here we have used the formula for the sum of a geometric series on the right-hand side; that is,

$$p(1 + i) = \sum_{j=1}^{n} (1 + i)^{j-1} = \frac{(1 + i)^n - 1}{(1 + i) - 1} = \frac{1}{i} [(1 + i)^n - 1]$$

Letting $a_{ij} = (1 + i)^{j-1}$ and $b_i = [(1 + i)^n - 1]/i$ in Equation (7), we have a linear system:

$$\sum_{j=1}^{n} a_{ij} x_j = b_i \qquad (1 \leq i \leq n) \tag{8}$$

EXAMPLE 1   Write a pseudocode that solves the system of Equation (8) for various values of n.

Solution   Since the naive Gaussian elimination procedure *ngauss* can be used, all that is needed is a calling program. We decide to use $n = 4, 5, 6, 7, 8, 9, 10$ for the test. Here is a suitable pseudocode:

program *main*
real array $(a_{ij})_{n \times n}, (b_i)_n, (x_i)_n$
integer i, j, n
for $n = 4$ **to** 10 **do**

```
        for i = 1 to n do
          for j = 1 to n do
            a_ij ← (i + 1)^(j-1)
          end do
          b_i ← [(i + 1)^n − 1]/i
        end do
        call ngauss(n, (a_ij), (b_i), (x_i))
        output n, (x_i)_n
      end do
    end program main
```

When this pseudocode was run on a machine that carries approximately seven decimal digits of accuracy, the solution was obtained with complete precision until n was changed to 9, and then the computed solution was worthless because one component exhibited a relative error of 16120%! □

The coefficient matrix for this linear system is an example of a well-known *ill-conditioned* matrix called the Vandermonde matrix, and this accounts for the fact that the system cannot be solved accurately using naive Gaussian elimination. What is amazing is that the trouble happens so suddenly! When $n \geqq 9$, the roundoff error present in computing x_i is propagated and magnified throughout the back substitution phase so that most of the computed values for x_i are worthless. Insert some intermediate print statements in the code to see for yourself what is going on here.

Residual and Error Vectors

Suppose that two students, each with a different home computer system, solve the same linear system $Ax = b$. What algorithm and what precision each used are not known. Each vehemently claims to have the correct answer, but the two computer solutions $\widetilde{x}$ and $\widehat{x}$ are totally different! How do we determine which, if either, computed solution is correct?

We can *check* them by substituting them into the original system, which is the same as computing the **residual vectors** $\widetilde{r} = A\widetilde{x} - b$ and $\widehat{r} = A\widehat{x} - b$. Of course, the computed solutions are not exact because they each must contain some roundoff errors. So we would want to accept the solution with the smaller residual vector. However, if we knew the exact solution x, then we would just compare the computed solutions with the exact solution, which is the same as computing the **error vectors** $\widetilde{e} = \widetilde{x} - x$ and $\widehat{e} = \widehat{x} - x$. Now the computed solution that produces the smaller error vector would most assuredly be the better answer.

Since the exact solution is usually not known in applications, one would tend to accept the computed solution with the smallest residual vector. But this may not be the best computed solution if the original problem is sensitive to roundoff errors—that is, ill-conditioned. In fact, the question of whether a computed solution to a linear system is a good solution is extremely difficult and beyond the scope of this book. Problem **6** may give some insight into the difficulty of assessing the accuracy of computed solutions of linear systems.

PROBLEMS
6.1

1. Solve each of the following systems using naive Gaussian elimination—that is, forward elimination and back substitution. Carry four significant figures.

a.
$$\begin{cases} 3x_1 + 4x_2 + 3x_3 = 16 \\ x_1 + 5x_2 - x_3 = -12 \\ 6x_1 + 3x_3 + 7x_3 = 102 \end{cases}$$

b.
$$\begin{cases} 3x + 2y - 5z = 4 \\ 2x - 3y + z = 8 \\ x + 4y - z = -3 \end{cases}$$

c.
$$\begin{bmatrix} 1 & -1 & 2 & 1 \\ 3 & 2 & 1 & 4 \\ 5 & 8 & 6 & 3 \\ 4 & 2 & 5 & 3 \end{bmatrix} \begin{bmatrix} x_1 \\ x_2 \\ x_3 \\ x_4 \end{bmatrix} = \begin{bmatrix} 1 \\ 1 \\ 1 \\ -1 \end{bmatrix}$$

2. Show that the system of equations

$$\begin{cases} x_1 + 4x_2 + \alpha x_3 = 6 \\ 2x_1 - x_2 + 2\alpha x_3 = 3 \\ \alpha x_1 + 3x_2 + x_3 = 5 \end{cases}$$

possesses a unique solution when $\alpha = 0$, no solution when $\alpha = -1$, and infinitely many solutions when $\alpha = 1$. Also, investigate the corresponding situation when the right-hand side is replaced by 0's.

3. For what values of α does naive Gaussian elimination produce erroneous answers for this system?

$$\begin{cases} x_1 + x_2 = 2 \\ \alpha x_1 + x_2 = 2 + \alpha \end{cases}$$

Explain what happens in the computer.

4. Apply naive Gaussian elimination to these examples and account for the failures. Solve the systems by other means if possible.

a.
$$\begin{cases} 3x_1 + 2x_2 = 4 \\ -x_1 - \frac{2}{3}x_2 = 1 \end{cases}$$

b.
$$\begin{cases} 6x_1 - 3x_2 = 6 \\ -2x_1 + x_2 = -2 \end{cases}$$

c.
$$\begin{cases} 0x_1 + 2x_2 = 4 \\ x_1 - x_2 = 5 \end{cases}$$

d.
$$\begin{cases} x_1 + x_2 + 2x_3 = 4 \\ x_1 + x_2 + 0x_3 = 2 \\ 0x_1 + x_2 + x_3 = 0 \end{cases}$$

5. Solve the following system of equations, retaining only four significant figures in each step of the calculation, and compare your answer with the solution obtained when eight significant figures are retained. Be consistent by either always rounding to the number of significant figures that are being carried or always chopping.

$$\begin{cases} 0.1036x + 0.2122y = 0.7381 \\ 0.2081x + 0.4247y = 0.9327 \end{cases}$$

6. Consider

$$A = \begin{bmatrix} 0.780 & 0.563 \\ 0.913 & 0.659 \end{bmatrix} \quad b = \begin{bmatrix} 0.217 \\ 0.254 \end{bmatrix} \quad \tilde{x} = \begin{bmatrix} 0.999 \\ -1.001 \end{bmatrix} \quad \hat{x} = \begin{bmatrix} 0.341 \\ -0.087 \end{bmatrix}$$

Compute residual vectors $\tilde{r} = A\tilde{x} - b$ and $\hat{r} = A\hat{x} - b$ and decide which of $\tilde{x}$ and $\hat{x}$ is the better solution vector. Now compute the error vectors $e = \tilde{x} - x$ and $\hat{e} = \hat{x} - x$, where $x = [1, -1]^T$ is the exact solution. Discuss the implications of this example.

7. Consider the system

$$\begin{cases} 10^{-4}x_1 + x_2 = b_1 \\ x_1 + x_2 = b_2 \end{cases}$$

where $b_1 \neq 0$ and $b_2 \neq 0$. Its exact solution is

$$x_1 = \frac{-b_1 + b_2}{1 - 10^{-4}} \quad \text{and} \quad x_2 = \frac{b_1 - 10^{-4}b_2}{1 - 10^{-4}}$$

a. Let $b_1 = 1$ and $b_2 = 2$. Solve this system using naive Gaussian elimination with three-digit (rounded) arithmetic and compare with the exact solution $x_1 = 1.00010\ldots$ and $x_2 = 0.999899\ldots$.

b. Find values of b_1 and b_2 so that naive Gaussian elimination does not give poor answers.

8. Using naive Gaussian elimination, solve the following systems.

a.
$$\begin{cases} 3x_1 + 2x_2 - x_3 = 7 \\ 5x_1 + 3x_2 + 2x_3 = 4 \\ -x_1 + x_2 - 3x_3 = -1 \end{cases}$$

b.
$$\begin{cases} x_1 + 3x_2 + 2x_3 + x_4 = -2 \\ 4x_1 + 2x_2 + x_3 + 2x_4 = 2 \\ 2x_1 + x_2 + 2x_3 + 3x_4 = 1 \\ x_1 + 2x_2 + 4x_3 + x_4 = -1 \end{cases}$$

COMPUTER PROBLEMS 6.1

1. Program and run the example in the text and insert some print statements to see what is happening.

2. Rewrite and test procedure *ngauss* so that it is column oriented; that is, the first index of a varies on the innermost loop.

3. Define an $n \times n$ matrix A by the equation $a_{ij} = i + j$. Define b by the equation $b_i = i + 1$. Solve $Ax = b$ by using procedure *ngauss*. What should x be?

4. Define an $n \times n$ array by $a_{ij} = -1 + 2\min\{i, j\}$. Then set up the array (b_i) in such a way that the solution of the system $\sum_{j=1}^{n} a_{ij}x_j = b_i (1 \leq i \leq n)$ is $x_j = 1$ $(1 \leq j \leq n)$. Test procedure *ngauss* on this system for a moderate value of n, say $n = 15$.

5. Write and test a version of procedure *ngauss* in which:

   a. An attempted division by 0 is signaled by an error return.

   b. The solution x is placed in array (b_i).

6. Write a complex-arithmetic version of *ngauss* by declaring certain variables complex and making other necessary changes to the code. Consider the complex linear system

$$Az = b$$

where

$$A = \begin{bmatrix} 5 + 9i & 5 + 5i & -6 - 6i & -7 - 7i \\ 3 + 3i & 6 + 10i & -5 - 5i & -6 - 6i \\ 2 + 2i & 3 + 3i & -1 + 3i & -5 - 5i \\ 1 + i & 2 + 2i & -3 - 3i & 4i \end{bmatrix}$$

Solve this system four times with the following right-hand sides:

$$\begin{bmatrix} -10 + 2i \\ -5 + i \\ -5 + i \\ -5 + i \end{bmatrix} \quad \begin{bmatrix} 2 + 6i \\ 4 + 12i \\ 2 + 6i \\ 2 + 6i \end{bmatrix} \quad \begin{bmatrix} 7 - 3i \\ 7 - 3i \\ 0 \\ 7 - 3i \end{bmatrix} \quad \begin{bmatrix} -4 - 8i \\ -4 - 8i \\ -4 - 8i \\ 0 \end{bmatrix}$$

Verify that the solutions are $z = \lambda^{-1}b$ for scalars λ. The numbers λ are called **eigenvalues**, and the solutions z are **eigenvectors** of A. Usually the b vector is not known and the solution of the problem $Az = \lambda z$ cannot be obtained by using a linear equation solver.

7. (Continuation) A common electrical engineering problem is to calculate currents in an electric circuit. For example, the circuit shown in the figure with R_i (ohms), C_i (microfarads), L (millihenries), and ω (hertz) leads to the system

$$(50 - 10i)I_1 + (50)I_2 + (50)I_3 = V_1$$
$$(10i)I_1 + (10 - 10i)I_2 + (10 - 20i)I_3 = 0$$
$$- (30i)I_2 + (20 - 50i)I_3 = -V_2$$

Select V_1 to be 100 millivolts and solve two cases:

   a. The two voltages are in phase; that is, $V_2 = V_1$.

   b. The second voltage is a quarter of a cycle ahead of the first; that is, $V_2 = iV_1$.

Use the complex-arithmetic version of *ngauss* and in each case solve the system for the amplitude (in milliamperes) and the phase (in degrees) for each current I_k. *Hint*: When $I_k = \Re(I_k) + i\Im(I_k)$, the amplitude is $|I_k|$, and the phase is $(180°/\pi)\arctan[\Im(I_k)/\Re(I_k)]$. Draw a diagram to show why this is so.

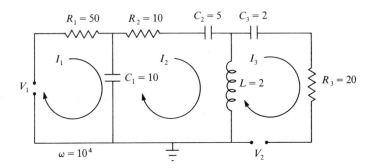

8. Select a reasonable value of n and generate a random $n \times n$ array a using a random-number generator. Define the array b so that the solution of the system

$$\sum_{j=1}^{n} a_{ij}x_j = b_i \qquad (1 \le i \le n)$$

is $x_j = j$, where $1 \le j \le n$. Test the naive Gaussian algorithm on this system. *Hint*: You may use the function *random*, which is developed in Chapter 9, to generate the random elements of the (a_{ij}) array.

9. Carry out the test described in the text for procedure *ngauss* but *reverse* the order of the equations. *Hint*: It suffices, in the code, to replace i by $n - i + 1$ in appropriate places.

10. Solve the linear system given in the lead-off example to this chapter using *ngauss*.

6.2 Gaussian Elimination with Scaled Partial Pivoting

Examples Where Naive Gaussian Elimination Fails

To see why the naive Gaussian elimination algorithm is unsatisfactory, consider the system

$$\begin{cases} 0x_1 + x_2 = 1 \\ x_1 + x_2 = 2 \end{cases} \tag{1}$$

The pseudocode constructed in Section 6.1 would attempt to subtract some multiple of the first equation from the second in order to produce a 0 as the coefficient for

x_1 in the second equation. This, of course, is impossible, so the algorithm fails if $a_{11} = 0$.

If a numerical procedure actually fails for some values of the data, then the procedure is probably untrustworthy for values of the data *near* the failing values. To test this dictum, consider the system

$$\begin{cases} \varepsilon x_1 + x_2 = 1 \\ x_1 + x_2 = 2 \end{cases} \tag{2}$$

in which ε is a small number different from 0. Now the naive algorithm of Section 6.1 works and produces first the system

$$\begin{cases} \varepsilon x_1 + \qquad x_2 = 1 \\ \qquad \left(1 - \dfrac{1}{\varepsilon}\right) x_2 = 2 - \dfrac{1}{\varepsilon} \end{cases} \tag{3}$$

In the back substitution, the arithmetic is as follows:

$$x_2 = \frac{2 - 1/\varepsilon}{1 - 1/\varepsilon} \qquad x_1 = \frac{1 - x_2}{\varepsilon}$$

Now $1/\varepsilon$ will be large, and so if this calculation is performed by a computer that has a fixed word length, then for small values of ε, both $(2 - 1/\varepsilon)$ and $(1 - 1/\varepsilon)$ would be computed as $-1/\varepsilon$.

For example, in an 8-digit decimal machine with a 16-digit accumulator, when $\varepsilon = 10^{-9}$, it follows that $1/\varepsilon = 10^9$. In order to subtract, the computer must interpret the numbers as

$$\frac{1}{\varepsilon} = 10^9 = 0.10000\,000 \times 10^{10} = 0.10000\,00000\,000000\,0 \times 10^{10}$$

$$2 = 0.20000\,000 \times 10^1 \quad = 0.00000\,00002\,000000\,0 \times 10^{10}$$

Thus, $(1/\varepsilon - 2)$ is computed initially as $0.09999\,99998\,000000\,0 \times 10^{10}$ and then rounded to $0.10000\,000 \times 10^{10} = 1/\varepsilon$.

We conclude that for values of ε sufficiently close to 0, the computer calculates x_2 as 1 and then x_1 as 0. Since the correct solution is

$$x_1 = \frac{1}{1 - \varepsilon} \approx 1 \qquad x_2 = \frac{1 - 2\varepsilon}{1 - \varepsilon} \approx 1$$

the relative error in the computed solution for x_1 is extremely large: 100%.

Actually, the naive Gaussian elimination algorithm works well on examples (1) and (2) if the equations are first permuted:

$$\begin{cases} x_1 + x_2 = 2 \\ \qquad x_2 = 1 \end{cases} \quad \text{or} \quad \begin{cases} x_1 + x_2 = 2 \\ \varepsilon x_1 + x_2 = 1 \end{cases}$$

Indeed, the second of these systems becomes

$$\begin{cases} x_1 + & x_2 = 2 \\ & (1 - \varepsilon)x_2 = 1 - 2\varepsilon \end{cases}$$

after the forward elimination. Then from the back substitution, the solution is computed as

$$x_2 = \frac{1 - 2\varepsilon}{1 - \varepsilon} \approx 1 \qquad x_1 = 2 - x_2 \approx 1$$

The difficulty in System (2) is not due simply to ε being small but rather to its being small relative to other coefficients in the same row. To verify this, consider

$$\begin{cases} x_1 + \dfrac{1}{\varepsilon}x_2 = \dfrac{1}{\varepsilon} \\ x_1 + x_2 = 2 \end{cases} \tag{4}$$

System (4) is mathematically equivalent to (2). The naive Gaussian elimination algorithm fails here, too, because it produces the triangular system

$$\begin{cases} x_1 + \dfrac{1}{\varepsilon}x_2 = \dfrac{1}{\varepsilon} \\ \left(1 - \dfrac{1}{\varepsilon}\right)x_2 = 2 - \dfrac{1}{\varepsilon} \end{cases}$$

and then, in the back substitution, it produces the erroneous result

$$x_2 = \frac{2 - \dfrac{1}{\varepsilon}}{1 - \dfrac{1}{\varepsilon}} \approx 1 \qquad x_1 = \frac{1}{\varepsilon} - \frac{1}{\varepsilon}x_2 \approx 0$$

Gaussian Elimination with Scaled Partial Pivoting

These simple examples should make it clear that the *order* in which we treat the equations significantly affects the accuracy of the elimination algorithm in the computer. In the naive Gaussian elimination algorithm, we use the first equation to eliminate x_1 from the following equations. Then we use the second equation to eliminate x_2 from the following equations, and so on. The order in which the equations are used as pivot equations is the *natural* order $\{1, 2, \ldots, n\}$. Note that the last equation (equation number n) is *not* used as an operating equation with the natural ordering: At no time are multiples of it subtracted from other equations in the naive algorithm.

The Gaussian elimination algorithm now to be described uses the equations in an order that is determined by the actual system being solved. For instance, if the algorithm were asked to solve System (1) or (2), the order in which the equations would be used as pivot equations would not be the natural order $\{1, 2\}$ but rather $\{2, 1\}$. This order is automatically determined by the computer program. The order in which the equations are employed is denoted by the row vector $[\ell_1, \ell_2, \ldots, \ell_n]$, with ℓ_n not actually being used in the forward elimination phase. Here the ℓ_i are integers from 1 to n in a possibly different order. We call $\ell = [\ell_1, \ell_2, \ldots, \ell_n]$ the **index vector**. The strategy to be described now for determining the index vector is termed **scaled partial pivoting**.

At the beginning, a scale factor must be computed for each equation in the system. Referring to the notation in Section 6.1, we define

$$s_i = \max_{1 \leq j \leq n} |a_{ij}| \qquad (1 \leq i \leq n)$$

These n numbers are recorded in the **scale vector** $\mathbf{s} = [s_1, s_2, \ldots, s_n]$.

In starting the forward elimination process, we do not arbitrarily use the first equation as the pivot equation. Instead we use the equation for which the ratio $|a_{i,1}|/s_i$ is greatest. Let ℓ_1 be the index for which this ratio is greatest. Now appropriate multipliers of equation ℓ_1 are subtracted from the other equations in order to create 0's as coefficients for each x_1 except in the pivot equation.

The best way of keeping track of the indices is as follows: At the beginning, define the index vector ℓ to be $[\ell_1, \ell_2, \ldots, \ell_n] = [1, 2, \ldots, n]$. Select j to be the index associated with the largest ratio in the set

$$\left\{ \frac{|a_{\ell_i 1}|}{s_{\ell_i}} : 1 \leq i \leq n \right\}$$

Now interchange ℓ_j with ℓ_1 in the index vector ℓ. Next, use multipliers

$$\frac{a_{\ell_i 1}}{a_{\ell_1 1}}$$

times row ℓ_1 and subtract from equations ℓ_i for $2 \leq i \leq n$. It is important to note that only entries in ℓ are being interchanged and *not* the equations. This eliminates the time-consuming and unnecessary process of moving the coefficients of equations around in the computer memory!

In the second step, the ratios

$$\left\{ \frac{|a_{\ell_i 2}|}{s_{\ell_i}} : 2 \leq i \leq n \right\}$$

are scanned. With j the index for the largest ratio, interchange ℓ_j with ℓ_2 in ℓ. Then multipliers

$$\frac{a_{\ell_i 2}}{a_{\ell_2 2}}$$

times equation ℓ_2 are subtracted from equations ℓ_i for $3 \leq i \leq n$.

At step k, select j to be the index corresponding to the largest of the ratios

$$\left\{ \frac{|a_{\ell_i k}|}{s_{\ell_i}} : k \leq i \leq n \right\}$$

and interchange ℓ_j and ℓ_k in index vector ℓ. Then multipliers

$$\frac{a_{\ell_i k}}{a_{\ell_k k}}$$

times pivot equation ℓ_k are subtracted from equations ℓ_i for $k + 1 \leq i \leq n$.

Notice that the scale factors are *not* changed after each pivot step. Intuitively, one might feel that after each pivot and elimination, the remaining (modified) coefficients should be used to recompute the scale factors instead of using the original scale vector. Of course, this could be done, but it is generally believed that the extra computations involved in this procedure are not worthwhile in the majority of linear systems. The reader is encouraged to explore this question. (See Computer Problem **16**.)

Numerical Example

We are not quite ready to write pseudocode, but let us consider what has been outlined in a concrete example. Consider this system:

$$\begin{bmatrix} 3 & -13 & 9 & 3 \\ -6 & 4 & 1 & -18 \\ 6 & -2 & 2 & 4 \\ 12 & -8 & 6 & 10 \end{bmatrix} \begin{bmatrix} x_1 \\ x_2 \\ x_3 \\ x_4 \end{bmatrix} = \begin{bmatrix} -19 \\ -34 \\ 16 \\ 26 \end{bmatrix}$$

The index vector is $\ell = [1, 2, 3, 4]$ at the beginning. The scale vector does not change throughout the procedure and is $s = [13, 18, 6, 12]$. To determine the first pivot row, we look at four ratios:

$$\left\{ \frac{|a_{\ell_i, 1}|}{s_{\ell_i}} : i = 1, 2, 3, 4 \right\} = \left\{ \frac{3}{13}, \frac{6}{18}, \frac{6}{6}, \frac{12}{12} \right\}$$

We select the index j as the *first* occurrence of the largest value of these ratios. In this example, the largest of these occurs for the index $j = 3$. So row 3 is to be the pivot equation in step 1 ($k = 1$) of the elimination process. In the index vector ℓ, entries ℓ_k and ℓ_j are interchanged so that the new index vector is $\ell = [3, 2, 1, 4]$. Thus, the pivot equation is ℓ_k, which is $\ell_1 = 3$. Now appropriate multiples of the third equation are subtracted from the other equations so as to create 0's as coefficients for x_1 in each of those equations. Explicitly, $\frac{1}{2}$ times row 3 is subtracted from row 1, -1 times row 3 is subtracted from row 2, and 2 times row 3 is subtracted from row 4. The result is

$$
\begin{bmatrix}
0 & -12 & 8 & 1 \\
0 & 2 & 3 & -14 \\
6 & -2 & 2 & 4 \\
0 & -4 & 2 & 2
\end{bmatrix}
\begin{bmatrix}
x_1 \\
x_2 \\
x_3 \\
x_4
\end{bmatrix}
=
\begin{bmatrix}
-27 \\
-18 \\
16 \\
-6
\end{bmatrix}
$$

In the next step ($k = 2$), we use the index vector $\ell = [3, 2, 1, 4]$ and scan the ratios

$$
\left\{ \frac{|a_{\ell_i,2}|}{s_{\ell_i}} : i = 2, 3, 4 \right\} = \left\{ \frac{2}{18}, \frac{12}{13}, \frac{4}{12} \right\}
$$

looking for the largest value. Hence, we set $j = 3$ and interchange ℓ_k with ℓ_j in the index vector. Thus, the index vector becomes $\ell = [3, 1, 2, 4]$. The pivot equation for step 2 in the elimination is now $\ell_2 = 1$. Next, multiples of the first equation are subtracted from the second equation and the fourth equation. The appropriate multiples are $-\frac{1}{6}$ and $\frac{1}{3}$, respectively. The result is

$$
\begin{bmatrix}
0 & -12 & 8 & 1 \\
0 & 0 & \frac{13}{3} & -\frac{83}{6} \\
6 & -2 & 2 & 4 \\
0 & 0 & -\frac{2}{3} & \frac{5}{3}
\end{bmatrix}
\begin{bmatrix}
x_1 \\
x_2 \\
x_3 \\
x_4
\end{bmatrix}
=
\begin{bmatrix}
-27 \\
-\frac{45}{2} \\
16 \\
3
\end{bmatrix}
$$

The third and final step ($k = 3$) is to examine the ratios

$$
\left\{ \frac{|a_{\ell_i,3}|}{s_{\ell_i}} : i = 3, 4 \right\} = \left\{ \frac{13/3}{18}, \frac{2/3}{12} \right\}
$$

with the index vector $\ell = [3, 1, 2, 4]$. The larger value is the first, so we set $j = 3$. Since this is step 3, interchanging ℓ_k with ℓ_j leaves the index vector as before, $\ell = [3, 1, 2, 4]$. The pivot equation is $\ell_3 = 2$, and we subtract $-\frac{2}{13}$ times the second equation from the fourth equation. So the forward elimination phase ends with the final system

$$
\begin{bmatrix}
0 & -12 & 8 & 1 \\
0 & 0 & \frac{13}{3} & -\frac{83}{6} \\
6 & -2 & 2 & 4 \\
0 & 0 & 0 & -\frac{6}{13}
\end{bmatrix}
\begin{bmatrix}
x_1 \\
x_2 \\
x_3 \\
x_4
\end{bmatrix}
=
\begin{bmatrix}
-27 \\
-\frac{45}{2} \\
16 \\
-\frac{6}{13}
\end{bmatrix}
$$

The order in which the pivot equations were selected is displayed in the final index vector $\ell = [3, 1, 2, 4]$.

Now, reading the entries in the index vector from the last to the first, we have the order in which the back substitution is to be performed. The solution is obtained by using equation $\ell_4 = 4$ to determine x_4, and then equation $\ell_3 = 2$ to find x_3, and so on. Clearly, we have

$$x_4 = \frac{1}{-6/13}[-6/13] = 1$$

$$x_3 = \frac{1}{13/3}[(-45/2) + (83/6)(1)] = -2$$

$$x_2 = \frac{1}{-12}[-27 - 8(-2) - 1(1)] = 1$$

$$x_1 = \frac{1}{6}[16 + 2(1) - 2(-2) - 4(1)] = 3$$

Hence, the solution is

$$x = \begin{bmatrix} 3 \\ 1 \\ -2 \\ 1 \end{bmatrix}$$

Pseudocode

The algorithm as it will be programmed carries out the forward elimination phase on the coefficient array (a_{ij}) only. The right-hand-side array (b_i) is treated in the next phase. This method is adopted because it is more efficient if several systems must be solved with the same array (a_{ij}) but differing (b_i)'s. Because we wish to treat (b_i) later, it is necessary to store not only the index array but also the various multipliers that are used. These multipliers are conveniently stored in array (a_{ij}) in the positions where the 0 entries would have been created.

We are now ready to write a procedure for forward elimination with scaled partial pivoting. Our approach is to modify procedure *ngauss* of Section 6.1 by introducing scaling and indexing arrays. The procedure that carries out Gaussian elimination with scaled partial pivoting on the square array (a_{ij}) is called *gauss*. Its calling sequence is $(n, (a_{ij}), (\ell_i))$, where (a_{ij}) is the $n \times n$ coefficient array and (ℓ_i) is the index array for ℓ. In the pseudocode, (s_i) is the scale array.

```
procedure gauss(n, (aij), (ℓi))
real array (aij)n×n, (ℓi)n
real array allocate (si)n
integer i, j, k, n
real r, rmax, smax, xmult
for i = 1 to n do
    ℓi ← i
    smax ← 0
    for j = 1 to n do
        smax ← max(smax, |aij|)
    end do
    si ← smax
end do
```

```
for k = 1 to n − 1 do
    rmax ← 0
    for i = k to n do
        r ← |a_{ℓ_i,k}/s_{ℓ_i}|
        if (r > rmax) then
            rmax ← r
            j ← i
        end if
    end do
    ℓ_j ↔ ℓ_k
    for i = k + 1 to n do
        xmult ← a_{ℓ_i,k}/a_{ℓ_k,k}
        a_{ℓ_i,k} ← xmult
        for j = k + 1 to n do
            a_{ℓ_i,j} ← a_{ℓ_i,j} − (xmult)a_{ℓ_k,j}
        end do
    end do
end do
end procedure gauss
```

A detailed explanation of the above procedure is now presented. In the first loop, the initial form of the index array is being established—namely, $\ell_i = i$. Then the scale array (s_i) is computed.

The statement **for** $k = 1$ **to** $n − 1$ **do** initiates the principal outer loop. The index k is the subscript of the variable whose coefficients will be made 0 in the array (a_{ij}); that is, k is the index of the column in which new 0's are to be created. Remember that the 0's in the array (a_{ij}) do not actually appear because those storage locations are used for the multipliers. This fact can be seen in the line of the procedure where *xmult* is stored in the array (a_{ij}).

Once k has been set, the first task is to select the correct pivot row, which is done by computing $|a_{\ell_i,k}|/s_{\ell_i}$ for $i = k, k + 1, \ldots, n$. The next set of lines in the pseudocode are calculating this greatest ratio, called *rmax* in the routine, and the index j where it occurs. Next, ℓ_k and ℓ_j are interchanged in the array (ℓ_i).

The arithmetic modifications in the array (a_{ij}) due to subtracting multiples of row ℓ_k from rows $\ell_{k+1}, \ell_{k+2}, \ldots, \ell_n$ all occur in the final lines. First the multiplier is computed and stored; then the subtraction occurs in a loop.

Caution: Values in array (a_{ij}) that result as *output* from procedure *gauss* are not the same as those in array (a_{ij}) at *input*. If the original array must be retained, one should store a duplicate of it in another array.

In the procedure *ngauss* for naive Gaussian elimination from Section 6.1, the right-hand side b was modified during the forward elimination phase; however, this was not done in the procedure *gauss*. Therefore, we need to update b before considering the back substitution phase. For simplicity, we discuss updating b for the naive forward elimination first. Stripping out the pseudocode from *ngauss* that involves the (b_i) array in the forward elimination phase, we obtain

real array $(a_{ij})_{n\times n}, (b_i)_n$
integer i, k, n

$\vdots$

for $k = 1$ **to** $n - 1$ **do**
   for $i = k + 1$ **to** n **do**
      $b_i = b_i - a_{ik}b_k$
   end do
end do

This updates the (b_i) array based on the stored multipliers from the (a_{ij}) array. When scaled partial pivoting is done in the forward elimination phase, such as in procedure *gauss*, the multipliers for each step are not one below another in the (a_{ij}) array but are jumbled around. To unravel this situation, all we have to do is introduce the index array (ℓ_i) into the above pseudocode:

real array $(a_{ij})_{n\times n}, (b_i)_n, (\ell_i)_n$
integer i, k, n

$\vdots$

for $k = 1$ **to** $n - 1$ **do**
   for $i = k + 1$ **to** n **do**
      $b_{\ell_i} = b_{\ell_i} - a_{\ell_i k}b_{\ell_k}$
   end do
end do

After the array b has been processed in the forward elimination, the back substitution process is carried out. It begins by solving the equation

$$a_{\ell_n,n}x_n = b_{\ell_n} \tag{5}$$

whence

$$x_n = \frac{b_{\ell_n}}{a_{\ell_n,n}}$$

Then the equation

$$a_{\ell_{n-1},n-1}x_{n-1} + a_{\ell_{n-1},n}x_n = b_{\ell_{n-1}}$$

is solved for x_{n-1}:

$$x_{n-1} = \frac{b_{\ell_{n-1}} - a_{\ell_{n-1},n}x_n}{a_{\ell_{n-1},n-1}}$$

After $x_n, x_{n-1}, \ldots, x_{i+1}$ have been determined, x_i is found from the equation

$$a_{\ell_i,i}x_i + a_{\ell_i,i+1}x_{i+1} + \cdots + a_{\ell_i,n}x_n = b_{\ell_i}$$

whose solution is

$$x_i = \frac{1}{a_{\ell_i,i}} \left(b_{\ell_i} - \sum_{j=i+1}^{n} a_{\ell_i,j} x_j \right) \qquad (6)$$

Except for the presence of the index array ℓ_i, this is similar to the back substitution formula obtained for naive Gaussian elimination.

The procedure for processing the array b and performing the back substitution phase is given next:

procedure *solve*$(n, (a_{ij}), (\ell_i), (b_i), (x_i))$
real array $(a_{ij})_{n \times n}, (\ell_i)_n, (b_i)_n, (x_i)_n$
integer i, k, n
real *sum*
for $k = 1$ **to** $n - 1$ **do**
  for $i = k + 1$ **to** n **do**
    $b_{\ell_i} \leftarrow b_{\ell_i} - a_{\ell_i,k} b_{\ell_k}$
  end do
end do
$x_n \leftarrow b_{\ell_n} / a_{\ell_n,n}$
for $i = n - 1$ **to** 1 **step** -1 **do**
  sum $\leftarrow b_{\ell_i}$
  for $j = i + 1$ **to** n **do**
    sum $\leftarrow$ *sum* $- a_{\ell_i,j} x_j$
  end do
  $x_i \leftarrow$ *sum*$/ a_{\ell_i,i}$
end do
end procedure *solve*

Here the first loop carries out the forward elimination process on array (b_i), using arrays (a_{ij}) and (ℓ_i) that result from procedure *gauss*. The next line carries out the solution of Equation (5). The final part carries out Equation (6). The variable *sum* is a temporary variable for accumulating the numerator in Equation (6).

As with most pseudocode in this book, those in this chapter contain only the basic ingredients for mathematical software. They are not suitable as *production* code for various reasons. For example, procedures for optimizing code are ignored. Furthermore, the procedures do not give warnings for difficulties that may be encountered, such as division by zero! *General* procedures should be **robust**; that is, anticipate every possible situation and deal with each in a prescribed way. (See Computer Problem **11**.)

Long Operation Count

Solving large systems of linear equations can be expensive on a computer. To understand why, let us perform an operation count on the two algorithms whose codes have been given. We count only multiplications and divisions (long operations) because they are more time consuming. Furthermore, we lump multiplications and divisions together even though division is slower than multiplication.

Consider first procedure *gauss*. In step 1, the choice of a pivot element requires the calculation of n ratios—that is, n divisions. Then for rows $\ell_2, \ell_3, \ldots, \ell_n$ we first compute a multiplier and then subtract from row ℓ_i that multiplier times row ℓ_1. The zero that is being created in this process is *not* computed. So the elimination requires $n - 1$ multiplications per row. If we include the calculation of the multiplier, there are n long operations (divisions or multiplications) per row. There are $n - 1$ rows to be processed for a total of $n(n - 1)$ operations. If we add the cost of computing the ratios, a total of n^2 operations is needed for step 1.

The next step is like step 1 except that row ℓ_1 is not affected; neither is the column of 0's created in step 1. So step 2 will require $(n - 1)^2$ multiplications or divisions because it operates on a system without row ℓ_1 and without column 1. Continuing this reasoning, we conclude that the total number of long operations for procedure *gauss* is

$$n^2 + (n - 1)^2 + (n - 2)^2 + \cdots + 4^2 + 3^2 + 2^2 = \frac{n}{6}(n + 1)(2n + 1) - 1 \approx \frac{n^3}{3}$$

(The derivation of this formula is outlined in Problem **15**.) Note that the number of long operations in this procedure grows like $n^3/3$, the dominant term.

Now consider procedure *solve*. The forward processing of the array (b_i) involves $n - 1$ steps. The first step contains $n - 1$ multiplications, the second $n - 2$ multiplications, and so on. The total of the forward processing of array (b_i) is thus

$$(n - 1) + (n - 2) + \cdots + 3 + 2 + 1 = \frac{n}{2}(n - 1)$$

(See Problem **14**.) In the back substitution procedure, one long operation is involved in the first step, two in the second step, and so on. The total is

$$1 + 2 + 3 + \cdots + n = \frac{n}{2}(n + 1)$$

Thus, procedure *solve* involves altogether n^2 long operations. To summarize:

LONG OPERATIONS

The forward elimination phase of the Gaussian elimination algorithm with scaled partial pivoting, if applied only to the $n \times n$ coefficient array, involves approximately $n^3/3$ long operations (multiplications or divisions). Solving for x requires an additional n^2 long operations.

Examples of Software Packages

There are a large number of computer programs and software packages for solving linear systems, each of which may use a slightly different pivoting strategy. We present just two—Maple and MATLAB.

A Maple program to solve the example of this section is as follows.

```
with(linalg):
A := array([[  3, -13,  9,   3, -19],
            [ -6,   4,  1, -18, -34],
            [  6,  -2,  2,   4,  16],
            [ 12,  -8,  6,  10,  26]]);
L := gausselim(A);
x := backsub(L);
```

Here the augmented matrix [A b] is used as the input array. Gaussian elimination with row pivoting on the augmented array A produces the array L:

$$
A := \begin{bmatrix} 3 & -13 & 9 & 3 & -19 \\ -6 & 4 & 1 & -18 & -34 \\ 6 & -2 & 2 & 4 & 16 \\ 12 & -8 & 6 & 10 & 26 \end{bmatrix}
$$

$$
L := \begin{bmatrix} 3 & -13 & 9 & 3 & -19 \\ 0 & -22 & 19 & -12 & -72 \\ 0 & 0 & 8 & -26 & -42 \\ 0 & 0 & 0 & 3/11 & 3/11 \end{bmatrix}
$$

```
x := [ 3, 1, -2, 1 ]
```

Repeating our example using MATLAB, we have the following input.

```
A = [ 3, -13,  9,   3
     -6,   4,  1, -18
      6,  -2,  2,   4
     12,  -8,  6,  10]

b = [-19
     -34
      16
      26]

x = A\b
```

MATLAB produces the following results.

```
x =
    3.0000
    1.0000
   -2.0000
    1.0000
```

In this latter example, we have not shown the forward elimination matrix. We will do so in Section 6.4.

**PROBLEMS
6.2**

1. Show how Gaussian elimination with scaled partial pivoting works on the systems in parts **a**, **b**, and **c** of Problem **1** in Section 6.1. What are the contents of the index array at each step?

2. Show how Gaussian elimination with scaled partial pivoting works on the following matrix A:

$$\begin{bmatrix} 2 & 3 & -4 & 1 \\ 1 & -1 & 0 & -2 \\ 3 & 3 & 4 & 3 \\ 4 & 1 & 0 & 4 \end{bmatrix}$$

3. Solve the following system using Gaussian elimination with scaled partial pivoting:

$$\begin{bmatrix} 1 & -1 & 2 \\ -2 & 1 & -1 \\ 4 & -1 & 2 \end{bmatrix} \begin{bmatrix} x_1 \\ x_2 \\ x_3 \end{bmatrix} = \begin{bmatrix} -2 \\ 2 \\ -1 \end{bmatrix}$$

Show intermediate matrices at each step.

4. Carry out Gaussian elimination with scaled partial pivoting on the matrix

$$\begin{bmatrix} 1 & 0 & 3 & 0 \\ 0 & 1 & 3 & -1 \\ 3 & -3 & 0 & 6 \\ 0 & 2 & 4 & -6 \end{bmatrix}$$

Show intermediate matrices.

5. Repeat Problem **7a** of Section 6.1 using Gaussian elimination with scaled partial pivoting.

6. Consider the matrix

$$\begin{bmatrix} -0.0013 & 56.4972 & 123.4567 & 987.6543 \\ 0. & -0.0145 & 8.8990 & 833.3333 \\ 0. & 102.7513 & -7.6543 & 69.6869 \\ 0. & -1.3131 & -9876.5432 & 100.0001 \end{bmatrix}$$

Identify the entry that will be used as the next pivot element of naive Gaussian elimination, of Gaussian elimination with partial pivoting (the scale vector is $[1., 1., 1., 1.]$), and of Gaussian elimination with scaled partial pivoting (the scale vector is $[987.6543, 46.79, 256.29, 1.096]$).

7. Without using the computer, determine the final contents of the array (a_{ij}) after procedure *gauss* has processed the following array. Indicate the multipliers by underlining.

$$\begin{bmatrix} 1 & 3 & 2 & 1 \\ 4 & 2 & 1 & 2 \\ 2 & 1 & 2 & 3 \\ 1 & 2 & 4 & 1 \end{bmatrix}$$

8. If the Gaussian elimination algorithm with scaled partial pivoting is used on the matrix shown, what is the scale vector? What is the second pivot row?

$$\begin{bmatrix} 4 & 7 & 3 \\ 1 & 3 & 2 \\ 2 & -4 & -1 \end{bmatrix}$$

9. If the Gaussian elimination algorithm with scaled partial pivoting is used on the example shown, which row will be selected as the third pivot row?

$$\begin{bmatrix} 8 & -1 & 4 & 9 & 2 \\ 1 & 0 & 3 & 9 & 7 \\ -5 & 0 & 1 & 3 & 5 \\ 4 & 3 & 2 & 2 & 7 \\ 3 & 0 & 0 & 0 & 9 \end{bmatrix}$$

10. Solve the system

$$\begin{cases} 2x_1 + 4x_2 - 2x_3 = 6 \\ x_1 + 3x_2 + 4x_3 = -1 \\ 5x_1 + 2x_2 \phantom{{}+ 4x_3} = 2 \end{cases}$$

using Gaussian elimination with scaled partial pivoting. Show intermediate results at each step and, in particular, display the scale and index vectors.

11. Consider the linear system

$$\begin{cases} 2x + 3y = 8 \\ -x + 2y - z = 0 \\ 3x + 2z = 9 \end{cases}$$

Solve for x, y, and z using Gaussian elimination with scaled partial pivoting. Show intermediate matrices and vectors.

12. Consider the linear system of equations

$$\begin{cases} -x_1 + x_2 & - 3x_4 = 4 \\ x_1 & + 3x_3 + x_4 = 0 \\ x_2 - x_3 & - x_4 = 3 \\ 3x_1 & + x_3 + 2x_4 = 1 \end{cases}$$

Solve this system using Gaussian elimination with scaled partial pivoting. Show all intermediate steps and write down the index vector at each step.

13. Consider Gaussian elimination with scaled partial pivoting applied to the coefficient matrix

$$\begin{bmatrix} x & x & x & x & 0 \\ x & x & x & 0 & x \\ 0 & x & x & x & 0 \\ 0 & x & 0 & x & 0 \\ x & 0 & 0 & x & x \end{bmatrix}$$

where each x denotes a different nonzero element. Circle the locations of elements in which multipliers will be stored and mark with an f those where fill-in will occur. The final index vector is $\ell = [2, 3, 1, 5, 4]$.

14. Derive the formula

$$\sum_{k=1}^{n} k = \frac{n}{2}(n + 1)$$

Hint: Set $S = \sum_{k=1}^{n} k$ and observe that

$$2S = (1 + 2 + \cdots + n) + [n + (n - 1) + \cdots + 2 + 1]$$
$$= (n + 1) + (n + 1) + \cdots$$

or use induction.

15. Derive the formula

$$\sum_{k=1}^{n} k^2 = \frac{n}{6}(n + 1)(2n + 1)$$

Hint: Induction is probably easiest.

16. Count the number of operations in the following pseudocode:

real array $(a_{ij})_{n \times n}, (x_{ij})_{n \times n}$
integer i, j, n
$\vdots$

real z
for $i = 1$ **to** n **do**
  for $j = 1$ **to** i **do**
    $z = z + a_{ij}x_{ij}$
  end do
end do

17. Count the number of divisions in procedure *gauss*. Count the number of multiplications. Count the number of additions or subtractions. Using execution times in microseconds (multiplication 1, division 2.9, addition 0.4, subtraction 0.4), write a function of n that represents the time used in these arithmetic operations.

18. Considering long operations only and assuming 1-microsecond execution time for all long operations, give the approximate execution times and costs for procedure *gauss* when $n = 10, 10^2, 10^3, 10^4$. Use only the dominant term in the operation count. Estimate costs at $500 per hour.

19. (Continuation) How much time would be used on the computer to solve 2000 equations using Gaussian elimination with scaled partial pivoting? How much would it cost? Give a rough estimate based on operation times.

20. After processing a matrix A by procedure *gauss*, how can the results be used to solve a system of equations of form $A^T x = b$?

21. What modifications would make procedure *gauss* more efficient if division were *much* slower than multiplication?

22. A matrix $A = (a_{ij})$ is **row-equilibrated** if it is scaled so that

$$\max_{1 \le j \le n} |a_{ij}| = 1 \qquad (1 \le i \le n)$$

In solving a system of equations, $Ax = b$, we can produce an equivalent system in which the matrix is row-equilibrated by dividing the ith equation by $\max_{1 \le j \le n} |a_{ij}|$.

a. Solve the system of equations

$$\begin{bmatrix} 1 & 1 & 2 \times 10^9 \\ 2 & -1 & 10^9 \\ 1 & 2 & 0 \end{bmatrix} \begin{bmatrix} x_1 \\ x_2 \\ x_3 \end{bmatrix} = \begin{bmatrix} 1 \\ 1 \\ 1 \end{bmatrix}$$

by Gaussian elimination with scaled partial pivoting.

b. Solve by using row-equilibrated naive Gaussian elimination. Are the answers the same? Why or why not?

COMPUTER PROBLEMS 6.2

1. Test the numerical example in the text using the naive Gaussian algorithm and the Gaussian algorithm with scaled partial pivoting.

2. Consider the system

$$\begin{bmatrix} 0.4096 & 0.1234 & 0.3678 & 0.2943 \\ 0.2246 & 0.3872 & 0.4015 & 0.1129 \\ 0.3645 & 0.1920 & 0.3781 & 0.0643 \\ 0.1784 & 0.4002 & 0.2786 & 0.3927 \end{bmatrix} \begin{bmatrix} x_1 \\ x_2 \\ x_3 \\ x_4 \end{bmatrix} = \begin{bmatrix} 0.4043 \\ 0.1550 \\ 0.4240 \\ 0.2557 \end{bmatrix}$$

Solve it by Gaussian elimination with scaled partial pivoting using procedures *gauss* and *solve*.

3. (Continuation) Assume that an error was made when the coefficient matrix in Computer Problem **2** was typed and that a single digit was mistyped—namely, 0.3645 became 0.3345. Solve this system and notice the effect of this small change. Explain.

4. The **Hilbert matrix** of order n is defined by $a_{ij} = (i + j - 1)^{-1}$ for $1 \leq i, j \leq n$. It is often used for test purposes because of its ill-conditioned nature. Define $b_i = \sum_{j=1}^{n} a_{ij}$. Then the solution of the system of equations $\sum_{j=1}^{n} a_{ij}x_j = b_i$ for $1 \leq i \leq n$ is $x = [1, 1, \ldots, 1]^T$. Verify this. Select some values of n in the range $2 \leq n \leq 15$, solve the system of equations for x using procedures *gauss* and *solve*, and see whether the result is as predicted. Do the case $n = 2$ by hand to see what difficulties occur in the computer.

5. Define the $n \times n$ array (a_{ij}) by $a_{ij} = -1 + 2 \max\{i, j\}$. Set up array (b_i) in such a way that the solution of the system $Ax = b$ is $x_i = 1$ for $1 \leq i \leq n$. Test procedures *gauss* and *solve* on this system for a moderate value of n, say $n = 30$.

6. Select a modest value of n, say $5 \leq n \leq 20$, and let $a_{ij} = (i - 1)^{j-1}$ and $b_i = i - 1$. Solve the system $Ax = b$ on the computer. By looking at the output, guess what the correct solution is. Establish algebraically that your guess is correct. Account for the errors in the computed solution.

7. For a fixed value of n from 2 to 4, let

$$a_{ij} = (i + j)^2 \quad \text{and} \quad b_i = ni(i + n + 1) + \frac{1}{6}n(1 + n(2n + 3))$$

Show that the vector $x = [1, 1, \ldots, 1]^T$ solves the system $Ax = b$. Test whether procedures *gauss* and *solve* can compute x correctly for $n = 2, 3, 4$. Explain what happens.

8. Using each value of n from 2 to 9, solve the $n \times n$ system $Ax = b$, where A and b are defined by

$$a_{ij} = (i + j - 1)^7 \quad \text{and} \quad b_i = p(n + i - 1) - p(i - 1)$$

where

$$p(x) = \frac{x^2}{24}(2 + x^2(-7 + n^2(14 + n(12 + 3n))))$$

Explain what happens.

9. Solve the following system using procedures *gauss* and *solve* and then using procedure *ngauss*. Compare the results and explain.

$$\begin{bmatrix} 2.908 & -2.253 & 6.775 & 3.970 \\ 1.212 & 1.995 & 2.266 & 8.008 \\ 4.552 & 5.681 & 8.850 & 1.302 \\ 5.809 & -5.030 & 0.099 & 7.832 \end{bmatrix} \begin{bmatrix} x_1 \\ x_2 \\ x_3 \\ x_4 \end{bmatrix} = \begin{bmatrix} 6.291 \\ 7.219 \\ 5.730 \\ 9.574 \end{bmatrix}$$

10. Without changing the parameter list, rewrite and test procedure *gauss* so that it does both forward elimination and back substitution. Increase the size of array (a_{ij}) and store the right-hand-side array (b_i) in the $n + 1$st column of (a_{ij}). Also, return the solution in this column.

11. Modify procedures *gauss* and *solve* so that they are more robust software. Two suggested changes are as follows: (i) Skip elimination if $a_{\ell_i,k} = 0$ and (ii) add an error parameter *ierr* to the parameter list and perform error checking (e.g., on division by zero or a row of zeros). Test the modified code on linear systems of varying sizes.

12. Rewrite procedures *gauss* and *solve* so that they are column oriented—that is, so that all inner loops vary the first index of (a_{ij}). On some computer systems, this implementation may avoid paging or swapping between high-speed and secondary memory and be more efficient for large matrices.

13. Computer memory can be minimized by using a different storage mode when the coefficient matrix is symmetric. An $n \times n$ symmetric matrix $A = (a_{ij})$ has the property that $a_{ij} = a_{ji}$ so that only the elements on and below the main diagonal need be stored in a vector of length $n(n + 1)/2$. The elements of the matrix A are placed in a vector $v = (v_k)$ in this order: $a_{11}, a_{21}, a_{22}, a_{31}, a_{32}, a_{33}, \ldots, a_{n,n}$. Storing a matrix in this way is known as **symmetric storage mode** and effects a savings of $n(n - 1)/2$ memory locations. Here $a_{ij} = v_k$, where $k = \frac{1}{2}i(i - 1) + j$ for $i \geq j$. Verify these statements.

    Write and test procedures *gausym*$(n, (v_i), (\ell_i))$ and *solsym*$(n, (v_i), (\ell_i), (b_i))$, which are analogous to procedures *gauss* and *solve* except that the coefficient matrix is stored in symmetric storage mode in a one-dimensional array (v_i) and the solution is returned in array (b_i).

14. The *determinant* of a square matrix can be easily computed with the help of procedure *gauss*. We require three facts about determinants. First, the determinant of a triangular matrix is the product of the elements on its diagonal. Second, if a multiple of one row is added to another row, the determinant of the matrix does not change. Third, if two rows in a matrix are interchanged, the determinant changes sign. Procedure *gauss* can be *interpreted* as a procedure for reducing a matrix to upper triangular form by interchanging rows and adding multiples of one row to another. Write a function det$(n, (a_{ij}))$ that computes the determinant of an $n \times n$ matrix. It will call procedure *gauss* and utilize the arrays (a_{ij}) and (ℓ_i) that result from that call. Numerically verify procedure det by using the following test matrices with several values of n:

    a. $a_{ij} = |i - j|$ $\det(A) = (-1)^{n-1}(n - 1)2^{n-2}$

b. $a_{ij} = \begin{cases} 1 & j \geq i \\ -j & j < i \end{cases}$ $\det(A) = n!$

c. $\begin{cases} a_{ij} = a_{j1} = n^{-1} & j \geq 1 \\ a_{ij} = a_{i-1,j} + a_{i,j-1} & i,j \geq 2 \end{cases}$ $\det(A) = n^{-n}$

15. (Continuation) Overflow and underflow may occur in evaluating determinants by this procedure. To avoid this, one can compute $\log |\det(A)|$ as the sum of terms $\log |a_{\ell,i}|$ and use the exponential function at the end. Repeat the numerical experiments in Computer Problem **14** using this idea.

16. Test a modification of procedure *gauss* in which the scale array is recomputed at each step (each new value of k) of the forward elimination phase. Try to construct an example for which this procedure would produce less roundoff error than the scaled partial pivoting method given in the text with fixed scale array. It is generally believed that the extra computations involved in this procedure are not worthwhile for most linear systems.

17. (Continuation) Modify and test procedure *gauss* so that the original system is initially row-equilibrated; that is, it is scaled so that the maximum element in every row is 1.

18. Modify and test procedures *gauss* and *solve* so that they carry out scaled *complete* pivoting; that is, the pivot element is selected from all elements in the submatrix, not just those in the kth column. Keep track of the order of the unknowns in the solution array in another index array because they will not be determined in the order $x_n, x_{n-1}, \ldots, x_1$.

19. Compare the computed numerical solutions of the following two linear systems:

$$\begin{bmatrix} 1 & \frac{1}{2} & \frac{1}{3} & \frac{1}{4} & \frac{1}{5} \\ \frac{1}{2} & \frac{1}{3} & \frac{1}{4} & \frac{1}{5} & \frac{1}{6} \\ \frac{1}{3} & \frac{1}{4} & \frac{1}{5} & \frac{1}{6} & \frac{1}{7} \\ \frac{1}{4} & \frac{1}{5} & \frac{1}{6} & \frac{1}{7} & \frac{1}{8} \\ \frac{1}{5} & \frac{1}{6} & \frac{1}{7} & \frac{1}{8} & \frac{1}{9} \end{bmatrix} \begin{bmatrix} x_1 \\ x_2 \\ x_3 \\ x_4 \\ x_5 \end{bmatrix} = \begin{bmatrix} 1 \\ 0 \\ 0 \\ 0 \\ 0 \end{bmatrix}$$

$$\begin{bmatrix} 1.0 & 0.5 & 0.333333 & 0.25 & 0.2 \\ 0.5 & 0.333333 & 0.25 & 0.2 & 0.166667 \\ 0.333333 & 0.25 & 0.2 & 0.166667 & 0.142857 \\ 0.25 & 0.2 & 0.166667 & 0.142857 & 0.125 \\ 0.2 & 0.166667 & 0.142857 & 0.125 & 0.111111 \end{bmatrix} \begin{bmatrix} x_1 \\ x_2 \\ x_3 \\ x_4 \\ x_5 \end{bmatrix} = \begin{bmatrix} 1 \\ 0 \\ 0 \\ 0 \\ 0 \end{bmatrix}$$

Compute the residual vector $\tilde{r} = A\tilde{x} - b$ and the error vector $\tilde{e} = \tilde{x} - x$, where $\tilde{x}$ is the computed solution and x is the true, or exact, solution. For the first system, the exact solution is $x = [25, -300, 1050, -1400, 630]^T$. To obtain the exact solution of the second system, solve it using double-precision arithmetic throughout. Do not change the input data of the second system to include more than the number of digits shown.

20. Repeat Computer Problem **19** but set $a_{ij} \leftarrow 7560a_{ij}$ and $b_i \leftarrow 7560b_i$ for each system before solving.

21. The fact that in Problems **6** and **7** of Section 6.1, solutions of complex linear systems were asked for may lead you to think that you *must* have complex versions of procedures *gauss* and *solve*. This is not the case. A complex system $Ax = b$ can also be written as a $2n \times 2n$ real system:

$$\sum_{j=1}^{n} \left[\Re(a_{ij})\Re(x_j) - \Im(a_{ij})\Im(x_j) \right] = \Re(b_i) \qquad (1 \leq i \leq n)$$

$$\sum_{j=1}^{n} \left[\Re(a_{ij})\Im(x_j) + \Im(a_{ij})\Re(x_j) \right] = \Im(b_i) \qquad (1 \leq i \leq n)$$

Repeat these two problems using this idea and the two procedures of this section. (Here $\Re$ denotes the real part and $\Im$ the imaginary part.)

6.3 Tridiagonal and Banded Systems

In many applications, including several considered later on, linear systems that have a **banded** structure are encountered. Banded matrices often occur in solving ordinary and partial differential equations.

Of practical importance is the **tridiagonal** system. Here all the nonzero elements in the coefficient matrix must be on the main diagonal or on the two diagonals just above and below the main diagonal (usually called **superdiagonal** and **subdiagonal**, respectively).

$$
\begin{bmatrix}
d_1 & c_1 & & & & & & \\
a_1 & d_2 & c_2 & & & & & \\
& a_2 & d_3 & c_3 & & & & \\
& & \ddots & \ddots & \ddots & & & \\
& & & a_{i-1} & d_i & c_i & & \\
& & & & \ddots & \ddots & \ddots & \\
& & & & & a_{n-2} & d_{n-1} & c_{n-1} \\
& & & & & & a_{n-1} & d_n
\end{bmatrix}
\begin{bmatrix}
x_1 \\ x_2 \\ x_3 \\ \vdots \\ x_i \\ \vdots \\ x_{n-1} \\ x_n
\end{bmatrix}
=
\begin{bmatrix}
b_1 \\ b_2 \\ b_3 \\ \vdots \\ b_i \\ \vdots \\ b_{n-1} \\ b_n
\end{bmatrix}
\qquad (1)
$$

(All elements not in the displayed diagonals are 0's.) A tridiagonal matrix is characterized by the condition $a_{ij} = 0$ if $|i - j| \geq 2$. In general, a matrix is said to have a **banded structure** if there is an integer k (less than n) such that $a_{ij} = 0$ whenever $|i - j| \geq k$.

The storage requirements for a banded matrix are less than those for a general matrix of the same size. Thus, an $n \times n$ diagonal matrix requires only n memory locations in the computer, and a tridiagonal matrix requires only $3n - 2$. This fact is important if banded matrices of very large order are being used.

For banded matrices, the Gaussian elimination algorithm can be made very efficient if it is known beforehand that pivoting is unnecessary. This situation occurs often enough to justify special procedures. Here we develop a code for the tridiagonal system and give a listing for the *pentadiagonal* system (in which $a_{ij} = 0$ if $|i - j| \geq 3$).

Tridiagonal Systems

The routine to be described now is called procedure *tri*. It is designed to solve a system of n linear equations in n unknowns, as shown in Equation (1). Both the forward elimination phase and the back substitution phase are incorporated in the procedure, and *no* pivoting is used; that is, the pivot equations are those given by the natural ordering $\{1, 2, \ldots, n\}$. Thus, naive Gaussian elimination is used.

In step 1, we subtract a_1/d_1 times row 1 from row 2, thus creating a 0 in the a_1 position. Only the entries b_2 and d_2 are altered. Observe that c_2 is *not* altered. In step 2 the process is repeated, using the new row 2 as the pivot row. Here is how the d_i's and b_i's are altered in each step:

$$
\begin{cases}
d_2 \leftarrow d_2 - \left(\dfrac{a_1}{d_1}\right)c_1 \\[3mm]
b_2 \leftarrow b_2 - \left(\dfrac{a_1}{d_1}\right)b_1
\end{cases}
$$

In general,

$$
\begin{cases}
d_i \leftarrow d_i - \left(\dfrac{a_{i-1}}{d_{i-1}}\right)c_{i-1} \\[3mm]
b_i \leftarrow b_i - \left(\dfrac{a_{i-1}}{d_{i-1}}\right)b_{i-1} \qquad (2 \leq i \leq n)
\end{cases}
$$

At the end of the forward elimination phase, the form of the system is as follows:

$$
\begin{bmatrix}
d_1 & c_1 & & & & & \\
 & d_2 & c_2 & & & & \\
 & & d_3 & c_3 & & & \\
 & & & \ddots & \ddots & & \\
 & & & & d_i & c_i & \\
 & & & & & \ddots & \ddots \\
 & & & & & & d_{n-1} & c_{n-1} \\
 & & & & & & & d_n
\end{bmatrix}
\begin{bmatrix}
x_1 \\ x_2 \\ x_3 \\ \vdots \\ x_i \\ \vdots \\ x_{n-1} \\ x_n
\end{bmatrix}
=
\begin{bmatrix}
b_1 \\ b_2 \\ b_3 \\ \vdots \\ b_i \\ \vdots \\ b_{n-1} \\ b_n
\end{bmatrix}
$$

Of course, the b_i's and d_i's are not as they were at the beginning of this process, but the c_i's are. The back substitution phase solves for $x_n, x_{n-1}, \ldots, x_1$ as follows:

$$x_n \leftarrow \frac{b_n}{d_n}$$

$$x_{n-1} \leftarrow \frac{b_{n-1} - c_{n-1}x_n}{d_{n-1}}$$

In general,

$$x_i \leftarrow \frac{b_i - c_i x_{i+1}}{d_i} \qquad (i = n - 1, n - 2, \ldots, 1)$$

In procedure *tri* for a tridiagonal system, we use single-dimensioned arrays (a_i), (d_i), and (c_i) for the diagonals in the coefficient matrix and array (b_i) for the right-hand side, and store the solution in array (x_i).

procedure *tri*$(n, (a_i), (d_i), (c_i), (b_i), (x_i))$
real array $(a_i)_n, (d_i)_n, (c_i)_n, (b_i)_n, (x_i)_n$
integer i, n
real *xmult*
for $i = 2$ **to** n **do**
   xmult $\leftarrow a_{i-1}/d_{i-1}$
   $d_i \leftarrow d_i - (xmult)c_{i-1}$
   $b_i \leftarrow b_i - (xmult)b_{i-1}$
end do
$x_n \leftarrow b_n/d_n$
for $i = n - 1$ **to** 1 **step** -1 **do**
   $x_i \leftarrow (b_i - c_i x_{i+1})/d_i$
end do
end procedure *tri*

A symmetric tridiagonal system arises in the cubic spline development of Chapter 7 and elsewhere. A general symmetric tridiagonal system has the form

$$\begin{bmatrix} d_1 & c_1 \\ c_1 & d_2 & c_2 \\ & c_2 & d_3 & c_3 \\ & & & \ddots & \ddots & \ddots \\ & & & & c_{i-1} & d_i & c_i \\ & & & & & \ddots & \ddots & \ddots \\ & & & & & & c_{n-2} & d_{n-1} & c_{n-1} \\ & & & & & & & c_{n-1} & d_n \end{bmatrix} \begin{bmatrix} x_1 \\ x_2 \\ x_3 \\ \vdots \\ x_i \\ \vdots \\ x_{n-1} \\ x_n \end{bmatrix} = \begin{bmatrix} b_1 \\ b_2 \\ b_3 \\ \vdots \\ b_i \\ \vdots \\ b_{n-1} \\ b_n \end{bmatrix} \qquad (2)$$

Such a system can be solved with a procedure call of the form

$$\textbf{call } tri(n, (c_i), (d_i), (c_i), (b_i), (x_i))$$

because *tri* arrays (a_i) and (c_i) are not changed and do not interact.

Diagonal Dominance

Since procedure *tri* does not involve pivoting, it is natural to ask whether it is likely to fail. Simple examples can be given to illustrate failure because of attempted division by zero even though the coefficient matrix in Equation (1) is nonsingular. On the other hand, it is not easy to give the weakest possible conditions on this matrix to guarantee the success of the algorithm. We content ourselves with one property that is easily checked and commonly encountered. If the tridiagonal coefficient matrix is diagonally dominant, then procedure *tri* will not encounter zero divisors. For a general matrix $A = (a_{ij})$, **diagonal dominance** is the condition:

$$|a_{ii}| > \sum_{\substack{j \neq i \\ j=1}}^{n} |a_{ij}| \qquad (1 \leq i \leq n)$$

In the case of the tridiagonal system of Equation (1), diagonal dominance means simply that

$$|d_i| > |c_i| + |a_{i-1}| \qquad (1 \leq i \leq n)$$

Let us verify that the forward elimination phase in procedure *tri* preserves diagonal dominance. The new coefficient matrix produced by Gaussian elimination has 0 elements where the a_i's originally stood, and new diagonal elements are determined recursively by

$$\widehat{d}_1 = d_1$$

$$\widehat{d}_i = d_i - c_{i-1}\left(\frac{a_{i-1}}{\widehat{d}_{i-1}}\right) \qquad (2 \leq i \leq n)$$

where $\widehat{d}_i$ denotes a new diagonal element. The c_i elements are unaltered. Now we assume that $|d_i| > |a_{i-1}| + |c_i|$, and we want to be sure that $|\widehat{d}_i| > |c_i|$. Obviously this is true for $i = 1$ because $\widehat{d}_1 = d_1$. If it is true for index $i - 1$ (that is, $|\widehat{d}_{i-1}| > |c_{i-1}|$) then it is true for index i because

$$|\widehat{d}_i| = \left|d_i - c_{i-1}\left(\frac{a_{i-1}}{\widehat{d}_{i-1}}\right)\right|$$

$$\geq |d_i| - |a_{i-1}|\frac{|c_{i-1}|}{|\widehat{d}_{i-1}|}$$

$$> |a_{i-1}| + |c_i| - |a_{i-1}| = |c_i|$$

Pentadiagonal Systems

The principles illustrated by procedure *tri* can be applied to matrices that have wider bands of nonzero elements. A procedure called *penta* is given here to solve the five-diagonal system:

$$
\begin{bmatrix}
d_1 & c_1 & f_1 \\
a_1 & d_2 & c_2 & f_2 \\
e_1 & a_2 & d_3 & c_3 & f_3 \\
& e_2 & a_3 & d_4 & c_4 & f_4 \\
& & \ddots & \ddots & \ddots & \ddots & \ddots \\
& & & e_{i-2} & a_{i-1} & d_i & c_i & f_i \\
& & & & \ddots & \ddots & \ddots & \ddots & \ddots \\
& & & & & e_{n-4} & a_{n-3} & d_{n-2} & c_{n-2} & f_{n-2} \\
& & & & & & e_{n-3} & a_{n-2} & d_{n-1} & c_{n-1} \\
& & & & & & & e_{n-2} & a_{n-1} & d_n
\end{bmatrix}
\begin{bmatrix}
x_1 \\ x_2 \\ x_3 \\ x_4 \\ \vdots \\ x_i \\ \vdots \\ x_{n-2} \\ x_{n-1} \\ x_n
\end{bmatrix}
=
\begin{bmatrix}
b_1 \\ b_2 \\ b_3 \\ b_4 \\ \vdots \\ b_i \\ \vdots \\ b_{n-2} \\ b_{n-1} \\ b_n
\end{bmatrix}
$$

In the pseudocode, the solution vector is placed in array (x_i). Also, the user should not use this routine if $n \leq 2$. (Why?)

procedure *penta*$(n, (e_i), (a_i), (d_i), (c_i), (f_i), (b_i), (x_i))$
real array $(e_i)_n, (a_i)_n, (d_i)_n, (c_i)_n, (f_i)_n, (b_i)_n, (x_i)_n$
integer i, n
real *xmult*
for $i = 2$ **to** $n - 1$ **do**
  $xmult \leftarrow a_{i-1}/d_{i-1}$
  $d_i \leftarrow d_i - (xmult)c_{i-1}$
  $c_i \leftarrow c_i - (xmult)f_{i-1}$
  $b_i \leftarrow b_i - (xmult)b_{i-1}$
  $xmult \leftarrow e_{i-1}/d_{i-1}$
  $a_i \leftarrow a_i - (xmult)c_{i-1}$
  $d_{i+1} \leftarrow d_{i+1} - (xmult)f_{i-1}$
  $b_{i+1} \leftarrow b_{i+1} - (xmult)b_{i-1}$
end do
$xmult \leftarrow a_{n-1}/d_{n-1}$
$d_n \leftarrow d_n - (xmult)c_{n-1}$
$x_n \leftarrow (b_n - (xmult)b_{n-1})/d_n$
$x_{n-1} \leftarrow (b_{n-1} - c_{n-1}x_n)/d_{n-1}$
for $i = n - 2$ **to** 1 **step** -1 **do**
  $x_i \leftarrow (b_i - f_i x_{i+2} - c_i x_{i+1})/d_i$
end do
end procedure *penta*

PROBLEMS 6.3

1. Count the long arithmetic operations involved in procedures:

   a. *tri*

   b. *penta*

2. How many storage locations are needed for a system of n linear equations if the coefficient matrix has banded structure in which $a_{ij} = 0$ for $|i - j| \geq k + 1$?

3. Give an example of a system of linear equations in tridiagonal form that cannot be solved without pivoting.

4. What is the appearance of a matrix A if its elements satisfy $a_{ij} = 0$ when:

a. $j < i - 2$

b. $j > i + 1$

5. Consider a diagonally dominant matrix A whose elements satisfy $a_{ij} = 0$ when $i > j + 1$. Does Gaussian elimination without pivoting preserve the diagonal dominance? Why or why not?

6. Let A be a matrix of form (1) such that $a_i c_i > 0$ for $1 \leq i \leq n - 1$. Find the general form of the diagonal matrix $D = \text{diag}(\alpha_i)$ with $\alpha_i \neq 0$ such that $D^{-1}AD$ is symmetric. What is the general form of $D^{-1}AD$?

7. What happens to the tridiagonal System (1) if Gaussian elimination with partial pivoting is used to solve it? In general, what happens to a banded system?

COMPUTER PROBLEMS 6.3

1. Rewrite procedure *tri* using only four arrays (a_i), (d_i), (c_i), and (b_i), storing the solution in the (b_i) array. Test the code with both a nonsymmetric and a symmetric tridiagonal system.

2. Repeat Problem **1** for procedure *penta* with six arrays (e_i), (a_i), (d_i), (c_i), (f_i), and (b_i). Use the example that begins this chapter as one of the test cases.

3. Write and test a special procedure to solve the tridiagonal system in which $a_i = c_i = 1$ for all i.

4. Use procedure *tri* to solve the following system of 100 equations. Compare the numerical solution to the obvious exact solution.

$$\begin{cases} x_1 & + \ 0.5x_2 & = 1.5 \\ 0.5x_{i-1} \ + & x_i \ + \ 0.5x_{i+1} & = 2.0 \qquad (2 \leq i \leq 99) \\ & 0.5x_{99} \ + \quad x_{100} & = 1.5 \end{cases}$$

What are the entries d_i after the algorithm is applied?

5. Solve the system

$$\begin{cases} 4x_1 & - \quad x_2 & = -20 \\ x_{j-1} \ - \ 4x_j & + \ x_{j+1} & = \quad 40 \qquad (2 \leq j \leq n - 1) \\ & - \ x_{n-1} \ + \ 4x_n & = -20 \end{cases}$$

using procedure *tri* with $n = 100$.

6. Let A be the 50×50 tridiagonal matrix

$$\begin{bmatrix} 5 & -1 \\ -1 & 5 & -1 \\ & -1 & 5 & -1 \\ & & \ddots & \ddots & \ddots \\ & & & -1 & 5 & -1 \\ & & & & -1 & 5 \end{bmatrix}$$

Consider the problem $Ax = b$ for 50 different vectors b of the form

$$[1, 2, \ldots, 49, 50]^T \qquad [2, 3, \ldots, 50, 1]^T \qquad [3, 4, \ldots, 50, 1, 2]^T \qquad \cdots$$

Write and test an efficient code for solving this problem. *Hint*: Rewrite procedure *tri*.

7. Rewrite and test procedure *tri* so that it performs Gaussian elimination with scaled partial pivoting. *Hint*: Additional temporary storage arrays may be needed.

8. Rewrite and test *penta* so that it does Gaussian elimination with scaled partial pivoting. Is this worthwhile?

9. Using the ideas illustrated in *penta*, write a procedure for solving seven-diagonal systems. Test it on several such systems.

10. Consider the system of equations ($n = 7$)

$$
\begin{bmatrix}
d_1 & & & & & & a_7 \\
 & d_2 & & & & a_6 & \\
 & & d_3 & a_5 & & & \\
 & & & d_4 & & & \\
 & & a_3 & d_5 & & & \\
 & a_2 & & & d_6 & & \\
a_1 & & & & & d_7
\end{bmatrix}
\begin{bmatrix}
x_1 \\ x_2 \\ x_3 \\ x_4 \\ x_5 \\ x_6 \\ x_7
\end{bmatrix}
=
\begin{bmatrix}
b_1 \\ b_2 \\ b_3 \\ b_4 \\ b_5 \\ b_6 \\ b_7
\end{bmatrix}
$$

For n odd, write and test

procedure $xgauss(n, (a_i), (d_i), (b_i))$

that does the forward elimination phase of Gaussian elimination (without scaled partial pivoting) and

procedure $xsolve(n, (a_i), (d_i), (b_i), (x_i))$

that does the back substitution for cross-systems of this form.

11. Consider the $n \times n$ lower-triangular system $Ax = b$, where $A = (a_{ij})$ and $a_{ij} = 0$ for $i < j$.

   a. Write an algorithm (in mathematical terms) for solving for x by forward substitution.

   b. Write

   procedure $forsub(n, (a_i), (b_i), (x_i))$

   which uses this algorithm.

   c. Determine the number of divisions, multiplications, and additions (or subtractions) in using this algorithm to solve for x.

   d. Should Gaussian elimination with partial pivoting be used to solve such a system?

12. (Normalized tridiagonal algorithm) Construct an algorithm for handling tridiagonal systems in which the normalized Gaussian elimination procedure without pivoting is used. In this process each pivot row is divided by the diagonal element before a multiple of the row is subtracted from the successive rows. Write the equations involved in the forward elimination phase and store the upper-diagonal entries back in array (c_i) and the right-hand-side entries back in array (b_i). Write the equations for the back substitution phase, storing the solution in array (b_i). Code and test this procedure. What are its advantages and disadvantages?

13. For a $(2n) \times (2n)$ tridiagonal system, write and test a procedure that proceeds as follows: In the forward elimination phase, the routine simultaneously eliminates the elements in the subdiagonal from the top to the middle and in the superdiagonal from the bottom to the middle. In the back substitution phase, the unknowns are determined two at a time from the middle outward.

14. (Continuation) Rewrite and test the procedure in Computer Problem **13** for a general $n \times n$ tridiagonal matrix.

15. Suppose

    procedure *trinor*$(n, (a_i), (d_i), (c_i), (b_i), (x_i))$

    performs the normalized Gaussian elimination algorithm of Computer Problem **12** and

    procedure *tri2n*$(n, (a_i), (d_i), (c_i), (b_i), (x_i))$

    does the algorithm outlined in Computer Problem **13**. Using a timing routine on your computer, compare *tri*, *trinor*, and *tri2n* to determine which of them is fastest for the tridiagonal system

    $$a_i = i(n - i + 1) \qquad c_i = (i + 1)(n - i - 1)$$
    $$d_i = (2i + 1)n - i - 2i \qquad b_i = i$$

    with a large even value of n. *Note:* Due to advances in computer architectures, mathematical algorithms may behave differently on parallel and vector computers. Generally speaking, parallel computations completely alter our conventional notions about what's best or most efficient.

16. Consider a special bidiagonal linear system of the form (illustrated with $n = 7$) with nonzero diagonal elements:

$$
\begin{bmatrix}
d_1 & & & & & & \\
a_1 & d_2 & & & & & \\
& a_2 & d_3 & & & & \\
& & a_3 & d_4 & a_4 & & \\
& & & & d_5 & a_5 & \\
& & & & & d_6 & a_6 \\
& & & & & & d_7
\end{bmatrix}
\begin{bmatrix}
x_1 \\ x_2 \\ x_3 \\ x_4 \\ x_5 \\ x_6 \\ x_7
\end{bmatrix}
=
\begin{bmatrix}
b_1 \\ b_2 \\ b_3 \\ b_4 \\ b_5 \\ b_6 \\ b_7
\end{bmatrix}
$$

Write and test

procedure $bidiag(n, (a_i), (d_i), (b_i))$

to solve the general system of order n (odd). Store the solution in array b and assume all arrays are of length n. *Do not* use forward elimination because the system can be solved quite easily without it.

17. Write and test

procedure $bktri(n, (a_i), (d_i), (c_i), (b_i), (x_i))$

for solving a backward tridiagonal system of linear equations of the form

$$
\begin{bmatrix}
 & & & a_1 & d_1 \\
 & & a_2 & d_2 & c_1 \\
 & a_3 & d_3 & c_2 \\
 & \ddots & \ddots & \ddots \\
a_{n-1} & d_{n-1} & c_{n-1} \\
d_n & c_{n-1}
\end{bmatrix}
\begin{bmatrix}
x_1 \\ x_2 \\ x_3 \\ \vdots \\ x_{n-1} \\ x_n
\end{bmatrix}
=
\begin{bmatrix}
b_1 \\ b_2 \\ b_3 \\ \vdots \\ b_{n-1} \\ b_n
\end{bmatrix}
$$

using Gaussian elimination without pivoting.

18. An upper **Hessenberg** matrix is of the form

$$
\begin{bmatrix}
a_{11} & a_{12} & a_{13} & \cdots & a_{1n} \\
a_{21} & a_{22} & a_{23} & \cdots & a_{2n} \\
 & a_{32} & a_{33} & \cdots & a_{3n} \\
 & & \ddots & \ddots & \vdots \\
 & & & a_{n,n-1} & a_{nn}
\end{bmatrix}
\begin{bmatrix}
x_1 \\ x_2 \\ x_3 \\ \vdots \\ x_n
\end{bmatrix}
=
\begin{bmatrix}
b_1 \\ b_2 \\ b_3 \\ \vdots \\ b_n
\end{bmatrix}
$$

Write a procedure for solving such a system and test it on at least a 10×10 system.

19. An $n \times n$ banded coefficient matrix with ℓ subdiagonals and m superdiagonals can be stored in **banded storage** mode in an $n \times (\ell + m + 1)$ array. The matrix is stored with the row and diagonal structure preserved with almost all 0 elements unstored. If the original $n \times n$ banded matrix had the form shown

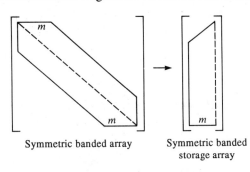

Symmetric banded array   Symmetric banded storage array

in the figure, then the $n \times (\ell + m + 1)$ array in banded storage mode would be as shown. The main diagonal would be the $\ell + 1$st column of the new array. Write and test a procedure for solving a linear system with the coefficient matrix stored in banded storage mode.

20. An $n \times n$ symmetric banded coefficient matrix with m subdiagonals and m superdiagonals can be stored in **symmetric banded storage** mode in an $n \times (m + 1)$ array. Only the main diagonal and subdiagonals are stored so that the main diagonal is the last column in the new array, shown in the figure. Write and test a procedure for solving a linear system with the coefficient matrix stored in symmetric banded storage mode.

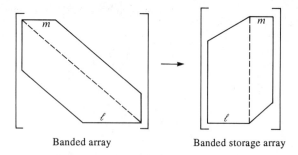

Banded array   Banded storage array

6.4 *LU* Factorization

This section requires some basic knowledge of linear algebra. Some of the notation and concepts of linear algebra are presented in Appendix A. This material can be skipped without loss of continuity.

An $n \times n$ system of linear equations can be written in matrix form

$$Ax = b \tag{1}$$

where the coefficient matrix A has the form

$$A = \begin{bmatrix} a_{11} & a_{12} & a_{13} & \cdots & a_{1n} \\ a_{21} & a_{22} & a_{23} & \cdots & a_{2n} \\ a_{31} & a_{32} & a_{33} & \cdots & a_{3n} \\ \vdots & \vdots & \vdots & \ddots & \vdots \\ a_{n1} & a_{n2} & a_{n3} & \cdots & a_{nn} \end{bmatrix}$$

Our main objective is to show that the naive Gaussian algorithm applied to A yields a factorization of A into a product of two simple matrices, one *unit lower triangular*:

$$L = \begin{bmatrix} 1 & & & & \\ \ell_{21} & 1 & & & \\ \ell_{31} & \ell_{32} & 1 & & \\ \vdots & \vdots & \vdots & \ddots & \\ \ell_{n1} & \ell_{n2} & \ell_{n3} & \cdots & 1 \end{bmatrix}$$

and the other upper triangular:

$$U = \begin{bmatrix} u_{11} & u_{12} & u_{13} & \cdots & u_{1n} \\ & u_{22} & u_{23} & \cdots & u_{2n} \\ & & u_{33} & \cdots & u_{3n} \\ & & & \ddots & \vdots \\ & & & & u_{nn} \end{bmatrix}$$

In short, we refer to this as an *LU* **factorization** of A; that is,

$$A = LU$$

Numerical Example

The system of Equations (2) of Section 6.1 can be written succinctly in matrix form:

$$\begin{bmatrix} 6 & -2 & 2 & 4 \\ 12 & -8 & 6 & 10 \\ 3 & -13 & 9 & 3 \\ -6 & 4 & 1 & -18 \end{bmatrix} \begin{bmatrix} x_1 \\ x_2 \\ x_3 \\ x_4 \end{bmatrix} = \begin{bmatrix} 16 \\ 26 \\ -19 \\ -34 \end{bmatrix} \tag{2}$$

Furthermore, the operations that led from this system to Equation (5) of Section 6.1—that is, the system

$$\begin{bmatrix} 6 & -2 & 2 & 4 \\ 0 & -4 & 2 & 2 \\ 0 & 0 & 2 & -5 \\ 0 & 0 & 0 & -3 \end{bmatrix} \begin{bmatrix} x_1 \\ x_2 \\ x_3 \\ x_4 \end{bmatrix} = \begin{bmatrix} 16 \\ -6 \\ -9 \\ -3 \end{bmatrix} \tag{3}$$

could be effected by an appropriate matrix multiplication. The forward elimination phase can be interpreted as starting from (1) and proceeding to

$$MAx = Mb \tag{4}$$

where M is a matrix chosen so that MA is the coefficient matrix for the System (3). Hence, we have

$$MA = \begin{bmatrix} 6 & -2 & 2 & 4 \\ 0 & -4 & 2 & 2 \\ 0 & 0 & 2 & -5 \\ 0 & 0 & 0 & -3 \end{bmatrix} \equiv U$$

which is an upper triangular matrix.

The first step of naive Gaussian elimination results in Equation (3) of Section 6.1 or the system

$$\begin{bmatrix} 6 & -2 & 2 & 4 \\ 0 & -4 & 2 & 2 \\ 0 & -12 & 8 & 1 \\ 0 & 2 & 3 & -14 \end{bmatrix} \begin{bmatrix} x_1 \\ x_2 \\ x_3 \\ x_4 \end{bmatrix} = \begin{bmatrix} 16 \\ -6 \\ -27 \\ -18 \end{bmatrix}$$

This step can be accomplished by multiplying (1) by a lower triangular matrix M_1:

$$M_1 A x = M_1 b$$

where

$$M_1 = \begin{bmatrix} 1 & 0 & 0 & 0 \\ -2 & 1 & 0 & 0 \\ -\frac{1}{2} & 0 & 1 & 0 \\ 1 & 0 & 0 & 1 \end{bmatrix}$$

Notice the special form of M_1. The diagonal elements are all 1's, and the only other nonzero elements are in the first column. These numbers are the *negatives of the multipliers* located in the positions where they created 0's as coefficients in step 1 of the forward elimination phase. To continue, step 2 resulted in Equation (4) of Section 6.1 or the system

$$\begin{bmatrix} 6 & -2 & 2 & 4 \\ 0 & -4 & 2 & 2 \\ 0 & 0 & 2 & -5 \\ 0 & 0 & 4 & -13 \end{bmatrix} \begin{bmatrix} x_1 \\ x_2 \\ x_3 \\ x_4 \end{bmatrix} = \begin{bmatrix} 16 \\ -6 \\ -9 \\ -21 \end{bmatrix}$$

which is equivalent to

$$M_2 M_1 A x = M_2 M_1 b$$

where

$$M_2 = \begin{bmatrix} 1 & 0 & 0 & 0 \\ 0 & 1 & 0 & 0 \\ 0 & -3 & 1 & 0 \\ 0 & \frac{1}{2} & 0 & 1 \end{bmatrix}$$

Again, M_2 differs from an identity matrix by the presence of the negatives of the multipliers in the second column from the diagonal down. Finally, step 3 gives System (3), which is equivalent to

$$M_3M_2M_1Ax = M_3M_2M_1b$$

where

$$M_3 = \begin{bmatrix} 1 & 0 & 0 & 0 \\ 0 & 1 & 0 & 0 \\ 0 & 0 & 1 & 0 \\ 0 & 0 & -2 & 1 \end{bmatrix}$$

Now the forward elimination phase is complete, and with

$$M = M_3M_2M_1 \tag{5}$$

we have the upper triangular coefficient System (3).

Using Equations (4) and (5), we can give a different interpretation of the forward elimination phase of naive Gaussian elimination. Now we see that

$$A = M^{-1}U$$
$$= M_1^{-1}M_2^{-1}M_3^{-1}U$$
$$= LU$$

Since each M_k has such a special form, its inverse is obtained by simply changing the signs of the negative multiplier entries! Hence, we have

$$L = \begin{bmatrix} 1 & 0 & 0 & 0 \\ 2 & 1 & 0 & 0 \\ \frac{1}{2} & 0 & 1 & 0 \\ -1 & 0 & 0 & 1 \end{bmatrix}\begin{bmatrix} 1 & 0 & 0 & 0 \\ 0 & 1 & 0 & 0 \\ 0 & 3 & 1 & 0 \\ 0 & -\frac{1}{2} & 0 & 1 \end{bmatrix}\begin{bmatrix} 1 & 0 & 0 & 0 \\ 0 & 1 & 0 & 0 \\ 0 & 0 & 1 & 0 \\ 0 & 0 & 2 & 1 \end{bmatrix}$$

$$= \begin{bmatrix} 1 & 0 & 0 & 0 \\ 2 & 1 & 0 & 0 \\ \frac{1}{2} & 3 & 1 & 0 \\ -1 & -\frac{1}{2} & 2 & 1 \end{bmatrix}$$

It is somewhat amazing that L is a unit lower triangular matrix composed of the multipliers. Notice that in forming L, we did not determine M first and then compute $M^{-1} = L$. (Why?)

It is easy to verify that

$$LU = \begin{bmatrix} 1 & 0 & 0 & 0 \\ 2 & 1 & 0 & 0 \\ \frac{1}{2} & 3 & 1 & 0 \\ -1 & -\frac{1}{2} & 2 & 1 \end{bmatrix}\begin{bmatrix} 6 & -2 & 2 & 4 \\ 0 & -4 & 2 & 2 \\ 0 & 0 & 2 & -5 \\ 0 & 0 & 0 & -3 \end{bmatrix} = \begin{bmatrix} 6 & -2 & 2 & 4 \\ 12 & -8 & 6 & 10 \\ 3 & -13 & 9 & 3 \\ -6 & 4 & 1 & -18 \end{bmatrix} = A$$

We see that A is **factored** or **decomposed** into a unit lower triangular matrix L and an upper triangular matrix U. The matrix L consists of the multipliers located in the positions of the elements they annihilated from A, of unit diagonal elements, and of 0 upper triangular elements. In fact, we now know the general form of L and can just write it down directly using the multipliers *without* forming the M_k's and the M_k^{-1}'s. The matrix U is upper triangular (not necessarily unit diagonal) and is the final coefficient matrix after the forward elimination phase is completed.

It should be noted that the pseudocode *ngauss* of Section 6.1 replaces the original coefficient matrix with its LU factorization. The elements of U are in the upper triangular part of the (a_{ij}) array including the diagonal. The entries below the main diagonal in L (that is, the multipliers) are found below the main diagonal in the (a_{ij}) array. Since it is known that L has a unit diagonal, nothing is lost by not storing the 1's. [In fact, we have run out of room in the (a_{ij}) array anyway!]

Formal Derivation

In order to see formally how the Gaussian elimination (in naive form) leads to an LU factorization, it is necessary to show that each row operation used in the algorithm can be effected by multiplying A on the left by an elementary matrix. Specifically, if we wish to subtract λ times row p from row q, we first apply this operation to the $n \times n$ identity matrix in order to create an elementary matrix M_{qp}. Then we form the matrix product $M_{qp}A$.

Before proceeding, let us verify that $M_{qp}A$ is obtained by subtracting λ times row p from row q in matrix A. Assume that $p < q$ (for in the naive algorithm this is always true). Then the elements of $M_{qp} = (m_{ij})$ are

$$m_{ij} = \begin{cases} 1 & \text{if } i = j \\ -\lambda & \text{if } i = q \text{ and } j = p \\ 0 & \text{in all other cases} \end{cases}$$

Therefore, the elements of $M_{qp}A$ are given by

$$(M_{pq}A)_{ij} = \sum_{s=1}^{n} m_{is}a_{sj} = \begin{cases} a_{ij} & \text{if } i \neq q \\ a_{qj} - \lambda a_{pj} & \text{if } i = q \end{cases}$$

The qth row of $M_{qp}A$ is the sum of the qth row of A and $-\lambda$ times the pth row of A, as was to be proved.

The kth step of Gaussian elimination corresponds to the matrix M_k, which is the product of $n - k$ elementary matrices:

$$M_k = M_{nk} M_{n-1,k} \cdots M_{k+1,k}$$

Notice that each elementary matrix M_{ik} here is lower triangular because $i > k$ and therefore M_k is also lower triangular. If we carry out the Gaussian forward elimination process on A, the result will be an upper triangular matrix U. On the other hand, the result is obtained by applying a succession of factors like M_k to the left of A. Hence, the entire process is summarized by writing

$$M_{n-1} \cdots M_2 M_1 A = U$$

Since each M_k is invertible, we have

$$A = M_1^{-1} M_2^{-1} \cdots M_{n-1}^{-1} U$$

Each M_k is lower triangular with 1's on its main diagonal (unit lower triangular). Each inverse M_k^{-1} has the same property, and the same is true of their product. Hence, the matrix

$$L = M_1^{-1} M_2^{-1} \cdots M_{n-1}^{-1} \tag{6}$$

is unit lower triangular, and we have

$$A = LU$$

This is the so-called **LU** factorization of A. Our construction of it depends upon *not* encountering any 0 divisors in the algorithm. It is easy to give examples of matrices that have no **LU** factorization; one of the simplest is $A = \begin{bmatrix} 0 & 1 \\ 1 & 1 \end{bmatrix}$. (See Problem **4**.)

LU FACTORIZATION THEOREM

Let $A = (a_{ij})$ be an $n \times n$ matrix. Assume that the forward elimination phase of the naive Gaussian algorithm is applied to A without encountering any 0 divisors. Let the resulting matrix be denoted by $\widetilde{A} = (\widetilde{a}_{ij})$. If

$$L = \begin{bmatrix} 1 & 0 & 0 & \cdots & 0 \\ \widetilde{a}_{21} & 1 & 0 & \cdots & 0 \\ \widetilde{a}_{31} & \widetilde{a}_{32} & 1 & \cdots & 0 \\ \vdots & \vdots & \ddots & \ddots & \vdots \\ \widetilde{a}_{n1} & \widetilde{a}_{n2} & \cdots & \widetilde{a}_{n,n-1} & 1 \end{bmatrix}$$

and

$$U = \begin{bmatrix} \widetilde{a}_{11} & \widetilde{a}_{12} & \widetilde{a}_{13} & \cdots & \widetilde{a}_{1n} \\ 0 & \widetilde{a}_{22} & \widetilde{a}_{23} & \cdots & \widetilde{a}_{2n} \\ 0 & 0 & \widetilde{a}_{33} & \cdots & \widetilde{a}_{3n} \\ \vdots & \vdots & \ddots & \ddots & \vdots \\ 0 & 0 & \cdots & 0 & \widetilde{a}_{nn} \end{bmatrix}$$

then $A = LU$.

Proof   We define the Gaussian algorithm formally as follows. Let $A^{(1)} = A$. Then we compute $A^{(2)}, A^{(3)}, \ldots, A^{(n)}$ recursively by the naive Gaussian algorithm, following these equations:

$$a_{ij}^{(k+1)} = a_{ij}^{(k)} \qquad \text{(if } i \leq k \text{ or } j < k) \tag{7}$$

$$a_{ij}^{(k+1)} = \frac{a_{ik}^{(k)}}{a_{kk}^{(k)}} \qquad \text{(if } i > k \text{ and } j = k) \tag{8}$$

$$a_{ij}^{(k+1)} = a_{ij}^{(k)} - \left(\frac{a_{ik}^{(k)}}{a_{kk}^{(k)}} \right) a_{kj}^{(k)} \qquad \text{(if } i > k \text{ and } j > k) \tag{9}$$

These equations describe in a precise form the forward elimination phase of the naive Gaussian elimination algorithm. For example, Equation (7) states that in proceeding from $A^{(k)}$ to $A^{(k+1)}$, we do not alter rows $1, 2, \ldots, k$ or columns $1, 2, \ldots, k - 1$. Equation (8) shows how the multipliers are computed and stored in passing from $A^{(k)}$ to $A^{(k+1)}$. Finally, Equation (9) shows how multiples of row k are subtracted from rows $k + 1, k + 2, \ldots, n$ in order to produce $A^{(k+1)}$ from $A^{(k)}$.

Notice that $A^{(n)}$ is the final result of the process. (It was referred to as $\widetilde{A}$ in the statement of the theorem.) The formal definitions of $L = (\ell_{ik})$ and $U = (u_{kj})$ are, therefore,

$$\ell_{ik} = 1 \qquad (i = k) \tag{10}$$

$$\ell_{ik} = a_{ik}^{(n)} \qquad (k < i) \tag{11}$$

$$\ell_{ik} = 0 \qquad (k > i) \tag{12}$$

$$u_{kj} = a_{kj}^{(n)} \qquad (j \geq k) \tag{13}$$

$$u_{kj} = 0 \qquad (j < k) \tag{14}$$

Now we draw some consequences of these equations. First, it follows immediately from Equation (7) that

$$a_{ij}^{(i)} = a_{ij}^{(i+1)} = \cdots = a_{ij}^{(n)} \tag{15}$$

Likewise, we have from Equation (7)

$$a_{ij}^{(j+1)} = a_{ij}^{(j+2)} = \cdots = a_{ij}^{(n)} \qquad (j < n) \tag{16}$$

From Equations (16) and (8), we have now

$$a_{ij}^{(n)} = a_{ij}^{(j+1)} = \frac{a_{ij}^{(j)}}{a_{jj}^{(j)}} \qquad (j < n) \tag{17}$$

From Equations (17) and (11), it follows that

$$\ell_{ik} = a_{ik}^{(n)} = \frac{a_{ik}^{(k)}}{a_{kk}^{(k)}} \qquad (k < i) \tag{18}$$

From Equations (13) and (15), we have

$$u_{kj} = a_{kj}^{(n)} = a_{kj}^{(k)} \qquad (k \leq j) \tag{19}$$

With the aid of all these equations, we can now prove that $LU = A$. First, consider the case $i \leq j$. Then

$$(LU)_{ij} = \sum_{k=1}^{n} \ell_{ik} u_{kj} \qquad \text{[definition of multiplication]}$$

$$= \sum_{k=1}^{i} \ell_{ik} u_{kj} \qquad \text{[by Equation (12)]}$$

$$= \sum_{k=1}^{i-1} \ell_{ik} u_{kj} + u_{ij} \qquad \text{[by Equation (10)]}$$

$$= \sum_{k=1}^{i-1} \left[\frac{a_{ik}^{(k)}}{a_{kk}^{(k)}} \right] a_{kj}^{(k)} + a_{ij}^{(i)} \qquad \text{[by Equations (18) and (19)]}$$

$$= \sum_{k=1}^{i-1} \left[a_{ij}^{(k)} - a_{ij}^{(k+1)} \right] + a_{ij}^{(i)} \qquad \text{[by Equation (9)]}$$

$$= a_{ij}^{(1)} = a_{ij}$$

In the remaining case, $i > j$, we have

$$(LU)_{ij} = \sum_{k=1}^{n} \ell_{ik} u_{kj} \qquad \text{[definition of multiplication]}$$

$$= \sum_{k=1}^{j} \ell_{ik} u_{kj} \qquad \text{[by Equation (14)]}$$

$$= \sum_{k=1}^{j} \left[\frac{a_{ik}^{(k)}}{a_{kk}^{(k)}} \right] a_{kj}^{(k)} \qquad \text{[by Equations (18) and (19)]}$$

$$= \sum_{k=1}^{j-1} \left[\frac{a_{ik}^{(k)}}{a_{kk}^{(k)}} \right] a_{kj}^{(k)} + a_{ij}^{(j)}$$

$$= \sum_{k=1}^{j-1} \left[a_{ij}^{(k)} - a_{ij}^{(k+1)} \right] + a_{ij}^{(j)} \qquad \text{[by Equation (9)]}$$

$$= a_{ij}^{(1)} = a_{ij} \qquad\qquad\blacksquare$$

Solving Linear Systems Using *LU* Factorization

Once the *LU* factorization of *A* is available, we can solve the system

$$Ax = b$$

by writing

$$LUx = b$$

Then we solve two triangular systems:

$$Lz = b \qquad\qquad\qquad (20)$$

for *z* and

$$Ux = z \qquad\qquad\qquad (21)$$

for *x*. This is particularly useful for problems that involve the same coefficient matrix *A* and many different right-hand vectors *b*.

Since *L* is unit lower triangular, *z* is obtained by the pseudocode

real array $(b_i)_n, (\ell_{ij})_{n \times n}, (z_i)_n$
integer i, n
   $\vdots$
$z_1 \leftarrow b_1$
for $i = 2$ **to** n **do**
   $z_i \leftarrow b_i - \sum_{j=1}^{i-1} \ell_{ij} z_j$
end do

Likewise, *x* is obtained by the pseudocode

real array $(u_{ij})_{n \times n}, (x_i)_n, (z_i)_n$
integer i, n
   $\vdots$
$x_n \leftarrow z_n / u_{nn}$
for $i = n - 1$ **to** 1 **step** -1 **do**
   $x_i \leftarrow \left(z_i - \sum_{j=i+1}^{n} u_{ij} x_j \right) \Big/ u_{ii}$
end do

The first of these two algorithms applies the forward phase of Gaussian elimination to the right-hand-side vector b. [Recall that the ℓ_{ij}'s are the *multipliers* that have been stored in the array (a_{ij}).] The easiest way to verify this assertion is to use Equation (6) and to rewrite the equation

$$Lz = b$$

in the form

$$M_1^{-1}M_2^{-1}\cdots M_{n-1}^{-1}z = b$$

From this we get immediately

$$z = M_{n-1}\cdots M_2M_1b$$

Thus, the same operations used to reduce A to U are to be used on b in order to produce z.

Another way to solve Equation (20) is to note that what must be done is to form

$$M_{n-1}M_{n-2}\cdots M_2M_1b$$

This can be accomplished using only the array (b_i) by putting the results back into b; that is,

$$b \leftarrow M_kb$$

We know what M_k looks like because it is made up of negative multipliers that have been saved in the array (a_{ij}). Clearly, we have

$$
M_kb = \begin{bmatrix} 1 & & & & & & \\ & \ddots & & & & & \\ & & 1 & & & & \\ & & -a_{k+1,k} & 1 & & & \\ & & \vdots & & \ddots & & \\ & & -a_{ik} & & & 1 & \\ & & \vdots & & & & \ddots \\ & & -a_{nk} & & & & 1 \end{bmatrix} \begin{bmatrix} b_1 \\ \vdots \\ b_k \\ \vdots \\ b_i \\ \vdots \\ b_n \end{bmatrix}
$$

The entries b_1 to b_k are not changed by this multiplication, while b_i (for $i \geq k + 1$) is replaced by $-a_{ik}b_k + b_i$. Hence, the following pseudocode updates the array (b_i) based on the stored multipliers in the array a:

real array $(a_{ij})_{n\times n}, (b_i)_n$
integer i, k, n

$$\vdots$$

> **for** $k = 1$ **to** $n - 1$ **do**
>   **for** $i = k + 1$ **to** n **do**
> $b_i \leftarrow b_i - a_{ik}b_k$
>   **end do**
> **end do**

This pseudocode should be familiar. It is the process for updating b from Section 6.2.

The algorithm for solving Equation (21) is the back substitution phase of the naive Gaussian elimination process.

Computing A^{-1}

In some applications, such as in statistics, it may be necessary to compute the inverse of a matrix A and explicitly display it as A^{-1}. This can be done by using procedures *gauss* and *solve*. If an $n \times n$ matrix A has an inverse, it is an $n \times n$ matrix X with the property that

$$AX = I \tag{22}$$

where I is the identity matrix. If $X^{(j)}$ denotes the jth column of X and $I^{(j)}$ denotes the jth column of I, then matrix Equation (22) can be written as

$$A[X^{(1)}, X^{(2)}, \ldots, X^{(n)}] = [I^{(1)}, I^{(2)}, \ldots, I^{(n)}]$$

This can be written as n linear systems of equations of the form

$$AX^{(j)} = I^{(j)} \qquad (1 \leq j \leq n)$$

Now use procedure *gauss* once to produce a factorization of A and use procedure *solve* n times with the vectors $I^{(j)}$ $(1 \leq j \leq n)$. This is equivalent to solving, one at a time, for the columns of A^{-1}, which are $X^{(j)}$. Hence,

$$A^{-1} = [X^{(1)}, X^{(2)}, \ldots, X^{(n)}]$$

A word of caution on computing the inverse of a matrix: When solving a linear system $Ax = b$, it is not advisable to determine A^{-1} and then compute $x = A^{-1}b$ because this requires many unnecessary calculations, compared to directly solving $Ax = b$ for x.

Example of a Software Package

The MATLAB input is as follows.

```
A = [ 6,   -2,   2,    4,
     12,   -8,   6,   10,
      3,  -13,   9,    3,
     -6,    4,   1,  -18]
```

```
[L, U, P] = lu(A)
```

The resulting MATLAB output is as follows.

```
L =
    1.0000         0         0         0
    0.2500    1.0000         0         0
   -0.5000         0    1.0000         0
    0.5000   -0.1818    0.0909    1.0000
```

```
U =
   12.0000   -8.0000    6.0000   10.0000
        0  -11.0000    7.5000    0.5000
        0         0    4.0000  -13.0000
        0         0         0    0.2727
```

```
P =
    0    1    0    0
    0    0    1    0
    0    0    0    1
    1    0    0    0
```

This corresponds to the factorization

$$PA = LU$$

where P is a permutation matrix corresponding to the pivoting strategy used.

PROBLEMS 6.4

1. Using naive Gaussian elimination, factor the following matrices so that $A = LU$, where L is a unit lower triangular matrix and U is an upper triangular matrix.

   a. $A = \begin{bmatrix} 3 & 0 & 3 \\ 0 & -1 & 3 \\ 1 & 3 & 0 \end{bmatrix}$

   b. $A = \begin{bmatrix} 1 & 0 & \frac{1}{3} & 0 \\ 0 & 1 & 3 & -1 \\ 3 & -3 & 0 & 6 \\ 0 & 2 & 4 & -6 \end{bmatrix}$

$$\text{c. } A = \begin{bmatrix} -20 & -15 & -10 & -5 \\ 1 & 0 & 0 & 0 \\ 0 & 1 & 0 & 0 \\ 0 & 0 & 1 & 0 \end{bmatrix}$$

2. Consider the matrix

$$A = \begin{bmatrix} 1 & 0 & 0 & 2 \\ 0 & 3 & 0 & 0 \\ 0 & 9 & 4 & 0 \\ 5 & 0 & 8 & 10 \end{bmatrix}$$

   a. Determine a unit lower triangular matrix M and an upper triangular matrix U such that $MA = U$.

   b. Determine a unit lower triangular matrix L and an upper triangular matrix U such that $A = LU$. Show that $ML = I$ so that $L = M^{-1}$.

3. Consider the matrix

$$A = \begin{bmatrix} 25 & 0 & 0 & 0 & 1 \\ 0 & 27 & 4 & 3 & 2 \\ 0 & 54 & 58 & 0 & 0 \\ 0 & 108 & 116 & 0 & 0 \\ 100 & 0 & 0 & 0 & 24 \end{bmatrix}$$

   a. Determine the unit lower triangular matrix M and the upper triangular matrix U such that $MA = U$.

   b. Determine $M^{-1} = L$ such that $A = LU$.

4. Consider the matrix

$$A = \begin{bmatrix} 2 & 2 & 1 \\ 1 & 1 & 1 \\ 3 & 2 & 1 \end{bmatrix}$$

   a. Show that A *cannot* be factored into the product of a unit lower triangular matrix and an upper triangular matrix.

   b. Interchange the rows of A so that this can be done.

5. Consider the matrix

$$A = \begin{bmatrix} a & 0 & 0 & z \\ 0 & b & 0 & 0 \\ 0 & x & c & 0 \\ w & 0 & y & d \end{bmatrix}$$

   a. Determine a unit lower triangular matrix M and an upper triangular matrix U such that $LU = U$.

b. Determine a lower triangular matrix L' and a unit upper triangular matrix U' such that $A = L'U'$.

6. Consider the matrix

$$A = \begin{bmatrix} 4 & -1 & -1 & 0 \\ -1 & 4 & 0 & -1 \\ -1 & 0 & 4 & -1 \\ 0 & -1 & -1 & 4 \end{bmatrix}$$

Factor A in the following ways:

a. $A = LU$, where L is unit lower triangular and U is upper triangular.

b. $A = LDU'$, where L is unit lower triangular, D is diagonal, and U' is unit upper triangular.

c. $A = L'U'$, where L' is lower triangular and U' is unit upper triangular.

d. $A = (L'')(L'')^T$, where L'' is lower triangular.

7. (Continuation) Evaluate the determinant of A.
  Hint: $\det(A) = \det(L)\det(D)\det(U') = \det(D)$.

8. Consider the 3×3 Hilbert matrix

$$A = \begin{bmatrix} 1 & \frac{1}{2} & \frac{1}{3} \\ \frac{1}{2} & \frac{1}{3} & \frac{1}{4} \\ \frac{1}{3} & \frac{1}{4} & \frac{1}{5} \end{bmatrix}$$

Repeat Problems **6** and **7** using this matrix.

9. Find the LU decomposition, where L is unit lower triangular for

$$A = \begin{bmatrix} 1 & 0 & 0 & 1 \\ 1 & 1 & 0 & -1 \\ -1 & 1 & 1 & 1 \\ 1 & -1 & 1 & -1 \end{bmatrix}$$

10. Consider

$$A = \begin{bmatrix} 2 & -1 & 2 \\ 2 & -3 & 3 \\ 6 & -1 & 8 \end{bmatrix}$$

a. Find the matrix factorization $A = LDU'$, where L is unit lower triangular, D is diagonal, and U' is unit upper triangular.

b. Use this decomposition of A to solve $Ax = b$, where $b = [-2, -5, 0]^T$.

11. Repeat Problem **10** for

$$A = \begin{bmatrix} -2 & 1 & -2 \\ -4 & 3 & -3 \\ 2 & 2 & 4 \end{bmatrix} \qquad b = \begin{bmatrix} 1 \\ 4 \\ 4 \end{bmatrix}$$

12. Consider the system of equations

$$\begin{cases} 6x_1 = & 12 \\ 6x_2 + 3x_1 = & -12 \\ 7x_3 - 2x_2 + 4x_1 = & 14 \\ 21x_4 + 9x_3 - 3x_2 + 5x_1 = & -2 \end{cases}$$

   a. Solve for x_1, x_2, x_3, and x_4 (in order) by forward substitution.

   b. Write this system in matrix notation $Ax = b$, where $x = [x_1, x_2, x_3, x_4]^T$. Determine the LU factorization $A = LU$, where L is unit lower triangular and U is upper triangular.

13. Given

$$A = \begin{bmatrix} 3 & 2 & -1 \\ 5 & 3 & 2 \\ -1 & 1 & -3 \end{bmatrix} \qquad L^{-1} = \begin{bmatrix} 1 & 0 & 0 \\ -\frac{5}{3} & 1 & 0 \\ -8 & 5 & 1 \end{bmatrix} \qquad U = \begin{bmatrix} 3 & 2 & -1 \\ 0 & -\frac{1}{3} & \frac{11}{3} \\ 0 & 0 & 15 \end{bmatrix}$$

   obtain the inverse of A by solving $UX^{(j)} = L^{-1}I^{(j)}$ for $j = 1, 2, 3$.

14. Using the system of Equations (2), form $M = M_3 M_2 M_1$ and determine M^{-1}. Verify that $M^{-1} = L$. Why is this, in general, not a good idea?

15. Consider the matrix

$$\begin{bmatrix} a_{11} & a_{12} & 0 & 0 \\ a_{21} & a_{22} & a_{23} & 0 \\ 0 & a_{32} & a_{33} & a_{34} \\ 0 & 0 & a_{43} & a_{44} \end{bmatrix}$$

   where $a_{ii} \neq 0$ for $i = 1, 2, 3, 4$.

   a. Establish the algorithm

   real array $(a_{ij})_{n \times n}, (\ell_{ij})_{n \times n}, (u_{ij})_{n \times n}$
   integer i, n

   $\vdots$

   $\ell_{11} \leftarrow a_{11}$
   for $i = 2$ **to** 4 **do**
     $\ell_{i,i-1} \leftarrow a_{i,i-1}$
     $u_{i-1,i} \leftarrow a_{i-1,i}/\ell_{i-1,i-1}$
     $\ell_{i,i} \leftarrow a_{i,i} - \ell_{i,i-1}u_{i-1,i}$
   end do

for determining the elements of a lower tridiagonal matrix $L = (\ell_{ij})$ and a *unit* upper tridiagonal matrix $U = (u_{ij})$ such that $A = LU$.

b. Establish the algorithm

real array $(a_{ij})_{n \times n}, (\ell_{i,j})_{n \times n}, (u_{i,j})_{n \times n}$
integer i, n

$\vdots$

$u_{11} \leftarrow a_{11}$
for $i = 2$ **to** 4 **do**
$\quad u_{i-1,i} \leftarrow a_{i-1,i}$
$\quad \ell_{i,i-1} \leftarrow a_{i,i-1} / u_{i-1,i-1}$
$\quad u_{i,j} \leftarrow a_{i,i} - \ell_{i,i-1} u_{i-1,i}$
end do

for determining the elements of a unit lower triangular matrix $L = (\ell_{ij})$ and an upper tridiagonal matrix $U = (u_{ij})$ such that $A = LU$.

By extending the loops, we can generalize these algorithms to $n \times n$ tridiagonal matrices.

16. Show that $AX = B$ can be solved by Gaussian elimination with scaled partial pivoting in $(n^3/3) + mn^2 + \mathcal{O}(n^2)$ multiplications and divisions, where A, X, and B are matrices of order $n \times n$, $n \times m$, and $n \times m$, respectively. Thus, if B is $n \times n$, then the $n \times n$ solution matrix X can be found by Gaussian elimination with scaled partial pivoting in $\frac{4}{3}n^3 + \mathcal{O}(n^2)$ multiplications and divisions. *Hint*: If $X^{(j)}$ and $B^{(j)}$ are the jth columns of X and B, respectively, then $AX^{(j)} = B^{(j)}$.

17. Let X be a square matrix that has the form

$$X = \begin{bmatrix} A & B \\ C & D \end{bmatrix}$$

where A and D are square matrices and A^{-1} exists. It is known that X^{-1} exists if and only if $(D - CA^{-1}B)^{-1}$ exists. Verify that X^{-1} is given by

$$X^{-1} = \begin{bmatrix} I & -A^{-1}B \\ 0 & I \end{bmatrix} \begin{bmatrix} A^{-1} & 0 \\ 0 & (D - CA^{-1}B)^{-1} \end{bmatrix} \begin{bmatrix} I & 0 \\ -CA^{-1} & I \end{bmatrix}$$

As an application, compute the inverse of the following.

a. $X = \begin{bmatrix} 1 & 0 & 0 & 1 \\ 0 & 1 & 1 & 0 \\ 1 & 0 & 1 & 2 \\ 0 & 0 & 0 & 1 \end{bmatrix}$

b. $X = \begin{bmatrix} 1 & 0 & 0 & 1 \\ 0 & 1 & 0 & 1 \\ 0 & 0 & 1 & 1 \\ 1 & 1 & 1 & 2 \end{bmatrix}$

18. Let A be an $n \times n$ complex matrix such that A^{-1} exists. Verify that

$$\begin{bmatrix} A & \bar{A} \\ -Ai & -\bar{A}i \end{bmatrix}^{-1} = \frac{1}{2} \begin{bmatrix} A^{-1} & A^{-1}i \\ \bar{A}^{-1} & -\bar{A}^{-1}i \end{bmatrix}$$

where $\bar{A}$ denotes the complex conjugate of A; that is, if $A = (a_{ij})$, then $\bar{A} = (\bar{a}_{ij})$. Recall that for a complex number $z = a + bi$, for real a and b, $\bar{z} = a - bi$.

COMPUTER PROBLEMS 6.4

1. Write and test a procedure for implementing the algorithms of Problem **15.**

2. The $n \times n$ factorization $A = LU$, where $L = (\ell_{ij})$ is lower triangular and $U = (u_{ij})$ is upper triangular, can be computed directly by the following algorithm (provided zero divisions are not encountered): Specify either ℓ_{11} or u_{11} and compute the other such that $\ell_{11}u_{11} = a_{11}$. Compute the first column in L by

$$\ell_{i1} = \frac{a_{i1}}{u_{11}} \qquad (1 \leq i \leq n)$$

and compute the first row in U by

$$u_{1j} = \frac{a_{1j}}{\ell_{11}} \qquad (1 \leq j \leq n)$$

Now suppose that columns $1, 2, \ldots, k-1$ have been computed in L and that rows $1, 2, \ldots, k-1$ have been computed in U. At the kth step, specify either ℓ_{kk} or u_{kk} and compute the other such that

$$\ell_{kk}u_{kk} = a_{kk} - \sum_{m=1}^{k-1} \ell_{km}u_{mk}$$

Compute the kth column in L by

$$\ell_{ik} = \frac{1}{u_{kk}} \left(a_{ik} - \sum_{m=1}^{k-1} \ell_{im}u_{mk} \right) \qquad (k \leq i \leq n)$$

and compute the kth row in U by

$$u_{kj} = \frac{1}{\ell_{kk}} \left(a_{kj} - \sum_{m=1}^{k-1} \ell_{km}u_{mj} \right) \qquad (k \leq j \leq n)$$

This algorithm is continued until all elements of U and L are completely determined. When $\ell_{ii} = 1$ $(1 \leq i \leq n)$, this procedure is called the Doolittle factorization, and when $u_{jj} = 1$ $(1 \leq j \leq n)$, it is known as the Crout factorization.

Define the test matrix

$$A = \begin{bmatrix} 5 & 7 & 6 & 5 \\ 7 & 10 & 8 & 7 \\ 6 & 8 & 10 & 9 \\ 5 & 7 & 9 & 10 \end{bmatrix}$$

Using the algorithm above, compute and print factorizations so that the diagonal entries of L and U are of the following forms:

diag(L)	diag(U)	
[1, 1, 1, 1]	[?, ?, ?, ?]	Doolittle
[?, ?, ?, ?]	[1, 1, 1, 1]	Crout
[1, ?, 1, ?]	[?, 1, ?, 1]	
[?, 1, ?, 1]	[1, ?, 1, ?]	
[?, ?, 7, 9]	[3, 5, ?, ?]	

Here ? means that the entry is to be computed. Write code to check the results by multiplying L and U together.

3. Write

procedure $poly(n, (a_{ij}), (c_i), k, (y_{ij}))$

for computing the $n \times n$ matrix $p_k(A)$ stored in array (y_{ij}):

$$y_k = p_k(A) = c_0 I + c_1 A + c_2 A^2 + \cdots + c_k A^k$$

where A is an $n \times n$ matrix and p_k is a kth-degree polynomial. Here (c_i) are real constants c_i for $0 \leq i \leq k$. Use nested multiplication and write efficient code. Test procedure *poly* on the following data:

Case 1

$$A = I_5 \qquad p_3(x) = 1 - 5x + 10x^3$$

Case 2

$$A = \begin{bmatrix} 1 & 2 \\ 3 & 4 \end{bmatrix} \qquad p_2(x) = 1 - 2x + x^2$$

Case 3

$$A = \begin{bmatrix} 0 & 2 & 4 \\ 0 & 0 & 8 \\ 0 & 0 & 0 \end{bmatrix} \qquad p_3(x) = 1 + 3x - 3x^2 + x^3$$

Case 4

$$A = \begin{bmatrix} 2 & -1 & 0 & 0 \\ -1 & 2 & -1 & 0 \\ 0 & -1 & 2 & -1 \\ 0 & 0 & -1 & 2 \end{bmatrix} \qquad p_5(x) = 10 + x - 2x^2 + 3x^3 - 4x^4 + 5x^5$$

Case 5

$$A = \begin{bmatrix} -20 & -15 & -10 & -5 \\ 1 & 0 & 0 & 0 \\ 0 & 1 & 0 & 0 \\ 0 & 0 & 1 & 0 \end{bmatrix} \qquad p_4(x) = 5 + 10x + 15x^2 + 20x^3 + x^4$$

Case 6

$$A = \begin{bmatrix} 5 & 7 & 6 & 5 \\ 7 & 10 & 8 & 7 \\ 6 & 8 & 10 & 9 \\ 5 & 7 & 9 & 10 \end{bmatrix} \qquad p_4(x) = 1 - 100x + 146x^2 - 35x^3 + x^4$$

4. Write and test a procedure for determining A^{-1} for a given square matrix A of order n. Your procedure should use procedures *gauss* and *solve*.

5. Write and test a procedure to solve the system $AX = B$ in which A, X, and B are matrices of order $n \times n$, $n \times m$, and $n \times m$, respectively. Verify that the procedure works on several test cases, one of which has $B = I$ so that the solution X is the inverse of A. *Hint*: See Problem **16**.

6. Write and test a procedure for directly computing the inverse of a tridiagonal matrix. Assume that pivoting is not necessary.

7. (Continuation) Test the procedure of Computer Problem **6** on the symmetric tridiagonal matrix A of order 10:

$$A = \begin{bmatrix} -2 & 1 & & & & \\ 1 & -2 & 1 & & & \\ & 1 & -2 & 1 & & \\ & & \ddots & \ddots & \ddots & \\ & & & 1 & -2 & 1 \\ & & & & 1 & -2 \end{bmatrix}$$

The inverse of this matrix is known to be

$$(A^{-1})_{ij} = (A^{-1})_{ji} = \frac{-i(n + 1 - j)}{(n + 1)} \qquad (i \leq j)$$

8. Investigate the numerical difficulties in inverting the following matrix.

$$A = \begin{bmatrix} -0.0001 & 5.096 & 5.101 & 1.853 \\ 0. & 3.737 & 3.740 & 3.392 \\ 0. & 0. & 0.006 & 5.254 \\ 0. & 0. & 0. & 4.567 \end{bmatrix}$$

7 APPROXIMATION BY SPLINE FUNCTIONS

By experimentation in a wind tunnel, an airfoil is constructed by trial and error so that it has certain desired characteristics. The cross section of the airfoil is then drawn as a curve on coordinate paper (see Figure 7.1). In order to study this airfoil by analytical methods or to manufacture it, it is essential to have a formula for this curve. To arrive at such a formula, one first obtains the coordinates of a finite set of points on the curve. Then a smooth curve called a **cubic interpolating spline** can be constructed to match these data points. This chapter discusses general polynomial spline functions and how they can be used in various numerical problems such as the data-fitting problem just described.

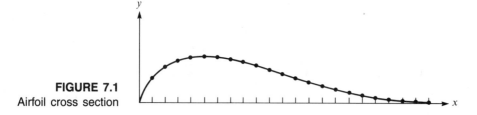

FIGURE 7.1
Airfoil cross section

7.1 First-Degree and Second-Degree Splines

First-Degree Spline

A **spline function** is a function that consists of polynomial pieces joined together with certain smoothness conditions. A simple example is the **polygonal** function (or spline of degree 1), whose pieces are linear polynomials joined together to achieve

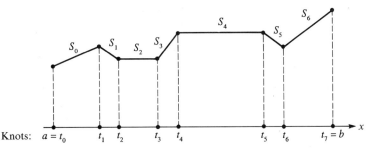

FIGURE 7.2
First-degree
spline function

continuity, as in Figure 7.2. The points $t_0, t_1, \ldots, t_n$ at which the function changes its character are termed **knots** in the theory of splines. Thus, the spline function shown in Figure 7.2 has eight knots.

Such a function is somewhat complicated to define in explicit terms. We are forced to write

$$
S(x) = \begin{cases}
S_0(x) & x \in [t_0, t_1] \\
S_1(x) & x \in [t_1, t_2] \\
\ \vdots & \quad \vdots \\
S_{n-1}(x) & x \in [t_{n-1}, t_n]
\end{cases}
\tag{1}
$$

where

$$
S_i(x) = a_i x + b_i \tag{2}
$$

because each piece of $S(x)$ is a linear polynomial. Such a function $S(x)$ is **piecewise linear**. If the knots $t_0, t_1, \ldots, t_n$ were given and if the coefficients $a_0, b_0, a_1, b_1, \ldots,$ a_{n-1}, b_{n-1} were all known, then the evaluation of $S(x)$ at a specific x would proceed by first determining the interval that contains x and then using the appropriate linear function for that interval.

If the function S defined by Equation (1) is continuous, we call it a **spline of degree 1**. It is characterized by the following three properties:

1. The domain of S is an interval $[a, b]$.

2. S is continuous on $[a, b]$.

3. There is a partitioning of the interval $a = t_0 < t_1 < \cdots < t_n = b$ such that S is a linear polynomial on each subinterval $[t_i, t_{i+1}]$.

Outside the interval $[a, b]$, $S(x)$ is usually defined to be the same function on the left of a as it is on the leftmost subinterval $[t_0, t_1]$ and the same on the right of b as it is on the rightmost subinterval $[t_{n-1}, t_n]$; namely, $S(x) = S_0(x)$ when $x < a$ and $S(x) = S_{n-1}(x)$ when $x > b$.

EXAMPLE 1   Determine whether this function is a first-degree spline function:

$$
f(x) = \begin{cases}
x & x \in [-1, 0] \\
1 - x & x \in (0, 1) \\
2x - 2 & x \in [1, 2]
\end{cases}
$$

Solution   The function is obviously piecewise linear but is not a spline of degree 1 because it is discontinuous at $x = 0$. Notice that $\lim_{x\to 0^+} f(x) = \lim_{x\to 0}(1 - x) = 1$, whereas $\lim_{x\to 0^-} f(x) = \lim_{x\to 0} x = 0$.   ☐

Continuity of a function f at a point s can be defined by the condition

$$\lim_{x\to s^+} f(x) = \lim_{x\to s^-} f(x) = f(s)$$

Here $\lim_{x\to s^+}$ means that the limit is taken over x values that converge to s from above s ; that is, $(x - s)$ is positive for all x values. Similarly, $\lim_{x\to s^-}$ means that the x values converge to s from below.

The spline functions of degree 1 can be used for interpolation. Suppose the following table of function values is given:

x	t_0	t_1	$\cdots$	t_n
y	y_0	y_1	$\cdots$	y_n

There is no loss of generality in supposing that $t_0 < t_1 < \cdots < t_n$ because this is only a matter of labeling the knots. The table can be represented by the set of $n + 1$ points in the plane, $(t_0, y_0), (t_1, y_1), \ldots, (t_n, y_n)$, and these points have distinct abscissas. Therefore, we can draw a polygonal line through the points without ever drawing a *vertical* segment. This polygonal line is the graph of a function, and this function is obviously a spline of degree 1. What are the equations of the individual line segments that make up this graph?

By referring to Figure 7.3 and using the point-slope form of a line, we obtain

$$S_i(x) = y_i + m_i(x - t_i) \tag{3}$$

on the interval $[t_i, t_{i+1}]$, where m_i is the slope of the line and is therefore given by the formula

$$m_i = \frac{y_{i+1} - y_i}{t_{i+1} - t_i}$$

The form of Equation (3) is better than that of Equation (2) for the practical evaluation of $S(x)$ because some of the quantities $x - t_i$ must be computed in any case simply to determine which subinterval contains x. The interval $[t_i, t_{i+1}]$ containing x is characterized by the fact that $x - t_i$ is the first of the quantities $x - t_{n-1}, x - t_{n-2}, \ldots, x - t_0$ that is *nonnegative*.

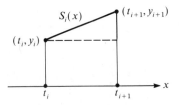

FIGURE 7.3

The following is a function procedure that utilizes $n + 1$ table values (t_i, y_i) in linear arrays (t_i) and (y_i), assuming that $a = t_0 < t_1 < \cdots < t_n = b$. Given an x value, the routine returns $S(x)$ using Equations (1) and (3). If $x < t_0$, then $S(x) = y_0 + m_0(x - t_0)$; if $x > t_n$, then $S(x) = y_{n-1} + m_{n-1}(x - t_{n-1})$.

real function *spline1* $(n, (t_i), (y_i), x)$
real array $(t_i)_{0:n}, (y_i)_{0:n}$
integer i, n
real x
for $i = n - 1$ **to** 1 **step** -1 **do**
  if $x - t_i \geqq 0$ **then** exit loop
end do
spline1 $\leftarrow y_i + (x - t_i)[(y_{i+1} - y_i)/(t_{i+1} - t_i)]$
end function *spline1*

Second-Degree Splines

Interpolating splines of degree higher than 1 are more complicated. We take up now the quadratic splines and show how to determine one that interpolates a given table:

x	t_0	t_1	t_2	$\cdots$	t_n
y	y_0	y_1	y_2	$\cdots$	y_n

We shall assume that the points $t_0, t_1, \ldots, t_n$ serve also as the knots for our spline function. Later, another quadratic spline interpolant is discussed in which the nodes for interpolation are different from the knots. Quadratic splines are not used in applications as often as are natural cubic splines, which are developed in the next section. However, the derivations of quadratic and cubic splines are similar enough that an understanding of the simpler second-degree spline theory will allow one to grasp easily the more complicated third-degree spline theory.

A function $S(x)$ is a **spline of degree 2** if S is a piecewise quadratic polynomial such that S and S' are continuous. A simple counting process shows us the number of conditions involved in defining such a **quadratic spline**. If there are $n + 1$ knots, then there are n subintervals and $n - 1$ interior knots. Since the spline $S(x)$ consists of quadratic polynomials of the form $a_i x^2 + b_i x + c_i$ over each subinterval $[t_i, t_{i+1}]$, there are $3n$ coefficients. We then expect that $3n$ conditions will fully define a quadratic spline function with n knots.

On each subinterval $[t_i, t_{i+1}]$, the quadratic spline function S_i must satisfy the interpolation conditions $S_i(t_i) = y_i$ and $S_i(t_{i+1}) = y_{i+1}$. Since there are n such subintervals, this imposes $2n$ conditions. The continuity of S does *not* add any additional conditions. (Why?) However, the continuity of S' at each of the interior knots gives $n - 1$ more conditions. Thus, we have $2n + n - 1 = 3n - 1$ conditions, or *one* condition short of the $3n$ conditions required. There are a variety of ways to impose this additional condition; for example, $S'(t_0) = 0$.

EXAMPLE 2   Determine whether the following function is a quadratic spline:

$$f(x) = \begin{cases} x^2 & x \le 0 \\ -x^2 & 0 \le x \le 1 \\ 1 - 2x & x \ge 1 \end{cases}$$

Solution   The function is obviously piecewise quadratic. Whether f and f' are continuous at the interior knots can be determined as follows:

$$\lim_{x \to 0^-} f(x) = \lim_{x \to 0^-} x^2 = 0 \qquad \lim_{x \to 0^+} f(x) = \lim_{x \to 0^+} (-x^2) = 0$$

$$\lim_{x \to 1^-} f(x) = \lim_{x \to 1^-} (-x^2) = -1 \qquad \lim_{x \to 1^+} f(x) = \lim_{x \to 1^+} (1 - 2x) = -1$$

$$\lim_{x \to 0^-} f'(x) = \lim_{x \to 0^-} 2x = 0 \qquad \lim_{x \to 0^+} f'(x) = \lim_{x \to 0^+} (-2x) = 0$$

$$\lim_{x \to 1^-} f'(x) = \lim_{x \to 1^-} (-2x) = -2 \qquad \lim_{x \to 1^+} f'(x) = \lim_{x \to 1^+} (-2) = -2$$

Consequently, $f(x)$ is a quadratic spline.   ◻

Quadratic Spline $Q(x)$

We now derive the equations for the interpolating quadratic spline, $Q(x)$. The value of $Q'(t_0)$ is prescribed as the additional condition. We seek a piecewise quadratic function

$$Q(x) = \begin{cases} Q_0(x) & t_0 \le x \le t_1 \\ Q_1(x) & t_1 \le x \le t_2 \\ \vdots & \vdots \\ Q_{n-1}(x) & t_{n-1} \le x \le t_n \end{cases} \tag{4}$$

which is continuously differentiable on the entire interval $[t_0, t_n]$ and which interpolates the table; that is, $Q(t_i) = y_i$ for $0 \le i \le n$.

Since Q' is continuous, we can put $z_i \equiv Q'(t_i)$. At present, we do not know the correct values of z_i, but nevertheless the following must be the formula for Q_i:

$$Q_i(x) = \frac{z_{i+1} - z_i}{2(t_{i+1} - t_i)}(x - t_i)^2 + z_i(x - t_i) + y_i \tag{5}$$

To see that this is correct, just verify that $Q_i(t_i) = y_i$, $Q_i'(t_i) = z_i$, and $Q_i'(t_{i+1}) = z_{i+1}$. These three conditions define the function Q_i uniquely on $[t_i, t_{i+1}]$ as given in Equation (5).

Now, in order for the quadratic spline function Q to be continuous and to interpolate the table of data, it is necessary and sufficient that $Q_i(t_{i+1}) = y_{i+1}$ for $i = 0, 1, \ldots, n - 1$ in Equation (5) with z_0 arbitrary. When this equation is written out in detail and simplified, the result is this:

$$z_{i+1} = -z_i + 2\left(\frac{y_{i+1} - y_i}{t_{i+1} - t_i}\right) \qquad (0 \leq i \leq n - 1) \qquad (6)$$

This equation can be used to obtain the vector $[z_0, z_1, \ldots, z_n]^T$, starting with an arbitrary value for z_0. We summarize with an algorithm:

Algorithm for Quadratic Spline Interpolation at the Knots

1. Determine $[z_0, z_1, \ldots, z_n]^T$ by selecting z_0 arbitrarily and computing $z_1, z_2, \ldots, z_n$ recursively by Formula (6).

2. The quadratic spline interpolating function Q is given by Formulas (4) and (5).

Subbotin Quadratic Spline

A useful approximation process, first proposed by Subbotin [1967], consists of interpolation with *quadratic* splines, where the nodes for interpolation are chosen to be the first and last knots and the midpoints between the knots. Remember that **knots** are defined as the points where the spline function is permitted to change in form from one polynomial to another. The **nodes** are the points where values of the spline are specified. In the Subbotin quadratic spline function, there are $n + 2$ interpolation conditions and $2(n - 1)$ conditions from the continuity of S and S'. Hence, we have the exact number of conditions needed, $3n$, in order to define the quadratic spline function completely.

   We outline the theory here, leaving details for the reader to fill in. Suppose that knots $a = t_0 < t_1 < \cdots < t_n = b$ have been specified; let the nodes be the points

$$\begin{cases} \tau_0 = t_0 \qquad \tau_{n+1} = t_n \\ \tau_i = \frac{1}{2}(t_i + t_{i-1}) \qquad (1 \leq i \leq n) \end{cases}$$

We seek a quadratic spline function S that has the given knots and takes prescribed values at the nodes:

$$S(\tau_i) = y_i \qquad (0 \leq i \leq n + 1)$$

as in Figure 7.4. The knots create n subintervals, and in each of them S can be a different quadratic polynomial. Let us say that on $[t_i, t_{i+1}]$, S is equal to the quadratic polynomial S_i. Since S is a quadratic spline, it and its first derivative should be continuous. Thus, $z_i \equiv S'(t_i)$ is well defined, although as yet we do not know its values. It is easy to see that on $[t_i, t_{i+1}]$, our quadratic polynomial can be represented in the form

$$S_i(x) = y_{i+1} + \frac{1}{2}(z_{i+1} + z_i)(x - \tau_{i+1}) + \frac{1}{2h_i}(z_{i+1} - z_i)(x - \tau_{i+1})^2 \qquad (7)$$

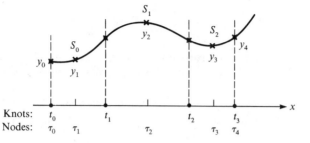

FIGURE 7.4
Subbotin quadratic
splines ($t_0 = \tau_0, t_3 = \tau_4$)

in which $h_i = t_{i+1} - t_i$. To verify the correctness of Equation (7), we must check that $S_i(\tau_{i+1}) = y_{i+1}, S_i'(t_i) = z_i$, and $S_i'(t_{i+1}) = z_{i+1}$. When the polynomial pieces $S_0, S_1, \ldots, S_{n-1}$ are joined together to form S, the result may be discontinuous. Hence, we impose continuity conditions at the interior knots:

$$\lim_{x \to t_i^-} S_{i-1}(x) = \lim_{x \to t_i^+} S_i(x) \qquad (1 \leq i \leq n-1)$$

The reader should carry out this analysis, which leads to

$$h_{i-1}z_{i-1} + 3(h_{i-1} + h_i)z_i + h_i z_{i+1} = 8(y_{i+1} - y_i) \qquad (1 \leq i \leq n-1) \quad \textbf{(8)}$$

The first and last interpolation conditions must also be imposed:

$$S(\tau_0) = y_0 \qquad S(\tau_{n+1}) = y_{n+1}$$

These two equations lead to

$$3h_0 z_0 + h_0 z_1 = 8(y_1 - y_0)$$

$$h_{n-1}z_{n-1} + 3h_{n-1}z_n = 8(y_{n+1} - y_n)$$

The system of equations governing the vector $\mathbf{z} = [z_0, z_1, \ldots, z_n]^T$ then can be written in the matrix form

$$\begin{bmatrix} 3h_0 & h_0 & & & & & \\ h_0 & 3(h_0 + h_1) & h_1 & & & & \\ & h_1 & 3(h_1 + h_2) & h_2 & & & \\ & & \ddots & \ddots & \ddots & & \\ & & & h_{n-2} & 3(h_{n-2} + h_{n-1}) & h_{n-1} \\ & & & & h_{n-1} & 3h_{n-1} \end{bmatrix} \begin{bmatrix} z_0 \\ z_1 \\ z_2 \\ \vdots \\ z_{n-1} \\ z_n \end{bmatrix}$$

$$= 8 \begin{bmatrix} y_1 - y_0 \\ y_2 - y_1 \\ y_3 - y_2 \\ \vdots \\ y_n - y_{n-1} \\ y_{n+1} - y_n \end{bmatrix}$$

This system of $n + 1$ equations in $n + 1$ unknowns can be conveniently solved by procedure *tri* in Chapter 6. After the z vector has been obtained, values of $S(x)$ can be computed from Equation (7). The writing of suitable code to carry out this interpolation method is left as a programming project.

PROBLEMS 7.1

1. Determine whether this function is a first-degree spline:

$$f(x) = \begin{cases} x & -1 \leq x \leq 0.5 \\ 0.5 + 2(x - 0.5) & 0.5 \leq x \leq 2 \\ x + 1.5 & 2 \leq x \leq 4 \end{cases}$$

2. The simplest type of spline function is the piecewise constant function, which could be defined as

$$S(x) = \begin{cases} c_0 & t_0 \leq x < t_1 \\ c_1 & t_1 \leq x < t_2 \\ \vdots & \vdots \\ c_{n-1} & t_{n-1} \leq x \leq t_n \end{cases}$$

Show that the indefinite integral of such a function is a polygonal function. What is the relationship between the piecewise constant functions and the rectangle rule of numerical integration? (See Problem **14** in Section 5.2.)

3. Show that $f(x) - p(x) = \frac{1}{2}f''(\xi)(x - a)(x - b)$ for some ξ in the interval (a, b), where p is a linear polynomial that interpolates f at a and b. *Hint*: Use a result from Section 4.2.

4. (Continuation) Show that $|f(x) - p(x)| \leq \frac{1}{8}M\ell^2$, where $\ell = b - a$, if $|f''(x)| \leq M$ on the interval (a, b).

5. (Continuation) Show that

$$f(x) - p(x) = \frac{(x - a)(x - b)}{b - a}\left[\frac{f(x) - f(b)}{x - b} - \frac{f(x) - f(a)}{x - a}\right]$$

6. (Continuation) If $|f'(x)| \leq Q$ on (a, b), show that $|f(x) - p(x)| \leq Q\ell/2$. *Hint*: Use the Mean-Value Theorem on the result of Problem 5.

7. (Continuation) Set S be a spline function of degree 1 that interpolates f at $t_0, t_1, \ldots, t_n$. Let $t_0 < t_1 < \cdots < t_n$ and let $\delta = \max_{0 \leq i \leq n-1}(t_{i+1} - t_i)$. Then $|f(x) - S(x)| \leq Q\delta/2$, where Q is an upper bound of $|f'(x)|$ on (t_0, t_n).

8. Let f be continuous on $[a, b]$. For a given $\varepsilon > 0$, let δ have the property that $|f(x) - f(y)| < \varepsilon$ whenever $|x - y| < \delta$ (uniform continuity principle). Let $n > 1 + (b - a)/\delta$. Show that there is a first-degree spline S having n knots such that $|f(x) - S(x)| < \varepsilon$ on $[a, b]$. *Hint*: Use Problem 5.

9. If the function $f(x) = \sin(100x)$ is to be approximated on the interval $[0, \pi]$ by an interpolating spline of degree 1, how many knots are needed to ensure that $|S(x) - f(x)| < 10^{-8}$?

10. Let $t_0 < t_1 < \cdots < t_n$. Construct first-degree spline functions $S_0, S_1, \ldots, S_n$ by requiring that S_i vanish at $t_0, t_1, \ldots, t_{i-1}, t_{i+1}, \ldots, t_n$ but that $S_i(t_i) = 1$. Show that the first-degree spline function that interpolates f at $t_0, t_1, \ldots, t_n$ is $\sum_{i=0}^{n} f(t_i) S_i(x)$.

11. Show that the trapezoid rule for numerical integration results from approximating f by a first-degree spline S and then using

$$\int_a^b f(x)\, dx \approx \int_a^b S(x)\, dx$$

12. Prove that the derivative of a quadratic spline is a first-degree spline.

13. If the knots t_i happen to be the integers $0, 1, \ldots, n$, find a good way to determine the index i for which $t_i \le x < t_{i+1}$?

14. Show that the indefinite integral of a first-degree spline is a second-degree spline.

15. Define $r(x) = 0$ if $x < 0$ and $r(x) = x^2$ if $x \ge 0$. Show that r and r' are continuous. Show that any quadratic spline with knots $t_0, t_1, \ldots, t_n$ is of the form

$$ax^2 + bx + c + \sum_{i=1}^{n-1} d_i r(x - t_i)$$

16. Define a function k by the equation

$$k(x) = \begin{cases} 0 & x < 0 \\ x & x \ge 0 \end{cases}$$

Prove that every first-degree spline function that has knots $t_0, t_1, \ldots, t_n$ can be written in the form

$$ax + b + \sum_{i=1}^{n-1} c_i k(x - t_i)$$

17. Find a quadratic spline interpolant for these data:

x	-1	0	$1/2$	1	2	$5/2$
y	2	1	0	1	2	3

18. (Continuation) Show that no quadratic spline S interpolates the table of Problem 17 and satisfies $S'(t_0) = S'(t_5)$.

19. What equations must be solved if a quadratic spline S that has knots $t_0, t_1, \ldots, t_n$ is required to take prescribed values at points $\frac{1}{2}(t_i + t_{i+1})$ for $0 \le i \le n - 1$?

20. Are these functions quadratic splines? Explain why or why not.

a. $S(x) = \begin{cases} 0.1x^2 & 0 \le x \le 1 \\ 9.3x^2 - 18.4x + 9.2 & 1 \le x \le 1.3 \end{cases}$

b. $S(x) = \begin{cases} -x^2 & x \le 0 \\ x & x > 0 \end{cases}$

$$\text{c.}\quad S(x) = \begin{cases} x & -\infty < x \leqq 1 \\ x^2 & 1 \leqq x \leqq 2 \\ 4 & 2 \leqq x < \infty \end{cases}$$

21. Is $f(x) = |x|$ a first-degree spline? Why or why not?

22. Verify that Formula (5) has the three properties $Q_i(t_i) = y_i$, $Q_i'(t_i) = z_i$, and $Q_i'(t_{i+1}) = z_{i+1}$.

23. (Continuation) Impose the continuity condition on Q and derive the system of Equation (6).

24. Show by induction that the recursive Formula (6) together with Equation (5) produces an interpolating quadratic spline function.

25. Verify the correctness of the equations in the text that pertain to Subbotin's spline interpolation process.

26. Analyze the Subbotin interpolation scheme in this alternative manner. First, let $v_i = S(t_i)$. Show that

$$S_i(x) = A_i(x - t_i)^2 + B_i(x - t_{i+1})^2 + C_i$$

where

$$C_i = 2y_i - \frac{1}{2}v_i - \frac{1}{2}v_{i+1} \qquad B_i = \frac{v_i - C_i}{h_i^2} \qquad A_i = \frac{v_{i+1} - C_i}{h_i^2} \qquad h_i = t_{i+1} - t_i$$

Hint: Show that $S_i(t_i) = v_i$, $S_i(t_{i+1}) = v_{i+1}$, and $S_i(\tau_i) = y_i$.

27. (Continuation) When continuity conditions on S' are imposed, show that the result is the following equation, in which $i = 1, 2, \ldots, n - 1$:

$$h_i v_{i-1} + 3(h_i + h_{i+1})v_i + h_{i-1}v_{i+1} = 4h_{i-1}y_i + 4h_i y_{i-1}$$

COMPUTER PROBLEMS 7.1

1. Rewrite procedure *spline1* so that ascending subintervals are considered instead of descending ones. Test the code on a table of 15 unevenly spaced data points.

2. Rewrite procedure *spline1* so that a *binary search* is used to find the desired interval. Test the revised code. What are the advantages and/or disadvantages of a binary search over the procedure in the text? A binary search is similar to the bisection method in that we choose t_k with $k = (i + j)/2$ or $k = (i + j + 1)/2$ and determine whether x is in $[t_i, t_k]$ or $[t_k, t_j]$.

3. A piecewise bilinear polynomial that interpolates points (x, y) specified in a rectangular grid is given by

$$p(x, y) = \frac{(\ell_{ij}z_{i+1,j+1} + \ell_{i+1,j+1}z_{ij}) - (\ell_{i+1,j}z_{i,j+1} + \ell_{i,j+1}z_{i+1,j})}{(x_{i+1} - x_i)(y_{j+1} - y_j)}$$

where $\ell_{ij} = (x_i - x)(y_j - y)$. Here $x_i \leqq x \leqq x_{i+1}$ and $y_j \leqq y \leqq y_{j+1}$. The given grid (x_i, y_j) is specified by strictly increasing arrays (x_i) and (y_j) of length n and

m, respectively. The given values z_{ij} at the grid points (x_i, y_j) are contained in the $n \times m$ array (z_{ij}), shown in the figure. Write

real function $bilinear((x_i), n, (y_j), m, (z_{ij}), x, y)$

to compute the value of $p(x, y)$. Test this routine on a set of 5×10 unequally spaced data points. Evaluate *bilinear* at four grid points and five nongrid points.

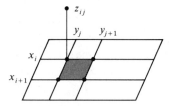

4. Write an adaptive spline interpolation procedure. The input should be a function f, an interval $[a, b]$, and a tolerance ε. The output should be a set of knots $a = t_0 < t_1 < \cdots < t_n = b$ and a set of function values $y_i = f(t_i)$ such that the first-degree spline interpolating function S satisfies $|S(x) - f(x)| \leq \varepsilon$ whenever x is any point $x_{ij} = t_i + j(t_{i+1} - t_j)/10$ for $0 \leq i \leq n - 1$ and $0 \leq j \leq 9$.

5. Write

procedure $spline2\_coef(n, t, (y_i), (z_i))$

that computes the (z_i) array in the quadratic spline interpolation process (interpolation at the knots). Then write

real function $spline2(n, (t_i), (y_i), (z_i), x)$

that computes values of $Q(x)$.

6. Carry out the programming project of Computer Problem **5** for the Subbotin quadratic spline.

7.2 Natural Cubic Splines

Introduction

The first- and second-degree splines discussed in the preceding section, though useful in certain applications, suffer an obvious imperfection: Their low-order derivatives are discontinuous. In the case of the first-degree spline (or polygonal line), this lack of smoothness is immediately evident because the slope of the spline may change abruptly from one value to another at each knot. For the quadratic spline, the discontinuity is in the second derivative and is therefore not so evident. But the *curvature* of the quadratic spline changes abruptly at each knot, and the curve may not be pleasing to the eye.

The general definition of spline functions of arbitrary degree is this: A function S is called a **spline of degree k** if:

1. The domain of S is an interval $[a, b]$.

2. $S, S', S'', \ldots, S^{(k-1)}$ are all continuous functions on $[a, b]$.

3. There are points t_i (the knots of S) such that $a = t_0 < t_1 < \cdots < t_n = b$ and such that S is a polynomial of degree $\leq k$ on each subinterval $[t_i, t_{i+1}]$.

Higher-degree splines are used whenever more smoothness is needed in the approximating function. From the definition of a spline function of degree k, we see that such a function will be continuous and have continuous derivatives $S', S'', \ldots, S^{(k-1)}$. If we want the approximating spline to have a continuous mth derivative, a spline of degree at least $m + 1$ is selected. To see why, consider a situation in which knots $t_0 < t_1 < \cdots < t_n$ have been prescribed. Suppose that a piecewise polynomial of degree m is to be defined, with its pieces joined at the knots in such a way that the resulting spline S has m continuous derivatives. At a typical interior knot t, we have the following circumstances: To the left of t, $S(x) = p(x)$; to the right of t, $S(x) = q(x)$, where p and q are mth-degree polynomials. The continuity of the mth derivative $S^{(m)}$ implies the continuity of the lower-order derivatives $S^{(m-1)}, S^{(m-2)}, \ldots, S', S$. Therefore, at the knot t,

$$\lim_{x \to t^-} S^{(k)}(x) = \lim_{x \to t^+} S^{(k)}(x) \qquad (0 \leq k \leq m)$$

from which we conclude that

$$\lim_{x \to t^-} p^{(k)}(x) = \lim_{x \to t^+} q^{(k)}(x) \qquad (0 \leq k \leq m) \qquad \textbf{(1)}$$

Since p and q are polynomials, their derivatives of all orders are continuous, and so Equation (1) is the same as

$$p^{(k)}(t) = q^{(k)}(t) \qquad (0 \leq k \leq m)$$

This condition forces p and q to be the *same* polynomial because by Taylor's Theorem,

$$p(x) = \sum_{k=0}^{m} \frac{1}{k!} p^{(k)}(t)(x - t)^k = \sum_{k=0}^{m} \frac{1}{k!} q^{(k)}(t)(x - t)^k = q(x)$$

This argument can be applied at each of the interior knots $t_1, t_2, \ldots, t_{n-1}$, and we see that S is simply one polynomial throughout the entire interval from t_0 to t_n. Thus, we need a piecewise polynomial of degree $m + 1$ with at most m continuous derivatives in order to have a spline function that is not just a single polynomial throughout the entire interval. (We already know that ordinary polynomials usually do not serve well in curve fitting. See Section 4.2.)

The choice of degree most frequently made for a spline function is 3. The resulting splines are termed **cubic splines**. In this case, we join cubic polynomials together in such a way that the resulting spline function has two continuous derivatives everywhere. At each knot, three continuity conditions will be imposed. Since S, S', and S'' are continuous, the graph of the function will appear smooth to the eye. Discontinuities, of course, will occur in the third derivative but cannot be detected visually, which is one reason for choosing degree 3. Experience has shown, moreover, that using splines of degree greater than 3 seldom yields any advantage. For technical reasons, odd-degree splines behave better than even-degree splines (when interpolating at the knots). Finally, a very elegant theorem, to be proved later, shows that in a certain precise sense the cubic interpolating spline function is the best interpolating function available. Thus, our emphasis on the cubic splines is well justified.

Natural Cubic Spline

We turn next to interpolating a given table of function values by a cubic spline whose knots coincide with the values of the independent variable in the table. As earlier, we start with the table:

x	t_0	t_1	$\cdots$	t_n
y	y_0	y_1	$\cdots$	y_n

The t_i's are the knots and are assumed to be arranged in ascending order.

The function S that we wish to construct consists of n cubic polynomial pieces:

$$S(x) = \begin{cases} S_0(x) & t_0 \leq x \leq t_1 \\ S_1(x) & t_1 \leq x \leq t_2 \\ \vdots & \vdots \\ S_{n-1}(x) & t_{n-1} \leq x \leq t_n \end{cases}$$

In this formula, S_i denotes the cubic polynomial that will be used on the subinterval $[t_i, t_{i+1}]$. The interpolation conditions are

$$S(t_i) = y_i \qquad (0 \leq i \leq n)$$

The continuity conditions are imposed only at the *interior* knots $t_1, t_2, \ldots, t_{n-1}$. (Why?) These conditions are written as

$$\lim_{x \to t_i^-} S^{(k)}(t_i) = \lim_{x \to t_i^+} S^{(k)}(t_i) \qquad (k = 0, 1, 2)$$

It turns out that two more conditions must be imposed in order to use all the degrees of freedom available. The choice that we make for these two extra conditions is

$$S''(t_0) = S''(t_n) = 0 \tag{2}$$

The resulting spline function is then termed a **natural cubic spline**.

We now verify that the number of conditions imposed equals the number of coefficients available. There are $n + 1$ knots and hence n subintervals. On each of these subintervals, we shall have a different cubic polynomial. Since a cubic polynomial has four coefficients, a total of $4n$ coefficients are available. As for the conditions imposed, we have specified that within each interval the interpolating polynomial must go through two points, which gives $2n$ conditions. The continuity adds no additional conditions. The first and second derivatives must be continuous at the $n - 1$ interior points, for $2(n - 1)$ more conditions. The second derivatives must vanish at the two endpoints for a total of $2n + 2(n - 1) + 2 = 4n$ conditions.

EXAMPLE 1   Determine the parameters a, b, c, d, e, f, g, and h so that $S(x)$ is a natural cubic spline, where

$$S(x) = \begin{cases} ax^3 + bx^2 + cx + d & x \in [-1, 0] \\ ex^3 + fx^2 + gx + h & x \in [0, 1] \end{cases}$$

with interpolation conditions $S(-1) = 1, S(0) = 2$, and $S(1) = -1$.

Solution   Let the two cubic polynomials be $S_0(x)$ and $S_1(x)$. From the interpolation conditions, we have $d = 2, h = 2, -a + b - c = -1$, and $e + f + g = -3$. Since

$$S'(x) = \begin{cases} 3ax^2 + 2bx + c \\ 3ex^2 + 2fx + g \end{cases}$$

we have $c = g$ from the continuity condition of S'. Also, since

$$S''(x) = \begin{cases} 6ax + 2b \\ 6ex + 2f \end{cases}$$

we have $b = f$ from the continuity condition of S''. In order for S to be a natural cubic spline, we must have $3a = b$ and $3e = -f$. From all of these equations, we obtain $a = -1, b = -3, c = -1, d = 2, e = 1, f = -3, g = -1$, and $h = 2$.   □

Algorithm for Natural Cubic Spline

From the previous example, it is evident that we need to develop a systematic procedure for determining the formula for a natural cubic spline, given a table of interpolation values. This is our objective in the material on the next several pages.

Since S'' is continuous, the numbers

$$z_i \equiv S''(t_i) \qquad (0 \leq i \leq n)$$

are unambiguously defined. We do not yet know the values $z_1, z_2, \ldots, z_{n-1}$, but, of course, $z_0 = z_n = 0$ by Equation (2).

If the z_i's were known, we could construct S as now described. On the interval $[t_i, t_{i+1}]$, S'' is a linear polynomial that takes the values z_i and z_{i+1} at the endpoints. Thus,

$$S_i''(x) = \frac{z_{i+1}}{h_i}(x - t_i) + \frac{z_i}{h_i}(t_{i+1} - x) \tag{3}$$

with $h_i = t_{i+1} - t_i$ for $0 \leq i \leq n - 1$. Clearly, $S_i''(t_i) = z_i$, $S_i''(t_{i+1}) = z_{i+1}$, and S_i'' is linear in x. If this is integrated twice, we obtain S_i itself:

$$S_i(x) = \frac{z_{i+1}}{6h_i}(x - t_i)^3 + \frac{z_i}{6h_i}(t_{i+1} - x)^3 + cx + d$$

where c and d are constants of integration. By adjusting the integration constants, we obtain a form for S_i that is easier to work with—namely,

$$S_i(x) = \frac{z_{i+1}}{6h_i}(x - t_i)^3 + \frac{z_i}{6h_i}(t_{i+1} - x)^3 + C(x - t_i) + D(t_{i+1} - x) \tag{4}$$

where C and D are constants. If we differentiate Equation (4) twice, we obtain Equation (3).

The interpolation conditions $S_i(t_i) = y_i$ and $S_i(t_{i+1}) = y_{i+1}$ can be imposed now to determine the appropriate values of C and D. The reader should do so (Problem **27**) and verify that the result is

$$S_i(x) = \frac{z_{i+1}}{6h_i}(x - t_i)^3 + \frac{z_i}{6h_i}(t_{i+1} - x)^3$$
$$+ \left(\frac{y_{i+1}}{h_i} - \frac{h_i}{6}z_{i+1}\right)(x - t_i) + \left(\frac{y_i}{h_i} - \frac{h_i}{6}z_i\right)(t_{i+1} - x) \tag{5}$$

When the values $z_0, z_1, \ldots, z_n$ have been determined, the spline function $S(x)$ is obtained from equations of this form for $S_0(x), S_1(x), \ldots, S_{n-1}(x)$.

We now show how to determine the z_i's. One condition remains to be imposed—namely, the continuity of S'. At the interior knots t_i for $1 \leq i \leq n - 1$, we must have $S_{i-1}'(t_i) = S_i'(t_i)$, as can be seen in Figure 7.5.

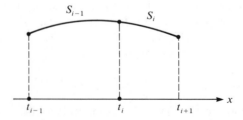

FIGURE 7.5

We have from Equation (5)

$$S'_i(x) = \frac{z_{i+1}}{2h_i}(x - t_i)^2 - \frac{z_i}{2h_i}(t_{i+1} - x)^2 + \frac{y_{i+1}}{h_i} - \frac{h_i}{6}z_{i+1} - \frac{y_i}{h_i} + \frac{h_i}{6}z_i$$

This gives

$$S'_i(t_i) = -\frac{h_i}{6}z_{i+1} - \frac{h_i}{3}z_i + b_i \tag{6}$$

where

$$b_i = \frac{1}{h_i}(y_{i+1} - y_i)$$

Analogously, we have

$$S'_{i-1}(t_i) = \frac{h_{i-1}}{6}z_{i-1} + \frac{h_{i-1}}{3}z_i + b_{i-1}$$

When these are set equal to each other, the resulting equation can be rearranged as

$$h_{i-1}z_{i-1} + 2(h_{i-1} + h_i)z_i + h_iz_{i+1} = 6(b_i - b_{i-1})$$

for $1 \leq i \leq n - 1$. By letting

$$u_i = 2(h_{i-1} + h_i)$$
$$v_i = 6(b_i - b_{i-1})$$

we obtain a tridiagonal system of equations:

$$\begin{cases} z_0 = 0 \\ h_{i-1}z_{i-1} + u_iz_i + h_iz_{i+1} = v_i \quad (1 \leq i \leq n - 1) \\ z_n = 0 \end{cases} \tag{7}$$

to be solved for the z_i's. The simplicity of the first and last equations is a result of the natural cubic spline conditions $S''(t_0) = S''(t_n) = 0$.

EXAMPLE 2   Derive the equations of the natural cubic interpolating spline for the following table:

x	−1	0	1
y	1	2	−1

Solution   First, we need to determine the tridiagonal system (7). From the table, we have $h_0 = 1, h_1 = 1, b_0 = 1, b_1 = -3, u_1 = 4$, and $v_1 = -24$. The system is

$$\begin{cases} z_0 & = 0 \\ z_0 + 4z_1 + z_2 = -24 \\ z_2 = 0 \end{cases}$$

Thus, $z_0 = 0, z_1 = -6$, and $z_2 = 0$, and from Equation (5) we have

$$S(x) = \begin{cases} -(x + 1)^3 + 3(x + 1) - x & x \in [-1, 0] \\ -(1 - x)^3 - x + 3(1 - x) & x \in [0, 1] \end{cases}$$

or

$$S(x) = \begin{cases} -x^3 - 3x^2 - x + 2 & x \in [-1, 0] \\ x^3 - 3x^2 - x + 2 & x \in [0, 1] \end{cases}$$

which agrees with the results from the previous example.

We can use Maple as follows to generate and plot this spline function.

```
readlib(spline):
spline([-1, 0, 1], [1, 2, -1], x, cubic);
S := 'spline/makeproc'(", x);
interface(plotdevice=postscript,plotoutput=myfile1);
plot(S, -1..1);                                                    □
```

Writing System (7) in matrix form, we have

$$\begin{bmatrix} 1 & 0 & & & & \\ h_0 & u_1 & h_1 & & & \\ & h_1 & u_2 & h_2 & & \\ & & \ddots & \ddots & \ddots & \\ & & & h_{n-2} & u_{n-1} & h_{n-1} \\ & & & & 0 & 1 \end{bmatrix} \begin{bmatrix} z_0 \\ z_1 \\ z_2 \\ \vdots \\ z_{n-1} \\ z_n \end{bmatrix} = \begin{bmatrix} 0 \\ v_1 \\ v_2 \\ \vdots \\ v_{n-1} \\ 0 \end{bmatrix} \tag{8}$$

On eliminating the first and last equations, we have

$$\begin{bmatrix} u_1 & h_1 & & & \\ h_1 & u_2 & h_2 & & \\ & \ddots & \ddots & \ddots & \\ & & h_{n-3} & u_{n-2} & h_{n-2} \\ & & & h_{n-2} & u_{n-1} \end{bmatrix} \begin{bmatrix} z_1 \\ z_2 \\ \vdots \\ z_{n-2} \\ z_{n-1} \end{bmatrix} = \begin{bmatrix} v_1 \\ v_2 \\ \vdots \\ v_{n-2} \\ v_{n-1} \end{bmatrix}$$

which is a symmetric tridiagonal system of order $n - 1$. We could use procedure *tridiagonal* developed in Section 6.3 to solve this system. However, we can design an algorithm specifically for it (based on the ideas in Section 6.3). In Gaussian elimination *without pivoting*, the forward elimination phase would modify the u_i's and v_i's as follows:

$$\begin{cases} u_i \leftarrow u_i - \dfrac{h_{i-1}^2}{u_{i-1}} \\[2ex] v_i \leftarrow v_i - \dfrac{h_{i-1}v_{i-1}}{u_{i-1}} \end{cases} \qquad (i = 2, 3, \ldots, n-1)$$

The back substitution phase yields

$$\begin{cases} z_{n-1} \leftarrow \dfrac{v_{n-1}}{u_{n-1}} \\[2ex] z_i \quad \leftarrow \dfrac{v_i - h_i z_{i+1}}{u_i} \qquad (i = n-2, n-3, \ldots, 1) \end{cases}$$

Putting all this together leads to the following algorithm, designed especially for the tridiagonal system (8).

Algorithm for Solving the Natural Cubic Spline Tridiagonal System Directly

Given the interpolation points (t_i, y_i) for $i = 0, 1, \ldots, n$:

1. Compute for $i = 0, 1, \ldots, n-1$

$$\begin{cases} h_i = t_{i+1} - t_i \\[2ex] b_i = \dfrac{1}{h_i}(y_{i+1} - y_i) \end{cases}$$

2. Set

$$\begin{cases} u_1 = 2(h_0 + h_1) \\ v_1 = 6(b_1 - b_0) \end{cases}$$

and compute inductively for $i = 2, 3, \ldots, n-1$

$$\begin{cases} u_i = 2(h_i + h_{i-1}) - \dfrac{h_{i-1}^2}{u_{i-1}} \\[2ex] v_i = 6(b_i - b_{i-1}) - \dfrac{h_{i-1}v_{i-1}}{u_{i-1}} \end{cases}$$

3. Set

$$\begin{cases} z_n = 0 \\ z_0 = 0 \end{cases}$$

and compute inductively for $i = n-1, n-2, \ldots, 1$

$$z_i = \dfrac{v_i - h_i z_{i+1}}{u_i}$$

This algorithm conceivably could fail because of divisions by zero in steps **2** and **3**. Therefore, let us prove that $u_i \neq 0$ for all i. It is clear that $u_1 > h_1 > 0$. If $u_{i-1} > h_{i-1}$, then $u_i > h_i$ because

$$u_i = 2(h_i + h_{i-1}) - \frac{h_{i-1}^2}{u_{i-1}} > 2(h_i + h_{i-1}) - h_{i-1} > h_i$$

Then by induction, $u_i > 0$ for $i = 1, 2, \ldots, n - 1$.

Equation (5) is not the best computational form for evaluating the cubic polynomial $S_i(x)$. We would prefer to have it in the form

$$S_i(x) = A_i + B_i(x - t_i) + C_i(x - t_i)^2 + D_i(x - t_i)^3 \qquad \textbf{(9)}$$

because nested multiplication can then be utilized.

Notice that Equation (9) is the Taylor expansion of S_i about the point t_i. Hence,

$$A_i = S_i(t_i) \qquad B_i = S_i'(t_i) \qquad C_i = \frac{1}{2}S_i''(t_i) \qquad D_i = \frac{1}{6}S_i'''(t_i)$$

Therefore, $A_i = y_i$ and $C_i = z_i/2$. The coefficient of x^3 in Equation (9) is D_i, whereas the coefficient of x^3 in Equation (5) is $(z_{i+1} - z_i)/6h_i$. Therefore,

$$D_i = \frac{1}{6h_i}(z_{i+1} - z_i)$$

Finally, Equation (6) provides the value of $S_i'(t_i)$, which is

$$B_i = -\frac{h_i}{6}z_{i+1} - \frac{h_i}{3}z_i + \frac{1}{h_i}(y_{i+1} - y_i)$$

Thus, the nested form of $S_i(x)$ is

$$S_i(x) = y_i + (x - t_i)\left(B_i + (x - t_i)\left(\frac{z_i}{2} + \frac{1}{6h_i}(x - t_i)(z_{i+1} - z_i)\right)\right) \qquad \textbf{(10)}$$

Pseudocode for Natural Cubic Splines

We now write routines for determining a natural cubic spline based on a table of values and for evaluating this function at a given value. First, we use the algorithm for directly solving the $(n + 1) \times (n + 1)$ tridiagonal system (8) for $z_0, z_1, \ldots, z_n$. This procedure, called *spline3_coef*, takes $n + 1$ table values (t_i, y_i) in arrays (t_i) and (y_i) and computes the z_i's, storing them in array (z_i). Intermediate (working) arrays (h_i), (b_i), (u_i), and (v_i) are needed.

procedure *spline3_coef* $(n, (t_i), (y_i), (z_i))$
real array $(t_i)_{0:n}, (y_i)_{0:n}, (z_i)_{0:n}$
allocate real array $(h_i)_{0:n-1}, (b_i)_{0:n-1}, (u_i)_{n-1}, (v_i)_{n-1}$
integer i, n
for $i = 0$ **to** $n - 1$ **do**
   $h_i \leftarrow t_{i+1} - t_i$
   $b_i \leftarrow (y_{i+1} - y_i)/h_i$
end do
$u_1 \leftarrow 2(h_0 + h_1)$
$v_1 \leftarrow 6(b_1 - b_0)$
for $i = 2$ **to** $n - 1$ **do**
   $u_i \leftarrow 2(h_i + h_{i+1}) - h_{i-1}^2/u_{i-1}$
   $v_i \leftarrow 6(b_i - b_{i-1}) - h_{i-1}v_{i-1}/u_{i-1}$
end do
$z_n \leftarrow 0$
for $i = n - 1$ **to** 1 **step** -1 **do**
   $z_i \leftarrow (v_i - h_i z_{i+1})/u_i$
end do
$z_0 \leftarrow 0$
end procedure *spline3_coef*

Now a procedure called *spline3_eval* is written for evaluating Equation (10), the natural cubic spline function $S(x)$, for x a given value. The procedure *spline3_eval* first determines the interval $[t_i, t_{i+1}]$ that contains x and then evaluates $S_i(x)$ using the nested form of this cubic polynomial.

real function *spline3_eval* $(n, (t_i), (y_i), (z_i), x)$
real array $(t_i)_{0:n}, (y_i)_{0:n}, (z_i)_{0:n}$
integer i
real h, tmp
for $i = n - 1$ **to** 1 **step** -1 **do**
   if $x - t_i \geqq 0$ **then** exit loop
end do
$h \leftarrow t_{i+1} - t_i$
$tmp \leftarrow (z_i/2) + (x - t_i)(z_{i+1} - z_i)/(6h)$
$tmp \leftarrow -(h/6)(z_{i+1} + 2z_i) + (y_{i+1} - y_i)/h + (x - t_i)(tmp)$
spline3_eval $\leftarrow y_i + (x - t_i)(tmp)$
end function *spline3_eval*

The function *spline3_eval* can be used repeatedly with different values of x after one call to procedure *spline3_coef*. For example, this would be the procedure when plotting a natural cubic spline curve. Since procedure *spline3_coef* stores the solution of the tridiagonal system corresponding to a particular spline function in the array (z_i), the arguments $n, (t_i), (y_i)$, and (z_i) must not be altered between repeated uses of *spline3_eval*.

Using Pseudocode for Interpolating and Curve Fitting

In order to illustrate the use of the natural cubic spline routines *spline3_coef* and *spline3_eval*, we rework an example from Section 4.1.

EXAMPLE 3   Write pseudocode for a program that determines the natural cubic spline interpolant for $\sin x$ at ten equidistant knots in the interval $[0, 1.6875]$. Print the value of $\sin x - S(x)$ at 37 equally spaced points in the same interval.

Solution   Here is a suitable pseudocode main program, which calls procedures *spline3_coef* and *spline3_eval*:

```
procedure main
integer parameter n ← 9
real parameter a ← 0, b ← 1.6875
real array (tᵢ)₀:ₙ, (yᵢ)₀:ₙ, (zᵢ)₀:ₙ
integer i
real d, h, x
h ← (b − a)/n
for i = 0 to n do
    tᵢ ← a + ih
    yᵢ ← sin(tᵢ)
end do
call spline3_coef (n, (tᵢ), (yᵢ), (zᵢ))
for i = 0 to 4n do
    x ← a + ih/4
    d ← sin(x) − spline3_eval(n, (tᵢ), (yᵢ), (zᵢ), x)
    output i, x, d
end do
end main
```

A typical result from the computer output is $i = 19$, $x = 0.890625$, and $d = 0.930E-05$. These values are *not* as accurate as those from the Newton interpolating polynomial! (Explain.)   □

We now illustrate the use of spline functions in fitting a curve to a set of data. Consider the following table:

x	0.0	0.6	1.5	1.7	1.9	2.1	2.3	2.6	2.8	3.0
y	−0.8	−0.34	0.59	0.59	0.23	0.1	0.28	1.03	1.5	1.44

3.6	4.7	5.2	5.7	5.8	6.0	6.4	6.9	7.6	8.0
0.74	−0.82	−1.27	−0.92	−0.92	−1.04	−0.79	−0.06	1.0	0.0

These 20 points were selected from a wiggly freehand curve drawn on graph paper. We intentionally selected more points where the curve bent sharply, and sought to

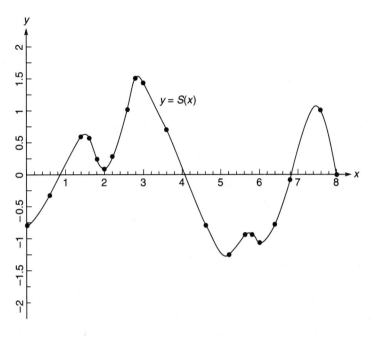

FIGURE 7.6
Natural cubic
spline curve

reproduce the curve using an automatic plotter. A visually pleasing curve is provided by using the cubic spline routines *spline3_coef* and *spline3_eval*. Figure 7.6 shows the resulting natural cubic spline curve.

Alternatively, we can use a Maple program as follows to plot the cubic spline function for this table.

```
readlib(spline):
spline([0,.6,1.5,1.7,1.9,2.1,2.3,2.6,2.8,3,
        3.6,4.7,5.2,5.7,5.8,6,6.4,6.9,7.6,8],
       [-.8,-.34,.59,.59,.23,.1,.28,1.03,1.5,
        1.44,.74,-.82,-1.27,-.92,-.92,-1.04,
        -.79,-.06,1,0], x, cubic);
S := 'spline/makeproc'(", x);
interface(plotdevice=postscript,plotoutput=myfile2);
plot(S, 0..8);
```

Here is another example of curve fitting using both the polynomial interpolation routines *coef* and *eval* from Chapter 4 and the cubic spline routines *spline3_coef* and *spline3_eval*. To compare the results, we select 13 points on the well-known *serpentine curve* given by

$$y = \frac{x}{1/4 + x^2}$$

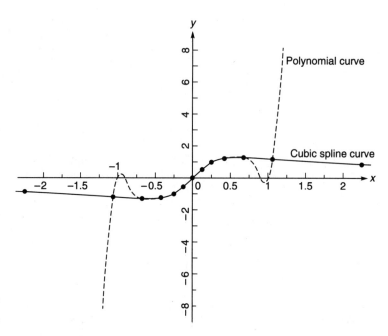

FIGURE 7.7
Serpentine curve

So that the knots will not be equally spaced, we write the curve in parametric form:

$$\begin{cases} x = \frac{1}{2}\cot\theta \\ y = \sin 2\theta \end{cases}$$

and take $\theta = i(\pi/14)$, where $i = -6, -5, \ldots, 6, 7$.

Figure 7.7 shows the resulting cubic spline curve and the polynomial curve (dashed line) from an automatic plotter. The polynomial becomes extremely erratic after the fourth knot from the origin and oscillates wildly, whereas the spline is a near perfect fit.

EXAMPLE 4   Use cubic spline functions to reproduce the curve shown in Figure 7.8.

Solution   The curve is continuous but its slope is not. Therefore, a single cubic spline is not suitable. Instead, we use two cubic spline interpolants, the first having knots $0, 1, 2, 3, 4$ and the second having knots $4, 5, 6, 7$. The data are

t	0	1	2	3	4	5	6	7
y	1.0	1.5	1.6	1.5	0.9	2.2	2.8	3.1

By carrying out two separate spline interpolation procedures, we obtain two cubic spline curves that meet at the point $(4, 0.9)$. At this point, the two curves have different slopes, as required by the curve in Figure 7.8.   ☐

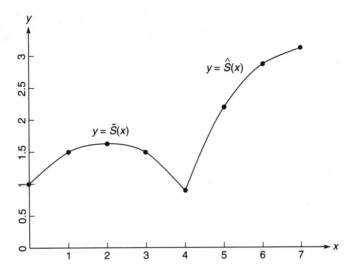

FIGURE 7.8
Two cubic splines

Smoothness Property

Why do spline functions serve the needs of data fitting better than ordinary polynomials? In order to answer this, one should understand that interpolation by polynomials of high degree is often unsatisfactory because polynomials may exhibit wild *oscillations*. Polynomials are smooth in the technical sense of possessing continuous derivatives of all orders, whereas in this sense spline functions are *not* smooth.

Wild oscillations in a function can be attributed to its derivatives being very large. Consider the function whose graph is shown in Figure 7.9. The slope of the chord that joins the points p and q is very large in magnitude. By the Mean-Value Theorem, the slope of that chord is the value of the derivative at some point between p and q. Thus, the derivative must attain large values. Indeed, somewhere on the curve between p and q there is a point where $f'(x)$ is large and negative. Similarly, between q and r there is a point where $f'(x)$ is large and positive. Hence, there is a point on the curve between p and r where $f''(x)$ is large. This reasoning can be continued to higher derivatives if there are more oscillations. This is the behavior that spline functions do *not* exhibit. In fact, the following result shows that, from a certain point of view, natural cubic splines are the *best* functions to use for curve fitting.

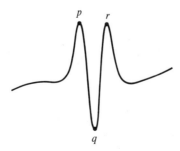

FIGURE 7.9

THEOREM ON CUBIC SPLINE SMOOTHNESS

If S is the natural cubic spline function that interpolates the twice-continuous function f at knots $a = t_0 < t_1 < \cdots < t_n = b$, then

$$\int_a^b [S''(x)]^2 \, dx \leqq \int_a^b [f''(x)]^2 \, dx$$

Proof   To verify the assertion about $[S''(x)]^2$, we let

$$g(x) = f(x) - S(x)$$

so that $g(t_i) = 0$ for $0 \leqq i \leqq n$, and

$$f'' = S'' + g''$$

Now

$$\int_a^b (f'')^2 \, dx = \int_a^b (S'')^2 \, dx + \int_a^b (g'')^2 \, dx + 2 \int_a^b S'' g'' \, dx$$

If the last integral were 0, we would be finished because then

$$\int_a^b (f'')^2 \, dx = \int_a^b (S'')^2 \, dx + \int_a^b (g'')^2 \, dx \geqq \int_a^b (S'')^2 \, dx$$

We apply the technique of integration by parts to the integral in question to show that it is 0.* We have

$$\int_a^b S'' g'' \, dx = S'' g' \Big|_a^b - \int_a^b S''' g' \, dx = - \int_a^b S''' g' \, dx$$

Here use has been made of the fact that S is a *natural* cubic spline—that is, $S''(a) = S''(b) = 0$. Continuing, we have

$$\int_a^b S''' g' \, dx = \sum_{i=0}^{n-1} \int_{t_i}^{t_{i+1}} S''' g' \, dx$$

Since S is a cubic polynomial in each interval $[t_i, t_{i+1}]$, its third derivative there is a

---

*The formula for integration by parts is:

$$\int u \, dv = uv - \int v \, du$$

constant, say c_i. So

$$\int_a^b S''' g' \, dx = \sum_{i=0}^{n-1} c_i \int_{t_i}^{t_{i+1}} g' \, dx = \sum_{i=0}^{n-1} c_i \left[g(t_{i+1}) - g(t_i) \right] = 0$$

because g vanishes at every knot.   ■

The interpretation of the integral inequality in the theorem is that the average value of $[S''(x)]^2$ on the interval $[a, b]$ is never larger than the average value of this expression with any twice-continuous function f that agrees with S at the knots. The quantity $[f''(x)]^2$ is related to the curvature of the function f.

PROBLEMS 7.2

1. Do there exist a, b, c, and d so that the function

$$S(x) = \begin{cases} ax^3 + x^2 + cx & -1 \leq x \leq 0 \\ bx^3 + x^2 + dx & 0 \leq x \leq 1 \end{cases}$$

is a natural cubic spline function that agrees with the absolute value function $|x|$ at the knots?

2. Do there exist a, b, c, and d so that the function

$$S(x) = \begin{cases} -x & x \leq -1 \\ ax^3 + bx^2 + cx + d & -1 \leq x \leq 1 \\ x & 1 \leq x \end{cases}$$

is a natural cubic spline function?

3. Determine the natural cubic spline that interpolates the function $f(x) = x^6$ over the interval $[0, 2]$ using knots 0, 1, and 2.

4. Determine the parameters a, b, c, d, and e so that S is a natural cubic spline:

$$S(x) = \begin{cases} a + b(x - 1) + c(x - 1)^2 + d(x - 1)^3 & x \in [0, 1] \\ (x - 1)^3 + ex^2 - 1 & x \in [1, 2] \end{cases}$$

5. Determine the values of a, b, c, and d so that f is a cubic spline and so that $\int_0^2 [f''(x)]^2 \, dx$ is a minimum:

$$f(x) = \begin{cases} 3 + x - 9x^3 & 0 \leq x \leq 1 \\ a + b(x - 1) + c(x - 1)^2 + d(x - 1)^3 & 1 \leq x \leq 2 \end{cases}$$

6. Determine whether f is a cubic spline with knots -1, 0, 1, and 2:

$$f(x) = \begin{cases} 1 + 2(x + 1) + (x + 1)^3 & -1 \leq x \leq 0 \\ 3 + 5x + 3x^2 & 0 \leq x \leq 1 \\ 11 + 1(x - 1) + 3(x - 1)^2 + (x - 1)^3 & 1 \leq x \leq 2 \end{cases}$$

7. List all the ways in which the following functions fail to be natural cubic splines.

a. $S(x) = \begin{cases} x + 1 & -2 \leq x \leq -1 \\ x^3 - 2x + 1 & -1 \leq x \leq 1 \\ x - 1 & 1 \leq x \leq 2 \end{cases}$

b. $f(x) = \begin{cases} x^3 + x - 1 & -1 \leq x \leq 0 \\ x^3 - x - 1 & 0 \leq x \leq 1 \end{cases}$

8. Suppose $S(x)$ is an mth-degree spline function over the interval $[a, b]$ with $n + 1$ knots $a = t_0 < t_1 < \cdots < t_n = b$.

   a. How many conditions are needed to define $S(x)$ uniquely over $[a, b]$?

   b. How many conditions are defined by the interpolation conditions at the knots?

   c. How many conditions are defined by the continuity of the derivatives?

   d. How many additional conditions are needed so that the total equals the number in part **a**?

9. Show that

$$S(x) = \begin{cases} 28 + 25x + 9x^2 + x^3 & -3 \leq x \leq -1 \\ 26 + 19x + 3x^2 - x^3 & -1 \leq x \leq 0 \\ 26 + 19x + 3x^2 - 2x^3 & 0 \leq x \leq 3 \\ -163 + 208x - 60x^2 + 5x^3 & 3 \leq x \leq 4 \end{cases}$$

is a natural cubic spline function.

10. Give an example of a cubic spline with knots 0, 1, 2, and 3 that is quadratic in $[0, 1]$, cubic in $[1, 2]$, and quadratic in $[2, 3]$.

11. Give an example of a cubic spline function S with knots 0, 1, 2, and 3 such that S is linear in $[0, 1]$ but of degree 3 in the other two intervals.

12. Determine a, b, and c so that S is a cubic spline function.

$$S(x) = \begin{cases} x^3 & 0 \leq x \leq 1 \\ \dfrac{1}{2}(x - 1)^3 + a(x - 1)^2 + b(x - 1) + c & 1 \leq x \leq 3 \end{cases}$$

13. Is there a choice of coefficients for which the following function is a natural cubic spline? Why or why not?

$$f(x) = \begin{cases} x + 1 & -2 \leq x \leq -1 \\ ax^3 + bx^2 + cx + d & -1 \leq x \leq 1 \\ x - 1 & 1 \leq x \leq 2 \end{cases}$$

14. Determine the coefficients in the function

$$S(x) = \begin{cases} x^3 - 1 & -9 \leq x \leq 0 \\ ax^3 + bx^2 + cx + d & 0 \leq x \leq 5 \end{cases}$$

so that it is a cubic spline that takes the value 2 when $x = 1$.

15. Determine the coefficients so that the function

$$S(x) = \begin{cases} x^2 + x^3 & 0 \leq x \leq 1 \\ a + bx + cx^2 + dx^3 & 1 \leq x \leq 2 \end{cases}$$

is a cubic spline and has the property $S_2'''(x) = 12$.

16. Assume that $a = x_0 < x_1 < \cdots < x_m = b$. Describe the function f that interpolates a table of values (x_i, y_i), where $0 \leq i \leq m$, and minimizes the expression $\int_a^b |f'(x)| dx$.

17. How many additional conditions are needed to specify uniquely a spline of degree 4 over n knots?

18. Let knots $t_0 < t_1 < \cdots < t_n$ and numbers y_i and z_i be given. Determine formulas for a piecewise cubic function f that has the given knots such that $f(t_i) = y_i$ ($0 \leq i \leq n$), $\lim_{x \to t_i^+} f''(x) = z_i$ ($0 \leq i \leq n - 1$), and $\lim_{x \to t_i^-} f''(x) = z_i$ ($1 \leq i \leq n$). Why is f not generally a cubic spline?

19. Define a function f by

$$f(x) = \begin{cases} x^3 + x - 1 & -1 \leq x \leq 0 \\ x^3 - x - 1 & 0 \leq x \leq 1 \end{cases}$$

Show that $\lim_{x \to 0^+} f(x) = \lim_{x \to 0^-} f(x)$ and that $\lim_{x \to 0^+} f''(x) = \lim_{x \to 0^-} f''(x)$. Are f and f'' continuous? Does it follow that f is a cubic spline? Explain.

20. Show that there is a unique cubic spline S having knots $t_0 < t_1 < \cdots < t_n$, interpolating data $S(t_i) = y_i$ ($0 \leq i \leq n$), and satisfying the two end conditions $S'(t_0) = S'(t_n) = 0$.

21. Describe explicitly the *natural cubic spline* that interpolates a table with only two entries:

x	t_0	t_1
y	y_0	y_1

Give a formula for it. Here t_0 and t_1 are the knots.

22. Suppose that $f(0) = 0$, $f(1) = 1.1752$, $f'(0) = 1$, and $f'(1) = 1.5431$. Determine the cubic interpolating polynomial $p_3(x)$ for these data. Is it a natural cubic spline?

23. A **periodic cubic spline** having knots $t_0, t_1, \ldots, t_n$ is defined as a cubic spline function $S(x)$ such that $S(t_0) = S(t_n)$, $S'(t_0) = S'(t_n)$, and $S''(t_0) = S''(t_n)$. It would be used to fit data known to be periodic. Carry out the analysis necessary to obtain a periodic cubic spline interpolant for a table

x	t_0	t_1	$\cdots$	t_n
y	y_0	y_1	$\cdots$	y_n

assuming that $y_n = y_0$.

24. The derivatives and integrals of polynomials are polynomials. State and prove a similar result about spline functions.

25. Given a differentiable function f and knots $t_0 < t_1 < \cdots < t_n$, show how to obtain a cubic spline S that interpolates f at the knots and satisfies the end conditions $S'(t_0) = f'(t_0)$ and $S'(t_n) = f'(t_n)$. *Note*: This procedure produces a better fit to f when applicable. If f' is not known, finite-difference approximations to $f'(t_0)$ and $f'(t_n)$ can be used.

26. Let S be a cubic spline that has knots $t_0 < t_1 < \cdots < t_n$. Suppose that on the two intervals $[t_0, t_1]$ and $[t_2, t_3]$, S reduces to linear polynomials. What can be said of S on $[t_1, t_2]$?

27. In the construction of the cubic interpolating spline, carry out the evaluation of constants C and D, and thus justify Equation (5).

28. Show that S_i can also be written in the form

$$S_i(x) = y_i + A_i(x - t_i) + \frac{1}{2}z_i(x - t_i)^2 + \frac{z_{i+1} - z_i}{6h_i}(x - t_i)^3$$

with

$$A_i = -\frac{h_i}{3}z_i - \frac{h_i}{6}z_{i+1} - \frac{y_i}{h_i} + \frac{y_{i+1}}{h_i}$$

29. Carry out the details in deriving Equation (7), starting with Equation (5).

30. Verify that the algorithm for computing the (z_i) array is correct by showing that if (z_i) satisfies Equation (7), then it satisfies the equation in step 3 of the algorithm.

31. Establish that $u_i > 2h_i + \frac{3}{2}h_{i-1}$ in the algorithm for determining the cubic spline interpolant.

32. By hand calculation, find the natural cubic spline interpolant for this table:

x	1	2	3	4	5
y	0	1	0	1	0

33. Find a cubic spline over knots $-1, 0$, and 1 such that the following conditions are satisfied: $S''(-1) = S''(1) = 0$, $S(-1) = S(1) = 0$, and $S(0) = 1$.

34. This problem and the next two lead to a more efficient algorithm for natural cubic spline interpolation in the case of equally spaced knots. Let $h_i = h$ in Equation (5), and replace the parameters z_i by $q_i = h^2 z_i/6$. Show that the new form of Equation (5) is then

$$S_i(x) = q_{i+1}\left(\frac{x - t_i}{h}\right)^3 + q_i\left(\frac{t_{i+1} - x}{h}\right)^3 + (y_{i+1} - q_{i+1})\left(\frac{x - t_i}{h}\right)$$
$$+ (y_i - q_i)\left(\frac{t_{i+1} - x}{h}\right)$$

35. (Continuation) Establish the new continuity conditions:

$$q_{i-1} + 4q_i + q_{i+1} = y_{i+1} - 2y_i + y_{i-1} \qquad (1 \leq i \leq n-1)$$

$$q_0 = q_n = 0$$

36. (Continuation) Show that the parameters q_i can be determined by backward recursion as follows:

$$q_n = 0 \qquad q_{n-1} = \beta_n \qquad q_i = \alpha_i q_{i+1} + \beta_i \qquad (i = n-2, n-3, \dots, 1)$$

where the coefficients α_i and β_i are generated by ascending recursion from the formulas

$$\alpha_1 = 0 \qquad \alpha_i = -(\alpha_{i-1} + 4)^{-1} \qquad (i = 1, 2, \dots, n)$$
$$\beta_1 = 0 \qquad \beta_i = -\alpha_i(y_{i+1} - 2y_i + y_{i-1} - \beta_{i-1}) \qquad (i = 1, 2, \dots, n)$$

(This algorithm, which is stable and efficient, is due to MacLeod [1973].)

37. Prove that if $S(x)$ is a spline of degree k on $[a, b]$, then $S'(x)$ is a spline of degree $k - 1$.

38. How many coefficients are needed to define a piecewise quartic (fourth-degree) function with $n + 1$ knots? How many conditions will be imposed if the piecewise quartic function is to be a quartic spline? Justify your answers.

39. Determine whether this function is a natural cubic spline:

$$S(x) = \begin{cases} x^3 + 3x^2 + 7x - 5 & -1 \leq x \leq 0 \\ -x^3 + 3x^2 + 7x - 5 & 0 \leq x \leq 1 \end{cases}$$

COMPUTER PROBLEMS 7.2

1. Rewrite and test procedure *spline3_coef* using procedure *tri* from Chapter 6. Use the symmetry of the $(n-1) \times (n-1)$ tridiagonal system.

2. The extra storage required in step 1 of the algorithm for solving the natural cubic spline tridiagonal system directly can be eliminated at the expense of a slight amount of extra computation—namely, by computing the h_i's and b_i's directly from the t_i's and y_i's in the forward elimination phase (step 2) and in the back substitution phase (step 3). Rewrite and test procedure *spline3_coef* using this idea.

3. Using at most twenty knots and the cubic spline routines *spline3_coef* and *spline3_eval*, plot on a computer plotter an outline of your:

a. school's mascot

b. signature

c. profile

4. Let S be the cubic spline function that interpolates the function $f(x) = (x^2 + 1)^{-1}$ at 41 equally spaced knots in the interval $[-5, 5]$. Evaluate $S(x) - f(x)$ at 101 equally spaced points on the interval $[0, 5]$.

5. Draw a free-form curve on graph paper, making certain that the curve is the graph of a function. Then read values of your function at a reasonable number of points, say 10 to 50, and compute the cubic spline function that takes those values. Compare the freely drawn curve to the graph of the cubic spline.

6. Draw a spiral (or other curve that is not a function) and reproduce it by spline functions as follows: Select points on the curve and label them $t = 0, 1, \ldots, n$. For each value of t, read off the x- and y-coordinates of the point, thus producing a table:

t	0	1	$\cdots$	n
x	x_0	x_1	$\cdots$	x_n
y	y_0	y_1	$\cdots$	y_n

Then fit $x = S(t)$ and $y = \overline{S}(t)$, where S and $\overline{S}$ are natural cubic spline interpolants. S and $\overline{S}$ give a parametric representation of the curve (see the figure).

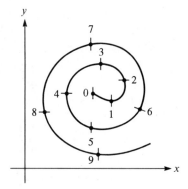

7. Write and test procedures, *as simple as possible*, to perform natural cubic spline interpolation with equally spaced knots. *Hint*: See Problems **34–36**.

8. Write a program to estimate $\int_a^b f(x)\, dx$, assuming that we know the values of f at only certain prescribed knots $a = t_0 < t_1 < \cdots < t_n = b$. Approximate f first by an interpolating cubic spline and then compute the integral of it using Equation (5).

9. Write a procedure to estimate $f'(x)$ for any x in $[a, b]$, assuming that we know only the values of f at knots $a = t_0 < t_1 < \cdots < t_n = b$.

7.3 B Splines

In this section, we give an introduction to the theory of **B splines**. These are special spline functions that are well adapted to numerical tasks and are being used more and more frequently in production-type programs for approximating data. Thus, the intelligent user of library code should have some familiarity with them. The

B splines were so named because they formed a *basis* for the set of all splines. (We prefer the more romantic description of *bell* splines because of their characteristic shape!)

Throughout this section, we suppose that an infinite set of knots $\{t_i\}$ has been prescribed in such a way that

$$\begin{cases} \cdots < t_{-2} < t_{-1} < t_0 < t_1 < t_2 < \cdots \\ \lim_{i \to \infty} t_i = \infty = -\lim_{i \to \infty} t_{-i} \end{cases} \tag{1}$$

The B splines to be defined now depend on this set of knots, although the notation does not show that dependence. The B splines of degree 0 are defined by

$$B_i^0(x) = \begin{cases} 1 & \text{if } t_i \le x < t_{i+1} \\ 0 & \text{otherwise} \end{cases} \tag{2}$$

The graph of B_i^0 is shown in Figure 7.10.

Obviously B_i^0 is discontinuous. However, it is continuous from the right at all points, even where the jumps occur. Thus,

$$\lim_{x \to t_i^+} B_i^0(x) = 1 = B_i^0(t_i) \quad \text{and} \quad \lim_{x \to t_{i+1}^+} B_i^0(x) = 0 = B_i^0(t_{i+1})$$

If the **support** of a function f is defined as the set of points x where $f(x) \ne 0$, then we can say that the support of B_i^0 is the half-open interval $[t_i, t_{i+1})$. Since B_i^0 is a piecewise constant function, it is a spline of degree 0.

Two further observations can be made:

$$B_i^0(x) \ge 0 \quad \text{for all } x \text{ and for all } i$$

$$\sum_{i=-\infty}^{\infty} B_i^0(x) = 1 \quad \text{for all } x$$

Although the second of these assertions contains an infinite series, there is no question of convergence because for each x only one term in the series is different from 0. Indeed, for fixed x, there is a unique integer m such that $t_m \le x < t_{m+1}$, and then

$$\sum_{i=-\infty}^{\infty} B_i^0(x) = B_m^0(x) = 1$$

FIGURE 7.10
B_i^0 spline

The reader should now see the reason for defining B_i^0 in the manner of Equation (2).

A final remark concerning these B splines of degree 0: Any spline of degree 0 that is continuous from the right and is based on the knots (1) can be expressed as a linear combination of the B splines B_i^0. Indeed, if S is such a function, then it can be specified by a rule such as

$$S(x) = b_i \quad \text{if } t_i \leqq x < t_{i+1} \qquad (i = 0, \pm 1, \pm 2, \ldots)$$

Then S can be written as

$$S = \sum_{i=-\infty}^{\infty} b_i B_i^0$$

With the functions B_i^0 as a starting point, we now generate all the higher-degree B splines by a simple recursive definition:

$$B_i^k(x) = \left(\frac{x - t_i}{t_{i+k} - t_i}\right) B_i^{k-1}(x) + \left(\frac{t_{i+k+1} - x}{t_{i+k+1} - t_{i+1}}\right) B_{i+1}^{k-1}(x) \qquad (k \geqq 1) \qquad \textbf{(3)}$$

Here $k = 1, 2, \ldots$ and $i = 0, \pm 1, \pm 2, \ldots$.

To illustrate Equation (3), let us determine B_i^1 in an alternative form:

$$B_i^1(x) = \left(\frac{x - t_i}{t_{i+1} - t_i}\right) B_i^0(x) + \left(\frac{t_{i+2} - x}{t_{i+2} - t_{i+1}}\right) B_{i+1}^0(x)$$

$$= \begin{cases} 0 & \text{if } x \geqq t_{i+2} \text{ or } x \leqq t_i \\[2mm] \dfrac{x - t_i}{t_{i+1} - t_i} & \text{if } t_i < x < t_{i+1} \\[2mm] \dfrac{t_{i+2} - x}{t_{i+2} - t_{i+1}} & \text{if } t_{i+1} \leqq x < t_{i+2} \end{cases}$$

The graph of B_i^1 is shown in Figure 7.11. The support of B_i^1 is the open interval (t_i, t_{i+2}). It is true, but perhaps not so obvious, that

$$\sum_{i=-\infty}^{\infty} B_i^1(x) = 1 \qquad \text{for all } x$$

and that every spline of degree 1 based on the knots (1) is a linear combination of B_i^1.

FIGURE 7.11
B_i^1 spline

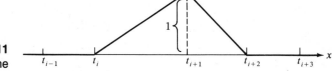

The functions B_i^k as defined by Equation (3) are called **B splines of degree k**. Since each B_i^k is obtained by applying linear factors to B_i^{k-1} and B_{i+1}^{k-1}, we see that the degrees actually increase by 1 at each step. So B_i^1 is piecewise linear, B_i^2 is piecewise quadratic, and so on.

It is also easily shown by induction that

$$B_i^k(x) = 0 \qquad x \notin [t_i, t_{i+k+1}) \qquad (k \geq 0)$$

To establish this, we start by observing that it is true when $k = 0$ because of definition (2). If it is true for index $k - 1$, then it is true for index k by the following reasoning. The inductive hypothesis tells us that $B_i^{k-1}(x) = 0$ if x is outside $[t_i, t_{i+k})$ and that $B_{i+1}^{k-1}(x) = 0$ if x is outside $[t_{i+1}, t_{i+k+1})$. If x is outside *both* intervals, it is outside their union, $[t_i, t_{i+k+1})$; then both terms on the right side of Equation (3) are 0. So $B_i^k(x) = 0$ outside $[t_i, t_{i+k+1})$. That $B_i^k(t_i) = 0$ follows directly from Equation (3), and so we know that $B_i^k(x) = 0$ for all x outside (t_i, t_{i+k+1}) if $k \geq 1$.

Complementary to the property just established, we can show, again by induction, that

$$B_i^k(x) > 0 \quad x \in (t_i, t_{i+k+1}) \qquad (k \geq 0)$$

By Equation (2), this assertion is true when $k = 0$. If it is true for index $k - 1$, then $B_i^{k-1}(x) > 0$ on (t_i, t_{i+k}) and $B_{i+1}^{k-1}(x) > 0$ on (t_{i+1}, t_{i+k+1}). In Equation (3), the factors that multiply $B_i^{k-1}(x)$ and $B_{i+1}^{k-1}(x)$ are positive when $t_i < x < t_{i+k+1}$. Thus, $B_i^k(x) > 0$ on this interval.

The principal use of the B splines $B_i^k (i = 0, \pm 1, \pm 2, \ldots)$ is as a basis for the set of all kth-degree splines that have the same knot sequence. Thus, linear combinations

$$\sum_{i=-\infty}^{\infty} c_i B_i^k$$

are important objects of study. (We use c_i for fixed k and C_i^k to emphasize the degree k of the corresponding B splines.) Our first task is to develop an efficient method to evaluate a function of the form

$$f(x) = \sum_{i=-\infty}^{\infty} C_i^k B_i^k(x) \tag{4}$$

under the supposition that the coefficients C_i^k are given (as well as the knot sequence t_i). Using definition (3) and some simple series manipulations, we have

$$f(x) = \sum_{i=-\infty}^{\infty} C_i^k \left[\left(\frac{x - t_i}{t_{i+k} - t_i} \right) B_i^{k-1}(x) + \left(\frac{t_{i+k+1} - x}{t_{i+k+1} - t_{i+1}} \right) B_{i+1}^{k-1}(x) \right]$$

$$= \sum_{i=-\infty}^{\infty} \left[C_i^k \left(\frac{x - t_i}{t_{i+k} - t_i} \right) + C_{i-1}^k \left(\frac{t_{i+k} - x}{t_{i+k} - t_i} \right) \right] B_i^{k-1}(x)$$

$$= \sum_{i=-\infty}^{\infty} C_i^{k-1} B_i^{k-1}(x) \tag{5}$$

where C_i^{k-1} is defined to be the appropriate coefficient from the line preceding Equation (5).

This algebraic manipulation shows how a linear combination of $B_i^k(x)$ can be expressed as a linear combination of $B_i^{k-1}(x)$. Repeating this process $k-1$ times, we eventually express $f(x)$ in the form

$$f(x) = \sum_{i=-\infty}^{\infty} C_i^0 B_i^0(x) \tag{6}$$

If $t_m \leqq x < t_{m+1}$, then $f(x) = C_m^0$. The formula by which the coefficients C_i^{j-1} are obtained is

$$C_i^{j-1} = C_i^j \left(\frac{x - t_i}{t_{i+j} - t_i} \right) + C_{i-1}^j \left(\frac{t_{i+j} - x}{t_{i+j} - t_i} \right) \tag{7}$$

A nice feature of Equation (4) is that only the $k+1$ coefficients $C_m^k, C_{m-1}^k, \ldots, C_{m-k}^k$ are needed to compute $f(x)$ if $t_m \leqq x < t_{m+1}$ (see **Problem 6**). Thus, if f is defined by Equation (4) and we want to compute $f(x)$, we use Equation (7) to calculate the entries in the following triangular array:

$$
\begin{array}{cccc}
C_m^k & C_m^{k-1} & \cdots & C_m^0 \\
C_{m-1}^k & C_{m-1}^{k-1} & \\
\vdots & \\
C_{m-k}^k &
\end{array}
$$

Although our notation does not show it, the coefficients in Equation (4) are independent of x, whereas the C_i^{j-1}'s calculated subsequently by Equation (7) do depend on x.

It is now a simple matter to establish that

$$\sum_{i=-\infty}^{\infty} B_i^k(x) = 1 \qquad \text{for all } x \text{ and all } k \geqq 0$$

If $k = 0$, we already know this. If $k > 0$, we use Equation (4) with $C_i^k = 1$ for all i. By Equation (7), all subsequent coefficients $C_i^k, C_i^{k-1}, C_i^{k-2}, \ldots, C_i^0$ are also equal to 1 (induction is needed here!). Thus at the end, Equation (6) is true with $C_i^0 = 1$, and so $f(x) = 1$. Therefore from Equation (4), the sum of all B splines of degree k is unity.

The smoothness of the B splines B_i^k increases with the index k. In fact, we can show by induction that B_i^k has a continuous $k-1$st derivative.

The B splines can be used as substitutes for complicated functions in many mathematical situations. Differentiation and integration are important examples. A basic result about the derivatives of B splines is

$$\frac{d}{dx}B_i^k(x) = \left(\frac{k}{t_{i+k} - t_i}\right)B_i^{k-1}(x) - \left(\frac{k}{t_{i+k+1} - t_{i+1}}\right)B_{i+1}^{k-1}(x) \qquad (8)$$

This equation can be proved by induction using the recursive Formula (3). Once Equation (8) is established, we get the useful formula

$$\frac{d}{dx}\sum_{i=-\infty}^{\infty} c_i B_i^k(x) = \sum_{i=-\infty}^{\infty} d_i B_i^{k-1}(x) \qquad (9)$$

where

$$d_i = k\left(\frac{c_i - c_{i-1}}{t_{i+k} - t_i}\right)$$

The verification is as follows: By Equation (8),

$$\frac{d}{dx}\sum_{i=-\infty}^{\infty} c_i B_i^k(x) = \sum_{i=-\infty}^{\infty} c_i \frac{d}{dx}B_i^k(x)$$

$$= \sum_{i=-\infty}^{\infty} c_i\left[\left(\frac{k}{t_{i+k} - t_i}\right)B_i^{k-1}(x) - \left(\frac{k}{t_{i+k+1} - t_{i+1}}\right)B_{i+1}^{k-1}(x)\right]$$

$$= \sum_{i=-\infty}^{\infty}\left[\left(\frac{c_i k}{t_{i+k} - t_i}\right) - \left(\frac{c_{i-1}k}{t_{i+k} - t_i}\right)\right]B_i^{k-1}(x)$$

$$= \sum_{i=-\infty}^{\infty} d_i B_i^{k-1}(x)$$

For numerical integration, the B splines are also recommended, especially for indefinite integration. Here is the basic result needed for integration:

$$\int_{-\infty}^{x} B_i^k(s)\,ds = \left(\frac{t_{i+k+1} - t_i}{k + 1}\right)\sum_{j=i}^{\infty} B_j^{k+1}(x) \qquad (10)$$

This equation can be verified by differentiating both sides with respect to x and simplifying by the use of Equation (9). In order to be sure that the two sides of Equation (10) do not differ by a constant, we note that for any $x < t_i$, both sides reduce to zero.

The basic result (10) produces this useful formula:

$$\int_{-\infty}^{x}\sum_{i=-\infty}^{\infty} c_i B_i^k(s)\,ds = \sum_{i=-\infty}^{\infty} e_i B_i^{k+1}(x) \qquad (11)$$

where

$$e_i = \frac{1}{k+1} \sum_{j=-\infty}^{i} c_j(t_{j+k+1} - t_j)$$

It should be emphasized that this formula gives an indefinite integral (antiderivative) of any function expressed as a linear combination of B splines. Any definite integral can be obtained by selecting a specific value of x. For example, if x is a knot, say $x = t_m$, then

$$\int_{-\infty}^{t_m} \sum_{i=-\infty}^{\infty} c_i B_i^k(s) \, ds = \sum_{i=-\infty}^{\infty} e_i B_i^{k+1}(t_m) = \sum_{i=m-k-1}^{m} e_i B_i^{k+1}(t_m)$$

PROBLEMS 7.3

1. Show that the functions $f_n(x) = \cos nx$ are generated by this recursive definition:

$$\begin{cases} f_0(x) = 1 \qquad f_1(x) = \cos x \\ f_{n+1}(x) = 2f_1(x)f_n(x) - f_{n-1}(x) \qquad (n \geq 1) \end{cases}$$

2. What functions are generated by the following recursive definition?

$$\begin{cases} f_0(x) = 1 \qquad f_1(x) = x \\ f_{n+1}(x) = 2xf_n(x) - f_{n-1}(x) \qquad (n \geq 1) \end{cases}$$

3. Find an expression for $B_i^2(x)$ and verify that it is piecewise quadratic. Show that $B_i^2(x)$ is zero at every knot except

$$B_i^2(t_{i+1}) = \frac{t_{i+1} - t_i}{t_{i+2} - t_i} \quad \text{and} \quad B_i^2(t_{i+2}) = \frac{t_{i+3} - t_{i+2}}{t_{i+3} - t_{i+1}}$$

4. Verify Equation (5).

5. Establish that $\sum_{i=-\infty}^{\infty} f(t_i)B_{i-1}^1(x)$ is a first-degree spline that interpolates f at every knot. What is the zero-degree spline that does so?

6. Show that if $t_m \leq x < t_{m+1}$, then

$$\sum_{i=-\infty}^{\infty} c_i B_i^k(x) = \sum_{i=m-k}^{m} c_i B_i^k(x)$$

7. Let $h_i = t_{i+1} - t_i$. Show that if

$$S(x) = \sum_{i=-\infty}^{\infty} c_i B_i^2(x) \quad \text{and if} \quad c_{i-1}h_{i-1} + c_{i-2}h_i = y_i(h_i + h_{i-1})$$

for all i, then $S(t_m) = y_m$ for all m. *Hint*: Use Problem **3**.

8. Show that the coefficients C_i^{j-1} generated by Equation (7) satisfy the condition $\min_i C_i^{j-1} \leq f(x) \leq \max_i C_i^{j-1}$.

9. For equally spaced knots, show that $k(k+1)^{-1}B_i^k(x)$ lies in the interval with endpoints $B_i^{k-1}(x)$ and $B_{i+1}^{k-1}(x)$.

10. Show that $B_i^k(x) = B_0^k(x - t_i)$ if the knots are the integers on the real line $(t_i = i)$.

11. Show that

$$\int_{-\infty}^{\infty} B_i^k(x)\, dx = \frac{t_{i+k+1} - t_i}{k+1}$$

12. Show that the class of all spline functions of degree m that have knots $x_0, x_1, \ldots, x_n$ includes the class of polynomials of degree m.

13. Establish Equation (8) by induction.

14. Which B splines B_i^k have a nonzero value on the interval (t_n, t_m)? Explain.

15. Show that on $[t_i, t_{i+1}]$ we have

$$B_i^k(x) = \frac{(x - t_i)^k}{(t_{i+1} - t_i)(t_{i+2} - t_i)\cdots(t_{i+k} - t_i)}$$

16. Is a spline of the form $S(x) = \sum_{i=-\infty}^{\infty} c_i B_i^k(x)$ *uniquely* determined by a finite set of interpolation conditions $S(t_i) = y_i$ $(0 \leq i \leq n)$? Why or why not?

17. If the spline function $S(x) = \sum_{i=-\infty}^{\infty} c_i B_i^k(x)$ vanishes at each knot, must it be identically zero? Why or why not?

18. What is the necessary and sufficient condition on the coefficients in order that $\sum_{i=-\infty}^{\infty} c_i B_i^k = 0$? State and prove.

19. Expand the function $f(x) = x$ in an infinite series $\sum_{i=-\infty}^{\infty} c_i B_i^1$.

20. Establish that $\sum_{i=-\infty}^{\infty} B_i^k$ is a constant function by means of Equation (9).

21. Show that if $k \geq 2$, then

$$\frac{d^2}{dx^2} \sum_{i=-\infty}^{\infty} c_i B_i^k$$

$$= k(k-1) \sum_{i=-\infty}^{\infty} \left[\frac{c_i - c_{i-1}}{(t_{i+k} - t_i)(t_{i+k-1} - t_i)} - \frac{c_{i-1} - c_{i-2}}{(t_{i+k-1} - t_{i-1})(t_{i+k-1} - t_i)} \right] B_i^{k-2}$$

COMPUTER PROBLEMS 7.3

1. Using an automatic plotter, graph B_0^k for $k = 0, 1, 2, 3, 4$. Use integer knots $t_i = i$ over the interval $[0, 5]$.

2. Let $t_i = i$ (so the knots are the integer points on the real line). Print a table of 100 values of the function $3B_7^1 + 6B_8^1 - 4B_9^1 + 2B_{10}^1$ on the interval $[6, 14]$. Using a plotter, construct the graph of this function on the given interval.

3. (Continuation) Repeat Computer Problem 2 for the function $3B_7^2 + 6B_8^2 - 4B_9^2 + 2B_{10}^2$.

4. Assuming that $S(x) = \sum_{i=0}^{n} c_i B_i^k(x)$, write a procedure to evaluate $S'(x)$ at a specified x. Input is $n, k, x, t_0, \ldots, t_{n+k+1}$ and $c_0, c_1, \ldots, c_n$.

5. Write a procedure to evaluate $\int_a^b S(x)\, dx$, assuming that $S(x) = \sum_{i=0}^{n} c_i B_i^k(x)$. Input will be $n, k, a, b, c_0, c_1, \ldots, c_n, t_0, \ldots, t_{n+k+1}$.

6. (March of the B splines) Produce graphs of several B splines of the same degree *marching* across the x-axis. Use an automatic plotter or a computer graphics package with a display screen.

7.4 Interpolation and Approximation by B Splines

In the preceding section, we developed a number of properties of B splines and showed how B splines are used in various numerical tasks. The problem of obtaining a B spline representation of a given function was not discussed. Here we consider the problem of interpolating a table of data; later a noninterpolatory method of approximation is described.

A basic question is how to determine the coefficients in the expression

$$S(x) = \sum_{i=-\infty}^{\infty} A_i B_{i-k}^k(x) \tag{1}$$

so that the resulting spline function interpolates a prescribed table:

x	t_0	t_1	$\cdots$	t_n
y	y_0	y_1	$\cdots$	y_n

We mean by *interpolate* that

$$S(t_i) = y_i \qquad (0 \leq i \leq n) \tag{2}$$

The natural starting point is with the simplest splines, corresponding to $k = 0$. Since

$$B_i^0(t_j) = \delta_{ij} = \begin{cases} 1 & i = j \\ 0 & i \neq j \end{cases}$$

the solution to the problem is immediate: Just set $A_i = y_i$ for $0 \leq i \leq n$. All other coefficients in Equation (1) are arbitrary. In particular, they can be zero. We arrive then at this result: The zero-degree B spline:

$$S(x) = \sum_{i=0}^{n} y_i B_i^0(x)$$

has the interpolation property (2).

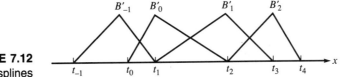

FIGURE 7.12
B_i^1 splines

The next case, $k = 1$, also has a simple solution. We use the fact that

$$B_{i-1}^1(t_j) = \delta_{ij}$$

Hence, the following is true: The first-degree B spline

$$S(x) = \sum_{i=0}^{n} y_i B_{i-1}^1(x)$$

has the interpolation property (2). So $A_i = y_i$ again.

If the table has four entries ($n = 3$), for instance, we use $B_{-1}^1, B_0^1, B_1^1,$ and B_2^1. They, in turn, require for their definition knots $t_{-1}, t_0, t_1, \ldots, t_4$. Knots t_{-1} and t_4 can be arbitrary. Figure 7.12 shows the graphs of the four splines. In such a problem, if t_{-1} and t_4 are not prescribed, it is natural to define them in such a way that t_0 is the midpoint of the interval $[t_{-1}, t_1]$ and t_3 is the midpoint of $[t_2, t_4]$.

In both elementary cases considered, the unknown coefficients $A_0, A_1, \ldots, A_n$ in Equation (1) were uniquely determined by the interpolation conditions (2). If terms were present in Equation (1) corresponding to values of i *outside* the range $\{0, 1, \ldots, n\}$, then they would have no influence on the values of $S(x)$ at $t_0, t_1, \ldots, t_n$.

For higher-degree splines, we shall see that some arbitrariness exists in choosing coefficients. In fact, *none* of the coefficients is uniquely determined by the interpolation conditions. This fact can be advantageous if other properties are desired of the solution. In the quadratic case, we begin with the equation

$$\sum_{i=-\infty}^{\infty} A_i B_{i-2}^2(t_j) = \frac{1}{t_{j+1} - t_{j-1}} \left[A_j(t_{j+1} - t_j) + A_{j+1}(t_j - t_{j-1}) \right] \qquad (3)$$

Its justification is left to Problem **3**. If the interpolation conditions (2) are now imposed, we obtain the following system of equations, which gives the necessary and sufficient conditions on the coefficients:

$$A_j(t_{j+1} - t_j) + A_{j+1}(t_j - t_{j-1}) = y_j(t_{j+1} - t_{j-1}) \qquad (0 \leq j \leq n) \qquad (4)$$

This is a system of $n + 1$ linear equations in $n + 2$ unknowns $A_0, A_1, \ldots, A_{n+1}$.

One way to solve Equation (4) is to assign any value to A_0 and then use Equation (4) to compute for $A_1, A_2, \ldots, A_{n+1}$, recursively. For this purpose, the equations could be rewritten as

$$A_{j+1} = \alpha_j + \beta_j A_j \qquad (0 \leq j \leq n) \qquad (5)$$

where these abbreviations have been used:

$$\begin{cases} \alpha_j = y_j \left(\dfrac{t_{j+1} - t_{j-1}}{t_j - t_{j-1}} \right) \\[4mm] \beta_j = \dfrac{t_j - t_{j+1}}{t_j - t_{j-1}} \end{cases} \qquad (0 \le j \le n)$$

To keep the coefficients small in magnitude, we recommend selecting A_0 so that the expression

$$\Phi = \sum_{i=0}^{n+1} A_i^2$$

will be a minimum. To determine this value of A_0, we proceed as follows: By successive substitution using Equation (5), we can show that

$$A_{j+1} = \gamma_j + \delta_j A_0 \qquad (0 \le j \le n) \tag{6}$$

where the coefficients γ_j and δ_j are obtained recursively by this algorithm:

$$\begin{cases} \gamma_0 = \alpha_0 & \delta_0 = \beta_0 \\ \gamma_j = \alpha_j + \beta_j \gamma_{j-1} & \delta_j = \beta_j \delta_{j-1} \end{cases} \qquad (1 \le j \le n) \tag{7}$$

Then Φ is a quadratic function of A_0 as follows:

$$\begin{aligned} \Phi &= A_0^2 + A_1^2 + \cdots + A_{n+1}^2 \\ &= A_0^2 + (\gamma_0 + \delta_0 A_0)^2 + (\gamma_1 + \delta_1 A_0)^2 + \cdots + (\gamma_n + \delta_n A_0)^2 \end{aligned}$$

To find the minimum of Φ, we take its derivative with respect to A_0 and set it equal to zero:

$$\frac{d\Phi}{dA_0} = 2A_0 + 2(\gamma_0 + \delta_0 A_0)\delta_0 + 2(\gamma_1 + \delta_1 A_0)\delta_1 + \cdots + 2(\gamma_n + \delta_n A_0)\delta_n = 0$$

This is equivalent to $qA_0 + p = 0$, where

$$\begin{cases} q = 1 + \delta_0^2 + \delta_1^2 + \cdots + \delta_n^2 \\ p = \gamma_0 \delta_0 + \gamma_1 \delta_1 + \cdots + \gamma_n \delta_n \end{cases}$$

Pseudocode and a Curve-Fitting Example

A procedure that computes coefficients $A_0, A_1, \ldots, A_{n+1}$ in the manner outlined above is given now. In its calling sequence, $(t_i)_{0:n}$ is the knot array, $(y_i)_{0:n}$ is the array of abscissa points, $(a_i)_{0:n+1}$ is the array of A_i coefficients, and $(h_i)_{0:n+1}$ is an array that contains $h_i = t_i - t_{i-1}$. Only n, (t_i), and (y_i) are input values. They are available unchanged when the routine is finished. Arrays (a_i) and (h_i) are computed and available as output.

```
procedure bspline2_coef (n, (tᵢ), (yᵢ), (aᵢ), (hᵢ))
real array (aᵢ)₀:ₙ₊₁, (hᵢ)₀:ₙ₊₁, (tᵢ)₀:ₙ, (yᵢ)₀:ₙ
integer i, n
real δ, γ, p, q
for i = 1 to n do
    hᵢ ← tᵢ − tᵢ₋₁
end do
h₀ ← h₁
hₙ₊₁ ← hₙ
δ ← −1
γ ← 2y₀
p ← δγ
q ← 2
for i = 1 to n do
    r ← hᵢ₊₁/hᵢ
    δ ← −rδ
    γ ← −rγ + (r + 1)yᵢ
    p ← p + γδ
    q ← q + δ²
end do
a₀ ← −p/q
for i = 1 to n + 1 do
    aᵢ ← [(hᵢ₋₁ + hᵢ)yᵢ₋₁ − hᵢaᵢ₋₁]/hᵢ₋₁
end do
end procedure bspline2_coef
```

Next we give a procedure function *bspline2_eval* for computing values of the quadratic spline given by $S(x) = \sum_{i=0}^{n+1} A_i B_{i-2}^2(x)$. Its calling sequence has some of the same variables as in the preceding pseudocode. The input variable x is a single real number that should lie between t_0 and t_n. The result of Problem **3** is used.

```
real function bspline2_eval (n, (tᵢ), (aᵢ), (hᵢ), x)
real array (aᵢ)₀:ₙ₊₁, (hᵢ)₀:ₙ₊₁, (tᵢ)₀:ₙ
integer i, n
real d, e, x
for i = n − 1 to 1 step −1 do
    if x − tᵢ ≧ 0 then exit loop
end do
i ← i + 1
d ← [aᵢ₊₁(x − tᵢ₋₁) + aᵢ(tᵢ − x + hᵢ₊₁)]/(hᵢ + hᵢ₊₁)
e ← [aᵢ(x − tᵢ₋₁ + hᵢ₋₁) + aᵢ₋₁(tᵢ₋₁ − x + hᵢ)]/(hᵢ₋₁ + hᵢ)
bspline2_eval ← [d(x − tᵢ₋₁) + e(tᵢ − x)]/hᵢ
end function bspline2_eval
```

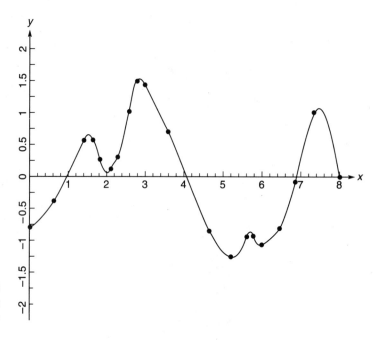

FIGURE 7.13
Quadratic
interpolating B spline

Using the table of 20 points from Section 7.2, we can compare the resulting natural cubic spline curve with the quadratic spline produced by the procedures *bspline2_coef* and *bspline2_eval*. The first of these curves is shown in Figure 7.6 and the second is in Figure 7.13. The latter is reasonable but perhaps not as pleasing as the former. These curves show once again that cubic natural splines are simple and elegant functions for curve fitting.

Schoenberg's Process

An efficient process due to Schoenberg [1967] can also be used to obtain B spline approximations to a given function. Its quadratic version is defined by

$$S(x) = \sum_{i=-\infty}^{\infty} f(\tau_i)B_i^2(x) \quad \text{where} \quad \tau_i = \frac{1}{2}(t_{i+1} + t_{i+2}) \tag{8}$$

Here, of course, the knots are $\{t_i\}_{i=-\infty}^{\infty}$, and the points where f must be evaluated are midpoints between the knots.

Equation (8) is useful in producing a quadratic spline function that approximates f. The salient properties of this process are as follows:

1. If $f(x) = ax + b$, then $S(x) = f(x)$.
2. If $f(x) \geq 0$ everywhere, then $S(x) \geq 0$ everywhere.
3. $\max_x |S(x)| \leq \max_x |f(x)|$

4. If f is continuous on $[a, b]$, if $\delta = \max_i |t_{i+1} - t_i|$, and if $\delta < b - a$, then for x in $[a, b]$,

$$|S(x) - f(x)| \leq \frac{3}{2} \max_{a \leq u \leq v \leq u + \delta \leq b} |f(u) - f(v)|$$

5. The graph of S does not cross any line in the plane a greater number of times than does the graph of f.

Some of these properties are elementary; others are more abstruse. Property **1** is outlined in Problem **6**. Property **2** is obvious because $B_i^2(x) \geq 0$ for all x. Property **3** follows easily from Equation (8) because if $|f(x)| \leq M$, then

$$|S(x)| \leq \left| \sum_{i=-\infty}^{\infty} f(\tau_i) B_i^2(x) \right| \leq \sum_{i=-\infty}^{\infty} |f(\tau_i)| B_i^2(x) \leq M \sum_{i=-\infty}^{\infty} B_i^2(x) = M$$

Properties **4** and **5** will be accepted without proof. Their significance, however, should not be overlooked. By **4**, we can make the function S close to a continuous function f simply by making the *mesh size* δ small. This is because $f(u) - f(v)$ can be made as small as we wish simply by imposing the inequality $|u - v| \leq \delta$ (uniform continuity property). Property **5** can be interpreted as a shape-preserving attribute of the approximation process. In a crude interpretation, S should not exhibit more undulations than f. Property **5** is exhibited in Computer Problem **5**.

Pseudocode

A pseudocode to obtain a spline approximation by means of Schoenberg's process is now developed. Suppose that f is defined on an interval $[a, b]$ and that the spline approximant of Equation (8) is wanted on the same interval. We define *nodes* $\tau_i = a + ih$, where $h = (b - a)/n$. Here i can be any integer, but the nodes in $[a, b]$ are only $\tau_0, \tau_1, \ldots, \tau_n$. In order to have $\tau_i = \frac{1}{2}(t_{i+1} + t_{i+2})$, we define the *knots* $t_i = a + (i - \frac{3}{2})h$. In Equation (8), the only B splines B_i^2 that are *active* on $[a, b]$ are $B_{-1}^2, B_0^2, \ldots, B_{n+1}^2$. Hence for our purposes, Equation (8) becomes

$$S(x) = \sum_{i=-1}^{n+1} f(\tau_i) B_i^2(x) \tag{9}$$

Thus, we require the values of f at $\tau_{-1}, \tau_0, \ldots, \tau_{n+1}$. Two of these nodes are outside the interval $[a, b]$, and therefore we furnish linearly extrapolated values in the code by defining

$$f(\tau_{-1}) = 2f(\tau_0) - f(\tau_1)$$
$$f(\tau_{n+1}) = 2f(t_n) - f(t_{n-1})$$

To use the formulas in Problem **3**, we write

$$S(x) = \sum_{i=1}^{n+3} D_i B_{i-2}^2(x) \qquad [D_i = f(\tau_{i-2})]$$

A pseudocode to compute $D_1, D_2, \ldots, D_{n+3}$ is given now. In the calling sequence for procedure *sch_coef*, f is an external function. After execution, the $n + 3$ desired coefficients are in the (d_i) array.

procedure *sch_coef* $(f, a, b, n, (d_i))$
real array $(d_i)_{1:n+3}$
integer i
real a, b, h
interface external function f
$h \leftarrow (b - a)/n$
for $i = 2$ **to** $n + 2$ **do**
$\quad d_i \leftarrow f(a + (i - 2)h)$
end do
$d_1 \leftarrow 2d_2 - d_3$
$d_{n+3} \leftarrow 2d_{n+2} - d_{n+1}$
end procedure *sch_coef*

After the coefficients D_i have been obtained by the procedure just given, we can recover values of the spline $S(x)$ in Equation (9). Here we use the algorithm of Problem **3**. Given an x, we first need to know where it is relative to the knots. To determine k such that $t_{k-1} \leq x \leq t_k$, we notice that k should be the largest integer such that $t_{k-1} \leq x$. This inequality is equivalent to the inequality $k \leq \frac{5}{2} + (x - a)/h$, as is easily verified. This explains the calculations of k in the pseudocode. The location of x is indicated in Figure 7.14. In the calling sequence for function *sch_eval*, a and b are the ends of the interval, and x is a point where the value of $S(x)$ is desired. The procedure determines knots t_i in such a way that the equally spaced points τ_i in the preceding procedure satisfy $\tau_i = \frac{1}{2}(t_{i+1} + t_{i+2})$.

real function *sch_eval* $(a, b, n, (d_i), x)$
real array $(d_i)_{1:n+3}$
integer k
real c, h, p, w
$h \leftarrow (b - a)/n$
$k \leftarrow$ **integer**$[(x - a)/h + 5/2]$
$p \leftarrow x - a - (k - 5/2)h$
$c \leftarrow [d_{k+1}p + d_k(2h - p)]/(2h)$
$e \leftarrow [d_k(p + h) + d_{k-1}(h - p)]/(2h)$
sch_eval $\leftarrow [cp + e(h - p)]/h$
end function *sch_eval*

FIGURE 7.14

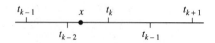

**PROBLEMS
7.4**

1. Establish formulas

$$B_{i-1}^2(t_i) = \frac{t_i - t_{i-1}}{t_{i+1} - t_{i-1}} = \frac{h_{i-1}}{h_i + h_{i-1}}$$

$$B_{i-2}^2(t_i) = \frac{t_{i+1} - t_i}{t_{i+1} - t_{i-1}} = \frac{h_i}{h_i + h_{i-1}}$$

where $h_i = t_{i+1} - t_i$.

2. Show by induction that if

$$A_j = \frac{1}{t_{j-1} - t_{j-2}} \left[y_{j-1}(t_j - t_{j-2}) - A_{j-1}(t_j - t_{j-1}) \right]$$

for $j = 2, 3, \ldots, n + 1$, then

$$\sum_{i=0}^{n+1} A_i B_{i-2}^2(t_j) = y_j \qquad (0 \le j \le n)$$

3. Show that if $S(x) = \sum_{i=-\infty}^{\infty} A_i B_{i-2}^2(x)$ and $t_{j-1} \le x \le t_j$, then

$$S(x) = \frac{1}{t_j - t_{j-1}} [d(x - t_{j-1}) + e(t_j - x)]$$

with

$$d = \frac{1}{t_{j+1} - t_{j-1}} [A_{j+1}(x - t_{j-1}) + A_j(t_{j+1} - x)]$$

and

$$e = \frac{1}{t_j - t_{j-2}} [A_j(x - t_{j-2}) + A_{j-1}(t_j - x)]$$

4. Verify Equations (6) and (7) by induction, using Equation (5).

5. If points $\tau_0 < \tau_1 < \cdots < \tau_n$ are given, can we always determine points t_i such that $t_i < t_{i+1}$ and $\tau_i = \frac{1}{2}(t_{i+1} + t_{i+2})$? Why or why not?

6. Show that if $f(x) = x$, then Schoenberg's process produces $S(x) = x$.

7. Show that $x^2 = \sum_{i=-\infty}^{\infty} t_{i+1} t_{i+2} B_i^2(x)$.

8. Let $f(x) = x^2$. Assume that $t_{i+1} - t_i \le \delta$ for all i. Show that the quadratic spline approximation to f given by Equation (8) differs from f by no more than $\delta^2/4$. *Hint:* Use Problem 7 and the fact that $\sum_{i=-\infty}^{\infty} B_i^2 \equiv 1$.

9. Verify (for $k > 0$) that $B_i^k(t_j) = 0$ if and only if $j \le i$ or $j \ge i + k + 1$.

10. What is the maximum value of B_i^2 and where does it occur?

COMPUTER PROBLEMS 7.4

1. The size of the Army of Flanders has been estimated by historians as follows:

Date	Sept. 1572	Dec. 1573	Mar. 1574	Jan. 1575	May 1576
Number	67,259	62,280	62,350	59,250	51,457

Feb. 1578	Sept. 1580	Oct. 1582	Apr. 1588	Nov. 1591	Mar. 1607
27,603	45,435	61,162	63,455	62,164	41,471

   Fit the table with a quadratic B spline and use it to find the average size of the army during the period given.

2. Rewrite procedures *bspline2_coef* and *bspline_eval* so that the array (h_i) is not used.

3. Rewrite procedures *bspline2_coef* and *bspline2_eval* for the special case of equally spaced knots, simplifying the code where possible.

4. Write a procedure to produce a spline approximation to $F(x) = \int_a^x f(t)\,dt$. Assume that $a \leq x \leq b$. Begin by finding a quadratic spline interpolant to f at the n points $t_i = a + i(b - a)/n$. Test your program on the following.

   a. $f(x) = \sin x$ $(0 \leq x \leq \pi)$

   b. $f(x) = e^x$ $(0 \leq x \leq 4)$

   c. $f(x) = (x^2 + 1)^{-1}$ $(0 \leq x \leq 2)$

5. Write a procedure to produce a spline function that approximates $f'(x)$ for a given f on a given interval $[a, b]$. Begin by finding a quadratic spline interpolant to f at $n + 1$ points evenly spaced in $[a, b]$, including endpoints. Test your procedure on the functions suggested in Computer Problem **4**.

6. Define f on $[0, 6]$ to be a polygonal line that joins points $(0, 0), (1, 2), (3, 3), (5, 3)$, and $(6, 0)$. Determine spline approximations to f, using Schoenberg's process and taking $7, 13, 19, 25$, and 31 knots.

7. Write suitable code to calculate $\sum_{i=-\infty}^{\infty} f(s_i)B_i^2(x)$ with $s_i = \frac{1}{2}(t_{i+1} + t_{i+2})$. Assume that f is defined on $[a, b]$ and that x will lie in $[a, b]$. Assume also that $t_1 < a < t_2$ and $t_{n+1} < b < t_{n+2}$. (Make no assumption about the spacing of knots.)

8. Write a procedure to carry out this approximation scheme:

$$S(x) = \sum_{i=-\infty}^{\infty} f(\tau_i)B_i^3(x) \qquad \tau_i = \frac{1}{3}(t_{i+1} + t_{i+2} + t_{i+3})$$

   Assume that f is defined on $[a, b]$ and that $\tau_i = a + ih$ for $0 \leq i \leq n$, where $h = (b - a)/n$.

9. Using a computer algebra system, compute and plot the quadratic B spline curve shown in Figure 7.13.

8 ORDINARY DIFFERENTIAL EQUATIONS

In a simple electrical circuit, the current I in amperes is a function of time: $I(t)$. The function $I(t)$ will satisfy an ordinary differential equation of the form

$$\frac{dI}{dt} = f(t, I)$$

Here the right side is a function of t and I that depends on the circuit and on the nature of the electromotive force supplied to the circuit. Using methods developed in this chapter, we can solve the differential equation numerically to produce a table of I as a function of t.

Initial-Value Problem: Analytical vs. Numerical Solution

An **ordinary differential equation** (ODE) is an equation that involves one or more derivatives of an unknown function. A **solution** of a differential equation is a specific function that satisfies the equation. Here are some examples of differential equations with their solutions. In each case, t is the independent variable and x is the dependent variable. Thus, x is the name of the unknown function of the independent variable t:

Equation	Solution
$x' - x = e^t$	$x(t) = te^t + ce^t$
$x'' + 9x = 0$	$x(t) = c_1 \sin 3t + c_2 \cos 3t$
$x' + \dfrac{1}{2x} = 0$	$x(t) = \sqrt{c - t}$

325

In these three examples, the letter c denotes an arbitrary constant. The fact that such constants appear in the solutions is an indication that a differential equation does not, in general, determine a unique solution function. When occurring in a scientific problem, a differential equation is usually accompanied by auxiliary conditions that (together with the differential equation) specify the unknown function precisely.

In this chapter, we concentrate on one type of differential equation and one type of auxiliary condition—the **initial-value problem** for a first-order differential equation. The standard form adopted is

$$\begin{cases} x' = f(t, x) \\ x(a) \text{ is given} \end{cases} \tag{1}$$

It is understood that x is a function of t, so that the differential equation written in more detail looks like this:

$$\frac{dx(t)}{dt} = f(t, x(t))$$

Problem (1) is termed an initial-value problem because t can be interpreted as time and $t = a$ can be thought of as the initial instant. We want to be able to determine the value of x at any time t before or after a.

Here are some examples of initial-value problems, together with their solutions:

Equation	Initial Value	Solution
$x' = x + 1$	$x(0) = 0$	$x = e^t - 1$
$x' = 6t - 1$	$x(1) = 6$	$x = 3t^2 - t + 4$
$x' = \dfrac{t}{x+1}$	$x(0) = 0$	$x = \sqrt{t^2 + 1} - 1$

Although many methods exist for obtaining analytical solutions of differential equations, they are primarily limited to special differential equations. When applicable, they produce a solution in the form of a formula, such as shown in the preceding examples. In practical problems, however, frequently a differential equation is not amenable to solution by special methods and a numerical solution must be sought. Even when a formal solution can be obtained, a numerical solution may be preferable, especially if the formal solution is very complicated. A numerical solution of a differential equation is usually obtained in the form of a table; the function remains unknown insofar as a specific formula is concerned.

The form of the differential equation adopted here permits the function f to depend on t and x. If f does not involve x, as in the second example above, then the differential equation can be solved by a direct process of indefinite integration. To illustrate, consider the initial-value problem

$$\begin{cases} x' = 3t^2 - 4t^{-1} + (1 + t^2)^{-1} \\ x(5) = 17 \end{cases}$$

The differential equation can be integrated to produce

$$x(t) = t^3 - 4 \ln t + \arctan t + C$$

The constant C can then be chosen so that $x(5) = 17$. We can use Maple to solve this differential equation explicitly and, thereby, find the value of this constant as follows.

```
ode := D(x)(t) = 3*t^2 -4/t + 1/(1 + t^2);
dsolve( {ode, x(5)=17}, x(t) );
```

The output confirms the given solution and indicates that $C = 4 \ln(5) - \arctan(5) - 108$. In such examples, a numerical solution may still be preferable because the function of t on the right side of differential Equation (1) may not be integrable in terms of elementary functions. Consider, for instance, the differential equation

$$x' = e^{-\sqrt{t^2 - \sin t}} + \ln|\sin t + \tanh t^3| \tag{2}$$

The solution is obtained by taking the integral or antiderivative of the right-hand side. It can be done in principle but not in practice. In other words, a function x exists for which dx/dt is the right member of Equation (2), but it is not possible to write $x(t)$ in terms of familiar functions.

Solving ordinary differential equations on a computer may require a large number of steps with small step size, so a great deal of roundoff error can accumulate. Consequently, multiple-precision computations may be necessary on small-word-length computers.

8.1 Taylor Series Methods

The method described first does not have the utmost generality, but it is natural and capable of high precision. Its principle is to represent the solution of a differential equation locally by a few terms of its Taylor series.

In what follows, we shall assume that our solution function x is represented by its Taylor series[*]

$$x(t + h) = x(t) + hx'(t) + \frac{1}{2!}h^2 x''(t) + \frac{1}{3!}h^3 x'''(t)$$

$$+ \frac{1}{4!}h^4 x^{(4)}(t) + \cdots + \frac{1}{n!}h^n x^{(n)}(t) + \cdots \tag{1}$$

For numerical purposes, the Taylor series truncated after $n + 1$ terms enables us to compute $x(t + h)$ rather accurately if h is small and if $x(t), x'(t), x''(t), \ldots, x^{(n)}(t)$ are

---

[*]Remember that the function e^{-1/x^2} is smooth but *not* represented by its Taylor series at 0.

known. When only terms through $h^n x^{(n)}(t)/n!$ are included in the Taylor series, the procedure to be discussed next is called the **Taylor series method of order n**.

Euler's Method Pseudocode

The Taylor series method of order 1 is known as **Euler's method**. To find approximate values of the solutions to the initial-value problem

$$\begin{cases} x' = f(t, x(t)) \\ x(a) = x_a \end{cases}$$

over the interval $[a, b]$, the first two terms in the Taylor series (1) are used:

$$x(t + h) \approx x(t) + hx'(t)$$

Hence, the formula

$$x(t + h) = x(t) + hf(t, x(t)) \tag{2}$$

can be used to step from $t = a$ to $t = b$ with n steps of size $h = (b - a)/n$. The pseudocode for Euler's method can be written as follows, where some prescribed values for n, a, b, and x_a are used:

```
program euler
integer parameter n ← 100
real parameter a ← 1, b ← 2, x ← −4
integer k
real h, t, x
h ← (b − a)/n
t ← a
output 0, t, x
for k = 1 to n do
    x ← x + hf(t, x)
    t ← t + h
    output k, t, x
end do
end program euler
```

To use this program, a procedure function for $f(t, x)$ is needed.

EXAMPLE 1   Using Euler's method, compute an approximate value for $x(2)$ for the differential equation $x' = 1 + x^2 + t^3$ with the initial value $x(1) = -4$ using 100 steps.

Solution   Use the pseudocode above with the initial values given and combine with the following function:

real function $f(t, x)$
real t, x
$f \leftarrow 1 + x^2 + t^3$
end function

The computed value of $x(2)$ is 4.23585.   □

Before accepting the result of this example and continuing, one should raise some questions such as: How accurate is the answer? Are higher-order Taylor series methods ever needed? Unfortunately, Euler's method is *not* very accurate because only two terms in the Taylor series (1) are used and therefore the truncation error is $\mathcal{O}(h^2)$.

Taylor Series Method of Higher Order

This same example can be used to explain the Taylor series method of higher order.
        Consider again the initial-value problem

$$\begin{cases} x' = 1 + x^2 + t^3 \\ x(1) = -4 \end{cases} \tag{3}$$

If the differential equation is differentiated several times with respect to t, the results are as follows. (Remember that a function of x must be differentiated with respect to t by using the chain rule.)

$$\begin{aligned} x' &= 1 + x^2 + t^3 \\ x'' &= 2xx' + 3t^2 \\ x''' &= 2xx'' + 2x'x' + 6t \\ x^{(4)} &= 2xx''' + 6x'x'' + 6 \end{aligned} \tag{4}$$

If t and $x(t)$ are known, these four formulas, applied in order, yield $x'(t)$; $x''(t)$, $x'''(t)$, and $x^{(4)}(t)$. Thus, it is possible from this work to use the first five terms in the Taylor series, Equation (1). Since $x(1) = -4$, we have a suitable starting point and we select $n = 100$, which determines h. Next, we can compute an approximation to $x(a + h)$ from Formulas (1) and (4). The same process can be repeated to compute $x(a + 2h)$ using $x(a + h), x'(a + h), \ldots, x^{(4)}(a + h)$. Here is the pseudocode:

program *taylor*
real parameter $a \leftarrow 1$, $b \leftarrow 2$, $x \leftarrow -4$
integer parameter $n \leftarrow 100$
integer k
real $h, t, x, x', x'', x''', x^{(4)}$
$h \leftarrow (b - a)/n$
$t \leftarrow a$

output $0, t, x$
for $k = 1$ **to** n **do**
$\quad x' \leftarrow 1 + x^2 + t^3$
$\quad x'' \leftarrow 2xx' + 3t^2$
$\quad x''' \leftarrow 2xx'' + 2(x')^2 + 6t$
$\quad x^{(4)} \leftarrow 2xx''' + 6x'x'' + 6$

$$x \leftarrow x + h\left[x' + \frac{h}{2}\left[x'' + \frac{h}{3}\left[x''' + \frac{h}{4}\left[x^{(4)}\right]\right]\right]\right]$$

$\quad t \leftarrow a + kh$
$\quad$ **output** k, t, x
end do
end program *taylor*

A few words of explanation may be helpful here. Before writing the pseudocode, determine the interval in which you want to compute the solution of the differential equation. In the example, this interval is chosen as $a = 1 \leq t \leq 2 = b$, and 100 steps are used. In each step, the current value of t is an integer multiple of the step size h. The assignment statements that define x', x'', x''', and $x^{(4)}$ are simply carrying out calculations of the derivatives according to Equation (4). The final calculation carries out the evaluation of the Taylor series in Equation (1) using five terms. Since this equation is a polynomial in h, it is evaluated most efficiently by using nested multiplication, which explains the formula for x in the pseudocode. The computation $t \leftarrow t + h$ may cause a small amount of roundoff error to accumulate in the value of t. This is avoided by using $t \leftarrow a + kh$.

As one might expect, the results of using only two terms in the Taylor series (Euler's method) are not at all as accurate as when five terms are used:

Euler's Method	Taylor Series Method (Order 4)
$x(2) \approx 4.2358541$	$x(2) \approx 4.3712096$

By further analysis, one can prove that the correct value of $x(2)$ to five significant figures is 4.3712. Here the computations were done with more precision just to show that lack of precision was not a contributing factor.

To verify these results, we use the Maple statements

```
ode := D(x)(t) = 1 + x(t)^2 + t^3;
f := dsolve({ode, x(1)=-4}, x(t), numeric);
f(0); f(1);
```

and obtain the results

```
{x(t) = -4., t = 1.}
{t = 2., x(t) = 4.371221866}
```

Types of Errors

When this pseudocode is programmed and run on a computer, what sort of accuracy can we expect? Are all the digits printed by the machine for the variable x accurate? Of course not! On the other hand, it is not easy to say how many digits *are* reliable. Here is a coarse assessment. Since terms up to $\frac{1}{24}h^4 x^{(4)}(t)$ are included, the first term *not* included in the Taylor series is $\frac{1}{120}h^5 x^{(5)}(t)$. The error may be larger than this, but the factor $h^5 = (10^{-2})^5 \approx 10^{-10}$ is affecting only the tenth decimal place. The printed solution is perhaps accurate to eight decimal places. Bridges or airplanes should not be built on such shoddy analysis, but for now our attention is focused on the general form of the procedure.

Actually, there are two types of errors to consider. At each step, if $x(t)$ is known and $x(t + h)$ is computed from the first few terms of the Taylor series, an error occurs because we have truncated the Taylor series. This error, then, is called the **truncation error** or, to be more precise, the **local truncation error**. In the preceding example, it is roughly $\frac{1}{120}h^5 x^{(5)}(\xi)$. In this situation, we say that the local truncation error is *of order* h^5, abbreviated by $\mathcal{O}(h^5)$.

The second type of error obviously present is due to the accumulated effects of all local truncation errors. Indeed, the calculated value of $x(t + h)$ is in error because $x(t)$ is already wrong (because of previous truncation errors) and because another local truncation error occurs in the computation of $x(t + h)$ by means of the Taylor series.

Additional sources of errors must be considered in a complete theory. One is **roundoff error**. Although not serious in any one step of the solution procedure, after hundreds or thousands of steps it may accumulate and contaminate the calculated solution seriously. Remember that an error made at a certain step is carried forward into all succeeding steps. Depending on the differential equation and the method used to solve it, such errors may be magnified by succeeding steps.

Mathematical manipulation languages have become available for carrying out various routine mathematical calculations of both a nonnumerical and a numerical type. Now differentiation and integration of expressions, even rather complicated ones, can be turned over to the computer! Of course, this applies only to a restricted class of functions, but this class is broad enough to include all the functions that one encounters in the typical calculus textbook. With the use of such a program for symbolic computations, the Taylor series method of high order can be carried out without difficulty. Here is a code showing how the preceding example can be worked using the algebraic manipulation potentialities of Maple:

```
# Using Taylor series method of order 4
y := array (1..4);
Digits := 24;
n := 100;
h := 0.01;
T := 1;
X := -4;
```

```
# f is a function of x and t
f :=   (x,t) ->   1 + (x(t))^2 + t^3;
# differentiate f with respect to t
one := 1 + (x(t))^2 + t^3;
two := diff(f(x,t), t);
three := diff(", t);
four := diff(", t);
first := diff(x(t), t);
second := diff(", t);
third := diff(", t);
for k from 1 to n do
y[1]  := subs(t = T, x(T) = X, one);
y[2]  := subs(first = y[1], t = T, x(T) = X, two);
y[3]  := subs({first = y[1], second = y[2]},
t = T, x(T) = X, three);
y[4]  := subs({first = y[1], second = y[2], third = y[3]},
t = T, x(T) = X, four);
X := X + sum(y[i]*h^i/factorial(i), i=1..4);
T := T + h;
od;
```

The final result is $X = 4.37121\,00522\,49692\,27234\,569$ at $T = 2.0000000$.

PROBLEMS 8.1

1. Give the solutions of these differential equations.

   a. $x' = t^3 + 7t^2 - t^{1/2}$

   b. $x' = x$

   c. $x' = -x$

   d. $x'' = -x$

   e. $x'' = x$

   f. $x'' + x' - 2x = 0$ *Hint:* Try $x = e^{at}$.

2. Give the solutions of these initial-value problems.

   a. $x' = t^2 + t^{1/3}$ $x(0) = 7$

   b. $x' = 2x$ $x(0) = 15$

   c. $x'' = -x$ $x(\pi) = 0$ $x'(\pi) = 3$

3. Solve the following differential equations.

   a. $x' = 1 + x^2$ *Hint:* $1 + \tan^2 t = \sec^2 t$

   b. $x' = \sqrt{1 - x^2}$ *Hint:* $\sin^2 t + \cos^2 t = 1$

   c. $x' = t^{-1} \sin t$ *Hint:* See Computer Problem 1 in Section 5.1.

   d. $x' + tx = t^2$ *Hint:* Multiply the equation by $f(t) = \exp(t^2/2)$.

4. Determine x'' when $x' = xt^2 + x^3 + e^x t$.

5. Find a polynomial p with the property $p - p' = t^3 + t^2 - 2t$.

6. The general first-order linear differential equation is $x' + px + q = 0$, where p and q are functions of t. Show that the solution is $x = -y^{-1}(z + c)$, where y and z are functions obtained as follows: Let u be an antiderivative of p. Put $y = e^u$ and let z be an antiderivative of yq.

7. Here is an example of an initial-value problem that has two solutions: $x' = x^{1/3}$, $x(0) = 0$. Verify that the two solutions are $x_1(t) = 0$ and $x_2(t) = \left(\frac{2}{3}t\right)^{3/2}$ for $t \geq 0$. If the Taylor series method is applied, what happens?

8. Solve Problem **3d** by substituting a power series $x(t) = \sum_{n=0}^{\infty} a_n t^n$ and then determining appropriate values of the coefficients.

9. Consider the problem $x' = x$. If the initial condition is $x(0) = c$, then the solution is $x(t) = ce^t$. If a roundoff error of ε occurs in reading the value of c into the computer, what effect is there on the solution at the point $t = 10$? At $t = 20$? Do the same for $x' = -x$.

10. If the Taylor series method is used on the initial-value problem $x' = t^2 + x^3$, $x(0) = 0$, and if we intend to use the derivatives of x up to and including $x^{(4)}$, what are the five main equations that must be programmed?

11. In solving the following differential equations by the Taylor series method of order n, what are the main equations in the algorithm?

   a. $x' = x + e^x$ $n = 4$

   b. $x' = x^2 - \cos x$ $n = 5$

12. Calculate an approximate value for $x(0.1)$ using one step of the Taylor series method of order 3 on the ordinary differential equation

$$\begin{cases} x'' = x^2 e^t + x' \\ x(0) = 1 \qquad x'(0) = 2 \end{cases}$$

13. Suppose that a differential equation is solved numerically on an interval $[a, b]$ and that the local truncation error is ch^p. Show that if all truncation errors have the same sign (the worst possible case), then the total truncation error is $(b - a)ch^{p-1}$, where $h = (b - a)/n$.

14. If we plan to use the Taylor series method with terms up to h^{20}, how should the computation $\sum_{n=0}^{20} x^{(n)}(t)h^n/n!$ be carried out? Assume that $x(t)$, $x^{(1)}(t)$, $x^{(2)}(t), \ldots$, and $x^{(20)}(t)$ are available. *Hint*: Only a few statements suffice.

COMPUTER PROBLEMS 8.1

1. Write and test a program for applying the Taylor series method to the initial-value problem

$$\begin{cases} x' = x + x^2 \\ x(1) = \dfrac{e}{16 - e} = 0.20466\,34172\,89155\,26943 \end{cases}$$

Generate the solution in the interval $[1, 2.77]$. Use derivatives to up to $x^{(5)}$ in the Taylor series. Use $h = 1/100$. Print out for comparison the values of the exact solution $x(t) = e^t/(16 - e^t)$. Verify that it is the exact solution.

2. Write a program to solve each problem on the indicated intervals. Use the Taylor series method with $h = 1/100$, and include terms to h^3. Account for any difficulties.

   a. $\begin{cases} x' = t + x^2 \\ x(0) = 1 \end{cases}$ interval $[0, 0.9]$

   b. $\begin{cases} x' = x - t \\ x(1) = 1 \end{cases}$ interval $[1, 1.75]$

   c. $\begin{cases} x' = tx + t^2 x^2 \\ x(2) = -0.63966\,25333 \end{cases}$ interval $[2, 5]$

3. Solve the differential equation $x' = x$ with initial value $x(0) = 1$ by the Taylor series method on the interval $[0, 10]$. Compare the result with the exact solution $x(t) = e^t$. Use derivatives up to and including the tenth. Use step size $h = 1/100$.

4. Solve for $x(1)$.

   a. $x' = 1 + x^2$ $x(0) = 0$

   b. $x' = (1 + t)^{-1}x$ $x(0) = 1$

   using the Taylor series method of order 5 with $h = 1/100$, and compare with the exact solutions, which are $\tan t$ and $1 + t$.

5. Solve the initial-value problem $x' = t + x + x^2$ on the interval $[0, 1]$ with initial condition $x(1) = 1$. Use the Taylor series method of order 5.

6. Solve the initial-value problem $x' = (x + t)^2$ with $x(0) = -1$ on the interval $[0, 1]$, using the Taylor series method with derivatives up to and including the fourth. Compare this to Taylor series methods of order 1, 2, and 3.

7. Write a program to solve on the interval $[0, 1]$ the initial-value problem

$$\begin{cases} x' = tx \\ x(0) = 1 \end{cases}$$

using the Taylor series method of order 20; that is, include terms in the Taylor series up to and including h^{20}. Observe that a simple recursive formula can be used to obtain $x^{(n)}$ for $n = 1, 2, \ldots, 20$.

8. Write a program to solve the initial-value problem $x' = \sin x + \cos t$, using the Taylor series method. Continue the solution from $t = 2$ to $t = 5$, starting with $x(2) = 0.32$. Include terms up to and including h^3.

9. Write a short program to solve the initial-value problem $x' = e^t x$ with $x(2) = 1$ on the interval $0 \leq t \leq 2$ using the Taylor series method. Include terms up to h^4.

10. Write a program to solve $x' = tx + t^4$ on the interval $0 \leq t \leq 5$ with $x(5) = 3$. Use the Taylor series method with terms to h^4.

11. Write a program to solve the initial-value problem of the example in this section over the interval $[1, 3]$. Explain.

12. Compute a table, at 101 equally spaced points in the interval $[0, 2]$, of the *Dawson integral*

$$f(x) = \exp(-x^2) \int_0^x \exp(t^2) \, dt$$

by numerically solving, with the Taylor series method of suitable order, an initial-value problem of which f is the solution. Make the table accurate to eight decimal places and print only eight decimal places. *Hint*: Find the relationship between $f'(x)$ and $xf(x)$. The Fundamental Theorem of Calculus is useful. *Check values*: $f(1) = 0.53807\,95069$ and $f(2) = 0.30134\,03889$.

13. Solve the initial-value problem $x' = t^3 + e^x$ with $x(3) = 7.4$ on the interval $0 \leq t \leq 3$ by means of the fourth-order Taylor series method.

8.2 Runge-Kutta Methods

The methods named after Carl Runge and Wilhelm Kutta are designed to imitate the Taylor series method without requiring analytic differentiation of the original differential equation. Recall that in using the Taylor series method on the initial-value problem

$$\begin{cases} x' = f(t, x) \\ x(a) = x_a \end{cases} \tag{1}$$

we need to obtain $x'', x''', \ldots$ by differentiating the function f. This requirement can be a serious obstacle to using the method. The user of this method must do some preliminary analytical work before writing a computer program. Ideally, a method for solving Equation (1) should involve nothing more than writing a code to evaluate f. The Runge-Kutta methods accomplish this.

For purposes of exposition, the Runge-Kutta method of order 2 is presented, although its low precision usually precludes its use in actual scientific calculations. Later the Runge-Kutta method of order 4 is given *without* a derivation. It is in common use. The order-2 Runge-Kutta procedure does find application in real-time calculations on small computers. For example, it is used in some aircraft by the on-board minicomputer.

At the heart of any method for solving an initial-value problem is a procedure for advancing the solution function one step at a time; that is, a formula must be given for $x(t + h)$ in terms of known quantities. As examples of known quantities, we can cite $x(t), x(t - h), x(t - 2h), \ldots$ if the solution process has gone through a number of steps. At the beginning, only $x(a)$ is known. Of course, we assume that $f(t, x)$ can be computed for any point (t, x).

Taylor Series for $f(x, y)$

Before explaining the Runge-Kutta method of order 2, let us present the **Taylor series in two variables**. The infinite series is

$$f(x + h, y + k) = \sum_{i=0}^{\infty} \frac{1}{i!} \left(h \frac{\partial}{\partial x} + k \frac{\partial}{\partial y} \right)^i f(x, y) \tag{2}$$

This series is analogous to the Taylor series in one variable given by Equation (11) in Section 1.2. The mysterious-looking terms in Equation (2) are interpreted as follows:

$$\left(h \frac{\partial}{\partial x} + k \frac{\partial}{\partial y} \right)^0 f(x, y) = f$$

$$\left(h \frac{\partial}{\partial x} + k \frac{\partial}{\partial y} \right)^1 f(x, y) = h \frac{\partial f}{\partial x} + k \frac{\partial f}{\partial y}$$

$$\left(h \frac{\partial}{\partial x} + k \frac{\partial}{\partial y} \right)^2 f(x, y) = h^2 \frac{\partial^2 f}{\partial x^2} + 2hk \frac{\partial^2 f}{\partial x \partial y} + k^2 \frac{\partial^2 f}{\partial y^2}$$

$$\vdots$$

where f and all partial derivatives are evaluated at (x, y). As in the one-variable case, if the Taylor series is truncated, an error term or remainder term is needed to restore the equality. Here is the appropriate equation:

$$f(x + h, y + k) = \sum_{i=0}^{n-1} \frac{1}{i!} \left(h \frac{\partial}{\partial x} + k \frac{\partial}{\partial y} \right)^i f(x, y) + \frac{1}{n!} \left(h \frac{\partial}{\partial x} + k \frac{\partial}{\partial y} \right)^n f(\bar{x}, \bar{y}) \tag{3}$$

The point $(\bar{x}, \bar{y})$ lies on the line segment that joins (x, y) to $(x + h, y + k)$ in the plane.

In applying Taylor series, we use subscripts to denote partial derivatives. So, for instance,

$$f_x = \frac{\partial f}{\partial x} \qquad f_t = \frac{\partial f}{\partial t} \qquad f_{xx} = \frac{\partial^2 f}{\partial x^2} \qquad f_{xt} = \frac{\partial^2 f}{\partial t \, \partial x} \tag{4}$$

We are dealing with functions for which the order of these subscripts is immaterial; for example, $f_{xt} = f_{tx}$. Thus, we have

$$f(x + h, y + k) = f + (hf_x + kf_y)$$

$$+ \frac{1}{2!} \left(h^2 f_{xx} + 2hk f_{xy} + k^2 f_{yy} \right)$$

$$+ \frac{1}{3!} \left(h^3 f_{xxx} + 3h^2 k f_{xxy} + 3hk^2 f_{xyy} + k^3 f_{yyy} \right)$$

$$+ \cdots$$

As special cases, we notice that

$$f(x + h, y) = f + hf_x + \frac{h^2}{2!}f_{xx} + \frac{h^3}{3!}f_{xxx} + \cdots$$

$$f(x, y + k) = f + kf_y + \frac{k^2}{2!}f_{yy} + \frac{k^3}{3!}f_{yyy} + \cdots$$

Runge-Kutta Method of Order 2

In the Runge-Kutta method of order 2, a formula is adopted that has two function evaluations of the form

$$\begin{cases} F_1 = hf(t, x) \\ F_2 = hf(t + \alpha h, x + \beta F_1) \end{cases}$$

and a linear combination of these is added to the value of x at t to obtain the value at $t + h$:

$$x(t + h) = x(t) + w_1 F_1 + w_2 F_2$$

or, equivalently,

$$x(t + h) = x(t) + w_1 hf(t, x) + w_2 hf(t + \alpha h, x + \beta hf(t, x)) \tag{5}$$

The objective is to determine constants w_1, w_2, α, and β so that Equation (5) is as accurate as possible. Explicitly, we want to reproduce as many terms as possible in the Taylor series

$$x(t + h) = x(t) + hx'(t) + \frac{1}{2!}h^2 x''(t) + \frac{1}{3!}h^3 x'''(t) + \cdots \tag{6}$$

Now compare Equation (5) with Equation (6). One way to force them to agree up through the term in h is to set $w_1 = 1$ and $w_2 = 0$ because $x' = f$. However, agreement up through the h^2 term is possible by a more adroit choice of parameters. To see how, apply the two-variable form of the Taylor series to the final term in Equation (5). We use the first three terms ($n = 2$) of the two-variable Taylor series given by Formula (3), with t, αh, x, and βhf playing the role of x, h, y, and k, respectively:

$$f(t + \alpha h, x + \beta hf) = f + \alpha hf_t + \beta hff_x + \frac{1}{2}\left(\alpha h\frac{\partial}{\partial t} + \beta hf\frac{\partial}{\partial x}\right)^2 f(\overline{x}, \overline{y})$$

Using the above equation results in a new form for Equation (5). We have

$$x(t + h) = x(t) + (w_1 + w_2)hf + \alpha w_2 h^2 f_t + \beta w_2 h^2 ff_x + \mathcal{O}(h^3) \tag{7}$$

Equation (6) is also given a new form by using differential Equation (1). Since $x' = f$, we have

$$x'' = \frac{dx'}{dt} = \frac{df(t, x)}{dt} = \left(\frac{\partial f}{\partial t}\right)\left(\frac{dt}{dt}\right) + \left(\frac{\partial f}{\partial x}\right)\left(\frac{dx}{dt}\right) = f_t + f_x f$$

So Equation (6) implies that

$$x(t + h) = x + hf + \frac{1}{2}h^2 f_t + \frac{1}{2}h^2 f f_x + \mathcal{O}(h^3) \tag{8}$$

Agreement between Equations (7) and (8) is achieved by stipulating that

$$w_1 + w_2 = 1 \qquad \alpha w_2 = \frac{1}{2} \qquad \beta w_2 = \frac{1}{2} \tag{9}$$

A convenient solution of these equations is

$$\alpha = 1 \qquad \beta = 1 \qquad w_1 = \frac{1}{2} \qquad w_2 = \frac{1}{2}$$

The resulting **Runge-Kutta method of order 2** is then, from Equation (5),

$$x(t + h) = x(t) + \frac{h}{2}f(t, x) + \frac{h}{2}f(t + h, x + hf(t, x))$$

or, equivalently,

$$x(t + h) = x(t) + \frac{1}{2}(F_1 + F_2) \tag{10}$$

where

$$\begin{cases} F_1 = hf(t, x) \\ F_2 = hf(t + h, x + F_1) \end{cases}$$

Formula (10) shows that the solution function at $t + h$ is computed at the expense of two evaluations of the function f.

Notice that other solutions for the nonlinear Equations (9) are possible. For example, α can be arbitrary and then

$$\beta = \alpha \qquad w_1 = 1 - \frac{1}{2\alpha} \qquad w_2 = \frac{1}{2\alpha}$$

One can show (see Problem **10**) that the error term for Runge-Kutta methods of order 2 is

$$\frac{h^3}{4}\left(\frac{2}{3} - \alpha\right)\left(\frac{\partial}{\partial t} + f\frac{\partial}{\partial x}\right)^2 f + \frac{h^3}{6}f_x\left(\frac{\partial}{\partial t} + f\frac{\partial}{\partial x}\right)f \tag{11}$$

Notice that the method with $\alpha = \frac{2}{3}$ is especially interesting. However, none of the second-order Runge-Kutta methods is widely used on large computers because the error is only $\mathcal{O}(h^3)$.

One algorithm in common use for the initial-value Problem (1) is the classical **Runge-Kutta method of order 4**. Its formulas are as follows:

$$x(t + h) = x(t) + \frac{1}{6}(F_1 + 2F_2 + 2F_3 + F_4) \tag{12}$$

where

$$\begin{cases} F_1 = hf(t, x) \\[2mm] F_2 = hf\left(t + \frac{1}{2}h, x + \frac{1}{2}F_1\right) \\[2mm] F_3 = hf\left(t + \frac{1}{2}h, x + \frac{1}{2}F_2\right) \\[2mm] F_4 = hf(t + h, x + F_3) \end{cases}$$

As can be seen, the solution at $x(t + h)$ is obtained at the expense of evaluating the function f four times. The final formula agrees with the Taylor expansion up to and including the term in h^4. The error therefore contains h^5 but no lower powers of h. Without knowing the coefficient of h^5 in the error, we cannot be precise about the local truncation error. In treatises devoted to this subject, these matters are further explored. See, for example, Butcher [1987] or Gear [1971].

Pseudocode

Here is a pseudocode to implement the Runge-Kutta method of order 4:

```
procedure rk4(f, t, x, h, n)
integer k, n
real F_1, F_2, F_3, F_4, h, t, t_a, x
interface external function f
output 0, t, x
t_a ← t
```

for $k = 1$ **to** n **do**
  $F_1 \leftarrow hf(t, x)$
  $F_2 \leftarrow hf(t + h/2, x + F_1/2)$
  $F_3 \leftarrow hf(t + h/2, x + F_2/2)$
  $F_4 \leftarrow hf(t + h, x + F_3)$
  $x \leftarrow x + (F_1 + 2F_2 + 2F_3 + F_4)/6$
  $t \leftarrow t_a + kh$
  output k, t, x
end do
end procedure *rk4*

To illustrate the use of the preceding pseudocode, consider the initial-value problem

$$\begin{cases} x' = 2 + (x - t - 1)^2 \\ x(1) = 2 \end{cases} \tag{13}$$

whose exact solution is $x(t) = 1 + t + \tan(t - 1)$. A pseudocode to solve this problem on the interval $[1, 1.5625]$ by the Runge-Kutta procedure follows. The step size needed is calculated by dividing the length of the interval by the number of steps, say $n = 72$.

program *main*
real parameter $a \leftarrow 1$, $b \leftarrow 1.5625$, $x \leftarrow 2$
integer $n \leftarrow 72$
real h, t, x
interface external function f
$h \leftarrow (b - a)/n$
$t \leftarrow a$
call $rk4(f, t, x, h, n)$
end program *main*

real function $f(t, x)$
real t, x
$f \leftarrow 2 + (x - t - 1)^2$
end function f

We include an interface-external-function statement both in the main program and in procedure *rk4* because the procedure f is passed in the argument list of *rk4*.

General-purpose routines incorporating the Runge-Kutta algorithm usually include additional programming to monitor the truncation error and make necessary adjustments in the step size as the solution progresses. In general terms, the step size can be large when the solution is slowly varying but should be small when it is rapidly varying. Such a program is presented in Section 8.3.

Alternatively, we can use Maple as follows

```
ode := D(x)(t) = 2 + (x(t) - t - 1)^2:
f := dsolve({ode, x(1)=2}, x(t), numeric);
f(1); f(1.5625);
F := t -> f(t)[2];
plot(F, 1..1.5625);
```

and obtain

```
{t = 1., x(t) = 2.}
{x(t) = 3.192937699, t = 1.562500000}
```

**PROBLEMS
8.2**

1. Derive the equations needed to apply the fourth-order Taylor series method to the differential equation $x' = tx^2 + x - 2t$. Compare them in complexity with the equations required for the fourth-order Runge-Kutta method.

2. Put these differential equations into a form suitable for numerical solution by the Runge-Kutta method.

   a. $x + 2xx' - x' = 0$

   b. $\log x' = t^2 - x^2$

   c. $(x')^2(1 - t^2) = x$

3. Solve the differential equation

$$\begin{cases} \dfrac{dx}{dt} = -tx^2 \\ x(0) = 2 \end{cases}$$

   at $t = -0.2$, correct to two decimal places, using one step of the Taylor series method of order 2 and one step of the Runge-Kutta method of order 2.

4. Consider the ordinary differential equation

$$\begin{cases} x' = (tx)^3 - \left(\dfrac{x}{t}\right)^2 \\ x(1) = 1 \end{cases}$$

   Take one step of the Taylor series method of order 2 with $h = 0.1$ and then use the Runge-Kutta method of order 2 to recompute $x(1.1)$. Compare answers.

5. In solving the following differential equations by using a Runge-Kutta procedure, it is necessary to write code for a function $f(t, x)$. Do so for each of the following.

   a. $x' = t^2 + tx' - 2xx'$

   b. $x' = e^t + x' \cos x + t^2$

6. Consider the ordinary differential equation $x' = t^3 x^2 - 2x^3/t^2$ with $x(1) = 0$. Determine the equations that would be used in applying the Taylor series method of order 3 and the Runge-Kutta method of order 4.

7. Consider the third-order Runge-Kutta method:

$$x(t + h) = x(t) + \frac{1}{9}(2F_1 + 3F_2 + 4F_3)$$

where

$$\begin{cases} F_1 = hf(t, x) \\ F_2 = hf\left(t + \frac{1}{2}h, x + \frac{1}{2}F_1\right) \\ F_3 = hf\left(t + \frac{3}{4}h, x + \frac{3}{4}F_2\right) \end{cases}$$

Show that it agrees with the Taylor series method of the same order for the differential equation $x' = x + t$.

8. Describe how the fourth-order Runge-Kutta method can be used to produce a table of values for the function

$$f(x) = \int_0^x e^{-t^2} dt$$

at 100 equally spaced points in the unit interval. *Hint*: Find an appropriate initial-value problem whose solution is f.

9. Show that the fourth-order Runge-Kutta formula reduces to a simple form when applied to an ordinary differential equation of the form

$$x' = f(t)$$

10. Establish the error term (11) for Runge-Kutta methods of order 2.

11. On a certain computer it was found that when the fourth-order Runge-Kutta method was used over an interval $[a, b]$ with $h = (b - a)/n$, the total error due to roundoff was about $36n2^{-50}$ and the total truncation error was $9nh^5$, where n is the number of steps and h is the step size. What is an optimum value of h? *Hint*: Minimize the total error: roundoff plus truncation.

12. How would you solve the initial-value problem

$$\begin{cases} x' = \sin x + \sin t \\ x(0) = 0 \end{cases}$$

on the interval $[0, 1]$ if ten decimal places of accuracy are required? Assume that you have a computer in which unit roundoff error is $\frac{1}{2} \times 10^{-14}$, and assume that the fourth-order Runge-Kutta method will involve local truncation errors of magnitude $100h^5$.

13. An important theorem of calculus states that the equation $f_{tx} = f_{xt}$ is true, provided that at least one of these two partial derivatives exists and is continuous. Test this equation on some functions, such as $f(t, x) = xt^2 + x^2t + x^3t^4$, $\log(x - t^{-1})$, and $e^x \sinh(t + x) + \cos(2x - 3t)$.

14. **a.** If $x' = f(t, y)$, then

$$x''' = f_{tt} + 2ff_{tx} + f^2 f_{xx} + f_x f_t + ff_x^2 = \left(\frac{\partial}{\partial t} + f \frac{\partial}{\partial x} \right)^2 f$$

$$= D^2 f + f_x Df$$

where

$$D = \frac{\partial}{\partial t} + f \frac{\partial}{\partial x} \qquad D^2 = \frac{\partial^2}{\partial t^2} + 2f \frac{\partial^2}{\partial x \, \partial t} + f^2 \frac{\partial^2}{\partial t^2}$$

Verify these equations.

b. Determine $x^{(4)}$ in a similar form.

15. Derive the two-variable form of the Taylor series from the one-variable form by considering the function of one variable $\phi(t) = f(x + th, y + tk)$ and expanding it by Taylor's Theorem.

16. The Taylor series expansion about point (a, b) in terms of two variables x and y is given by

$$f(x, y) = \sum_{i=0}^{\infty} \frac{1}{i!} \left((x - a) \frac{\partial}{\partial x} + (y - b) \frac{\partial}{\partial y} \right)^i f(a, b)$$

Show that Formula (2) can be obtained from this form by a change of variables.

17. (Continuation) Using the form given in Problem **16**, determine the first four nonzero terms in the Taylor series for $f(x, y) = \sin x + \cos y$ about the point $(0, 0)$. Compare the result to the known series for $\sin x$ and $\cos y$. Make a conjecture about the Taylor series for functions that have the special form $f(x, y) = g(x) + h(y)$.

18. For the function $f(x, y) = y^2 - 3 \ln x$, write the first six terms in the Taylor series of $f(1 + h, 0 + k)$.

19. Using the truncated Taylor series about $(1, 1)$, give a three-term approximation to $e^{(1-xy)}$. *Hint*: Use Problem **16**.

20. The function $f(x, y) = xe^y$ can be approximated by the Taylor series in two variables by $f(x + h, y + k) \approx (Ax + B)e^y$. Determine A and B when terms through the second partial derivatives are used in the series.

21. For $f(x, y) = (y - x)^{-1}$, the Taylor series can be written as

$$f(x + h, y + k) = Af + Bf^2 + Cf^3 + \cdots$$

where $f = f(x, y)$. Determine the coefficients A, B, and C.

22. Consider the function e^{x^2+y}. Determine its Taylor series about the point $(0, 1)$ through second-partial-derivative terms. Use this result to obtain an approximate value for $f(0.001, 0.998)$.

COMPUTER PROBLEMS 8.2

1. Run the sample code given in the text for differential Equation (13) to illustrate the Runge-Kutta method.

2. Solve the initial-value problem $x' = x/t + t \sec(x/t)$ with $x(0) = 0$ by the fourth-order Runge-Kutta method. Continue the solution to $t = 1$ using step size $h = 2^{-7}$. Compare the numerical solution with the exact solution, which is $x(t) = t \arcsin t$. Define $f(0, 0) = 0$, where $f(t, x) = x/t + t \sec(x/t)$.

3. Select one of the following initial-value problems and compare the numerical solutions obtained with fourth-order Runge-Kutta formulas and fourth-order Taylor series. Use different values of $h = 2^{-n}$, for $n = 2, 3, \ldots, 7$, to compute the solution on the interval $[1, 2]$.

   a. $x' = 1 + x/t$ $x(1) = 1$
   b. $x' = 1/x^2 - xt$ $x(1) = 1$
   c. $x' = 1/t^2 - x/t - x^2$ $x(1) = -1$

4. Select a Runge-Kutta routine from your program library and test it on the initial-value problem $x' = (2 - t)x$ with $x(2) = 1$. Compare with the exact solution, $x = \exp[-(\frac{1}{2})(t - 2)^2]$.

5. (A *stiff* ODE) Solve the ordinary differential equation $x' = 10x + 11t - 5t^2 - 1$ with initial value $x(0) = 0$. Continue the solution from $x = 0$ to $x = 3$, using the fourth-order Runge-Kutta method with $h = 2^{-8}$. Print the numerical solution and the exact solution $(t^2/2 - t)$ at every tenth step and draw a graph of the two solutions. Verify that the solution of the same differential equation with initial value $x(0) = \varepsilon$ is $\varepsilon e^{10t} + t^2/2 - t$ and thus account for the discrepancy between the numerical and exact solutions of the original problem.

6. Solve the initial-value problem $x' = x\sqrt{x^2 - 1}$ with $x(0) = 1$ by the Runge-Kutta method on the interval $0 \leq t \leq 1.6$ and account for any difficulties. Then using negative h, solve the same differential equation on the same interval with initial value $x(1.6) = 1.0$.

7. The following pathological example has been given by Dahlquist and Björck [1974]. Consider the differential equation $x' = 100(\sin t - x)$ with initial value $x(0) = 0$. Integrate it with the fourth-order Runge-Kutta method on the interval $[0, 3]$, using $h = 0.015, 0.020, 0.025, 0.030$. Observe the numerical instability!

8. Consider the differential equation

$$\begin{cases} x' = \begin{cases} x + t & -1 \leq t \leq 0 \\ x - t & 0 \leq t \leq 1 \end{cases} \\ x(-1) = 1 \end{cases}$$

Using the Runge-Kutta procedure *rk4* with step size $h = 0.1$, solve this problem over the interval $[-1, 1]$. Now solve by using $h = 0.09$. Which numerical solution is more accurate and why? *Hint:* The true solution is given by $x = e^{(t+1)} - (t + 1)$ if $t \leq 0$ and $x = e^{(t+1)} - 2e^t + (t + 1)$ if $t \geq 0$.

9. Solve $t - x' + 2xt = 0$ with $x(0) = 0$ on the interval $[0, 10]$ using the Runge-Kutta formulas with $h = 0.1$. Compare with the true solution, which is $\frac{1}{2}(e^{t^2} - 1)$. Draw a graph or have one created by an automatic plotter. Then graph the logarithm of the solution.

10. Write a program to solve $x' = \sin(xt) + \arctan t$ on $1 \leq t \leq 7$ with $x(2) = 4$ using the Runge-Kutta procedure *rk4*.

11. The general form of Runge-Kutta methods of order 2 is given by Equations (5) and (9). Write and test **procedure** $rk2(f, t, x, h, \alpha, n)$ for carrying out n steps with step size h and initial conditions t and x for several given α values.

12. We want to solve

$$\begin{cases} x' = e^t x^2 + e^3 \\ x(2) = 4 \end{cases}$$

at $x(5)$ with step size 0.5. Solve it in the following two ways.

a. Write the procedure function $f(t, x)$ that is needed and use procedure *rk4*.

b. Write a short program that uses the Taylor series method including terms up to h^4.

13. Another fourth-order Runge-Kutta method is given by

$$x(t + h) = x(t) + w_1 F_1 + w_2 F_2 + w_3 F_3 + w_4 F_4$$

where

$$\begin{cases} F_1 = hf(t, x) \\ F_2 = hf\left(t + \frac{2}{5}h, \ x + \frac{2}{5}F_1\right) \\ F_3 = hf\left(t + \frac{1}{16}(14 - 3\sqrt{5})h, \ x + c_{31}F_1 + c_{32}F_2\right) \\ F_4 = hf(t + h, \ x + c_{41}F_1 + c_{42}F_2 + c_{43}F_3) \end{cases}$$

Here the appropriate constants are

$$c_{31} = \frac{3(-963 + 476\sqrt{5})}{1024} \qquad c_{32} = \frac{5(757 - 324\sqrt{5})}{1024}$$

$$c_{41} = \frac{-3365 + 2094\sqrt{5}}{6040} \qquad c_{42} = \frac{-975 - 3046\sqrt{5}}{2552}$$

$$c_{43} = \frac{32(14595 + 6374\sqrt{5})}{2\,40845}$$

$$w_1 = \frac{263 + 24\sqrt{5}}{1812} \qquad w_2 = \frac{125(1 - 8\sqrt{5})}{3828}$$

$$w_3 = \frac{1024(3346 + 1623\sqrt{5})}{59\,24787} \qquad w_4 = \frac{2(15 - 2\sqrt{5})}{123}$$

Select a differential equation with a known solution and compare the two fourth-order Runge-Kutta methods. Print the errors at each step. Is the ratio of the two errors a constant at each step? What are the advantages and/or disadvantages of each method? *Note*: There are any number of Runge-Kutta methods of any order. The higher the order, the more complicated the formulas. Since the one given by Equation (12) has error $\mathcal{O}(h^5)$ and is rather simple, it is the most popular fourth-order Runge-Kutta method. The error term for the method of this problem is also $\mathcal{O}(h^5)$, and it is optimum in a certain sense. (See Ralston [1965] for details.)

14. A fifth-order Runge-Kutta method is given by

$$x(t + h) = x(t) + \frac{1}{24}F_1 + \frac{5}{48}F_4 + \frac{27}{56}F_5 + \frac{125}{336}F_6$$

where

$$\begin{cases}
F_1 = hf(t, x) \\[2mm]
F_2 = hf\left(t + \frac{1}{2}h, \ x + \frac{1}{2}F_1\right) \\[2mm]
F_3 = hf\left(t + \frac{1}{2}h, \ x + \frac{1}{4}F_1 + \frac{1}{4}F_2\right) \\[2mm]
F_4 = hf(t + h, \ x - F_2 + F_3) \\[2mm]
F_5 = hf\left(t + \frac{2}{3}h, \ x + \frac{7}{27}F_1 + \frac{10}{27}F_2 + \frac{1}{27}F_4\right) \\[2mm]
F_6 = hf\left(t + \frac{1}{5}h, \ x + \frac{28}{625}F_1 - \frac{1}{5}F_2 + \frac{546}{625}F_3 + \frac{54}{625}F_4 - \frac{378}{625}F_5\right)
\end{cases}$$

Write and test a procedure that uses this formula.

8.3   Stability and Adaptive Runge-Kutta Methods

Stability Analysis

Let us now resume the discussion of errors that inevitably occur in the numerical solution of an initial-value problem

$$\begin{cases} x' = f(t, x) \\ x(a) = s \end{cases} \tag{1}$$

The exact solution is a function $x(t)$. It depends on the initial value s, and in order to show this, we write $x(t, s)$. The differential equation gives rise, therefore, to a family of solution curves, each corresponding to one value of the parameter s. For example, the differential equation

$$\begin{cases} x' = x \\ x(a) = s \end{cases}$$

gives rise to the family of solution curves $x = se^{(t-a)}$ that differ in their initial values $x(a) = s$. A few such curves are shown in Figure 8.1. The fact that the curves there diverge from one another as t increases has important numerical significance. Suppose, for instance, that initial value s is read into the computer with some roundoff error. Then even if all subsequent calculations are precise and *no truncation errors* occur, the computed solution will be wrong. An error made at the beginning has the effect of selecting the wrong *curve* from the family of all solution curves. Since these curves diverge from one another, the minute error made at the beginning is responsible for an eventual complete loss of accuracy. This phenomenon is not restricted to errors made in the first step because each point in the numerical solution can be interpreted as the initial value for succeeding points.

For an example in which this difficulty does not arise, consider

$$\begin{cases} x' = -x \\ x(a) = s \end{cases}$$

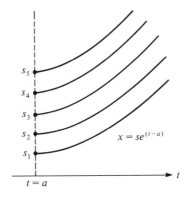

FIGURE 8.1
Solution curves to
$x' = x$ with $x(a) = s$

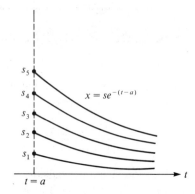

FIGURE 8.2
Solution curves to
$x' = -x$ with $x(a) = s$

Its solutions are $x = se^{-(t-a)}$. As t increases, these curves come closer together, as in Figure 8.2. Thus, errors made in the numerical solution still result in selecting the wrong curve, but the effect is not as serious because the curves coalesce.

For the general differential Equation (1), how can the two modes of behavior just discussed be distinguished? It is simple. If $f_x > \delta$ for some positive δ, the curves diverge. However, if $f_x < -\delta$, they converge. To see why, consider two nearby solution curves that correspond to initial values s and $s + h$. By Taylor series, we have

$$x(t, s + h) = x(t, s) + h\frac{\partial}{\partial s}x(t, s) + \frac{1}{2}h^2\frac{\partial^2}{\partial s^2}x(t, s) + \cdots$$

whence

$$x(t, s + h) - x(t, s) \approx h\frac{\partial}{\partial s}x(t, s)$$

Thus, the condition of divergence of the curves, which means that

$$\lim_{t\to\infty}|x(t, s + h) - x(t, s)| = \infty$$

can be written as

$$\lim_{t\to\infty}\left|\frac{\partial}{\partial s}x(t, s)\right| = \infty$$

To calculate this partial derivative, start with the differential equation satisfied by $x(t, s)$:

$$\frac{\partial}{\partial t}x(t, s) = f(t, x(t, s))$$

and differentiate partially with respect to s:

$$\frac{\partial}{\partial s}\frac{\partial}{\partial t}x(t,s) = \frac{\partial}{\partial s}f(t,x(t,s))$$

Hence,

$$\frac{\partial}{\partial t}\frac{\partial}{\partial s}x(t,s) = f_x(t,x(t,s))\frac{\partial}{\partial s}x(t,s) + f_t(t,x(t,s))\frac{\partial t}{\partial s} \tag{2}$$

But s and t are independent variables (a change in s produces no change in t), so $\partial t/\partial s = 0$. If s is now fixed and if we put $u(t) = (\partial/\partial s)x(t,s)$ and $q(t) = f_x(t,x(t,s))$, then Equation (2) becomes

$$u' = qu \tag{3}$$

This is a linear differential equation with solution $u(t) = ce^{Q(t)}$, where Q is the indefinite integral (antiderivative) of q. Obviously, the condition $\lim_{t\to\infty}|u(t)| = \infty$ is met if $\lim_{t\to\infty}Q(t) = \infty$. This situation, in turn, occurs if $q(t)$ is positive and bounded away from zero because then

$$Q(t) = \int_a^t q(\theta)\,d\theta > \int_a^t \delta\,d\theta = \delta(t-a) \to \infty$$

as $t \to \infty$ if $f_x = q > \delta > 0$.

To illustrate, consider the differential equation $x' = t + \tan x$. The solution curves diverge from one another as $t \to \infty$ because $f_x(t,x) = \sec^2 x > 1$.

An Adaptive Runge-Kutta-Fehlberg Method

In realistic situations involving the numerical solution of initial-value problems, there is always a need to estimate the precision attained in the computation. Usually a tolerance is prescribed, and the numerical solution must not deviate from the true solution beyond this tolerance. Once a method is selected, the error tolerance dictates the allowable step size. Even if we consider only the local truncation error, determining an appropriate step size may be difficult. Moreover, often a small step size is needed on one portion of the solution curve, whereas a larger one may suffice elsewhere.

For the reasons given, various methods have been developed for *automatically* adjusting the step size in algorithms for the initial-value problem. One simple procedure is now described. Consider the fourth-order Runge-Kutta method discussed in Section 8.2. To advance the solution curve from t to $t + h$, we can take one step of size h using the Runge-Kutta formulas. But we can also take *two* steps of size $h/2$ to arrive at $t + h$. If there were no truncation error, the value of the numerical solution $x(t + h)$ would be the same for both procedures. The difference in the numerical results can be taken as an estimate of the local truncation error. So, in practice, if this difference is within the prescribed tolerance, the current step size h is satisfactory. If this difference exceeds the tolerance, the step size is halved. If the difference is very much less than the tolerance, the step size is doubled.

The procedure just outlined is easily programmed but rather wasteful of computing time and is not recommended. A more sophisticated method was developed by Fehlberg [1969]. The Fehlberg method of order 4 is of Runge-Kutta type and uses these formulas:

$$x(t + h) = x(t) + \frac{25}{216}F_1 + \frac{1408}{2565}F_3 + \frac{2197}{4104}F_4 - \frac{1}{5}F_5$$

where

$$
\begin{cases}
F_1 = hf(t, x) \\[2mm]
F_2 = hf\left(t + \frac{1}{4}h, \ x + \frac{1}{4}F_1\right) \\[2mm]
F_3 = hf\left(t + \frac{3}{8}h, \ x + \frac{3}{32}F_1 + \frac{9}{32}F_2\right) \\[2mm]
F_4 = hf\left(t + \frac{12}{13}h, \ x + \frac{1932}{2197}F_1 - \frac{7200}{2197}F_2 + \frac{7296}{2197}F_3\right) \\[2mm]
F_5 = hf\left(t + h, \ x + \frac{439}{216}F_1 - 8F_2 + \frac{3680}{513}F_3 - \frac{845}{4104}F_4\right)
\end{cases}
$$

Since this scheme requires one more function evaluation than the standard Runge-Kutta method of order 4, it is of questionable value alone. However, with an additional function evaluation

$$F_6 = hf\left(t + \frac{1}{2}h, \ x - \frac{8}{27}F_1 + 2F_2 - \frac{3544}{2565}F_3 + \frac{1859}{4104}F_4 - \frac{11}{40}F_5\right)$$

we can obtain a Runge-Kutta method of order 5, namely,

$$x(t + h) = x(t) + \frac{16}{135}F_1 + \frac{6656}{12825}F_3 + \frac{28561}{56430}F_4 - \frac{9}{50}F_5 + \frac{2}{55}F_6$$

The difference between the values of $x(t + h)$ obtained from the fourth- and fifth-order procedures is an estimate of the local truncation error in the fourth-order procedure. So six function evaluations give a fourth-order approximation, together with an error estimate!

A pseudocode for the Runge-Kutta-Fehlberg method is given in procedure *rk45*. The coefficients are computed and the numerical values assigned so that they need not be recomputed each time the procedure is called. The error estimate ε can tell us when to adjust the step size in order to control the single-step error.

We now describe a simple adaptive procedure. In the *rk45* procedure, the fourth- and fifth-order approximations for $x(t + h)$, say x_4 and x_5, are computed from six function evaluations and the error estimate $\varepsilon = |x_4 - x_5|$ is known. From user-specified bounds on the allowable error estimate ($\varepsilon_{min} \leq \varepsilon \leq \varepsilon_{max}$), the step size h is doubled or halved as needed to keep ε within these bounds. A range for the allowable step size h is also specified by the user ($h_{min} \leq h \leq h_{max}$). Clearly, the user must set the bounds ($\varepsilon_{min}, \varepsilon_{max}, h_{min}, h_{max}$) carefully so that the adaptive procedure does not get caught in a loop, trying repeatedly to halve and double the step size from the same point in order to meet error bounds that are too restrictive for the given differential equation.

Basically, our adaptive process is as follows:

1. Given a step size h and an initial value $x(t)$, the *rk45* routine computes the value $x(t + h)$ and an error estimate ε.

2. If $\varepsilon_{min} \leq \varepsilon \leq \varepsilon_{max}$, then the step size h is not changed and the next step is taken by repeating step 1 with initial value $x(t + h)$.

3. If $\varepsilon < \varepsilon_{min}$, then h is replaced by $2h$.

4. If $\varepsilon > \varepsilon_{max}$, then h is replaced by $h/2$.

5. If $h_{min} \leq |h| \leq h_{max}$, then the step is repeated by returning to step 1 with $x(t)$ and the new h value.

procedure *rk45*$(f, t, x, h, \varepsilon)$
real parameter $c_{20} \leftarrow 0.25$, $c_{21} \leftarrow 0.25$
real parameter $c_{30} \leftarrow 0.375$, $c_{31} \leftarrow 0.09375$, $c_{32} \leftarrow 0.28125$
real parameter $c_{40} \leftarrow 12./13.$, $c_{41} \leftarrow 1932./2197.$
real parameter $c_{42} \leftarrow -7200./2197.$, $c_{43} \leftarrow 7296./2197.$
real parameter $c_{51} \leftarrow 439./216.$, $c_{52} \leftarrow -8.$
real parameter $c_{53} \leftarrow 439./216.$, $c_{54} \leftarrow -845./4104.$
real parameter $c_{60} \leftarrow 0.5$, $c_{61} \leftarrow -8./27.$, $c_{62} \leftarrow 2.$
real parameter $c_{63} \leftarrow -3544./2565.$, $c_{64} \leftarrow 1859./4104.$
real parameter $c_{65} \leftarrow -0.275$
real parameter $a_1 \leftarrow 25./216.$, $a_2 \leftarrow 0.$, $a_3 \leftarrow 1408./2565.$
real parameter $a_4 \leftarrow 2197./4104.$, $a_5 \leftarrow -0.2$
real parameter $b_1 \leftarrow 16./135.$, $b_2 \leftarrow 0.$, $b_3 \leftarrow 6656./12825.$
real parameter $b_4 \leftarrow 28561./56430.$, $b_5 \leftarrow -0.18$
real parameter $b_6 \leftarrow 2./55.$
real $\varepsilon, F_1, F_2, F_3, F_4, F_5, F_6, h, t, x, x_4$
interface external function f
$F_1 \leftarrow hf(t, x)$
$F_2 \leftarrow hf(t + c_{20}h, x + c_{21}F_1)$
$F_3 \leftarrow hf(t + c_{30}h, x + c_{31}F_1 + c_{32}F_2)$
$F_4 \leftarrow hf(t + c_{40}h, x + c_{41}F_1 + c_{42}F_2 + c_{43}F_3)$
$F_5 \leftarrow hf(t + h, x + c_{51}F_1 + c_{52}F_2 + c_{53}F_3 + c_{54}F_4)$
$F_6 \leftarrow hf(t + c_{60}h, x + c_{61}F_1 + c_{62}F_2 + c_{63}F_3 + c_{64}F_4 + c_{65}F_5)$

$$x_4 \leftarrow x + a_1F_1 + a_3F_3 + a_4F_4 + a_5F_5$$
$$x \leftarrow x + b_1F_1 + b_3F_3 + b_4F_4 + b_5F_5 + b_6F_6$$
$$t \leftarrow t + h$$
$$\varepsilon \leftarrow |x - x_4|$$
end procedure *rk45*

The programmer may wish to consider various optimization techniques such as assigning numerical values to the coefficients with decimal expansions corresponding to the precision of the computer being used so that the fractions do not need to be recomputed at each call to the procedure.

The procedure for this adaptive scheme is *rk45ad*. In the parameter list, f is the function $f(t, x)$ for the differential equation, t and x contain the initial values, h is the initial step size, t_b is the final value for t, *itmax* is the maximum number of steps to be taken in going from $a = t_a$ to $b = t_b$, ε_{min} and ε_{max} are lower and upper bounds on the allowable error estimate ε, h_{min} and h_{max} are bounds on the step size h, and *iflag* is an error flag that returns one of the following values:

iflag	Meaning
0	Successful march from t_a to t_b
1	Maximum number of iterations reached

On return, t and x are the exit values and h is the final step size value considered or used.

procedure *rk45ad*$(f, t, x, h, t_b, itmax, \varepsilon_{max}, \varepsilon_{min}, h_{min}, h_{max}, iflag)$
real parameter $\delta \leftarrow \frac{1}{2} \times 10^{-5}$
integer *iflag, itmax, n*
real $\varepsilon, \varepsilon_{max}, \varepsilon_{min}, d, h, h_{min}, h_{max}, t, t_b, x, x_{save}, t_{save}$
output "$n \qquad h \qquad t \qquad x$"
interface external function f
output $0, h, t, x$
iflag $\leftarrow 1$
$k \leftarrow 0$
while $k \leq itmax$ **do**
   $k \leftarrow k + 1$
   if $|h| < h_{min}$ **then** $h \leftarrow \text{sign}(h)h_{min}$
   if $|h| > h_{max}$ **then** $h \leftarrow \text{sign}(h)h_{max}$
   $d \leftarrow |t_b - t|$
   if $d \leq |h|$ **then**
      iflag $\leftarrow 0$
      if $d \leq \delta \max\{|t_b|, |t|\}$ **then** exit loop
      $h \leftarrow \text{sign}(h)d$
   end if
   $x_{save} \leftarrow x$
   $t_{save} \leftarrow t$

call $rk45(f, t, x, h, \varepsilon)$
output n, h, t, x, ε
if $iflag = 0$ **then** exit loop
if $\varepsilon < \varepsilon_{\min}$ **then** $h \leftarrow 2h$
if $\varepsilon > \varepsilon_{\max}$ **then**
 $h \leftarrow h/2$
 $x \leftarrow x_{\text{save}}$
 $t \leftarrow t_{\text{save}}$
 $k \leftarrow k - 1$
end if
end do
end procedure $rk45ad$

As an illustration, the reader should repeat the computer example in the previous section using $rk45ad$, which allows variable step size, instead of $rk4$. Compare the accuracy of these two computed solutions.

Solving Differential Equations and Integration

There is a close connection between solving differential equations and integration. Consider the differential equation

$$\begin{cases} \dfrac{dx}{dr} = f(r, x) \\[2mm] x(a) = s \end{cases}$$

Integrating from t to $t + h$, we have

$$\int_t^{t+h} dx = \int_t^{t+h} f(r, x(r)) \, dr$$

Hence,

$$x(t + h) = x(t) + \int_t^{t+h} f(r, x(r)) \, dr$$

Replacing the integral with one of the numerical integration rules from Chapter 5, we obtain a formula for solving the differential equation. For example, Euler's method, Equation (2) in Section 8.1, is obtained from the left rectangle approximation (see Problem **14** in Section 5.2):

$$\int_t^{t+h} f(r, x(r)) \, dr \approx hf(t, x(t))$$

The trapezoid rule:

$$\int_t^{t+h} f(r, x(r)) \, dr \approx \frac{h}{2}[f(t, x(t)) + f(t + h, x(t + h))]$$

gives the formula

$$x(t + h) = x(t) + \frac{h}{2}[f(t, x(t)) + f(t + h, x(t + h))]$$

Since $x(t + h)$ appears on both sides of this equation, it is called an **implicit formula**. If Euler's method

$$x(t + h) = x(t) + hf(t, x(t))$$

is used for the $x(t + h)$ on the right side, then we obtain the Runge-Kutta formula of order 2—namely, Equation (10) in Section 8.2.

Using the Fundamental Theorem of Calculus, we can easily show that an approximate numerical value for the integral

$$\int_a^b f(r, x(r)) \, dr$$

can be computed by solving the following initial-value problem for $x(b)$:

$$\begin{cases} \dfrac{dx}{dr} = f(r, x) \\ x(a) = 0 \end{cases}$$

PROBLEMS 8.3

1. Establish Equation (3).

2. The initial-value problem $x' = (1 + t^2)x$ with $x(0) = 1$ is to be solved on the interval $[0, 9]$. How sensitive is $x(9)$ to perturbations in the initial value $x(0)$?

3. For each differential equation, determine regions in which the solution curves tend to diverge from one another as t increases.

   a. $x' = \sin t + e^x$

   b. $x' = x + te^{-t}$

   c. $x' = xt$

   d. $x' = x^3(t^2 + 1)$

   e. $x' = \cos t - e^x$

   f. $x' = (1 - x^3)(1 + t^2)$

4. Solve the problem

$$\begin{cases} x' = -x \\ x(0) = 1 \end{cases}$$

by using the trapezoid rule, as discussed in the preceding subsection. Compare the true solution at $t = 1$ to the approximate solution obtained with n steps. Show, for example, that for $n = 5$, the error is 0.00123.

5. For the differential equation $x' = t(x^3 - 6x^2 + 15x)$, determine whether the solution curves diverge from one another as $t \to \infty$.

6. Determine whether the solution curves of the differential equation $x' = (1 + t^2)^{-1}x$ diverge from one another as $t \to \infty$.

7. Derive a formula of the form

$$x(t + h) = ax(t) + bx(t - h) + h[cx'(t + h) + dx''(t) + ex'''(t - h)]$$

that is accurate for polynomials of as high a degree as possible. *Hint*: Use polynomials $1, t, t^2$, and so on.

8. Determine the coefficients of an implicit, one-step, ordinary differential equation method of the form

$$x(t + h) = ax(t) + bx'(t) + cx'(t + h)$$

so that it is exact for polynomials of as high a degree as possible. What is the order of the error term?

9. Derive an implicit multistep formula based on Simpson's rule, involving uniformly spaced points $x(t - h)$, $x(t)$, and $x(t + h)$, for numerically solving the ordinary differential equation $x' = f$.

COMPUTER PROBLEMS 8.3

1. Design and carry out a numerical experiment to verify that a slight perturbation in an initial-value problem can cause *catastrophic* errors in the numerical solution.

2. Solve

$$\begin{cases} x' = \dfrac{3x}{t} + \dfrac{9}{2}t - 13 \\ x(3) = 6 \end{cases}$$

at $x(\frac{1}{2})$ using procedure *rk45ad* to obtain the desired solution to nine decimal places. Compare with the true solution:

$$x = t^3 - \frac{9}{2}t^2 + \frac{13}{2}t$$

3. (Continuation) Repeat Problem **2** for $x(-\frac{1}{2})$.

4. It is known that the fourth-order Runge-Kutta method described in Equation (12), Section 8.2, has a local truncation error that is $\mathcal{O}(h^5)$. Devise and carry out a numerical experiment to test this. *Suggestions*: Take just one step in the numerical solution of a nontrivial differential equation whose solution is known beforehand. However, use a variety of values for h, such as 2^{-n}, where $1 \leq n \leq 24$. Test whether the ratio of errors to h^5 remains bounded as $h \to 0$. A multiple-precision calculation may be needed. Print the indicated ratios.

5. Compute and print a table of the function

$$f(\phi) = \int_0^\phi \sqrt{1 - \frac{1}{4}\sin^2 \theta}\, d\theta$$

by solving an appropriate initial-value problem. Cover the interval $[0, 90°]$ with steps of $1°$ and use the Runge-Kutta method of order 4. *Check values*: Use $f(30°) = 0.51788\,193$, and $f(90°) = 1.46746\,221$. *Note*: This is an example of an elliptic integral of the second kind. It arises in finding an arc length on an ellipse and in many engineering problems.

6. Compute the numerical solution of

$$\begin{cases} x' = -x \\ x(0) = 1 \end{cases}$$

using the midpoint formula

$$x_{n+1} = x_{n-1} + 2hx_n'$$

with $x_0 = 1$ and $x_1 = -h + \sqrt{1 + h^2}$. Are there any difficulties in using this method for this problem? Carry out an analysis of the stability of this method. *Hint*: Consider fixed h and assume $x_n = \lambda^n$.

7. Determine the numerical value of $2\pi \int_4^5 (e^s/s)\, ds$ using a procedure for solving an ordinary differential equation.

8. Tabulate and graph the function $[1 - \ln v(x)]v(x)$ on $[0, e]$, where $v(x)$ is the solution of the initial-value problem $(dv/dx)[\ln v(x)] = 2x, v(0) = 1$. *Check value*: $v(1) = e$.

9. By solving an appropriate initial-value problem, make a table of the function

$$f(x) = \int_{1/x}^\infty \frac{dt}{te^t}$$

on the interval $[0, 1]$. Determine how well f is approximated by $xe^{-1/x}$. *Hint*: Let $t = -\ln s$.

10. By solving an appropriate initial-value problem, make a table of the function

$$f(x) = \frac{2}{\sqrt{\pi}} \int_0^x e^{-t^2} \, dt$$

on the interval $0 \leq x \leq 2$. Determine how accurately $f(x)$ is approximated on this interval by the function

$$g(x) = 1 - (ay + by^2 + cy^3) \frac{2}{\sqrt{\pi}} e^{-x^2}$$

where

$$a = 0.30842\,84 \qquad b = -0.08497\,13$$
$$c = 0.66276\,98 \qquad y = (1 + 0.47047x)^{-1}$$

11. The Adams-Moulton method of order 2 is given by

$$\tilde{x}(t + h) = x(t) + \frac{h}{2}[3f(t, x(t)) - f(t - h, x(t - h))]$$

$$x(t + h) = x(t) + \frac{h}{2}[f(t + h, \tilde{x}(t + h)) + f(t, x(t))]$$

The approximate single-step error is $\varepsilon \equiv K|x(t + h) - \tilde{x}(t + h)|$, where $K = \frac{1}{6}$. Using ε to monitor the convergence, write and test an adaptive procedure for solving an ODE of your choice using these formulas.

12. (Continuation) Carry out the instructions of Computer Problem **11** for the Adams-Moulton method of order 3:

$$\tilde{x}(t + h) = x(t) + \frac{h}{12}[23f(t, x(t)) - 16f(t - h, x(t - h))$$

$$+ 5f(t - 2h, x(t - 2h))]$$

$$x(t + h) = x(t) + \frac{h}{12}[5f(t + h, \tilde{x}(t + h)) + 8f(t, x(t))$$

$$- f(t - h, x(t - h))]$$

where $K = \frac{1}{10}$ in the expression for the approximate single-step error.

13. (A predictor-corrector scheme) Using the Adams-Bashforth-Moulton formulas in Problem **15** of Section 5.5, derive the predictor-corrector scheme given by the following equations:

$$\tilde{x}(t + h) = x(t) + \frac{h}{24}[55f(t, x(t)) - 59f(t - h, x(t - h))$$

$$+ 37f(t - 2h, x(t - 2h)) - 9f(t - 3h, x(t - 3h))]$$

$$x(t + h) = x(t) + \frac{h}{24}[9f(t + h, \tilde{x}(t + h)) + 19f(t, x(t))$$

$$- 5f(t - h, x(t - h)) + f(t - 2h, x(t - 2h))]$$

Write and test a procedure for the Adams-Bashforth-Moulton method. *Note*: This is a multistep process because values of x at t, $t - h$, $t - 2h$, and $t - 3h$ are used to determine the *predicted* value $\tilde{x}(t + h)$, which, in turn, is used with values of x at t, $t - h$, and $t - 2h$ to obtain the *corrected* value $x(t + h)$. The error terms for these formulas are $(251/720)h^5 f^{(4)}(\xi)$ and $-(19/720)h^5 f^{(4)}(\eta)$, respectively. (See Section 9.3 for additional discussion of these methods.)

14. Use the Runge-Kutta method to compute $\int_0^1 \sqrt{1 + s^3}\, ds$.

15. Write and run a program to print an accurate table of the function

$$Si(t) = \int_0^t \frac{\sin r}{r}\, dr$$

The table should cover the interval $0 \leq t \leq 1$ in steps of size 0.01. [Use $\sin(0)/0 = 1$.]

16. Compute a table of the function

$$Shi(x) = \int_0^x \frac{\sinh t}{t}\, dt$$

by finding an initial-value problem that it satisfies and then solving the initial-value problem. Your table should be accurate to nearly machine precision. [Use $\sinh(0)/0 = 1$.]

9 SYSTEMS OF ORDINARY DIFFERENTIAL EQUATIONS

A simple model to account for the way in which two different animal species sometimes react is the *predator-prey* model. If $u(t)$ is the number of individuals in the predator species and $v(t)$ the number of individuals in the prey species, then under suitable simplifying assumptions and with appropriate constants a, b, c, and d,

$$\begin{cases} \dfrac{du}{dt} = a(v + b)u \\[2mm] \dfrac{dv}{dt} = c(u + d)v \end{cases}$$

This is a pair of nonlinear ordinary differential equations that govern the populations of the two species (as functions of time t). In this chapter, numerical procedures are developed for solving such problems.

9.1 Methods for First-Order Systems

In Chapter 8, ordinary differential equations were considered in the simplest context; that is, we restricted our attention to a single differential equation of the first order with an accompanying auxiliary condition. Scientific and technological problems often lead to more complicated situations, however. The next degree of complication occurs with **systems** of several first-order equations.

Uncoupled and Coupled Systems

For example, the sun and the nine planets form a system of *particles* moving under the jurisdiction of Newton's law of gravitation. The position vectors of the planets constitute a system of 27 functions, and the Newtonian laws of motion can be written, then, as a system of 54 first-order ordinary differential equations. In principle, the past and future positions of the planets can be obtained by solving these equations numerically.

Taking an example of more modest scope, we consider two equations with two auxiliary conditions. Let $x = x(t)$ and $y = y(t)$ be two functions subject to the system

$$\begin{cases} x'(t) = x(t) - y(t) + 2t - t^2 - t^3 \\ y'(t) = x(t) + y(t) - 4t^2 + t^3 \end{cases} \tag{1}$$

with initial conditions

$$\begin{cases} x(0) = 1 \\ y(0) = 0 \end{cases}$$

This is an example of an initial-value problem that involves a system of two first-order differential equations. The reader is invited to verify that the analytic solution is

$$\begin{cases} x(t) = e^t \cos(t) + t^2 \\ y(t) = e^t \sin(t) - t^3 \end{cases}$$

Alternatively, one can use Maple as follows to confirm this solution.

```
sysode := { D(x)(t) = x(t) - y(t) + 2*t - t^2 - t^3,
                 D(y)(t) = x(t) + y(t) - 4*t^2 + t^3 };
fcns := {x(t), y(t)};
ic := {x(0) = 1, y(0) = 0};
f := dsolve( sysode union ic, fcns );
```

Observe that System (1) can be written in vector notation:

$$\begin{bmatrix} x' \\ y' \end{bmatrix} = \begin{bmatrix} x - y + 2t - t^2 - t^3 \\ x + y - 4t^2 + t^3 \end{bmatrix}$$

with initial conditions

$$\begin{bmatrix} x(0) \\ y(0) \end{bmatrix} = \begin{bmatrix} 1 \\ 0 \end{bmatrix}$$

This is a special case of a more general problem that can be written as

$$\begin{cases} X' = F(t, X) \\ X(a) = S \quad \text{given} \end{cases}$$

where

$$X = \begin{bmatrix} x \\ y \end{bmatrix} \qquad X' = \begin{bmatrix} x' \\ y' \end{bmatrix}$$

and F is the vector whose two components are given by the right-hand sides in Equation (1). Since F depends on t and X, we write $F(t, X)$.

Note that in the example given, it is not possible to solve either of the two differential equations by itself because the first equation governing x' involves the unknown function y, and the second equation governing y' involves the unknown function x. In this situation, we say that the two differential equations are **coupled**.

Let us look at another example that is superficially similar to the first but is actually simpler:

$$\begin{cases} x'(t) = x(t) + 2t - t^2 - t^3 \\ y'(t) = y(t) - 4t^2 + t^3 \end{cases} \qquad (2)$$

with

$$\begin{cases} x(0) = 1 \\ y(0) = 0 \end{cases}$$

These two equations are *not* coupled and can be solved separately as two unrelated initial-value problems (using, for instance, the methods of Chapter 8). Naturally, our concern here is with systems that are coupled, although methods that solve coupled systems also solve those that are not. The procedures discussed in Chapter 8 extend to systems whether coupled or uncoupled.

Taylor Series Method

We illustrate the Taylor series method for System (1) and begin by differentiating the equations constituting it:

$$\begin{cases} x' = x - y + 2t - t^2 - t^3 \\ y' = x + y - 4t^2 + t^3 \end{cases}$$

$$\begin{cases} x'' = x' - y' + 2 - 2t - 3t^2 \\ y'' = x' + y' - 8t + 3t^2 \end{cases}$$

$$\begin{cases} x''' = x'' - y'' - 2 - 6t \\ y''' = x'' + y'' - 8 + 6t \end{cases}$$

$$\begin{cases} x^{(4)} = x''' - y''' - 6 \\ y^{(4)} = x''' + y''' + 6 \end{cases}$$

etc.

A program to proceed from $x(t)$ to $x(t + h)$ and from $y(t)$ to $y(t + h)$ is easily written by using a few terms of the Taylor series:

$$x(t + h) = x + hx' + \frac{h^2}{2}x'' + \frac{h^3}{6}x''' + \frac{h^4}{24}x^{(4)} + \cdots$$

$$y(t + h) = y + hy' + \frac{h^2}{2}y'' + \frac{h^3}{6}y''' + \frac{h^4}{24}y^{(4)} + \cdots$$

together with equations for the various derivatives. Here $x = x(t)$, $y = y(t)$, $x' = x'(t)$, $y'' = y''(t)$, and so on.

A pseudocode program that generates and prints a numerical solution from 0 to 1 in 100 steps is as follows. Terms up to h^4 have been used in the Taylor series.

program *taylorsys*
integer parameter *nsteps* ← 100
real parameter $a \leftarrow 0$, $b \leftarrow 1$
integer i, k
real $h, x, y, x', y', x'', y'', x''', y''', x^{(4)}, y^{(4)}$
$x \leftarrow 1; y \leftarrow 0$
$t \leftarrow a$
output $0, t, x, y$
$h \leftarrow (b - a)/nsteps$
for $k = 1$ **to** *nsteps* **do**
   $x' \leftarrow x - y + t(2 - t(1 + t))$
   $y' \leftarrow x + y + t^2(-4 + t)$
   $x'' \leftarrow x' - y' + 2 - t(2 + 3t)$
   $y'' \leftarrow x' + y' + t(-8 + 3t)$
   $x''' \leftarrow x'' - y'' - 2 - 6t$
   $y''' \leftarrow x'' + y'' - 8 + 6t$
   $x^{(4)} \leftarrow x''' - y''' - 6$
   $y^{(4)} \leftarrow x''' + y''' + 6$
   $x \leftarrow x + h[x' + (h/2)[x'' + (h/3)[x''' + (h/4)[x^{(4)}]]]]$
   $y \leftarrow y + h[y' + (h/2)[y'' + (h/3)[y''' + (h/4)[y^{(4)}]]]]$
   $t \leftarrow a + kh$
   output k, t, x, y
end do
end program *taylorsys*

Simplification

Before describing how other methods of Chapter 8 can be used for systems of equations, we introduce a slight simplification. When we wrote the system of differential equations in vector form:

$$X' = F(t, X)$$

we assumed that the variable t was explicitly separated from the other variables x and y and treated differently. It is not necessary to do so. Indeed, we can introduce a new variable x_0 that is t in disguise and add a new differential equation $x_0' = 1$.

A new initial condition must also be provided, $x_0(a) = a$. In this way, we increase the number of differential equations by 1 and obtain a system written in the more elegant vector form

$$\begin{cases} X' = F(X) \\ X(a) = S \quad \text{given} \end{cases}$$

Consider the system of two equations given by Equation (1). We write it as a system with three variables by letting

$$x_0 = t \qquad x_1 = x \qquad x_2 = y$$

Thus, we have

$$\begin{bmatrix} x_0' \\ x_1' \\ x_2' \end{bmatrix} = \begin{bmatrix} 1 \\ x_1 - x_2 + 2x_0 - x_0^2 - x_0^3 \\ x_1 + x_2 - 4x_0^2 + x_0^3 \end{bmatrix} \tag{3}$$

The auxiliary condition for the vector X is $X(0) = [0, 1, 0]^T$.

As a result of the preceding remarks, we sacrifice no generality in considering a system of $n + 1$ first-order differential equations written as:

$$\begin{cases} x_0' = f_0(x_0, x_1, x_2, \ldots, x_n) \\ x_1' = f_1(x_0, x_1, x_2, \ldots, x_n) \\ \quad \vdots \\ x_n' = f_n(x_0, x_1, x_2, \ldots, x_n) \\ x_0(a) = s_0, \ x_1(a) = s_1, \ldots, \ x_n(a) = s_n \quad \text{all given} \end{cases}$$

We can write this system in general vector notation as

$$\begin{cases} X' = F(X) \\ X(a) = S \quad \text{given} \end{cases} \tag{4}$$

where

$$X = \begin{bmatrix} x_0 \\ x_1 \\ x_2 \\ \vdots \\ x_n \end{bmatrix} \qquad X' = \begin{bmatrix} x_0' \\ x_1' \\ x_2' \\ \vdots \\ x_n' \end{bmatrix} \qquad F = \begin{bmatrix} f_0 \\ f_1 \\ f_2 \\ \vdots \\ f_n \end{bmatrix} \qquad S = \begin{bmatrix} s_0 \\ s_1 \\ s_2 \\ \vdots \\ s_n \end{bmatrix}$$

A differential equation without the t variable explicitly present is said to be *autonomous*. The numerical methods that we discuss do not require that $x_0 = t$ or $f_0 = 1$ or $s_0 = a$.

Taylor Series Method: Vector Notation

The Taylor series method of order m would be written as

$$X(t + h) = X + hX' + \frac{h^2}{2}X'' + \cdots + \frac{h^m}{m!}X^{(m)} \tag{5}$$

where $X = X(t)$, $X' = X'(t)$, $X'' = X''(t)$, and so on.

A pseudocode for the Taylor series method of order 4 applied to the preceding problem can be easily rewritten by a simple change of variables and the introduction of an array and an inner loop.

```
program taylorsys
integer parameter n ← 2, nsteps ← 100
real parameter a ← 0, b ← 1
real array (x_i)_{0:n}
integer i, k
real h
(x_i) ← (0, 1, 0)
output 0, (x_i)
h ← (b − a)/nsteps
x'_0 ← 1
x''_0 ← 0
x'''_0 ← 0
x_0^{(4)} ← 0
for k = 1 to nsteps do
    x'_1 ← x_1 − x_2 + x_0(2 − x_0(1 + x_0))
    x'_2 ← x_1 + x_2 + x_0^2(−4 + x_0)
    x''_1 ← x'_1 − x'_2 + 2 − x_0(2 + 3x_0)
    x''_2 ← x'_1 + x'_2 + x_0(−8 + 3x_0)
    x'''_1 ← x''_1 − x''_2 − 2 − 6x_0
    x'''_2 ← x''_1 + x''_2 − 8 + 6x_0
    x_1^{(4)} ← x'''_1 − x'''_2 − 6
    x_2^{(4)} ← x'''_1 + x'''_2 + 6
    for i = 0 to n do
        x_i ← x_i + h[x'_i + (h/2)[x''_i + (h/3)[x'''_i + (h/4)[x_i^{(4)}]]]]
    end do
    output k, (x_i)
end do
end program taylorsys
```

A two-dimensional array can be used instead of all these different variables, say $x_{ij} \leftrightarrow x_i^{(j)}$. In fact, this and other methods in this chapter become particularly easy to program if the computer language supports vector operations.

Runge-Kutta Method

The Runge-Kutta methods of Chapter 8 also extend to systems of differential equations. The classical fourth-order Runge-Kutta method for System (4) uses these formulas:

$$X(t + h) = X + \frac{h}{6}(F_1 + 2F_2 + 2F_3 + F_4) \tag{6}$$

where

$$
\begin{cases}
F_1 = F(X) \\[2mm]
F_2 = F\left(X + \frac{1}{2}hF_1\right) \\[2mm]
F_3 = F\left(X + \frac{1}{2}hF_2\right) \\[2mm]
F_4 = F(X + hF_3)
\end{cases}
$$

Here $X = X(t)$ and all quantities are vectors with $n + 1$ components except t and h.

A procedure for carrying out the Runge-Kutta procedure is given next. It is assumed that the system to be solved is in the form of Equation (4) and that there are $n + 1$ equations in the system. The user furnishes the initial value of t, the initial value of X, the step size h, and the number of steps to be taken, *nsteps*. Furthermore, procedure $xpsys((x_i), (f_i))$ is needed, which evaluates the right-hand side of Equation (4) for a given value of array (x_i) and stores the result in array (f_i). (The name *xpsys* is chosen as an abbreviation of *x-prime for a system*.)

```
procedure rk4sys(n, h, (xᵢ), nsteps)
real array (xᵢ)₀:ₙ
allocate real array (yᵢ)₀:ₙ, (Fᵢⱼ)₀:ₙ×₄
integer i, k, n
real h
output 0, (xᵢ)
for k = 1 to nsteps do
    call xpsys(n, (xᵢ), (Fᵢ,₁))
    for i = 0 to n do
        yᵢ ← xᵢ + ½hFᵢ,₁
    end do
    call xpsys(n, (yᵢ), (Fᵢ,₂))
    for i = 0 to n do
        yᵢ ← xᵢ + ½hFᵢ,₂
    end do
```

```
call xpsys(n, (yᵢ), (Fᵢ,₃))
for i = 0 to n do
    yᵢ ← xᵢ + hFᵢ,₃
end do
call xpsys(n, (y)ᵢ, (Fᵢ,₄))
for i = 0 to n do
    xᵢ ← xᵢ + (h/6)[Fᵢ,₁ + 2Fᵢ,₂ + 2Fᵢ,₃ + Fᵢ,₄]
end do
    output k, (xᵢ)
end do
end procedure rk4sys
```

To illustrate the use of this procedure, we again use System (1) for our example. Of course, it must be rewritten in the form of Equation (3). A suitable main program and a procedure for computing the right side of Equation (3) follow.

```
program main
integer parameter n ← 2, nsteps ← 100
real parameter a ← 0, b ← 1
real array (xᵢ)₀:ₙ
(xᵢ) ← (0, 1, 0)
h ← (b − a)/nsteps
call rk4sys(n, h, (xᵢ), nsteps)
end program main
```

```
procedure xpsys(n, (xᵢ), (fᵢ))
real array (xᵢ)₀:ₙ, (fᵢ)₀:ₙ
integer n
f₀ ← 1
f₁ ← x₁ − x₂ + x₀(2 − x₀(1 + x₀))
f₂ ← x₁ + x₂ − x₀²(4 − x₀)
end procedure
```

A numerical experiment to compare the results of the Taylor series method and the Runge-Kutta method with the analytic solution of System (1) is suggested in Computer Problem **1**. At the point $t = 1.0$, the results are as follows:

	Taylor Series	Runge-Kutta	Analytic Solution
$x(1.0) \approx$	2.46869 40	2.46869 42	2.46869 39399
$y(1.0) \approx$	1.28735 46	1.28735 61	1.28735 52872

Alternatively, we can solve this system directly using the previous Maple program with the following additional commands.

```
f := dsolve( sysode union ic, fcns, numeric );
f(0); f(1);
```

We obtain the following results.

```
{x(t) = 1., y(t) = 0, t = 0}
{t = 1., y(t) = 1.287355325, x(t) = 2.468693912}
```

PROBLEMS 9.1

1. Write an equivalent system of first-order differential equations without t appearing on the right-hand side.

$$\begin{cases} x' = x^2 + \log(y) + t^2 \\ y' = e^y - \cos(x) + \sin(tx) - (xy)^7 \\ x(0) = 1 \quad y(0) = 3 \end{cases}$$

2. Consider

$$\begin{cases} x' = y \\ y' = x \end{cases} \quad \text{with} \quad \begin{cases} x(0) = -1 \\ y(0) = 0 \end{cases}$$

Write down the equations, without derivatives, to be used in the Taylor series method of order 5.

3. How would you solve this differential equation numerically?

$$\begin{cases} x_1' = x_1^2 + e^t - t^2 \\ x_2' = x_2 - \cos t \\ x_1(0) = 0 \quad x_2(1) = 0 \end{cases}$$

4. How would you solve the initial-value problem

$$\begin{cases} x_1'(t) = x_1(t)e^t + \sin t - t^2 \\ x_2'(t) = [x_2(t)]^2 - e^t + x_2(t) \\ x_1(1) = 2 \quad x_2(1) = 4 \end{cases}$$

if a computer program were available to solve an initial-value problem of the form $x' = f(t, x)$ involving a single unknown function $x = x(t)$?

COMPUTER PROBLEMS 9.1

1. Solve the system of differential equations (1) by using the two methods given in the text and compare the results with the analytic solution.

2. Solve the initial-value problem

$$\begin{cases} x' = t + x^2 - y \\ y' = t^2 - x + y^2 \\ x(0) = 3 \quad y(0) = 2 \end{cases}$$

by means of the Taylor series method using $h = 1/128$ on the interval $[0, 0.38]$. Include terms involving three derivatives in x and y. How accurate are the computed function values?

3. Write the Runge-Kutta procedure to solve

$$\begin{cases} x_1' = -3x_2 \\ x_2' = \frac{1}{3}x_1 \\ x_1(0) = 0 \quad x_2(0) = 1 \end{cases}$$

on the interval $0 \leq t \leq 4$. Plot the solution.

4. Write a driver program for procedure *rk4sys* that solves the ordinary differential equation system given by Equation (2). Use $h = -10^{-2}$ and print out the values of x_0, x_1, and x_2, together with the true solution on the interval $[-1, 0]$. Verify that the true solution is $x(t) = e^t + 6 + 6t + 4t^2 + t^3$ and $y(t) = e^t - t^3 + t^2 + 2t + 2$.

5. Using the Runge-Kutta procedure, solve the following initial-value problem on the interval $0 \leq t \leq 2\pi$. Plot the resulting curves $(x_1(t), x_2(t))$ and $(x_3(t), x_4(t))$. They should be circles.

$$\begin{cases} \mathbf{X}' = \begin{bmatrix} x_3 \\ x_4 \\ -x_1(x_1^2 + x_2^2)^{-3/2} \\ -x_2(x_1^2 + x_2^2)^{-3/2} \end{bmatrix} \\ \mathbf{X}(0) = [1, 0, 0, 1]^T \end{cases}$$

6. Solve the problem

$$\begin{cases} x_0' = 1 \\ x_1' = -x_2 + \cos x_0 \\ x_2' = \quad x_1 + \sin x_0 \\ x_0(1) = 1 \quad x_1(1) = 0 \quad x_2(1) = -1 \end{cases}$$

Use the Runge-Kutta method and the interval $-1 \leq t \leq 2$.

7. Write and test a program, using the Taylor series method of order 5, to solve the system

$$\begin{cases} x' = tx - y^2 + 3t \\ y' = x^2 - ty - t^2 \\ x(5) = 2 \quad y(5) = 3 \end{cases}$$

on the interval $[5, 6]$ using $h = 10^{-3}$. Print values of x and y at steps of 0.1.

8. Print a table of $\sin t$ and $\cos t$ on the interval $[0, \pi/2]$ by numerically solving the system

$$\begin{cases} x' = y \\ y' = -x \\ x(0) = 0 \quad y(0) = 1 \end{cases}$$

9. Recode and test procedure *rk4sys* using a computer language that supports vector operations.

10. Write a program for using the Taylor series method of order 3 to solve the system

$$\begin{cases} x' = tx + y' - t^2 \\ y' = ty + 3t \\ z' = tz - y' + 6t^3 \\ x(0) = 1 \quad y(0) = 2 \quad z(0) = 3 \end{cases}$$

on the interval $[0, 0.75]$ using $h = 0.01$.

11. Write and test a short program for solving the system of differential equations

$$\begin{cases} y' = x^3 - t^2 y - t^2 \\ x' = tx^2 - y^4 + 3t \\ y(2) = 5 \quad x(2) = 3 \end{cases}$$

over the interval $[2, 5]$ with $h = 0.25$. Use the Taylor series method of order 4.

9.2 Higher-Order Equations and Systems

Consider the initial-value problem for ordinary differential equations of order higher than 1. A differential equation of order n is normally accompanied by n auxiliary conditions. This many initial conditions are needed to specify the solution of the differential equation precisely (assuming certain smoothness conditions are present). Take, for example, a particular second-order initial-value problem:

$$\begin{cases} x''(t) = -3\cos^2(t) + 2 \\ x(0) = 0 \quad x'(0) = 0 \end{cases} \tag{1}$$

Without the auxiliary conditions, the general analytic solution is

$$x(t) = \frac{1}{4}t^2 + \frac{3}{8}\cos(2t) + c_1 t + c_2$$

where c_1 and c_2 are arbitrary constants. To select one specific solution, c_1 and c_2 must be fixed, and two initial conditions allow this to be done. In fact, $x(0) = 0$ yields $c_2 = -\frac{3}{8}$ and $x'(0) = 0$ forces $c_1 = 0$.

To verify these conclusions, we use the Maple statement

```
dsolve( {(D@@2)(x)(t) = -3*cos(t)^2 +2,
          x(0) = 0, D(x)(0) = 0}, x(t) );
```

and obtain the results

$$x(t) = \frac{1}{4}t^2 + \frac{3}{4}\cos^2(t) - \frac{3}{4}$$

Higher-Order Differential Equations

In general, higher-order problems can be much more complicated than this simple example because System (1) has the special property that the function on the right side of the differential equation does not involve x. The most general form of an ordinary differential equation with initial conditions that we shall consider is

$$\begin{cases} x^{(n)} = f(t, x, x', x'', \dots, x^{(n-1)}) \\ x(a), x'(a), x''(a), \dots, x^{(n-1)}(a) \quad \text{all given} \end{cases} \tag{2}$$

This can be solved numerically by turning it into a system of *first-order* differential equations. To do so, we define new variables $x_0, x_1, \dots, x_n$ as follows:

$$x_0 = t \quad x_1 = x \quad x_2 = x' \quad x_3 = x'' \quad \dots \quad x_{n-1} = x^{(n-2)} \quad x_n = x^{(n-1)}$$

Consequently, the original initial-value Problem (2) is equivalent to

$$\begin{cases} x_0' = 1 \\ x_1' = x_2 \\ x_2' = x_3 \\ \quad \vdots \\ x_{n-1}' = x_n \\ x_n' = f(x_0, x_1, x_2, \dots, x_n) \\ x_0(a), x_1(a), \dots, x_n(a) \quad \text{all given} \end{cases}$$

or, in vector notation,

$$\begin{cases} X' = F(X) \\ X(a) \quad \text{given} \end{cases} \tag{3}$$

where

$$X = [x_0, x_1, \dots, x_n]^T$$
$$X' = [x_0', x_1', \dots, x_n']^T$$
$$F = [1, x_2, x_3, x_4, \dots, x_n, f]^T$$

and

$$X(a) = [a, x_1(a), x_2(a), \ldots, x_n(a)]$$

Whenever a problem must be transformed by introducing new variables, it is recommended that a *dictionary* be given to show the relationship between the new and the old variables. At the same time, this information, together with the differential equations and the initial values, can be displayed in a chart. Such systematic bookkeeping can be helpful in a complicated situation.

To illustrate, let us transform the initial-value problem

$$\begin{cases} x''' = \cos x + \sin x' - e^{x''} + t^2 \\ x(0) = 3 \quad x'(0) = 7 \quad x''(0) = 13 \end{cases}$$

into a form suitable for solution by the Runge-Kutta procedure. We notice that t is present on the right-hand side and so the equations $x_0 = t$ and $x_0' = 1$ are necessary. A chart summarizing the transformed problem is as follows:

Old Variable	New Variable	Initial Value	Differential Equation
t	x_0	0	$x_0' = 1$
x	x_1	3	$x_1' = x_2$
x'	x_2	7	$x_2' = x_3$
x''	x_3	13	$x_3' = \cos x_1 + \sin x_2 - e^{x_3} + x_0^2$

So the corresponding first-order system is

$$X' = \begin{bmatrix} 1 \\ x_2 \\ x_3 \\ \cos x_1 + \sin x_2 - e^{x_3} + x_0^2 \end{bmatrix}$$

and $X(0) = [0, 3, 7, 13]^T$.

Systems of Differential Equations

By systematically introducing new variables, we can transform a system of differential equations of various orders into a larger system of first-order equations. For instance, the system

$$\begin{cases} x'' = x - y - (3x')^2 + (y')^3 + 6y'' + 2t \\ y''' = y'' - x' + e^x - t \\ x(1) = 2 \quad x'(1) = -4 \quad y(1) = -2 \quad y'(1) = 7 \quad y''(1) = 6 \end{cases}$$

can be solved by the Runge-Kutta procedure if we first transform it according to the following chart:

Old Variable	New Variable	Initial Value	Differential Equation
t	x_0	1	$x_0' = 1$
x	x_1	2	$x_1' = x_2$
x'	x_2	-4	$x_2' = x_1 - x_3 - 9x_2^2 + x_4^3 + 6x_5 + 2x_0$
y	x_3	-2	$x_3' = x_4$
y'	x_4	7	$x_4' = x_5$
y''	x_5	6	$x_5' = x_5 - x_2 + e^{x_1} - x_0$

Hence, we have

$$X' = \begin{bmatrix} 1 \\ x_2 \\ x_1 - x_3 - 9x_2^2 + x_4^3 + 6x_5 + 2x_0 \\ x_4 \\ x_5 \\ x_5 - x_2 + e^{x_1} - x_0 \end{bmatrix}$$

and $X(1) = [1, 2, -4, -2, 7, 6]^T$.

**PROBLEMS
9.2**

1. Rewrite the following system of first-order differential equations without t appearing on the right-hand side.

$$\begin{cases} x^{(4)} = (x''')^2 + \cos(x'x'') - \sin(tx) + \log\left(\dfrac{x}{t}\right) \\ x(0) = 1 \quad x'(0) = 3 \quad x''(0) = 4 \quad x'''(0) = 5 \end{cases}$$

2. Turn this differential equation into a system of first-order equations suitable for applying the Runge-Kutta method.

$$\begin{cases} x''' = 2x' + \log(x'') + \cos(x) \\ x(0) = 1 \quad x'(0) = -3 \quad x''(0) = 5 \end{cases}$$

3. **a.** Assuming that a program is available for solving initial-value problems of the form (3), how can it be used to solve the following differential equation?

$$\begin{cases} x''' = t + x + 2x' + 3x'' \\ x(1) = 3 \quad x'(1) = -7 \quad x''(1) = 4 \end{cases}$$

   b. How would this problem be solved if the initial conditions were $x(1) = 3$, $x'(1) = -7$, and $x'''(1) = 0$?

4. How would you solve this differential equation problem numerically?

$$\begin{cases} x_1'' = x_1' + x_1^2 - \sin t \\ x_2'' = x_2 - (x_2')^{1/2} + t \\ x_1(0) = 1 \quad x_2(1) = 3 \quad x_1'(0) = 0 \quad x_2'(1) = -2 \end{cases}$$

5. Convert to a first-order system the orbital equations

$$\begin{cases} x'' + x(x^2 + y^2)^{-3/2} = 0 \\ y'' + y(x^2 + y^2)^{-3/2} = 0 \end{cases}$$

with initial conditions

$$x(0) = 0.5 \quad x'(0) = 0.75 \quad y(0) = 0.25 \quad y'(0) = 1.0$$

6. Express the system of ordinary differential equations

$$\begin{cases} \dfrac{d^2z}{dt^2} - 2t\dfrac{dz}{dt} = 2te^{xz} \\[2mm] \dfrac{d^2x}{dt^2} - 2xz\dfrac{dx}{dt} = 3x^2yt^2 \\[2mm] \dfrac{d^2y}{dt^2} - e^y\dfrac{dy}{dt} = 4xt^2z \\[2mm] z(1) = x'(1) = y'(1) = 2 \quad z'(1) = x(1) = y(1) = 3 \end{cases}$$

as a system of first-order ordinary differential equations.

7. Determine a system of first-order equations equivalent to each of the following.

a. $x''' + x'' \sin x + tx' + x = 0$

b. $x^{(4)} + x'' \cos x' + txx' = 0$

c. $\begin{cases} x'' = 3x^2 - 7y^2 + \sin t + \cos(x'y') \\ y''' = y + x^2 - \cos t - \sin(xy'') \end{cases}$

8. Consider

$$\begin{cases} x'' = x' = x \\ x(0) = 0 \quad x'(0) = 1 \end{cases}$$

Determine the associated first-order system and its auxiliary initial conditions.

9. The problem

$$\begin{cases} x''(t) = x + y - 2x' + 3y' + \log t \\ y''(t) = 2x - 3y + 5x' + ty' - \sin t \\ x(0) = 1 \quad x'(0) = 2 \\ y(0) = 3 \quad y'(0) = 4 \end{cases}$$

is to be put into the form of a system of five first-order equations. Give the resulting system and the appropriate initial values.

10. Write procedure *xpsys* for use with the fourth-order Runge-Kutta routine *rk4sys* for the following differential equation.

$$\begin{cases} x''' = 10e^{x''} - x''' \sin(x'x) - (xt)^{10} \\ x(2) = 6.5 \quad x'(2) = 4.1 \quad x''(2) = 3.2 \end{cases}$$

11. If we are going to solve the initial-value problem

$$\begin{cases} x''' = x' - tx'' + x + \ln t \\ x(1) = x'(1) = x''(1) = 1 \end{cases}$$

using Runge-Kutta formulas, how should the problem be transformed?

**COMPUTER
PROBLEMS
9.2**

1. Use *rk4sys* to solve each of the following for $0 \leq t \leq 1$. Use $h = 2^{-k}$ with $k = 5, 6,$ and 7 and compare results.

a. $\begin{cases} x'' = 2(e^{2t} - x^2)^{1/2} \\ x(0) = 0 \quad x'(0) = 1 \end{cases}$

b. $\begin{cases} x'' = x^2 - y + e^t \\ y'' = x - y^2 - e^t \\ x(0) = 0 \quad x'(0) = 0 \\ y(0) = 1 \quad y'(0) = -2 \end{cases}$

2. Solve the Airy differential equation

$$\begin{cases} x'' = tx \\ x(0) = 0.35502\,80538\,87817 \\ x'(0) = -0.25881\,94037\,92807 \end{cases}$$

on the interval $[0, 4.5]$ using the Runge-Kutta method. *Check value*: The value $x(4.5) = 0.00033\,02503$ is correct.

3. The differential equation

$$\begin{cases} x^2 x'' + 2(x')^2 + 2\left(\dfrac{1+x}{2+x}\right) x' = \left(\dfrac{2}{2+x}\right) \\ x(0) = 0.01 \quad x'(0) = 0.5 \end{cases}$$

describes ice buildup on a frozen lake in terms of thermal resistance $x(t)$ at time t. Write and test a main program and procedure *xpsys* that use the fourth-order Runge-Kutta routine *rk4sys* to solve this differential equation from time 0 to 1.5 with steps of 0.05.

4. Solve

$$\begin{cases} x'' + x' + x^2 - 2t = 0 \\ x(0) = 0 \quad x'(0) = 0.1 \end{cases}$$

on $[0, 3]$ by any convenient method. If a plotter is available, graph the solution.

5. Solve

$$\begin{cases} x'' = 2x' - 5x \\ x(0) = 0 \quad x'(0) = 0.4 \end{cases}$$

on the interval $[-2, 0]$.

9.3   Adams-Moulton Methods

The procedures explained so far have solved the initial-value problem

$$\begin{cases} X' = F(X) \\ X(a) \quad \text{given} \end{cases} \tag{1}$$

by means of **single-step** numerical methods. In other words, if the solution $X(t)$ is known at a particular point t, then $X(t + h)$ can be computed with no knowledge of the solution at points earlier than t. The Runge-Kutta and Taylor series methods compute $X(t + h)$ in terms of $X(t)$ and various values of F.

A Predictor-Corrector Scheme

More efficient methods can be devised if several of the values $X(t)$, $X(t - h)$, $X(t - 2h), \ldots$ are used in computing $X(t + h)$. Such methods are called **multistep** methods. They have the obvious drawback that at the beginning of the numerical solution, no prior values of X are available. So it is usual to start a numerical solution with a single-step method, such as the Runge-Kutta procedure, and transfer to a multistep procedure for efficiency as soon as enough starting values have been computed.

One example of a multistep formula is known by the name of **Adams-Bashforth**. (See Problem **15** of Section 5.5.) It is

$$X(t + h) = X(t) + \frac{h}{24}\{55F[X(t)] - 59F[X(t - h)] + 37F[X(t - 2h)]$$

$$- 9F[X(t - 3h)]\} \tag{2}$$

If the solution X has been computed at the four points $t, t - h, t - 2h,$ and $t - 3h$, then Formula (2) can be used to compute $X(t + h)$. If this is done systematically, then only *one* evaluation of F is required for each step. This represents a considerable saving over the fourth-order Runge-Kutta procedure; the latter requires *four* evaluations of F per step. (Of course, a consideration of truncation error and stability might permit a larger step size in the Runge-Kutta method and make it much more competitive.)

In practice, Formula (2) is never used by itself. Instead, it is used as a *predictor* and then another formula is used as a *corrector*. The corrector usually used with Formula (2) is the **Adams-Moulton formula**:

$$X(t + h) = X(t) + \frac{h}{24}\{9F[\widetilde{X}(t + h)] + 19F[X(t)] - 5F[X(t - h)]$$

$$+ F[X(t - 2h)]\} \tag{3}$$

Here $\widetilde{X}(t + h)$ is the predicted value of $X(t + h)$ computed from Formula (2). Thus, Equation (2) predicts a tentative value of $X(t + h)$, and Equation (3) computes this X value more accurately. The combination of the two formulas results in a **predictor-corrector scheme**.

With initial values of X specified at a, three steps of a Runge-Kutta method can be performed to determine enough X values so that the Adams-Bashforth-Moulton procedure can begin. The fourth-order Adams-Bashforth and Adams-Moulton formulas, started with the fourth-order Runge-Kutta method, are referred to as the **Adams-Moulton method**. Predictor and corrector formulas of the same order are used so that only one application of the corrector formula is needed. Some suggest iterating the corrector formula, but research has demonstrated that the best overall approach is only *one* application per step.

Pseudocode

Storage of the approximate solution at previous steps in the Adams-Moulton method is usually handled either by storing in an array of dimension larger than the total number of steps to be taken or by physically shifting data after each step (discarding the oldest data and storing the newest in their place). If an adaptive process is used, the total number of steps to be taken cannot be determined beforehand. Physical shifting of data can be eliminated by cycling the indices of a storage array of fixed dimension. For the Adams-Moulton method, the x_i data for $X(t)$ are stored in a two-dimensional array with entries z_{im} in locations $m = 1, 2, 3, 4, 5, 1, 2, \ldots$ for $t = a, a + h, a + 2h, a + 3h, a + 4h, a + 5h, a + 6h, \ldots$, respectively. The sketch in Figure 9.1 shows the first several t values with corresponding m values and abbreviations for the formulas used.

An error analysis can be conducted after each step of the Adams-Moulton method. If $x_i^{(p)}$ is the numerical approximation of the ith equation in System (1) at $t + h$ obtained by predictor Formula (2) and x_i is that from corrector Formula (3) at $t + h$, then it can be shown that the single-step error for the ith component at $t + h$ is given approximately by

$$\varepsilon_i = \frac{19}{270} \frac{|x_i - x_i^{(p)}|}{|x_i|}$$

So we compute

$$est = \max_{1 \le i \le n} |\varepsilon_i|$$

in the Adams-Moulton procedure *amsys* to obtain an estimate of the maximum single-step error at $t + h$.

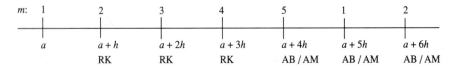

FIGURE 9.1

A control procedure is needed that calls the Runge-Kutta procedure three times and then calls the Adams-Moulton predictor-corrector scheme to compute the remaining steps. Such a procedure for doing *nsteps* steps with a fixed step size h follows.

```
procedure amrk(n, h, (xi), nsteps)
real array (xi)0:n
allocate real array (fij)0:n×0:4, (zij)0:n×0:4
integer i, k, m, n
real est, h
m ← 0
output h
output 0, (xi)
for i = 0 to n do
    zim ← xi
end do
for k = 1 to 3 do
    call rksys(m, n, h, (zij), (fij))
    output k, (zim)
end do
for k = 4 to nsteps do
    call amsys(m, n, h, est, (zij), (fij), )
    output k, (zim)
    output est
end do
for i = 0 to n do
    xi ← zim
end do
deallocate array (f, z)
end procedure amrk
```

The Adams-Moulton method for a system and the computation of the single-step error are accomplished in the following pseudocode.

```
procedure amsys(m, n, h, est, (zij), (fij))
real parameter array (ai)1:4 ← (55, -59, 37, -9)
real parameter array (bi)1:4 ← (9, 19, -5, 1)
real array (xij)0:n×4, (fij)0:n×4
allocate real array (si)0:n, (yi)0:n
integer i, j, k, m, mp1
real d, dmax, est, h
mp1 ← (1 + m) mod 5
call xpsys(n, (zim), (fim))
for i = 0 to n do
    si ← 0
end do
```

for $k = 1$ **to** 4 **do**
   $j \leftarrow (m - k + 6) \bmod 5$
   for $i = 0$ **to** n **do**
      $s_i \leftarrow s_i + a_k f_{ij}$
   end do
end do
for $i = 0$ **to** n **do**
   $y_i \leftarrow z_{im} + h s_i / 24$
end do
call $xpsys(n, (y_i), (f_{i,mp1}))$
for $i = 0$ **to** n **do**
   $s_i \leftarrow 0$
end do
for $k = 1$ **to** 4 **do**
   $j \leftarrow (mp1 - k + 6) \bmod 5$
   for $i = 0$ **to** n **do**
      $s_i \leftarrow s_i + b_k f_{ij}$
   end do
end do
for $i = 0$ **to** n **do**
   $z_{i,mp1} \leftarrow z_{im} + h s_i / 24$
end do
$m \leftarrow mp1$
$d_{\max} \leftarrow 0$
for $i = 0$ **to** n **do**
   $d \leftarrow |z_{im} - y_i| / |z_{im}|$
   if $d > d_{\max}$ **then**
      $d_{\max} \leftarrow d$
      $j \leftarrow i$
   end if
end do
$est \leftarrow 19 d_{\max} / 270$
deallocate array (s, y)
end procedure *amsys*

Here the function evaluations are stored cyclically in f_{im} for use by Formulas (2) and (3). Various optimization techinques are possible in this pseudocode. For example, the programmer may wish to move the computation of $\frac{1}{24} h$ outside of the loops.

    A companion Runge-Kutta procedure is needed, which is a modification of procedure *rk4sys* from Section 9.1.

procedure $rksys(m, n, h, (z_{ij}), (f_{ij}))$
real array $(z_{ij})_{0:n \times 0:4}, (f_{ij})_{0:n \times 0:4}$
allocate real array $(g_{ij})_{0:n \times 3}, (y_i)_{0:n}$
integer $i, m, mp1, n$
real h
$mp1 \leftarrow (1 + m) \bmod 5$

call $xpsys(n, (z_{im}), (f_{im}))$
for $i = 0$ **to** n **do**
$\quad y_i \leftarrow z_{im} + \frac{1}{2}hf_{im}$
end do
call $xpsys(n, (y_i), (g_{i,1}))$
for $i = 0$ **to** n **do**
$\quad y_i \leftarrow z_{im} + \frac{1}{2}hg_{i,1}$
end do
call $xpsys(n, (y_i), (g_{i,2}))$
for $i = 0$ **to** n **do**
$\quad y_i \leftarrow z_{im} + hg_{i,2}$
end do
call $xpsys(n, (y_i), (g_{i,3}))$
for $i = 0$ **to** n **do**
$\quad z_{i,mp1} \leftarrow z_{im} + h[f_{im} + 2g_{i,1} + 2g_{i,2} + g_{i,3}]/6$
end do
$m \leftarrow mp1$
deallocate array (g, y)
end procedure *rksys*

As before, the programmer may wish to move $\frac{1}{6}h$ out of the loop.

To use the Adams-Moulton pseudocode, we supply the procedure *xpsys* that defines the system of ordinary differential equations and write a driver program with a call to procedure *amrk*. The complete program then consists of the following five parts: the main program and procedures *xpsys*, *amrk*, *rksys*, and *amsys*.

As an illustration, the pseudocode for the last example in Section 9.2 is as follows:

program *main*
integer parameter $n \leftarrow 5$, *nsteps* $\leftarrow 100$
real parameter $a \leftarrow 0$, $b \leftarrow 1$
real array $(x_i)_{0:n}$
real h
$(x_i) \leftarrow (1, 2, -4, -2, 7, 6)$
$h \leftarrow (b - a)/nsteps$
call $amrk(n, h, (x_i), nsteps)$
end program *main*

procedure $xpsys(n, (x_i), (f_i))$
real array $(x_i)_{0:n}, (f_i)_{0:n}$
integer n
$f_0 \leftarrow 1$
$f_1 \leftarrow x_2$
$f_2 \leftarrow x_1 - x_3 - 9x_2^2 + x_4^3 + 6x_5 + 2x_0$
$f_3 \leftarrow x_4$
$f_4 \leftarrow x_5$
$f_5 \leftarrow x_5 - x_2 + e^{x_1} - x_0$
end procedure *xpsys*

An Adaptive Scheme

Since an estimate of the error is available from the Adams-Moulton method, it is natural to replace procedure *amrk* with one that employs an adaptive scheme—that is, one that changes the step size. A procedure similar to the one used in Section 8.3 is outlined here. The Runge-Kutta method is used to compute the first three steps and then the Adams-Moulton method is used. If the error test determines that halving or doubling of the step size is necessary in the first step using the Adams-Moulton method, then the step size is halved or doubled and the whole process starts again with the initial values—so at least one step of the Adams-Moulton method must take place. If during this process, the error test indicates that halving is required at some point within the interval $[a, b]$, then the step size is halved. A retreat is made to an appropriate previous value, and after a suitable number of Runge-Kutta steps have been computed, the process continues using the Adams-Moulton method again but with the new step size. In other words, the point at which the error was too large should be computed by the Adams-Moulton method, not the Runge-Kutta method. Doubling the step size is handled in an analogous manner. To simplify this process (whether halving or doubling the step size), one could always back up two steps with the *old* step size and then use this as the beginning point of a *new* initial-value problem. Other more complicated procedures can be designed and can be the subject of numerical experimentation. (See Computer Problem 3.)

COMPUTER PROBLEMS 9.3

1. Test the procedures of this section on the system given in Computer Problem **2** of Section 9.2.

2. The single-step error is closely controlled by using fourth-order formulas; however, the roundoff error in performing the computations in Equations (2) and (3) can be large. It is logical to carry out these in what is known as **partial double-precision** arithmetic. The function F would be evaluated in single precision at the desired points $X(t + ih)$, but the inner product $\sum_i c_i F(X(t + ih))$ would be accumulated in double precision. Also, the addition of $X(t)$ to this result is done in double precision. Recode the Adams-Moulton method so that partial double-precision arithmetic is used. Compare this code with that in the text for a system with a known solution. How do they compare with regard to roundoff error at each step?

3. Write and test an adaptive process similar to *rk45ad* in Section 8.3 with calling sequence

   procedure $amrkad(n, h, t_a, t_b, (x_i), itmax, \varepsilon_{min}, \varepsilon_{max}, h_{min}, h_{max}, iflag)$

   This routine should carry out the adaptive procedure outlined at the end of this section and be used in place of the *amrk* procedure.

4. Rewrite and test the code in this section using a computer language that supports vector operations.

5. Solve the predator-prey problem in the example at the beginning of this chapter with $a = -10^{-2}$, $b = -\frac{1}{4} \times 10^2$, $c = 10^{-2}$, $d = -10^2$, and with initial values $u(0) = 80$, $v(0) = 30$. Plot u (the prey) and v (the predator) as functions of time t.

10 SMOOTHING OF DATA AND THE METHOD OF LEAST SQUARES

Surface tension S in a liquid is known to be a linear function $S = aT + b$ of temperature T. For a particular liquid, measurements have been made of the surface tension at certain temperatures. The results were

T	0	10	20	30	40	80	90	95
S	68.0	67.1	66.4	65.6	64.6	61.8	61.0	60.0

How can the most probable values of the constants in the equation

$$S = aT + b$$

be determined? Methods for solving such problems are developed in this chapter.

10.1 The Method of Least Squares

Linear Least Squares

In experimental, social, and behavioral sciences, an experiment or survey often produces a mass of data. To interpret the data, the investigator may resort to graphical methods. For instance, an experiment in physics might produce a numerical table of the form

x	x_0	x_1	$\cdots$	x_m
y	y_0	y_1	$\cdots$	y_m

(1)

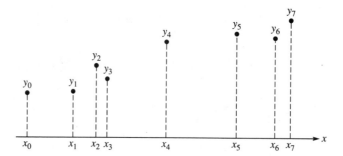

FIGURE 10.1

and from it, $m + 1$ points on a graph could be plotted. Suppose that the resulting graph looks like Figure 10.1. A reasonable tentative conclusion is that the underlying function is *linear* and that the failure of the points to fall *precisely* on a straight line is due to experimental error. If one proceeds on this assumption—or if theoretical reasons exist for believing that the function in indeed linear—the next step is to determine the correct function. Assuming that

$$y = ax + b$$

what are the coefficients a and b? Thinking geometrically, we ask, *What line most nearly passes through the ten points plotted?*

To answer this question, suppose that a guess is made about the correct values of a and b. This is equivalent to deciding on a specific line to represent the data. In general, the data points will not fall on the line $y = ax + b$. If by chance the kth datum falls on the line, then

$$ax_k + b - y_k = 0$$

If it does not, then there is a discrepancy or *error* of magnitude

$$|ax_k + b - y_k|$$

The total absolute error for all $m + 1$ points is therefore

$$\sum_{k=0}^{m} |ax_k + b - y_k|$$

This is a function of a and b, and it would be reasonable to choose a and b so that the function assumes its minimum value. This problem is an example of ℓ_1 **approximation** and can be solved by the techniques of linear programming, a subject dealt with in Chapter 15.

In practice, it is common to minimize a different error function of a and b:

$$\phi(a, b) = \sum_{k=0}^{m} (ax_k + b - y_k)^2 \tag{2}$$

This function is suitable because of statistical considerations. Explicitly, if the errors follow a *normal probability distribution,* then the minimization of ϕ produces a best estimate of a and b. This is called an ℓ_2 **approximation**. Another advantage is that the methods of calculus can be used on Equation (2).

The ℓ_1 and ℓ_2 approximations are related to specific cases of the ℓ_p norm defined by

$$\|x\|_p = \left\{ \sum_{i=1}^{n} |x_i|^p \right\}^{1/p} \qquad (1 \leqq p < \infty)$$

for the vector $x = (x_1, x_2, \ldots, x_n)^T$.

Let us try to make $\phi(a, b)$ a minimum. By calculus, the conditions

$$\frac{\partial \phi}{\partial a} = 0 \qquad \frac{\partial \phi}{\partial b} = 0$$

(partial derivatives of ϕ with respect to a and b, respectively) are *necessary* at the minimum. Taking the derivatives in Equation (2), we obtain

$$\begin{cases} \displaystyle\sum_{k=0}^{m} 2(ax_k + b - y_k)x_k = 0 \\[2mm] \displaystyle\sum_{k=0}^{m} 2(ax_k + b - y_k) = 0 \end{cases}$$

This is a pair of simultaneous linear equations in the unknowns a and b. They are called the **normal equations** and can be written as

$$\begin{cases} \left(\displaystyle\sum_{k=0}^{m} x_k^2 \right) a + \left(\displaystyle\sum_{k=0}^{m} x_k \right) b = \displaystyle\sum_{k=0}^{m} y_k x_k \\[4mm] \left(\displaystyle\sum_{k=0}^{m} x_k \right) a + \ (m + 1)b = \displaystyle\sum_{k=0}^{m} y_k \end{cases} \qquad (3)$$

Here, of course, $\sum_{k=0}^{m} 1 = m + 1$. The explicit solution of Equation (3) is

$$a = \frac{1}{d} \left((m + 1) \sum_{k=0}^{m} x_k y_k - \sum_{k=0}^{m} x_k \sum_{k=0}^{m} y_k \right)$$

$$b = \frac{1}{d} \left(\sum_{k=0}^{m} x_k^2 \sum_{k=0}^{m} y_k - \sum_{k=0}^{m} x_k \sum_{k=0}^{m} x_k y_k \right)$$

where

$$d = (m + 1) \sum_{k=0}^{m} x_k^2 - \left(\sum_{k=0}^{m} x_k \right)^2$$

Linear Example

The preceding analysis illustrates the **least squares** procedure in the simple linear case. As a concrete example, the table

x	1.0	2.0	2.5	3.0
y	3.7	4.1	4.3	5.0

leads to the system of two equations

$$\begin{cases} 20.25a + 8.5b = 37.65 \\ 8.5a + 4b = 17.1 \end{cases}$$

whose solution is $a = 0.6$ and $b = 3.0$. The value of $\phi(a, b)$, which ideally should be 0, is 0.1.

Nonpolynomial Example

The method of least squares is not restricted to linear (first-degree) polynomials or to any specific functional form. Suppose, for instance, that we want to fit a table of values (x_k, y_k), where $k = 0, 1, \ldots, m$, by a function of the form

$$a \ln x + b \cos x + c e^x$$

in the least squares sense. We consider the function

$$\phi(a, b, c) = \sum_{k=0}^{m} (a \ln x_k + b \cos x_k + c e^{x_k} - y_k)^2$$

and set $\partial \phi / \partial a = 0$, $\partial \phi / \partial b = 0$, and $\partial \phi / \partial c = 0$. This results in the following three normal equations:

$$\begin{cases} a \sum_{k=0}^{m} (\ln x_k)^2 + b \sum_{k=0}^{m} (\ln x_k)(\cos x_k) + c \sum_{k=0}^{m} (\ln x_k)e^{x_k} = \sum_{k=0}^{m} y_k \ln x_k \\[2mm] a \sum_{k=0}^{m} (\ln x_k)(\cos x_k) + b \sum_{k=0}^{m} (\cos x_k)^2 + c \sum_{k=0}^{m} (\cos x_k)e^{x_k} = \sum_{k=0}^{m} y_k \cos x_k \\[2mm] a \sum_{k=0}^{m} (\ln x_k)e^{x_k} + b \sum_{k=0}^{m} (\cos x_k)e^{x_k} + c \sum_{k=0}^{m} (e^{x_k})^2 = \sum_{k=0}^{m} y_k e^{x_k} \end{cases}$$

For the table

x	0.24	0.65	0.95	1.24	1.73	2.01	2.23	2.52	2.77	2.99
y	0.23	-0.26	-1.10	-0.45	0.27	0.10	-0.29	0.24	0.56	1.00

we obtain the 3×3 system

$$\begin{cases} 6.79410a - 5.34749b + 63.25889c = 1.61627 \\ -5.34749a + 5.10842b - 49.00859c = -2.38271 \\ 63.25889a - 49.00859b + 1002.50650c = 26.77277 \end{cases}$$

which has the solution $a = -1.04103$, $b = -1.26132$, and $c = 0.03073$. So the curve

$$y = -1.04103 \ln x - 1.26132 \cos x + 0.03073 e^x$$

has the required form and fits the table in the least squares sense. The value of $\phi(a, b, c)$ is 0.92557.

Basis Functions $\{g_0, g_1, \ldots, g_n\}$

The principle of least squares, illustrated in these two simple cases, can be extended to general linear families of functions without involving any new ideas. Suppose that the data (1) are thought to conform to a relationship like

$$y = \sum_{j=0}^{n} c_j g_j(x) \tag{4}$$

in which the functions $g_0, g_1, \ldots, g_n$ (called *basis functions*) are known and held fixed. The coefficients $c_0, c_1, \ldots, c_n$ are to be determined according to the principle of least squares. In other words, we define the expression

$$\phi(c_0, c_1, \ldots, c_n) = \sum_{k=0}^{m} \left[\sum_{j=0}^{n} c_j g_j(x_k) - y_k \right]^2 \tag{5}$$

and select the coefficients to make it as small as possible. Of course, the expression $\phi(c_0, c_1, \ldots, c_n)$ is the sum of the squares of the errors associated with each entry (x_k, y_k) in the given table.

Proceeding as before, we write down as necessary conditions for the minimum the n equations

$$\frac{\partial \phi}{\partial c_i} = 0 \qquad (0 \leq i \leq n)$$

These partial derivatives are obtained from Equation (5). Indeed,

$$\frac{\partial \phi}{\partial c_i} = \sum_{k=0}^{m} 2 \left[\sum_{j=0}^{n} c_j g_j(x_k) - y_k \right] g_i(x_k) \qquad (0 \leq i \leq n)$$

When set equal to zero, the resulting equations can be rearranged as

$$\sum_{j=0}^{n}\left[\sum_{k=0}^{m} g_i(x_k)g_j(x_k)\right] c_j = \sum_{k=0}^{m} y_k g_i(x_k) \qquad (0 \leq i \leq n). \tag{6}$$

These are the **normal equations** in this situation and serve to determine the best values of the parameters $c_0, c_1, \ldots, c_n$. The normal equations are linear in $c_0, c_1, \ldots, c_n$, and thus, in principle, they can be solved by the method of Gaussian elimination (see Chapter 6).

In practice, the normal equations may be difficult to solve if care is not taken in choosing the basis functions $g_0, g_1, \ldots, g_n$. First, the set $\{g_0, g_1, \ldots, g_n\}$ should be **linearly independent**. This means that no linear combination $\sum_{i=0}^{n} c_i g_i$ can be the zero function (except in the trivial case when $c_0 = c_1 = \cdots = c_n = 0$). Second, the functions $g_0, g_1, \ldots, g_n$ should be *appropriate* to the problem at hand. Finally, one should choose a set of basis functions that is *well conditioned* for numerical work. We elaborate on this aspect of the problem in the next section.

PROBLEMS 10.1

1. Using the method of least squares, find the constant function that best fits the data

x	-1	2	3
y	5/4	4/3	5/12

2. Determine the *constant* function c that is produced by the least squares theory applied to Table (1) in the text. Does the resulting formula involve the points x_k in any way? Apply your general formula to Problem **1**.

3. Find an equation of the form $y = ae^{x^2} + bx^3$ that best fits the points $(-1, 0)$, $(0, 1)$, and $(1, 2)$ in the least squares sense.

4. Suppose that the x points in Table (1) are symmetrically situated about 0 on the x-axis. In this case, there is an especially simple formula for the line that best fits the points. Find it.

5. Find the equation of a parabola of form $y = ax^2 + b$ that best represents these data:

x	-1	0	1
y	3.1	0.9	2.9

   Use the method of least squares.

6. Suppose that Table (1) is known to conform to a function like $y = x^2 - x + c$. What value of c is obtained by the least squares theory?

7. Suppose that Table (1) is thought to be represented by a function $y = c \log x$. If so, what value for c emerges from the least squares theory?

8. Show that the solution of Equation (3) is as asserted in the text.

9. (Continuation) How do we know that divisor d is not zero? In fact, show that d is positive for $m \geq 1$. *Hint:* Show that

$$d = \sum_{k=0}^{m} \sum_{l=0}^{k-1} (x_k - x_l)^2$$

by induction on m. The Cauchy-Schwarz inequality can also be used to prove that $d > 0$.

10. (Continuation) Show that a and b can also be computed as follows.

$$\hat{x} = \frac{1}{m+1} \sum_{k=0}^{m} x_k \qquad \hat{y} = \frac{1}{m+1} \sum_{k=0}^{m} y_k$$

$$c = \sum_{k=0}^{m} (x_k - \hat{x})^2 \qquad a = \frac{1}{c} \sum_{k=0}^{m} (x_k - \hat{x})(y_k - \hat{y})$$

$$b = \hat{y} - a\hat{x}$$

Hint: Show that $d = (m+1)c$.

11. How do we know that the coefficients $c_0, c_1, \ldots, c_n$ that satisfy the normal equations (6) do not lead to a maximum in the function defined by Equation (5)?

12. If Table (1) is thought to conform to a relationship $y = \log(cx)$, what is the value of c obtained by the method of least squares?

13. What straight line best fits the data

x	1	2	3	4
y	0	1	1	2

in the least squares sense?

14. In analytic geometry, we learn that the distance from a point (x_0, y_0) to a line represented by the equation $ax + by = c$ is $(ax_0 + by_0 - c)(a^2 + b^2)^{-1/2}$. Determine a straight line that fits a table of data points (x_i, y_i), for $0 \leq i \leq m$, in such a way that the sum of the squares of the distances from the points to the line is minimized.

15. Show that if a straight line is fitted to a table (x_i, y_i) by the method of least squares, then the line will pass through the point (x^*, y^*), where x^* and y^* are the averages of the x_i's and y_i's, respectively.

16. The viscosity V of a liquid is known to vary with temperature according to a quadratic law $V = a + bT + cT^2$. Find the best values of a, b, and c for this table.

T	1	2	3	4	5	6	7
V	2.31	2.01	1.80	1.66	1.55	1.47	1.41

17. An experiment involves two independent variables x and y and one dependent variable z. How can a function $z = a + bx + cy$ be fitted to the table of points (x_k, y_k, z_k)? Give the normal equations.

18. Find the best function (in the least squares sense) that fits these data points and is of the form $f(x) = a \sin \pi x + b \cos \pi x$.

x	−1	−1/2	0	1/2	1
y	−1	0	1	2	1

19. Find the quadratic polynomial that best fits the following data in the sense of least squares.

x	−2	−1	0	1	2
y	2	1	1	1	2

20. What line best represents these data in the least squares sense?

x	0	1	2
y	5	−6	7

21. What constant c makes the equation

$$\sum_{i=0}^{m} [f(x_i) - ce^{x_i}]^2$$

as small as possible?

COMPUTER PROBLEMS 10.1

1. Write a procedure that sets up the normal equations (6). Using that procedure and other routines, such as *gauss* and *solve* from Chapter 6, verify the solution given for the problem involving $\ln x$, $\cos x$, and e^x in the subsection *Nonpolynomial Example*.

2. Write a procedure that fits a straight line to Table (1). Use this procedure to find the constants in the equation $S = aT + b$ for the table in the example that begins this chapter. Also, verify the results obtained for the problem in the subsection *Linear Example*.

3. Write and test a program that takes $m + 1$ points in the plane (x_i, y_i), where $0 \leq i \leq m$ with $x_0 < x_1 < \cdots < x_m$, and computes the best linear fit by the method of least squares. Then the program should create a plot of the points and the best line determined by the least squares method.

10.2 Orthogonal Systems and Chebyshev Polynomials

Orthonormal Basis Functions $\{g_0, g_1, \ldots, g_n\}$

Once the functions $g_0, g_1, \ldots g_n$ of Equation (4) in Section 10.1 have been chosen, the least squares problem can be interpreted as follows: The set of all functions g that can be expressed as linear combinations of $g_0, g_1, \ldots, g_n$ is a vector space G. (Familiarity with vector spaces is not essential to understanding the discussion here.) In symbols,

$$G = \left\{ g : \text{there exist } c_0, c_1, \ldots, c_n \text{ such that } g(x) = \sum_{j=0}^{n} c_j g_j(x) \right\}$$

The function sought in the least squares problem is thus an element of the vector space G. The functions $g_0, g_1, \ldots, g_n$ form a **basis** for G if they are not linearly dependent. However, a given vector space has many different bases, and they can differ drastically in their numerical properties.

Let us turn our attention away from the given basis $\{g_0, g_1, \ldots, g_n\}$ to the vector space G generated by that basis. Without changing G, we ask, *What basis for G should be chosen for numerical work?* In the present problem, the principal numerical task is to solve the normal equations—that is, Equation (6) in Section 10.1:

$$\sum_{j=0}^{n} \left[\sum_{k=0}^{m} g_i(x_k) g_j(x_k) \right] c_j = \sum_{k=0}^{m} y_k g_i(x_k) \qquad (0 \leq i \leq n) \tag{1}$$

The nature of this system obviously depends on the basis $\{g_0, g_1, \ldots, g_n\}$. We want these equations to be *easily* solved or to be capable of being *accurately* solved. The ideal situation occurs when the coefficient matrix in Equation (1) is the identity matrix. This happens if the basis $\{g_0, g_1, \ldots, g_n\}$ has the property of **orthonormality**:

$$\sum_{k=0}^{m} g_i(x_k) g_j(x_k) = \delta_{ij} = \begin{cases} 1 & \text{if } i = j \\ 0 & \text{if } i \neq j \end{cases}$$

In the presence of this property, Equation (1) simplifies dramatically to

$$c_j = \sum_{k=0}^{m} y_k g_j(x_k) \qquad (0 \leq j \leq n)$$

which is no longer a system of equations to be solved but rather an explicit formula for the coefficients c_j.

Under rather general conditions, the space G has a basis that is orthonormal in the sense just described. A procedure known as the **Gram-Schmidt process** can be used to obtain such a basis. There are some situations in which the effort of obtaining an orthonormal basis is justified, but simpler procedures often suffice. We describe one procedure now.

Remember that our goal is to make Equation (1) well disposed for numerical solution. We want to avoid any matrix of coefficients that involves the difficulties encountered in connection with the Hilbert matrix (see Computer Problem 4, Section 6.2). This objective can be met if the basis for the space G is well chosen.

We now consider the space G that consists of all polynomials of degree $\leq n$, which is an important example of the least squares theory. It may seem natural to use the following $n + 1$ functions as a basis for G:

$$g_0(x) = 1 \quad g_1(x) = x \quad g_2(x) = x^2 \quad \cdots \quad g_n(x) = x^n$$

Using this basis, we write a typical element of the space G in the form

$$g(x) = \sum_{j=0}^{n} c_j g_j(x) = \sum_{j=0}^{n} c_j x^j = c_0 + c_1 x + c_2 x^2 + \cdots + c_n x^n$$

This basis, however natural, is almost always a *poor* choice for numerical work. For many purposes, the Chebyshev polynomials (suitably defined for the interval involved) do form a *good* basis.

Figure 10.2 gives an indication of why the simple polynomials x^j do not form a good basis for numerical work: These functions are too much alike. If a function g is given and we wish to analyze it into components, $g(x) = \sum_{j=0}^{n} c_j x^j$, it is difficult to determine the coefficients c_j precisely. Figure 10.2 also shows a few of the Chebyshev polynomials; they are quite different from one another.

For simplicity, assume that the points in our least squares problem have the property

$$-1 = x_0 < x_1 < \cdots < x_m = 1$$

Then the *Chebyshev polynomials* for the interval $[-1, 1]$ can be used. The traditional notation is

$$T_0(x) = 1 \qquad\qquad T_1(x) = x \qquad\qquad T_2(x) = 2x^2 - 1$$
$$T_3(x) = 4x^3 - 3x \qquad T_4(x) = 8x^4 - 8x^2 + 1 \qquad \text{etc.}$$

A recursive formula for these polynomials is

$$T_j(x) = 2x T_{j-1}(x) - T_{j-2}(x) \qquad (j \geq 2) \tag{2}$$

This formula, together with the equations $T_0(x) = 1$ and $T_1(x) = x$, provides a formal definition of the Chebyshev polynomials. Alternatively, $T_k(x) = \cos(k \arccos x)$.

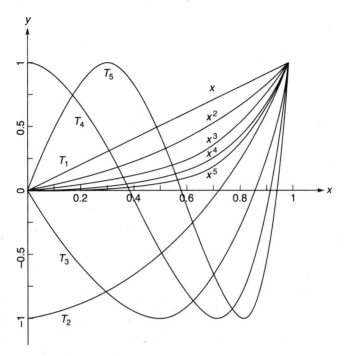

FIGURE 10.2
Polynomials x^k
and Chebyshev
polynomials T_k

Linear combinations of Chebyshev polynomials are easy to evaluate because a special nested multiplication algorithm applies. To describe this procedure, consider an arbitrary linear combination of $T_0, T_1, T_2, \ldots, T_n$:

$$g(x) = \sum_{j=0}^{n} c_j T_j(x)$$

An algorithm to compute $g(x)$ for any given x goes as follows:

$$\begin{cases} w_{n+2} = w_{n+1} = 0 \\ w_j = c_j + 2xw_{j+1} - w_{j+2} \qquad (n \geq j \geq 0) \\ g(x) = w_0 - xw_1 \end{cases} \qquad (3)$$

To see that this algorithm actually produces $g(x)$, we write down the series for g, shift some indices, and use Formulas (2) and (3):

$$\begin{aligned} g(x) &= \sum_{j=0}^{n} c_j T_j(x) \\ &= \sum_{j=0}^{n} (w_j - 2xw_{j+1} + w_{j+2}) T_j \\ &= \sum_{j=0}^{n} w_j T_j - 2x \sum_{j=0}^{n} w_{j+1} T_j + \sum_{j=0}^{n} w_{j+2} T_j \end{aligned}$$

$$= \sum_{j=0}^{n} w_j T_j - 2x \sum_{j=1}^{n+1} w_j T_{j-1} + \sum_{j=2}^{n+2} w_j T_{j-2}$$

$$= \sum_{j=0}^{n} w_j T_j - 2x \sum_{j=1}^{n} w_j T_{j-1} + \sum_{j=2}^{n} w_j T_{j-2}$$

$$= w_0 T_0 + w_1 T_1 + \sum_{j=2}^{n} w_j T_j - 2xw_1 T_0 - 2x \sum_{j=2}^{n} w_j T_{j-1} + \sum_{j=2}^{n} w_j T_{j-2}$$

$$= w_0 + xw_1 - 2xw_1 + \sum_{j=2}^{n} w_j (T_j - 2xT_{j-1} + T_{j-2})$$

$$= w_0 - xw_1$$

In general, it is best to arrange the data so that all the abscissas $\{x_i\}$ lie in the interval $[-1, 1]$. Then, if the first few Chebyshev polynomials are used as a basis for the polynomials, the normal equations should be reasonably well conditioned. We have not given a technical definition of this term; it can be interpreted informally to mean that Gaussian elimination with pivoting produces an accurate solution to the normal equations.

If the original data do not satisfy $\min\{x_k\} = -1$ and $\max\{x_k\} = 1$ but lie instead in another interval $[a, b]$, then the change of variable

$$x = \frac{1}{2}(b - a)z + \frac{1}{2}(a + b)$$

produces a variable z that traverses $[-1, 1]$ as x traverses $[a, b]$.

Outline of Algorithm

Here is an outline of a procedure, based on the preceding discussion, that produces a polynomial of degree $\leq (n + 1)$ that best fits a given table of values (x_k, y_k) $(0 \leq k \leq m)$. Here m is usually much greater than n.

1. Find the smallest interval $[a, b]$ that contains all the x_k. Thus, let $a = \min\{x_k\}$ and $b = \max\{x_k\}$.

2. Make a transformation to interval $[-1, 1]$ by defining

$$z_k = \frac{2x_k - a - b}{b - a} \qquad (0 \leq k \leq m)$$

3. Decide on the value of n to be used. In this situation, 8 or 10 would be a large value for n.

4. Using Chebyshev polynomials as a basis, generate the $(n + 1) \times (n + 1)$ normal equations

$$\sum_{j=0}^{n} \left[\sum_{k=0}^{m} T_i(z_k) T_j(z_k) \right] c_j = \sum_{k=0}^{m} y_k T_i(z_k) \qquad (0 \leq i \leq n) \qquad \textbf{(4)}$$

5. Use an equation-solving routine to solve the normal equations for coefficients $c_0, c_1, \ldots, c_n$ in the function

$$g(x) = \sum_{j=0}^{n} c_j T_j(x)$$

6. The polynomial sought is

$$g\left(\frac{2x - a - b}{b - a}\right)$$

The details of step **4** are as follows: Begin by introducing a double-subscripted variable:

$$t_{jk} = T_j(z_k) \qquad (0 \le k \le m,\ 0 \le j \le n)$$

The matrix $T = (t_{jk})$ can be computed efficiently by using the recursive definition of the Chebyshev polynomials, Equation (2), as in the following segment of pseudocode:

real array $(t_{ij})_{0:n \times 0:m}, (z_i)_{0:n}$
integer j, k, m
for $k = 0$ **to** m **do**
  $t_{0k} \leftarrow 1$
  $t_{1k} \leftarrow z_k$
  for $j = 2$ **to** n **do**
    $t_{jk} \leftarrow 2z_k t_{j-1,k} - t_{j-2,k}$
  end do
end do

The normal equations have a coefficient matrix $A = (a_{ij})_{0:n \times 0:n}$ and a right-hand side $b = (b_i)_{0:n}$ given by

$$a_{ij} = \sum_{k=0}^{m} T_i(z_k) T_j(z_k) = \sum_{k=0}^{m} t_{ik} t_{jk} \qquad (0 \le i,\ j \le n)$$

$$b_i = \sum_{k=0}^{m} y_k T_i(z_k) = \sum_{k=0}^{m} y_k t_{ik} \qquad (0 \le i \le n)$$

(5)

The pseudocode to calculate A and b follows:

real array $(b_i)_{0:n}, (t_{ij})_{0:n \times 0:m}, (y_i)_{0:n}$
integer i, j, m, n
real s
$\vdots$

for $i = 0$ **to** n **do**
  $s \leftarrow 0$
  for $k = 0$ **to** m **do**
    $s \leftarrow s + y_k t_{ik}$
  end do
  $b_i \leftarrow s$
  for $j = i$ **to** n **do**
    $s \leftarrow 0$
    for $k = 0$ **to** m **do**
      $s \leftarrow s + t_{ik} t_{jk}$
    end do
    $a_{ij} \leftarrow s$
    $a_{ji} \leftarrow s$
  end do
end do
end do

To fit data with polynomials, other methods exist that employ systems of polynomials tailor-made for a given set of abscissas. The method outlined above is, however, simple and direct.

Smoothing Data: Polynomial Regression

One of the important applications of the least squares procedure is in the smoothing of data. In this context, **smoothing** refers to the removal of experimental errors to whatever degree possible. If one knows the type of function to which the data should conform, then the least squares procedure can be used to compute any unknown parameters in the function. This has been amply illustrated in the examples given previously. If, however, one simply wishes to smooth the data by fitting them with any convenient function, then polynomials of increasing degree can be used until a reasonable balance between good fit and smoothness is obtained.

This idea will be illustrated by the experimental data depicted in the table, which shows 20 points (x_i, y_i):

x	-1.0	-0.92	-0.84	-0.8	-0.72	-0.64	-0.56	-0.48	-0.36
y	4.0	1.0	5.0	7.0	6.0	3.0	2.0	2.0	5.0

-0.24	-0.12	0.0	0.12	0.2	0.32	0.4	0.52	0.64	0.76	0.92
10.0	13.0	11.0	7.0	4.0	-2.0	-6.0	-8.0	-2.0	4.0	9.0

Of course, a polynomial of degree 19 can be determined that passes through these points *exactly*. But if the points are contaminated by experimental errors, our purposes are better served by some lower-degree polynomial that fits the data *approximately* in the least squares sense. In statistical parlance, this is the problem of *curvilinear regression.* A good software library will contain code for the polynomial fitting of empirical data using a least squares criterion. Such programs will determine the fitting polynomials of degrees $0, 1, 2, \ldots$ with a minimum of computing effort and with high precision. One can, of course, use the techniques illustrated already in this chapter, although they are not at all streamlined. Thus, with the Chebyshev polynomials as a basis, we can set up and solve the normal equations for $n = 0, 1, 2, \ldots$ and plot the resulting functions. Some of the polynomials obtained in this way for the data of the table are shown in Figure 10.3.

An efficient procedure for polynomial regression is now explained. This procedure uses a system of orthogonal polynomials tailor-made for the problem at hand. We begin with a table of experimental values:

x	x_0	x_1	$\ldots$	x_m
y	y_0	y_1	$\ldots$	y_m

The ultimate objective is to replace this table by a suitable polynomial of modest degree, with the experimental errors of the table somehow suppressed. We do not know what degree of polynomial should be used.

For statistical purposes, a reasonable hypothesis is that there is a polynomial

$$p_N(x) = \sum_{i=0}^{N} a_i x^i$$

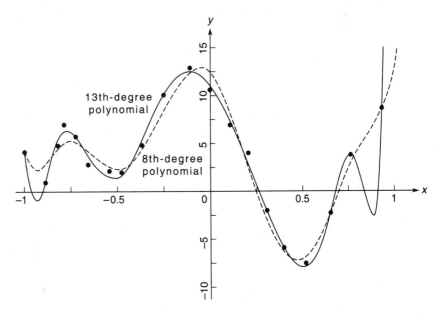

FIGURE 10.3
Polynomial of degree 8 (dashed line) and polynomial of degree 13 (solid line)

that represents the trend of the table and that the given tabular values obey the equation

$$y_i = p_N(x_i) + \varepsilon_i \qquad (0 \le i \le m)$$

In this equation, ε_i represents an observational error present in y_i. A further reasonable hypothesis is that these errors are independent random variables that are normally distributed.

For a fixed value of n, we have already discussed a method of determining p_n by the method of least squares. Thus, a system of normal equations can be set up to determine the coefficients of p_n. Once these are known, a quantity called the **variance** can be computed from the formula

$$\sigma_n^2 = \frac{1}{m-n} \sum_{i=0}^{m} [y_i - p_n(x_i)]^2 \qquad (m > n) \tag{6}$$

Statistical theory tells us that if the trend of the table is truly a polynomial of degree N, then

$$\sigma_0^2 > \sigma_1^2 > \cdots > \sigma_N^2 = \sigma_{N+1}^2 = \sigma_{N+2}^2 = \cdots = \sigma_{m-1}^2$$

This fact suggests the following strategy for dealing with the case in which N is not known: Compute $\sigma_0^2, \sigma_1^2, \ldots$ in succession. As long as these are decreasing significantly, continue the calculation. When an integer N is reached for which $\sigma_N^2 \approx \sigma_{N+1}^2 \approx \sigma_{N+2}^2 \approx \cdots$, stop and declare p_n to be the polynomial sought.

If $\sigma_0^2, \sigma_1^2, \ldots$ are to be computed directly from the definition in Equation (6), then each of the polynomials $p_0, p_1, \ldots$ will have to be determined. The procedure described below can avoid the determination of all but the one desired polynomial.

In the remainder of the discussion, the abscissas x_i are to be held fixed. These points are assumed to be distinct, although the theory can be extended to include cases in which some points repeat. If f and g are two functions whose domains include the points x_i, then the following notation is used:

$$\langle f, g \rangle = \sum_{i=0}^{m} f(x_i)g(x_i) \tag{7}$$

This quantity is called the **inner product** of f and g. Much of our discussion does not depend on the exact form of the inner product but only on certain of its properties:

1. $\langle f, g \rangle = \langle g, f \rangle$

2. $\langle f, f \rangle > 0$ unless $f(x_i) = 0$ for all i

3. $\langle af, g \rangle = a \langle f, g \rangle$ where $a \in \mathbb{R}$

4. $\langle f, g + h \rangle = \langle f, g \rangle + \langle f, h \rangle$

The reader should verify that the inner product defined in Equation (7) has the properties listed.

A set of functions is now said to be **orthogonal** if $\langle f, g \rangle = 0$ for any two different functions f and g in that set. An orthogonal set of polynomials can be generated recursively by the following formulas:

$$
\begin{cases}
q_0(x) = 1 \\[2mm]
q_1(x) = x - \alpha_0 \\[2mm]
q_{n+1}(x) = xq_n(x) - \alpha_n q_n(x) - \beta_n q_{n-1}(x)
\end{cases}
$$

where

$$
\begin{cases}
\alpha_n = \dfrac{\langle xq_n, q_n \rangle}{\langle q_n, q_n \rangle} \\[4mm]
\beta_n = \dfrac{\langle xq_n, q_{n-1} \rangle}{\langle q_{n-1}, q_{n-1} \rangle}
\end{cases}
$$

In these formulas a slight abuse of notation occurs where "xq_n" is used to denote the function whose value at x is $xq_n(x)$.

To understand how this definition leads to an orthogonal system, let's examine a few cases. First,

$$
\langle q_1, q_0 \rangle = \langle x - \alpha_0, q_0 \rangle = \langle xq_0 - \alpha_0 q_0, q_0 \rangle = \langle xq_0, q_0 \rangle - \alpha_0 \langle q_0, q_0 \rangle = 0
$$

Notice that several properties of an inner product listed previously have been used here. Also, the definition of α_0 was used. Another of the first few cases is this:

$$
\langle q_2, q_1 \rangle = \langle xq_1 - \alpha_1 q_1 - \beta_1 q_0, q_1 \rangle
$$
$$
= \langle xq_1, q_1 \rangle - \alpha_1 \langle q_1, q_1 \rangle - \beta_1 \langle q_0, q_1 \rangle = 0
$$

Here the definition of α_1 has been used, as well as the fact (established above) that $\langle q_1, q_0 \rangle = 0$. The next step in a formal proof is to verify that $\langle q_2, q_0 \rangle = 0$. Then an inductive proof completes the argument.

One part of this proof consists in showing that the coefficients α_n and β_n are well defined. This means that the denominators $\langle q_n, q_n \rangle$ are not zero. To verify that this is the case, suppose that $\langle q_n, q_n \rangle = 0$. Then $\sum_{i=0}^{m} [q_n(x_i)]^2 = 0$, and consequently $q_n(x_i) = 0$ for *each* value of i. This means that the polynomial q_n has $m + 1$ roots, $x_0, x_1, \ldots, x_m$. Since the degree n is less than m, we conclude that q_n is the zero polynomial. However, this is not possible because obviously

$$q_0(x) = 1$$

$$q_1(x) = x - \alpha_0$$

$$q_2(x) = x^2 + \text{(lower-order terms)}$$

and so on. Observe that this argument requires $n < m$.

The system of orthogonal polynomials $\{q_0, q_1, \ldots, q_{m-1}\}$ generated by the above algorithm is a basis for the vector space $\prod_{m-1}$ of all polynomials of degree $\leq m - 1$. It is clear from the algorithm that each q_n starts with the highest term x^n. If it is desired to express a given polynomial p of degree n ($n \leq m - 1$) as a linear combination of $q_0, q_1, \ldots, q_n$, this can be done as follows: Set

$$p = \sum_{i=0}^{n} a_i q_i \tag{8}$$

On the right-hand side, only one summand contains x^n. It is the term $a_n q_n$. On the left-hand side, there is also a term in x^n. One chooses a_n so that $a_n x^n$ on the right is equal to the corresponding term in p. Now write

$$p - a_n q_n = \sum_{i=0}^{n-1} a_i q_i$$

On both sides of this equation there are polynomials of degree at most $n - 1$ (because of the choice of a_n). Hence, we can now choose a_{n-1} in the way we chose a_n; that is, choose a_{n-1} so that the terms in x^{n-1} are the same on both sides. By continuing in this way, we discover the unique values that the coefficients a_i must have. This establishes that $\{q_0, q_1, \ldots, q_n\}$ is a basis for $\prod_n$, for $n = 0, 1, \ldots, m - 1$.

Another way of determining the coefficients a_i (once we know that they exist!) is to take the inner product of both sides of Equation (8) with q_j. The result is

$$\langle p, q_j \rangle = \sum_{i=0}^{n} a_i \langle q_i, q_j \rangle \qquad (0 \leq j \leq n)$$

Since the set $q_0, q_1, \ldots, q_n$ is orthogonal, $\langle q_i, q_j \rangle = 0$ for each i different from j. Hence,

$$\langle p, q_j \rangle = a_j \langle q_j, q_j \rangle$$

This gives a_j as a quotient of two inner products.

Now we return to the least squares problem. Let F be a function that we wish to fit by a polynomial p_n of degree n. We shall find the polynomial that minimizes the expression

$$\sum_{i=0}^{m} [F(x_i) - p_n(x_i)]^2$$

The solution is given by the formulas

$$p_n = \sum_{i=0}^{n} c_i q_i \qquad c_i = \frac{\langle F, q_i \rangle}{\langle q_i, q_i \rangle} \qquad \text{(9)}$$

It is especially noteworthy that c_i does *not* depend on n. This implies that the various polynomials $p_0, p_1, \ldots$ that we are seeking can all be obtained by simply truncating *one* series—namely, $\sum_{i=0}^{m-1} c_i q_i$. To prove that p_n, as given in Equation (9), solves our problem, we return to the normal equations, Equation (1). The basic functions now being used are $q_0, q_1, \ldots, q_n$. Thus, the normal equations are

$$\sum_{j=0}^{n} \left[\sum_{k=0}^{m} q_i(x_k) q_j(x_k) \right] c_j = \sum_{k=0}^{m} y_k q_i(x_k) \qquad (0 \leq i \leq n)$$

Using the inner product notation, we get

$$\sum_{j=0}^{n} \langle q_i, q_j \rangle c_j = \langle F, q_i \rangle \qquad (0 \leq i \leq n)$$

where F is some function such that $F(x_k) = y_k$ for $0 \leq k \leq m$. Next, apply the orthogonality property $\langle q_i, q_j \rangle = 0$ when $i \neq j$. The result is

$$\langle q_i, q_i \rangle c_i = \langle F, q_i \rangle \qquad (0 \leq i \leq n) \qquad \text{(10)}$$

Now we return to the variance numbers $\sigma_0^2, \sigma_1^2, \ldots$ and show how they can be easily computed. First, an important observation: The set $\{q_0, q_1, \ldots, q_n, F - p_n\}$ is orthogonal! The only new fact here is that $\langle F - p_n, q_i \rangle = 0$ for $0 \leq i \leq n$. To check this, write

$$\langle F - p_n, q_i \rangle = \langle F, q_i \rangle - \langle p_n, q_i \rangle$$

$$= \langle F, q_i \rangle - \left\langle \sum_{j=0}^{n} c_j q_j, q_i \right\rangle$$

$$= \langle F, q_i \rangle - \sum_{j=0}^{n} c_j \langle q_j, q_i \rangle$$

$$= \langle F, q_i \rangle - c_i \langle q_i, q_i \rangle = 0$$

In this computation, we used Equations (9) and (10). Since p_n is a linear combination of $q_0, q_1, \ldots, q_n$, it follows easily that

$$\langle F - p_n, p_n \rangle = 0$$

Now recall that the variance σ_n^2 was defined by

$$\sigma_n^2 = \frac{\rho_n}{m - n} \qquad \rho_n = \sum_{i=0}^{m} [y_i - p_n(x_i)]^2$$

The quantities ρ_n can be written in another way:

$$\begin{aligned}
\rho_n &= \langle F - p_n, F - p_n \rangle \\
&= \langle F - p_n, F \rangle \\
&= \langle F, F \rangle - \langle F, p_n \rangle \\
&= \langle F, F \rangle - \sum_{i=0}^{n} c_i \langle F, q_i \rangle \\
&= \langle F, F \rangle - \sum_{i=0}^{n} \frac{\langle F, q_i \rangle^2}{\langle q_i, q_i \rangle}
\end{aligned}$$

Thus, the numbers $\rho_0, \rho_1, \ldots$ can be generated recursively by the algorithm

$$\begin{cases}
\rho_0 = \langle F, F \rangle - \dfrac{\langle F, q_0 \rangle^2}{\langle q_0, q_0 \rangle} \\[2ex]
\rho_n = \rho_{n-1} - \dfrac{\langle F, q_n \rangle^2}{\langle q_n, q_n \rangle} \qquad (n \geq 1)
\end{cases}$$

PROBLEMS 10.2

1. Let $g_0, g_1, \ldots, g_n$ be a set of functions such that $\sum_{k=0}^{m} g_i(x_k) g_j(x_k) = 0$ if $i \neq j$. What linear combination of these functions best fits the data of Table (1) in Section 10.1?

2. Consider polynomials $g_0, g_1, \ldots, g_n$ defined by $g_0(x) = 1$, $g_1(x) = x - 1$, and $g_j(x) = 3x g_{j-1}(x) + 2g_{j-2}(x)$. Develop an efficient algorithm for computing values of the function $f(x) = \sum_{j=0}^{n} c_j g_j(x)$.

3. Show that $\cos n\theta = 2 \cos \theta \cos(n-1)\theta - \cos(n-2)\theta$. *Hint:* Use the familiar identity $\cos(A \mp B) = \cos A \cos B \pm \sin A \sin B$.

4. (Continuation) Show that if $f_n(x) = \cos(n \arccos x)$, then $f_0(x) = 1$, $f_1(x) = x$, and $f_n(x) = 2x f_{n-1}(x) - f_{n-2}(x)$.

5. (Continuation) Show that an alternate definition of Chebyshev polynomials is $T_n(x) = \cos(n \arccos x)$ for $-1 \leq x \leq 1$.

6. (Continuation) Give a one-line proof that $T_n(T_m(x)) = T_{nm}(x)$.

7. (Continuation) Show that $|T_n(x)| \leq 1$ for x in the interval $[-1, 1]$.

8. Define $g_k(x) = T_k(\frac{1}{2}x + \frac{1}{2})$. What recursive relation do these functions satisfy?

9. Show that $T_0, T_2, T_4, \ldots$ are even and that $T_1, T_3, \ldots$ are odd functions. Recall that an even function satisfies the equation $f(x) = f(-x)$; an odd function satisfies the equation $f(x) = -f(-x)$.

10. Count the number of operations involved in the algorithm used to compute $g(x) = \sum_{j=0}^{n} c_j T_j(x)$.

11. Show that the algorithm for computing $g(x) = \sum_{j=0}^{n} c_j T_j(x)$ can be modified to read

$$\begin{cases} w_{n-1} = c_{n-1} + 2xc_n \\ \quad w_k = c_k + 2xw_{k+1} - w_{k-2} \qquad (n - 2 \geq k \geq 1) \\ \quad g(x) = c_0 + xw_1 - w_2 \end{cases}$$

thus making w_{n+2}, w_{n+1}, and w_0 unnecessary.

12. (Continuation) Count the operations for the algorithm in Problem **11**.

13. Determine $T_6(x)$ as a polynomial in x.

14. Verify the four properties of an inner product that were listed in the text using Definition (7).

15. Verify these formulas.

$$p_0(x) = \frac{1}{m+1} \sum_{i=0}^{m} y_i$$

$$\beta_n = \frac{\langle q_n, q_n \rangle}{\langle q_{n-1}, q_{n-1} \rangle}$$

$$c_n = \frac{p_{n-1} - p_n}{\langle F, q_n \rangle}$$

16. Complete the proof that the algorithm for generating the orthogonal system of polynomials works.

17. There is a function f of the form

$$f(x) = \alpha x^{12} + \beta x^{13}$$

for which $f(0.1) = 6 \times 10^{-13}$ and $f(0.9) = 3 \times 10^{-2}$. What is it? Are α and β sensitive to perturbations in the two given values of $f(x)$?

COMPUTER PROBLEMS 10.2

1. Carry out an experiment in data smoothing as follows: Start with a polynomial of modest degree, say 7. Compute 100 values of this polynomial at random points in the interval $[-1, 1]$. Perturb these values by adding random numbers chosen from a small interval, say $[-\frac{1}{8}, \frac{1}{8}]$. Try to recover the polynomial from these perturbed values by using the method of least squares.

2. Write

real function *cheb*(n, x)

for evaluating $T_n(x)$. Use the recursive formula satisfied by Chebyshev polynomials. Do not use a subscripted variable. Test the program on these 15 cases: $n = 0, 1, 3, 6, 12$ and $x = 0, -1, 0.5$.

3. Write

real function *theb*$(n, x, (y_i))$

to calculate $T_0(x), T_1(x), \ldots, T_n(x)$ and store these numbers in the array (y_i). Use your routine, together with suitable plotting routines, to obtain graphs of $T_0, T_1, T_2, \ldots, T_8$ on $[-1, 1]$.

4. Write

   real function $F(n, (c_i), x)$

   for evaluating $f(x) = \sum_{j=0}^{n} c_j T_j(x)$. Test your routine by means of the formula $\sum_{k=0}^{\infty} t^k T_k(x) = (1 - tx)/(1 - 2tx + t^2)$, valid for $|t| < 1$. If $|t| \leq \frac{1}{2}$, then only a few terms of the series are needed to give full machine precision. Add terms in ascending order of magnitude.

5. Obtain a graph of T_n for some reasonable value of n by means of the following idea: Generate 100 equally spaced angles θ_i in the interval $[0, \pi]$. Define $x_i \cos \theta_i$ and $y_i = T_n(x_i) = \cos(n \arccos x_i) = \cos n\theta_i$. Send the points (x_i, y_i) to a suitable plotting routine.

6. Write suitable code to carry out the procedure outlined in the text for fitting a table with a linear combination of Chebyshev polynomials. Test it in the manner of Computer Problem **1**, first by using an unperturbed polynomial. Find out experimentally how large n can be in this process before roundoff errors become serious.

7. Select a modest value of n. Select $m > 2n$ and define $x_k = \cos[(2k - 1)\pi/(2m)]$. Compute and print the matrix A whose elements are

$$a_{ij} = \sum_{k=0}^{m} T_i(x_k) T_j(x_k) \qquad (0 \leq i, \, j \leq n)$$

   Interpret the results in terms of the least-squares-polynomial fitting problem.

8. Program the algorithm for finding $\sigma_0^2, \sigma_1^2, \ldots$ in the polynomial regression problem.

9. Program the complete polynomial regression algorithm. The output should be $\alpha_n, \beta_n, \alpha_n^2$, and c_n for $0 \leq n \leq N$, where N is determined by the condition $\sigma_{N-1}^2 > \sigma_N^2 \approx \sigma_{N+1}^2$.

10.3 Other Examples of the Least Squares Principle

The principle of least squares is also used in other situations. In one of these, we attempt to *solve* an inconsistent system of linear equations of the form

$$\sum_{j=0}^{n} a_{kj} x_j = b_k \qquad (0 \leq k \leq m) \tag{1}$$

in which $m > n$. Here there are $m + 1$ equations but only $n + 1$ unknowns. If a given $n + 1$-tuple $(x_0, x_1, \ldots, x_n)$ is substituted on the left, the discrepancy between the two sides of the kth equation is termed the **kth residual**. Ideally, of course, all residuals should be zero. If it is not possible to select $(x_0, x_1, \ldots, x_n)$ so as to make all residuals zero, System (1) is said to be **inconsistent** or **incompatible**. In this case, an alternative is to minimize the sum of the squares of the residuals. So we are led to minimize the expression

$$\phi(x_0, x_1, \ldots, x_n) = \sum_{k=0}^{m} \left(\sum_{j=0}^{n} a_{kj}x_j - b_k \right)^2 \tag{2}$$

by making an appropriate choice of $(x_0, x_1, \ldots, x_n)$. Proceeding as before, we take partial derivatives with respect to x_i and set them equal to zero, arriving, thereby, at the normal equations

$$\sum_{j=0}^{n} \left(\sum_{k=0}^{m} a_{ki}a_{kj} \right) x_j = \sum_{k=0}^{m} b_k a_{ki} \qquad (0 \leq i \leq n) \tag{3}$$

This is a linear system of just $n + 1$ equations involving unknowns $x_0, x_1, \ldots, x_n$. It can be shown that this system is consistent, provided that the column vectors in the original coefficient array are linearly independent. System (3) can be solved, for instance, by Gaussian elimination. The solution of System (3) is then a best approximate solution of Equation (1) in the least squares sense.

Special methods have been devised for the problem just discussed. Generally they gain in precision over the simple approach outlined above. One such algorithm for solving System (1),

$$Ax = b$$

begins by factoring

$$A = QR$$

where Q is an $(m + 1) \times (n + 1)$ matrix satisfying $Q^T Q = I$ and R is an $(n + 1) \times (n + 1)$ matrix satisfying $r_{ii} > 0$ and $r_{ij} = 0$ for $j < i$. Then the least squares solution is obtained by an algorithm called the **modified Gram-Schmidt process**.

A more elaborate (and more versatile) algorithm depends on the singular value decomposition of the matrix A. This is a factoring, $A = U\Sigma V^T$, in which $U^T U = I_{m+1}$, $V^T V = I_{n+1}$, and Σ is an $(m + 1) \times (n + 1)$ diagonal matrix that has nonnegative entries. For these more modern procedures, the reader is referred to Stewart [1973] and Lawson and Hanson [1974] for details.

Weight Function $w(x)$

Another important example of the principle of least squares occurs in fitting or approximating functions on *intervals* rather than discrete sets. For example, a given function f defined on an interval $[a, b]$ may be required to be approximated by a function like

$$g(x) = \sum_{j=0}^{n} c_j g_j(x)$$

It is natural, then, to attempt to minimize the expression

$$\phi(c_0, c_1, \ldots, c_n) = \int_a^b [g(x) - f(x)]^2 \, dx \qquad \text{(4)}$$

by choosing coefficients appropriately. In some applications, it is desirable to force functions g and f into better agreement in certain parts of the interval. For this purpose, we can modify Equation (4) by including a positive weight function $w(x)$, which can, of course, be $w(x) \equiv 1$ if all parts of the interval are to be treated the same. The result is

$$\phi(c_0, c_1, \ldots, c_n) = \int_a^b [g(x) - f(x)]^2 w(x) \, dx$$

The minimum of ϕ is again sought by differentiating with respect to each c_i and setting the partial derivatives equal to zero. The result is a system of normal equations:

$$\sum_{j=0}^{n} \left[\int_a^b g_i(x) g_j(x) w(x) \, dx \right] c_j = \int_a^b f(x) g_i(x) w(x) \, dx \qquad (0 \leq i \leq n) \quad \text{(5)}$$

Again it is a system of $n + 1$ linear equations in $n + 1$ unknowns $c_0, c_1, \ldots, c_n$ and can be solved by Gaussian elimination. Earlier remarks about choosing a good basis apply here also. The ideal situation is to have functions $g_0, g_1, \ldots, g_n$ that have the orthogonality property:

$$\int_a^b g_i(x) g_j(x) w(x) \, dx = 0 \qquad (i \neq j) \qquad \text{(6)}$$

Many such orthogonal systems have been developed over the years. For example, Chebyshev polynomials form one such system—namely,

$$\int_{-1}^{1} T_i(x) T_j(x) (1 - x^2)^{-1/2} \, dx = \begin{cases} 0 & \text{if } i \neq j \\ \dfrac{\pi}{2} & \text{if } i = j > 0 \\ \pi & \text{if } i = j = 0 \end{cases}$$

The weight function $(1 - x^2)^{-1/2}$ assigns heavy weight to the ends of the interval $[-1, 1]$.

If a sequence of nonzero functions $g_0, g_1, \ldots, g_n$ is orthogonal according to Equation (6), then the sequence $\lambda_0 g_0, \lambda_1 g_1, \ldots, \lambda_n g_n$ is orthonormal for appropriate positive real numbers λ_j; namely,

$$\lambda_j = \left\{ \int_a^b [g_j(x)]^2 w(x) \, dx \right\}^{-1/2}$$

Nonlinear Example

As another example of the least squares principle, here is a nonlinear problem. Suppose that a table of points (x_k, y_k) is to be fitted by a function of the form

$$y = e^{cx}$$

Proceeding as before leads to the problem of minimizing the function

$$\phi(c) = \sum_{k=0}^{m} (e^{cx_k} - y_k)^2$$

The minimum occurs for a value of c such that

$$0 = \frac{\partial \phi}{\partial c} = \sum_{k=0}^{m} 2(e^{cx_k} - y_k)e^{cx_k} x_k$$

This equation is nonlinear in c. One could contemplate solving it by Newton's method or the secant method. On the other hand, the problem of minimizing $\phi(c)$ could be attacked directly. Since there can be multiple roots in the normal equation and local minima in ϕ itself, a direct minimization of ϕ would be safer. This type of difficulty is typical of nonlinear least squares problems. Consequently, other methods of curve fitting are often preferred if the unknown parameters do not occur linearly in the problem.

Alternatively, this particular example can be linearized by a change of variables $z = \ln y$ and by considering

$$z = cx$$

The problem of minimizing the function

$$\phi(c) = \sum_{k=0}^{m} (cx_k - z_k)^2 \qquad (z_k = \ln y_k)$$

is easy and leads to

$$c = \frac{\displaystyle\sum_{k=0}^{m} z_k x_k}{\displaystyle\sum_{k=0}^{m} x_k^2}$$

This value of c is *not* the solution of the original problem but may be satisfactory in some applications.

Linear/Nonlinear Example

The final example contains elements of linear and nonlinear theory. Suppose that an (x_k, y_k) table is given with $m + 1$ entries and that a functional relationship like

$$y = a \sin(bx)$$

is suspected. Can the least squares principle be used to obtain the appropriate values of the parameters a and b?

Notice that parameter b enters this function in a nonlinear way, creating some difficulty, as will be seen. According to the principle of least squares, the parameters should be so chosen that the expression

$$\sum_{k=0}^{m} [a \sin(bx_k) - y_k]^2$$

has a minimum value. The minimum value is sought by differentiating this expression with respect to a and b and setting these partial derivatives equal to zero. The results are

$$
\begin{cases}
\displaystyle\sum_{k=0}^{m} 2[a \sin(bx_k) - y_k] \sin(bx_k) & = 0 \\
\displaystyle\sum_{k=0}^{m} 2[a \sin(bx_k) - y_k] a x_k \cos(bx_k) = 0
\end{cases}
$$

If b were known, a could be obtained from either equation. The correct value of b is the one for which these corresponding two a values are identical. So each of the preceding equations should be solved for a and the results set equal to each other. This process leads to the equation

$$\frac{\displaystyle\sum_{k=0}^{m} y_k \sin bx_k}{\displaystyle\sum_{k=0}^{m} (\sin bx_k)^2} = \frac{\displaystyle\sum_{k=0}^{m} x_k y_k \cos bx_k}{\displaystyle\sum_{k=0}^{m} x_k \sin bx_k \cos bx_k}$$

which can now be solved for parameter b, using, for example, the bisection or the secant method. Then either side of this equation can be evaluated as the value of a.

PROBLEMS 10.3

1. Analyze the least squares problem of fitting data by a function of the form $y = x^c$.

2. Show that the Hilbert matrix (Computer Problem **4**, Section 6.2) arises in the normal equations when we minimize

$$\int_0^1 \left[\sum_{j=0}^n c_j x^j - f(x) \right]^2 dx$$

3. Find a function of the form $y = e^{cx}$ that best fits this table.

x	0	1
y	1/2	1

4. (Continuation) Repeat Problem **3** for the following table.

x	0	1
y	a	b

5. (Continuation) Repeat Problem **4** under the supposition that b is negative.

6. Show that the normal equation for the problem of fitting $y = e^{cx}$ to points $(1, -12)$ and $(2, 7.5)$ has two real roots: $c = \ln 2$ and $c = 0$. Which value is correct for the fitting problem?

7. Consider the inconsistent System (1). Suppose that each equation has associated with it a positive number w_i indicating its relative importance or reliability. How should Equations (2) and (3) be modified to reflect this?

8. Determine the best approximate solution of the inconsistent system of linear equations

$$\begin{cases} 2x + 3y = 1 \\ x - 4y = -9 \\ 2x - y = -1 \end{cases}$$

in the least squares sense.

9. **a.** Find the constant c for which cx is the best approximation in the sense of least squares to the function $\sin x$ on interval $[0, \pi/2]$.

   b. Do the same for e^x on $[0,1]$.

10. Analyze the problem of fitting a function $y = (c - x)^{-1}$ to a table of $m + 1$ points.

11. Show that the normal equations for the least squares solution of $Ax = b$ can be written $(A^T A)x = A^T b$.

12. Derive the normal equations given by System (5).

13. A table of values (x_k, y_k), where $k = 0, 1, \ldots, m$, is obtained from an experiment. When plotted on semilogarithmic graph paper, the points lie nearly on a straight line, implying that $y \approx e^{ax+b}$. Suggest a simple procedure for obtaining parameters a and b.

14. In fitting a table of values to a function of the form $a + bx^{-1} + cx^{-2}$, we try to make each point lie on the curve. This leads to $a + bx_k^{-1} + cx_k^{-2} = y_k$ for $0 \leq k \leq m$. An equivalent equation is $ax_k^2 + bx_k + c = y_k x_k^2$ for $0 \leq k \leq m$. Are the least squares problems for these systems of equations equivalent?

15. A table of points (x_k, y_k) is plotted and appears to lie on a hyperbola of the form $y = (a + bx)^{-1}$. How can the *linear* theory of least squares be used to obtain good estimates of a and b?

16. Consider $f(x) = e^{2x}$ over $[0, \pi]$. We wish to approximate the function by a trigonometric polynomial of the form $p(x) = a + b\cos(x) + c\sin(x)$. Determine the linear system to be solved for determining the least squares fit of p to f.

17. Find the constant c that makes the expression $\int_0^1 (e^x - cx)^2 dx$ a minimum?

18. Show that in every least squares matrix problem, the normal equations have a symmetric coefficient matrix.

19. Verify that the following steps produce the least squares solution of $Ax = b$.

   a. Factor $A = QR$, where Q and R have the properties described in the text.

   b. Define $y = Q^T b$.

   c. Solve the lower triangular system $Rx = y$.

20. What value of c should be used if a table of experimental data (x_i, t_i) for $0 \leq i \leq m$ is to be represented by the formula $y = c \sin x$? An explicit usable formula for c is required. Use the principle of least squares.

COMPUTER PROBLEMS 10.3

1. Using the method suggested in the text, fit the data

x	0.1	0.2	0.3	0.4	0.5	0.6	0.7	0.8
y	0.6	1.1	1.6	1.8	2.0	1.9	1.7	1.3

by a function $y = a \sin bx$.

2. (Prony's method, $n = 1$) To fit a table of the form

x	1	2	$\cdots$	m
y	y_1	y_2	$\cdots$	y_m

by the function $y = ab^x$, we can proceed as follows: If y is actually ab^x, then $y_k = ab^k$ and $y_{k+1} = by_k$ for $k = 1, 2, \ldots, m - 1$. So we determine b by solving this system of equations using the least squares method. Having found b, we find a by solving the equations $y_k = ab^k$ in the least squares sense. Write a program to carry out this procedure and test it on an artificial example.

3. (Continuation) Modify the procedure of Computer Problem 2 to handle any case of equally spaced points.

4. A quick way of fitting a function of the form

$$f(x) \approx \frac{a + bx}{1 + cx}$$

is to apply the least squares method to the problem $(1 + cx)f(x) \approx a + bx$. Use this technique to fit the world population data given here.

Year	Population (in billions)
1000	0.340
1650	0.545
1800	0.907
1900	1.61
1950	2.51
1970	3.65
1980	4.20
1990	5.30

Determine when the world population will become infinite! *Project*: Explore the question of whether the least squares method should be used to predict. For example, study the variances in this problem to determine whether a polynomial of any degree would be satisfactory.

5. Write a procedure that takes as input an $(m + 1) \times (n + 1)$ matrix A and an $m + 1$ vector b and returns the least squares solution of the system $Ax = b$.

11 MONTE CARLO METHODS AND SIMULATION

A highway engineer wishes to simulate the flow of traffic in a proposed design for a major freeway intersection. The information obtained will then be used to determine the capacity of *storage lanes* (in which cars must slow down to yield the right of way). The intersection has the form shown in Figure 11.1, and various flows (cars per minute) are postulated at the points where arrows are drawn. By running the simulation program, the engineer can study the effect of different speed limits, determine which flows lead to saturation (bottlenecks), and so on. Some techniques for constructing such programs are developed in this chapter.

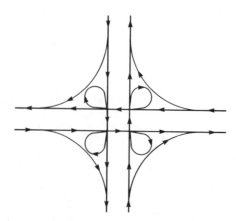

FIGURE 11.1
Traffic flow

ITP HIGHER EDUCATION

The enclosed materials are sent to you for your review by
DAVID HART 800 225-4904

LOCATION	QTY.	ISBN	AUTHOR / TITLE
K-18C-008-21	1	0-534-20112-1	CHENEY/KINCAID NUMERICAL MATH & COMPUTING

Account   Contact   Date
Account   Contact   Date

11.1   Random Numbers

This chapter differs from most of the others in its point of view. Instead of addressing clear-cut mathematical problems, it attempts to develop methods for simulating complicated processes or phenomena. If the computer can be made to imitate an experiment or a process, then by repeating the computer simulation with different data, we can draw statistical conclusions. In such an approach, the conclusions may lack a high degree of mathematical precision but still be sufficiently accurate to enable us to understand the process being simulated.

Particular emphasis is given to problems in which the computer simulation involves an element of chance. The whimsical name of *Monte Carlo methods* was applied some years ago by Stanislaw M. Ulam (1909–1984) to this way of imitating reality by a computer. Since chance or randomness is part of the method, we begin with the elusive concept of **random numbers**.

Consider a sequence of real numbers $x_1, x_2, \ldots$ all lying in the unit interval $(0, 1)$. Expressed informally, the sequence is **random** if the numbers seem to be distributed haphazardly throughout the interval and if there seems to be no pattern in the progression $x_1, x_2, \ldots$. For example, if all the numbers in decimal form begin with the digit 3, then the numbers are clustered in the subinterval $0.3 \leqq x < 0.4$ and are not randomly distributed in $(0, 1)$. If the numbers are monotonically increasing, they are not random. If each x_i is obtained from its predecessor by a simple continuous function, say $x_i = f(x_{i-1})$, then the sequence is not random (although it might appear so). A precise definition of *randomness* is quite difficult to formulate, and the interested reader may wish to consult an article by Chaitlin [1975], in which randomness is related to the complexity of computer algorithms! Thus, it seems best, at least in introductory material, to accept intuitively the notion of a random sequence of numbers in an interval and to accept certain algorithms for generating sequences that are more or less random.

A recommended reference is the book of Niederreiter [1992].

Random-Number Algorithms/Generators

Most computer systems have **random-number generators**, which are procedures that produce either a single random number or an entire array of random numbers with each call. In this chapter, we call such a procedure *random*. The reader can use a random-number generator available on his or her own computing system, within the computer language being used, or one of the generators described below.*

Random-number generators are often available in mathematical program libraries. For the problems in this chapter, one should select a routine to provide random numbers uniformly distributed in the interval $(0, 1)$. A sequence of numbers

---

*In Fortran 90, the procedure RANDOM_NUMBER returns one or an array of uniformly distributed pseudorandom numbers in the interval $[0, 1)$ depending on whether the argument is a scalar variable or an array. The procedure RANDOM_SEED restarts or queries the pseudorandom number generator used by RANDOM_NUMBER.

is **uniformly distributed** in the interval $(0, 1)$ if no subset of the interval contains more than its share of the numbers. In particular, the probability that an element x drawn from the sequence falls within the subinterval $[a, a + h]$ should be h and hence independent of the number a. Similarly, if $p_i = (x_i, y_i)$ are random points in the plane uniformly distributed in some rectangle, then the number of these points that fall inside a small square of area k should depend only on k and not on where the square is located inside the rectangle.

Random numbers produced by a computer code cannot be truly random because the manner in which they are produced is completely *deterministic*; that is, no element of chance is actually present. But the sequences produced by these routines appear to be random, and they do pass certain tests for randomness. Some authors prefer to emphasize this point by calling such computer-generated sequences **pseudorandom** numbers.

If the reader wishes to program a random-number generator, the following one should be satisfactory on a machine that has 32-bit word length. This algorithm generates random numbers $x_1, x_2, \ldots, x_n$ uniformly distributed in the open interval $(0, 1)$ by means of the following recursive algorithm:

integer array $(\ell_i)_n$
real array $(x_i)_n$
$\ell_0 \leftarrow$ any integer such that $1 < \ell_0 < 2^{31} - 1$
for $i = 1$ **to** n **do**
   $\ell_i \leftarrow$ remainder when $7^5 \ell_{i-1}$ is divided by $2^{31} - 1$
   $x_i \leftarrow \ell_i / (2^{31} - 1)$
end do

Here all ℓ_i's are integers in the range $1 < \ell_i < 2^{31} - 1$. The initial integer ℓ_0 is called the **seed** for the sequence and is selected as any integer between 1 and the Mersenne prime number $2^{31} - 1 = 2147483647$.[*]

The interested reader should consult the article by Schrage [1979] on portable random-number generators for more information on this subject. A fast *normal* random-number generator can be written in only a few lines of code as presented in Leva [1992]. It is based on the ratio of uniform deviates method of Kinderman and Monahan [1977].

An external function procedure to generate a new array of pseudorandom numbers per call could be based on the following pseudocode:

---

[*]In 1644, Mersenne (a French friar) conjectured that $2^n - 1$ was a prime number for $n = 17, 19, 31, 67, 127, 257$ and for no other n in the range $1 \leqq n \leqq 257$. Lehmer in 1937 showed, however, that $2^{257} - 1$ was *not* prime. Lucas in 1876 proved that $2^{127} - 1$ was prime. Until 1952, that was the largest known prime. Then it was shown that $2^{521} - 1$ was prime. As a means of testing new computer systems, the search for ever-larger Mersenne primes continues. In 1992, a Cray 2 supercomputer using the Lucas-Lehmer test determined after a 19-hour computation that the 227832-decimal-digit number $2^{756839} - 1$ was a prime number. The previous largest known Mersenne prime was identified in 1985 as $2^{216091} - 1$ (*Focus*, June 1992, Mathematical Association of American, Washington, D.C.).

```
procedure random( (xᵢ) )
integer parameter k ← 16807, j ← 21474 83647
real array (xᵢ)
integer seed
seed ← select initial value for seed
n ← size( (xᵢ) )
for i = 1 to n do
    seed ← mod(k · seed, j)
    xᵢ ← real(seed)/real(j)
end do
end procedure random
```

To allow adequate representation of the numbers involved in procedure *random*, it must be written using double or extended precision for use on a 32-bit computer; otherwise, it will produce nonrandom numbers.

Recall that here and elsewhere $\text{mod}(n, m)$ is the remainder when n is divided by m; that is, it results in $n - [\text{integer}(n/m)]m$ where $\text{integer}(n/m)$ is the integer resulting from the truncation of n/m. Thus, $\text{mod}(44, 7)$ is 2, $\text{mod}(3, 11)$ is 3, and $\text{mod}(n, m)$ is 0 whenever m divides n evenly. On the other hand, $x \equiv y$ modulo (z) means that $x - y$ is divisible by z.

Two other random-number routines that can be easily programmed follow:

1. Let the seed x_0 be any number in the interval $(0, 1)$. For $i = 1, 2, \ldots$, let x_i be the fractional part of $(\pi + x_{i-1})^5$.

2. Let $u_0 = 1$. For $i = 1, 2, \ldots$, let $u_i \equiv (8t - 3)u_{i-1}$ modulo (28) and $x_i = u_i/2^i$. Here t can be any large integer. In computing u_i, we retain only q binary bits in the indicated product. This algorithm is suitable for a binary machine with word length q.

A few words of caution about random-number generators in computing systems are needed. The fact that the sequences produced by these programs are not truly random has already been noted. In some simulations, the failure of randomness can lead to erroneous conclusions. Here are three specific points and examples to remember:

1. The algorithms of the type illustrated here by *random* and those above produce **periodic** sequences; that is, the sequences eventually repeat themselves. The period is of the order 2^{30} for *random*, which is quite large.

2. If a random-number generator is used to produce random points in n-dimensional space, these points lie on a relatively small number of planes or hyperplanes. As Marsaglia [1968] reports, points obtained in this way in three-space lie on a set of only 119086 planes for computers with integer storage of 48 bits. In ten-space they lie on a set of 126 planes.

3. The individual digits that make up random numbers generated by routines such as *random* are not, in general, independent random digits. For example, it might happen that the digit 3 follows the digit 5 more (or less) often than would be expected.

Examples

An example of a pseudocode to compute and print ten random numbers using procedure *random* follows:

program *test_random*
integer parameter $n \leftarrow 10$
real array $(x_i)_n$
call *random* $((x_i))$
output (x_i)
end program *test_random*

The computer results from a typical run are as follows:

$$0.31852\,29, \ 0.53260\,59, \ 0.50676\,22, \ 0.15271\,48, \ 0.67687\,93,$$
$$0.31067\,89, \ 0.57963\,66, \ 0.95331\,68, \ 0.39584\,57, \ 0.97879\,35$$

Maple has a collection of random-number generators with various distributions. For example, the following code generates ten uniformly distributed random numbers.

```
with(stats):
x := RandUniform(0..1):
seq(x(),i=1..10);
```

As a coarse check on the random-number generator, let us compute a long sequence of random numbers and determine what proportion of them lie in the interval $(0, \frac{1}{2}]$. The computed answer should be approximately 50%. The results with different sequence lengths are tabulated. Here is the pseudocode to carry out this experiment:

program *coarse_check*
integer parameter $n \leftarrow 10000$
integer i, m
real *per*
real array $(r_i)_n$
$m \leftarrow 0$
call *random* $((r_i))$
for $i = 1$ **to** n **do**
  if $r_i \leq 1/2$ **then** $m \leftarrow m + 1$
  if $\mathrm{mod}(i, 1000) = 0$ **then**
     $per \leftarrow 100\,\mathrm{real}(m)/\mathrm{real}(n)$
     output i, per
  end if
end do
end program *coarse_check*

In this pseudocode, a sequence of 10000 random numbers is generated. Along the way, the current proportion of numbers less than $\frac{1}{2}$ is computed at the 1000th step and then at multiples of 1000. Some of the computer results of the experiment are 49.5, 50.2, 51.0, and 50.625.

The experiment described can also be interpreted as a computer simulation of the tossing of a coin. A single toss corresponds to the selection of a random number x in the interval $(0, 1)$. We arbitrarily associate heads with event $0 < x \leq \frac{1}{2}$ and tails with event $\frac{1}{2} < x < 1$. One thousand tosses of the coin correspond to 1000 choices of random numbers. The results show the proportion of heads that result from repeated tossing of the coin. Random integers can be used to simulate coin tossing as well.

Observe that (at least in this experiment) reasonable precision is attained with only a moderate number of random numbers (4000). Repeating the experiment 10000 times has only a marginal influence on the precision. Of course, theoretically, if the random numbers were truly random, the limiting value as the number of random numbers used increases without bound would be exactly 50%.

In this pseudocode and others in the chapter, all of the random numbers are generated initially, stored in an array, and used later in the program as needed. This is an efficient way to obtain these numbers because it minimizes the number of procedure calls but at the cost of storage space. If memory space is at a premium, the call to the random-number generator can be moved closer to its use (inside the loop(s)) so that it returns a single random-number with each call.

Now we consider some basic questions about generating random points in various geometric configurations. Assume that procedure *random* is used to obtain a random number r. First, if uniformly distributed random points are needed on some interval (a, b), the statement

$$x \leftarrow (b - a)r + a$$

accomplishes this. Second, the pseudocode

$$i \leftarrow \text{integer}((n + 1)r)$$

produces random integers in the set $\{0, 1, \ldots, n\}$. Third, for random integers from j to k $(j \leq k)$, use the assignment statement

$$i \leftarrow \text{integer}((k - j + 1)r + j)$$

Finally, the following statements can be used to obtain the first four digits in a random number:

integer parameter $n \leftarrow 4$
integer array $(m_i)_n$
integer i
real r, x

$\vdots$

```
call random( r )
for i = 1 to n do
    x ← 10r
    mᵢ ← integer(x)
    x ← x − real(mᵢ)
end do
output (mᵢ)
```

$$\vdots$$

Uses of Pseudocode *random*

We now illustrate both correct and incorrect uses of procedure *random* for producing uniformly distributed points.

Consider the problem of generating 1000 random points uniformly distributed inside the ellipse $x^2 + 4y^2 = 4$. One way to do so is to generate random points in the rectangle $-2 \leq x \leq 2$, $-1 \leq y \leq 1$, and discard those that do not lie in the ellipse (see Figure 11.2).

```
program ellipse
integer parameter n ← 1000, npts ← 2000
real array (xᵢ)ₙ, (yᵢ)ₙ, (rᵢⱼ)npts×2
integer i, j
real u, v
call random ( (rᵢⱼ) )
j ← 1
for i = 1 to npts do
    u ← 4rᵢ,₁ − 2
    v ← 2rᵢ,₂ − 1
    if u² + 4v² ≤ 4 then
        xⱼ ← u
        yⱼ ← v
```

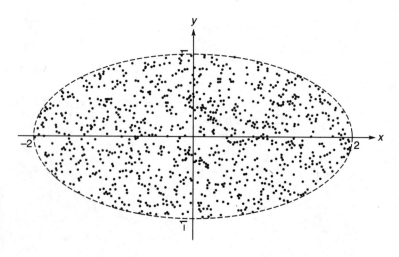

FIGURE 11.2
Uniformly distributed
random points in
ellipse $x^2 + 4y^2 = 4$

        if $j = n$ **then** exit loop
           $j \leftarrow j + 1$
      end if
    end do
    $\vdots$

    end program *ellipse*

    To be less wasteful, we can *force* the $|y|$ value to be less than $\frac{1}{2}\sqrt{4 - x^2}$, as in the following pseudocode, *which produces erroneous results* (see Figure 11.3):

    program *ellipse_erroneous*
    integer parameter $n \leftarrow 1000$
    real array$(x_i)_n, (y_i)_n, (r_{ij})_{n \times 2}$
    integer i
    call *random* $(\,(r_{ij})\,)$
    for $i = 1$ **to** n **do**
      $x_i \leftarrow 4r_{i,1} - 2$
      $y_i \leftarrow [(2r_{i,2} - 1)/2]\sqrt{4 - x_i^2}$
    end do
    $\vdots$

    end program *ellipse_erroneous*

    This pseudocode does *not* produce uniformly distributed points inside the ellipse. To be convinced of this, consider two vertical strips taken inside the ellipse (see Figure 11.4). If each strip is of width h, then approximately $1000(h/4)$ of the random points lie in each strip because the random variable x is uniformly distributed in $(-2, 2)$, and with each x a corresponding y is generated by the program so that (x, y) is inside the ellipse. But the two strips shown should *not* contain approximately the same number of points because they do not have the

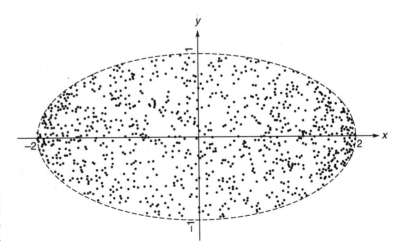

FIGURE 11.3
Nonuniformly distributed random points in the ellipse $x^2 + 4y^2 = 4$

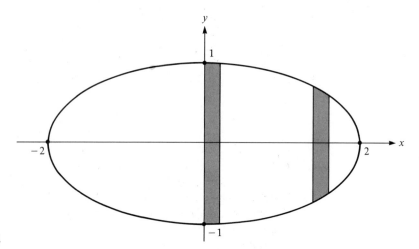

FIGURE 11.4

same area. The points generated by the second program tend to be clustered at the left and right extremities of the ellipse in Figure 11.3.

For the same reasons, the following pseudocode does *not* produce uniformly distributed random points in the circle $x^2 + y^2 = 1$ (see Figure 11.5):

```
program circle_erroneous
integer parameter n ← 1000
real array (xᵢ)ₙ, (yᵢ)ₙ, (rᵢⱼ)ₙ×₂
integer i
call random ( (rᵢⱼ) )
for i = 1 to n do
    xᵢ ← rᵢ,₁ cos(2πrᵢ,₂)
    yᵢ ← rᵢ,₁ sin(2πrᵢ,₂)
end do
    ⋮
end program circle_erroneous
```

In this pseudocode, $2\pi r_{i,2}$ is uniformly distributed in $(0, 2\pi)$ and $r_{i,1}$ is uniformly distributed in $(0, 1)$. However, in the transfer from polar to rectangular coordinates by the equations $x = r_{i,1} \cos(2\pi r_{i,2})$ and $y = r_{i,1} \sin(2\pi r_{i,2})$, the uniformity is lost. The random points are strongly clustered near the origin in Figure 11.5.

**PROBLEMS
11.1**

1. Taking the seed to be 123456, compute by hand the first three random numbers produced by procedure *random*.

2. Show that if the seed ℓ is less than or equal to 12777, then the first random number produced by procedure *random* is less than $\frac{1}{10}$.

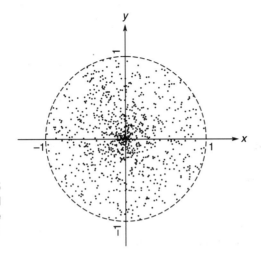

FIGURE 11.5
Nonuniformly distributed
random points in the
circle $x^2 + y^2 = 1$

3. Show that the numbers produced by procedure *random* are not random because their products with $2^{31} - 1$ are integers.

4. Describe in what ways this algorithm for random numbers differs from procedure *random*.

$$\begin{cases} x_0 \text{ arbitrary in } (0, 1) \\ x_i = \text{fractional part of } 7^5 x_{i-1} \quad (i \ge 1) \end{cases}$$

COMPUTER PROBLEMS 11.1

1. Write a program to generate 1000 random points uniformly distributed in the cardioid $r = 2 - \cos \theta$.

2. Using procedure *random*, write code for procedure *ranrec*(x, y), which generates a pseudorandom point (x, y) inside or on the trapezoid formed by the points $(1, 3)$, $(2, 5)$, $(4, 3)$, and $(3, 5)$.

3. Without using any procedures, write a program to generate and print 100 random numbers uniformly distributed in $(0, 1)$. Eight statements suffice.

4. Test the two algorithms in the text as random-number generators.

5. Test the random-number generator on your computer system in the following way: Generate 1000 random numbers $x_1, x_2, \ldots, x_{1000}$.

   a. In any small interval of width h, approximately $1000h$ of the x_i's should lie in that interval. Count the number of random numbers in each of ten intervals $[0, 1/n]$, where $n = 1, 2, \ldots, 10$.

   b. Inequality $x_i < x_{i+1}$ should occur approximately 500 times. Count them in your sample.

6. Write a procedure to generate with each call a random vector of the form $x = [x_1, x_2, \ldots, x_{20}]^T$, where each x_i is an integer from 1 to 100 and no two components of x are the same.

7. Write a program to generate $n = 1000$ random points uniformly distributed in the

   a. equilateral triangle in the figure below

   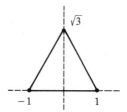

   b. diamond in the figure below

   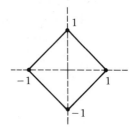

   Store the random points (x_i, y_i) in arrays $(x_i)_n$ and $(y_i)_n$.

8. If $x_1, x_2, \ldots$ is a random sequence of numbers uniformly distributed in the interval $(0, 1)$, what proportion would you expect to satisfy the inequality $40x^2 + 7 > 43x$? Write a program to test this on 1000 random numbers.

9. Write a program to generate and print 1000 points uniformly and randomly distributed in the circle $(x - 3)^2 + (y + 1)^2 \leq 9$.

10. Generate 1000 random numbers x_i according to a uniform distribution in the interval $(0, 1)$. Define a function f on $(0, 1)$ as follows: $f(t)$ is the number of random numbers $x_1, x_2, \ldots, x_{1000}$ less than t. Compute $f(t)/1000$ for 200 points t uniformly distributed in $(0, 1)$. What do you expect $f(t)/1000$ to be? Is this expectation borne out by the experiment? If a plotter is available, plot $f(t)/1000$.

11. Let n_i $(1 \leq i \leq 1000)$ be a sequence of integers that satisfies $0 \leq n_i \leq 9$. Write a program to test the given sequence for periodicity. (The sequence is **periodic** if there is an integer k such that $n_i = n_{i+k}$ for all i.)

12. Generate in the computer 1000 random numbers in the interval $(0, 1)$. Print and examine them for evidence of nonrandom behavior.

13. Generate 1000 random numbers x_i ($1 \leq i \leq 1000$) on your computer. Let n_i denote the eighth decimal digit in x_i. Count how many 0's, 1's, ..., 9's there are among the 1000 numbers n_i. How many of each would you expect? This code can be written with nine statements.

14. (Continuation) Using a random-number generator, generate 1000 random numbers and count how many times the digit i occurs in the jth decimal place. Print a table of these values—that is, frequency of digit versus decimal place. By examining the table, determine which decimal place seems to produce the best uniform distribution of random digits. *Hint*: Use the routine from Computer Problem **6** of Section 1.1 to compute the arithmetic mean, variance, and standard deviations of the table entries.

15. Using random integers, write a short program to simulate five persons matching coin flips. Print the percentage of matchups (five of a kind) after 125 flips.

16. Write a program to generate 1600 random points uniformly distributed in the sphere defined by $x^2 + y^2 + z^2 \leq 1$. Count the number of random points in the first octant.

17. Write a program to simulate 1000 flips of three coins. Print the number of times that two of the three coins come up heads.

18. Compute 1000 triples of random numbers drawn from a uniform distribution. For each triple (x, y, z), compute the leading significant digit of the product xyz. (The leading significant digit is one of $1, 2, \ldots, 9$.) Determine the frequencies with which the digits 1 through 9 occur among the 1000 cases. Try to account for the fact that these digits do not occur with the same frequency. (For example, 1 occurs approximately 7 times more often than 9.) If you are intrigued by this, you may wish to consult the articles by Flehinger [1966], Raimi [1969], and Turner [1982].

19. Run the example programs in this section and see whether similar results are obtained on your computer system.

20. Write a program to generate and plot 1000 pseudorandom points with the following *exponential* distribution inside the figure: $x = -\ln(1 - r)/\lambda$ for $r \in [0, 1)$ and $\lambda = \frac{1}{30}$.

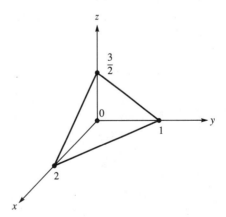

11.2 Estimation of Areas and Volumes by Monte Carlo Techniques

Numerical Integration

Now we turn to applications, the first being the approximation of a definite integral by the Monte Carlo method. If we select the first n elements $x_1, x_2, \ldots, x_n$ from a random sequence in the interval $(0, 1)$, then

$$\int_0^1 f(x)\,dx \approx \frac{1}{n} \sum_{i=1}^n f(x_i)$$

So the integral is approximated by the average of n numbers $f(x_1), f(x_2), \ldots, f(x_n)$. When this is actually carried out, the error is of order $1/\sqrt{n}$, which is not at all competitive with good algorithms, such as the Romberg method. However, in higher dimensions, the Monte Carlo method can be quite attractive. For example,

$$\int_0^1 \int_0^1 \int_0^1 f(x, y, z)\,dx\,dy\,dz \approx \frac{1}{n} \sum_{i=1}^n f(x_i, y_i, z_i)$$

where (x_i, y_i, z_i) is a random sequence of n points in the unit cube $0 \leq x \leq 1$, $0 \leq y \leq 1$, and $0 \leq z \leq 1$. To obtain random points in the cube, we assume that we have a random sequence in $(0, 1)$ denoted by $\xi_1, \xi_2, \xi_3, \xi_4, \xi_5, \xi_6, \ldots$. To get our first random point p_1 in the cube, just let $p_1 = (\xi_1, \xi_2, \xi_3)$. The second is, of course, $p_2 = (\xi_4, \xi_5, \xi_6)$, and so on.

If the interval (in a one-dimensional integral) is not of length 1, but, say, is the general case (a, b), then the average of f over n random points in (a, b) is not simply an approximation for the integral but rather for

$$\frac{1}{b - a} \int_a^b f(x)\,dx$$

which agrees with our intention that the function $f(x) = 1$ have an average of 1. Similarly, in higher dimensions, the average of f over a region is obtained by integrating and dividing by the area, volume, or measure of that region. For instance,

$$\frac{1}{8} \int_1^3 \int_{-1}^1 \int_0^2 f(x, y, z)\,dx\,dy\,dz$$

is the average of f over the parallelepiped described by the following three inequalities: $0 \le x \le 2, -1 \le y \le 1, 1 \le z \le 3$.*

So if (x_i, y_i) denote random points with appropriate uniform distribution, the following examples illustrate Monte Carlo techniques:

$$\int_0^5 f(x)\, dx \approx \frac{5}{n} \sum_{i=1}^n f(x_i)$$

$$\int_2^5 \int_1^6 f(x, y)\, dx\, dy \approx \frac{15}{n} \sum_{i=1}^n f(x_i, y_i)$$

In each case, the random points should be uniformly distributed in the regions involved.

In general, we have

$$\int_A f \approx (\text{measure of } A) \times (\text{average of } f \text{ over } n \text{ random points in } A)$$

Here we are using the fact that the average of a function on a set is equal to the integral of the function over the set divided by the measure of the set.

Example and Pseudocode

Let us consider the problem of obtaining the numerical value of the integral

$$\iint_\Omega \sin \sqrt{\ln(x + y + 1)}\, dx\, dy = \iint_\Omega f(x, y)\, dx\, dy$$

over the disk in xy-space, defined by the inequality

$$\Omega = \left\{ (x, y) : \left(x - \frac{1}{2}\right)^2 + \left(y - \frac{1}{2}\right)^2 \le \frac{1}{4} \right\}$$

A sketch of this domain, with a surface above it, is shown in Figure 11.6. We proceed by generating random points in the square and discarding those that do not lie in the disk. We take $n = 1000$ points in the disk. If the points are $p_i = (x_i, y_i)$, then the integral is estimated to be

---

*To keep the limits of integration straight, recall that

$$\int_a^b \int_c^d f(x, y)\, dx\, dy = \int_a^b \left[\int_c^d f(x, y)\, dx \right] dy$$

and

$$\int_{a_1}^{a_2} \int_{b_1}^{b_2} \int_{c_1}^{c_2} f(x, y, z)\, dx\, dy\, dz = \int_{a_1}^{a_2} \left\{ \int_{b_1}^{b_2} \left[\int_{c_1}^{c_2} f(x, y, z)\, dx \right] dy \right\} dz$$

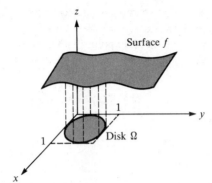

FIGURE 11.6

$$\iint_{\Omega} f(x, y) \, dx \, dy \approx (\text{area of disk } \Omega) \times (\text{average height of } f \text{ over } n \text{ random points})$$

$$= (\pi r^2) \left[\frac{1}{n} \sum_{i=1}^{n} f(p_i) \right]$$

$$= \frac{\pi}{4n} \sum_{i=1}^{n} f(p_i)$$

The pseudocode for this example follows. Intermediate estimates of the integral are printed when n is a multiple of 1000. This gives us some idea of how the correct value is being approached by our averaging process. We obtain an approximate numerical value of 0.57 for the integral.

```
program double_integral
integer parameter n ← 5000, iprt ← 1000
real array (r_ij)_n×2
integer i, j
real sum, vol, x, y
call random ( (r_ij) )
j ← 1; sum ← 0
for i = 1 to n do
    x = r_i,1; y = r_i,2
    if (x − 1/2)² + (y − 1/2)² ≤ 1/4 then
        j ← j + 1
        sum ← sum + f(x, y)
        if mod(j, iprt) = 0 then
            vol ← (π/4)sum/real(j)
            output j, vol
        end if
    end if
end do
vol ← (π/4)sum/real(j)
output j, vol
end program double_integral
```

```
real function f(x, y)
real x, y
f ← sin (√(ln(x + y + 1)))
end function
```

We obtain an approximate value of 0.57 for the integral.

Computing Volumes

The volume of a complicated region in 3-space can be computed by a Monte Carlo technique. Taking a simple case, let us determine the volume of the region whose points satisfy the inequalities

$$\begin{cases} 0 \leq x \leq 1 \quad 0 \leq y \leq 1 \quad 0 \leq z \leq 1 \\ x^2 + \sin y \leq z \\ x - z + e^y \leq 1 \end{cases}$$

The first line defines a cube whose volume is 1. The region defined by *all* the given inequalities is, therefore, a subset of this cube. If we generate n random points in the cube and determine that m of them satisfy the last two inequalities, then the volume of the desired region is approximately m/n. Here is a pseudocode that carries out this procedure:

```
program volume_region
integer parameter n ← 5000, iprt ← 1000
real array (r_{ij})_{n×3}
integer i, m
real vol, x, y, z
call random ( (r_{ij}) )
for i = 1 to n do
    x ← r_{i,1}
    y ← r_{i,2}
    z ← r_{i,3}
    if x² + sin y ≤ z, x − z + e^y ≤ 1 then m ← m + 1
    if mod(i, iprt) = 0 then
        vol ← real(m)/real(i)
        output i, vol
    end if
end do
end program volume_region
```

Observe that intermediate estimates are printed out when we reach 1000, 2000, . . . , 5000 points. An approximate value of 0.14 is determined for the volume of the region.

Ice Cream Cone Example

Consider the problem of finding the volume above the cone $z^2 = x^2 + y^2$ and inside the sphere $x^2 + y^2 + (z - 1)^2 = 1$ as shown in Figure 11.7. Clearly, the volume is contained in the box bounded by $-1 \le x \le 1$, $-1 \le y \le 1$, and $0 \le z \le 2$, which has volume 8. Thus, we want to generate random points inside this box and multiply by 8 the ratio of those inside the desired volume to the total number generated. A pseudocode for doing this follows:

```
program cone
integer parameter n ← 5000, iprt ← 1000
real array (r_ij)_n×3
integer i, m
real vol, x, y, z
m ← 0
call random ( (r_ij) )
for i = 1 to n do
    x ← 2r_{i,1} − 1; y ← 2r_{i,2} − 1; z ← 2r_{i,3}
    if x² + y² ≤ z², x² + y² ≤ z(2 − z) then m ← m + 1
    if mod(i, iprt) = 0 then
        vol ← 8 real(m)/real(i)
        output i, vol
    end if
end do
end program cone
```

The volume of the cone is approximately 3.3.

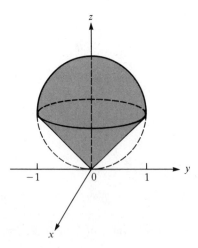

FIGURE 11.7
Ice cream cone region

COMPUTER PROBLEMS 11.2

1. Run the codes given in this section on your computer system and verify that they produce reasonable answers.

2. Write and test a program to evaluate the integral $\int_0^1 e^x \, dx$ by the Monte Carlo method, using $n = 25, 50, 100, 200, 400, 800, 16000$, and 32000. Observe that 32000 random numbers are needed and that the work in each case can be used in the next case. Print the exact answer.

3. Write a program to verify numerically that $\pi = \int_0^2 (4 - x^2)^{1/2} \, dx$. Use the Monte Carlo method and 2500 random numbers.

4. Use the Monte Carlo method to approximate the integral

$$\int_{-1}^{1} \int_{-1}^{1} \int_{-1}^{1} (x^2 + y^2 + z^2) \, dx \, dy \, dz$$

Compare with the correct answer.

5. Write a program to estimate

$$\int_{0}^{2} \int_{3}^{6} \int_{-1}^{1} (yx^2 + z \log y + e^x) \, dx \, dy \, dz$$

6. Using the Monte Carlo technique, write a pseudocode to approximate the integral

$$\iiint_{\Omega} (e^x \sin y \log z) \, dx \, dy \, dz$$

where Ω is the circular cylinder that has height 3 and circular base $x^2 + y^2 \leq 4$.

7. Estimate the area under the curve $y = e^{-(x+1)^2}$ and inside the triangle that has vertices $(1, 0)$, $(0, 1)$, and $(-1, 0)$ by writing and testing a short program.

8. Using the Monte Carlo approach, find the area of the irregular figure defined by

$$\begin{cases} 1 \leq x \leq 3 \quad (-1 \leq y \leq 4) \\ x^3 + y^3 \leq 29 \\ y \geq e^x - 2 \end{cases}$$

9. Use the Monte Carlo method to estimate the volume of the solid whose points (x, y, z) satisfy

$$\begin{cases} 0 \leq x \leq y \quad (1 \leq y \leq 2) \quad (-1 \leq z \leq 3) \\ e^x \leq y \\ (\sin z)y \geq 0 \end{cases}$$

10. Using a Monte Carlo technique, estimate the area of the region determined by the inequalities $0 \leq x \leq 1$, $10 \leq y \leq 13$, $y \geq 12 \cos x$, and $y \geq 10 + x^3$. Print intermediate answers.

11. Use the Monte Carlo method to approximate the following integrals.

    a. $\displaystyle\int_{-1}^{1}\int_{-1}^{1}\int_{-1}^{1} (x^2 - y^2 - z^2)\, dx\, dy\, dz$

    b. $\displaystyle\int_{1}^{4}\int_{2}^{5} (x^2 - y^2 + xy - 3)\, dx\, dy$

    c. $\displaystyle\int_{2}^{3}\int_{1+y}^{\sqrt{y}} (x^2 y + xy^2)\, dx\, dy$

    d. $\displaystyle\int_{0}^{1}\int_{y^2}^{\sqrt{y}}\int_{0}^{y+z} xy\, dx\, dy\, dz$

12. The value of the integral

$$\int_{0}^{\pi/4}\int_{0}^{2\cos\phi}\int_{0}^{2\pi} \rho^2 \sin\phi\, d\theta\, d\rho\, d\phi$$

using spherical coordinates is the volume above the cone $z^2 = x^2 + y^2$ and inside the sphere $x^2 + y^2 + (z - 1)^2 = 1$. Use the Monte Carlo method to approximate this integral and compare the results with that from the example in the text.

13. Let R denote the region in the xy-plane defined by the inequalities

$$\begin{cases} \dfrac{1}{3} \leq 3x \leq 9 - y \\ \sqrt{x} \leq y \leq 3 \end{cases}$$

Estimate the integral

$$\iint_{R} (e^x + \cos xy)\, dx\, dy$$

14. Using a Monte Carlo technique, estimate the area of the region defined by the inequalities $4x^2 + 9y^2 \leq 36$ and $y \leq \arctan(x + 1)$.

15. Write a program to estimate the area of the region defined by the inequalities

$$\begin{cases} x^2 + y^2 \leq 4 \\ |y| \leq e^x \end{cases}$$

16. An integral can be estimated by the formula

$$\int_{0}^{1} f(x)\, dx \approx \frac{1}{n}\sum_{i=1}^{n} f(x_i)$$

even if the x_i's are not random numbers; in fact, some nonrandom sequences may be better. Use the sequence $x_i = \left(\text{fractional part of } i\sqrt{2}\,\right)$ and test the corresponding numerical integration scheme. Test whether the estimates converge at the rate $1/n$ or $1/\sqrt{n}$ by using some simple examples, such as $\int_0^1 e^x\, dx$ and $\int_0^1 (1 + x^2)^{-1}\, dx$.

17. Consider the ellipsoid

$$\frac{x^2}{4} + \frac{y^2}{16} + \frac{z^2}{4} = 1$$

a. Write a program to generate and store 5000 random points uniformly distributed in the first octant of this ellipsoid.

b. Write a program to estimate the volume of this ellipsoid in the first octant.

18. A Monte Carlo method for estimating $\int_a^b f(x)\, dx$ if $f(x) \geq 0$ is as follows: Let $c \geq \max_{a \leq x \leq b} f(x)$. Then generate n random points (x, y) in the rectangle $a \leq x \leq b, 0 \leq y \leq c$. Count the number k of these random points (x, y) that satisfy $y \leq f(x)$. Then $\int_a^b f(x)\, dx \approx kc(b - a)/n$. Verify this and test the method on $\int_1^2 x^2\, dx$, $\int_0^1 (2x^2 - x + 1)\, dx$, and $\int_0^1 (x^2 + \sin 2x)\, dx$.

19. (Continuation) Use the method of Computer Problem **18** to estimate the value of $\pi = 4 \int_0^1 \sqrt{1 - x^2}\, dx$. Generate random points in $0 \leq x \leq 1, 0 \leq y \leq 1$. Use $n = 1000, 2000, \ldots, 10000$ and try to determine whether the error is behaving like $1/\sqrt{n}$.

20. (Continuation) Modify the method outlined in Computer Problem **19** to handle the case when f takes positive and negative values on $[a, b]$. Test the method on $\int_{-1}^1 x^3\, dx$.

21. Another Monte Carlo method for evaluating $\int_a^b f(x)\, dx$ is as follows: Generate an odd number of random numbers in (a, b). Reorder these points so that $a < x_1 < x_2 < \cdots < x_n < b$. Now compute

$$f(x_1)(x_2 - a) + f(x_3)(x_4 - x_2) + f(x_5)(x_6 - x_4) + \cdots + f(x_n)(b - x_{n-1})$$

Test this method on

$$\int_0^1 (1 + x^2)^{-1}\, dx \qquad \int_0^1 (1 - x^2)^{-1/2}\, dx \qquad \int_0^1 x^{-1} \sin x\, dx$$

22. What is the expected value of the volume of a tetrahedron formed by four points chosen randomly inside the tetrahedron whose vertices are $(0, 0, 0)$, $(0, 1, 0)$, $(0, 0, 1)$, and $(1, 0, 0)$? (The precise answer is unknown!)

23. Write a program to compute the area under the curve $y = \sin x$ and above the curve $y = \ln(x + 2)$. Use the Monte Carlo method and print intermediate results.

24. Estimate the integral

$$\int_{3.2}^{5.9} \left(\frac{e^{\sin x + x^2}}{\ln x} \right) dx$$

by the Monte Carlo method.

11.3 Simulation

We next illustrate the idea of **simulation**. We consider a physical situation in which an element of chance is present and try to imitate the situation on the computer. Statistical conclusions can be drawn if the experiment is performed many times.

Loaded Die Problem

In simulation problems, we must often produce random variables with a prescribed distribution. Suppose, for example, that we want to simulate the throw of a loaded die and that the probabilities of various outcomes have been determined as shown:

Outcome	1	2	3	4	5	6
Probability	0.2	0.14	0.22	0.16	0.17	0.11

If the random variable x is uniformly distributed in the interval $(0, 1)$, then by breaking this interval into six subintervals of lengths given by the table, we can simulate the throw of this loaded die. For example, we agree that if x is in $(0, 0.2)$, the die shows 1; if x is in $[0.2, 0.34)$, the die shows 2, and so on. A pseudocode to count the outcome of 5000 throws of this die might be written as follows:

```
program loaded_die
real parameter n ← 5000
real array (y_i)_6, (m_i)_6, (r_i)_n
integer i, j
(y_i)_6 ← (0.2, 0.34, 0.56, 0.72, 0.89, 1.0)
(m_i)_6 ← (0.0, 0.0, 0.0, 0.0, 0.0, 0.0)
call random ( (r_i) )
for i = 1 to n do
    for j = 1 to 6 do
        if r_i < y_j then
            m_j ← m_j + 1
            exit loop j
        end if
    end do
end do
output (m_i)
end program loaded_die
```

The computer output is $1012, 672, 1126, 800, 867$, and 523.

Birthday Problem

An interesting problem that can be solved using simulation is the famous **birthday problem**. Suppose that in a room of n persons, each of the 365 days of the year is equally likely to be someone's birthday. From probability theory, it can be shown that, contrary to intuition, only 23 persons need be present for the chances to be better than fifty-fifty that at least two of them will have the same birthday! (It is always fun to try this experiment at a large party or in class to see it work in practice.)

Many people are curious about the theoretical reasoning behind this result, so we discuss it briefly before solving the simulation problem. After someone is asked his or her birthday, the chances that the next person asked will not have the same birthday are 364/365. The chances that the third person's birthday will not match those of the first two people are 363/365. The chances of two successive independent events occurring is the product of the probability of the separate events. (The sequential nature of the explanation does not imply that the events are dependent.) In general, the probability that the nth person asked will have a birthday different from that of anyone already asked is

$$\left(\frac{365}{365}\right)\left(\frac{364}{365}\right)\left(\frac{363}{365}\right)\cdots\left(\frac{365-(n-1)}{365}\right)$$

The probability that the nth person asked will provide a match is 1 minus this value. A table of the quantity $1 - (365)(364)\cdots[365 - (n-1)]/365^n$ shows that with 23 persons the chances are 50.7%; with 55 or more persons, the chances are 98.6% or almost theoretically certain that at least two out of 55 people will have the same birthday. (See Table 11.1.)

Without using probability theory, we can write a routine that uses the random-number generator to compute the approximate chances for groups of n persons. Clearly, all that is needed is to select n random integers from the set $\{1, 2, 3, \ldots, 365\}$ and to examine them in some way to determine whether there is a match. By repeating this experiment a large number of times, we can compute the probability of at least one match in any gathering of n persons.

TABLE 11.1 Birthday problem

n	Theoretical	Simulation
5	0.027	0.028
10	0.117	0.110
15	0.253	0.255
20	0.411	0.412
22	0.476	0.462
23	0.507	0.520
25	0.569	0.553
30	0.706	0.692
35	0.814	0.819
40	0.891	0.885
45	0.941	0.936
50	0.970	0.977
55	0.986	0.987

One way of writing a routine for simulating the birthday problem follows. In it we use the approach of checking off days on a calendar to find out whether there is a match. Of course, there are many other ways of approaching this problem.

Function procedure *probably* calculates the probability of repeated birthdays.

real function *probably*(*n*, *npts*)
integer *i*, *npts*
logical *birthday*
real *sum*
sum ← 0
for *i* = 1 **to** *npts* **do**
    if *birthday* (*n*) **then** *sum* ← *sum* + 1
end do
probably ← *sum*/real(*npts*)
end function *probably*

Logical function *birthday* generates *n* random numbers and compares them. It returns a value of `true` if these numbers contain at least one repetition and `false` if all *n* numbers are different.

logical function *birthday* (*n*)
real array $(r_i)_n$
logical array $(days_i)_{365}$
integer *i*, *n*, *number*
call *random* ((r_i))
for *i* = 1 **to** 365 **do**
    days(*i*) ← `false`
end do
birthday ← `false`
for *i* = 1 **to** *n* **do**
    number ← integer($365r_i$ + 1)
    if *days*(*number*) **then**
        birthday ← `true`
        exit loop *i*
    end if
    days(*number*) ← `true`
end do
end function *birthday*

The results of the theoretical calculations and the simulation are given in Table 11.1.

Buffon's Needle Problem

The next example of a simulation is a very old problem known as **Buffon's needle problem**. Imagine that a needle of unit length is dropped onto a sheet of paper ruled by parallel lines 1 unit apart. What is the probability that the needle intersects one of the lines?

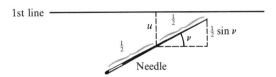

1st line

FIGURE 11.8
Buffon's needle problem   2nd line

To make the problem precise, assume that the center of the needle lands between the lines at a random point. Assume further that the angular orientation of the needle is another random variable. Finally, assume that our random variables are drawn from a uniform distribution. Figure 11.8 shows the geometry of the situation.

Let the distance of the center of the needle from the nearest of the two lines be u and the angle from the horizontal be v. Here u and v are the two random variables. The needle intersects one of the lines if and only if $u \leq \frac{1}{2} \sin v$. We perform the experiment many times, say 5000. Because of the problem's symmetry, we select u from a uniform random distribution on the interval $(0, \frac{1}{2})$ and v from a uniform random distribution on the interval $(0, \pi/2)$, and we determine the number of times that $2u \leq \sin v$. We let $w = 2u$ and test $w \leq \sin v$, where w is a random variable in $(0, 1)$. In this program, intermediate answers are printed out so that their progression can be observed. Also, the theoretical answer, $t = 2/\pi \approx 0.63662$, is printed for comparison.

```
program needle
integer parameter n ← 5000, iprt ← 1000
real array (r_ij)_{n×2}
integer i, m
real prob, v, w
m ← 0
call random ( (r_ij) )
for i = 1 to n do
    w ← r_{i1}
    v ← (π/2)r_{i,2}
    if w ≤ sin v then m ← m + 1
    if mod(i, iprt) = 0 then
        prob ← real(m)/real(i)
        output i, prob, (2/π)
    end if
end do
end program needle
```

Two Dice Problem

Our next example again has an analytic solution. This is advantageous for us because we wish to compare the results of Monte Carlo simulations with theoretical solutions. Consider the experiment of tossing two dice. For an (unloaded) die, the numbers 1, 2, 3, 4, 5, and 6 are equally likely to occur. We ask, *What is the probability of throwing a 12 (i.e., 6 appearing on each die) in 24 throws of the dice?*

There are six possible outcomes from each die for a total of 36 possible combinations. Only one of these combinations is a double 6, so 35 out of the 36 combinations are not correct. With 24 throws, we have $(35/36)^{24}$ as the probability of a wrong outcome. Hence, $1 - (35/36)^{24} = 0.49140$ is the answer. Not all problems of this type can be analyzed like this, so we model the situation using a random-number generator.

If we simulate this process, a single experiment consists of throwing the dice 24 times, and this experiment must be repeated a large number of times, say 1000. For the outcome of the throw of a single die, we need random integers uniformly distributed in the set $\{1, 2, 3, 4, 5, 6\}$. If x is a random variable in $(0, 1)$, then $6x + 1$ is a random variable in $(1, 7)$ and the integer part is a random integer in $\{1, 2, 3, 4, 5, 6\}$. Here is a pseudocode:

```
program two_dice
integer parameter n ← 5000, iprt ← 1000
real array (r_{ijk})_{n×24×2}
integer i, j, i_1, i_2, m
real prob
call random ( (r_{ijk}) )
m ← 0
for i = 1 to n do
   for j = 1 to 24 do
      i_1 ← integer(6r_{ij1} + 1)
      i_2 ← integer(6r_{ij2} + 1)
      if i_1 + i_2 = 12 then
         m ← m + 1
         exit loop j
      end if
   end do
   if mod(i, 1000) = 0 then
      prob ← real(m)/real(i)
      output i, prob
   end if
end do
end program two_dice
```

This program computes the probability of throwing a 12 in 24 throws of the dice at approximately *even money*—that is, 0.487.

Neutron Shielding

Our final example concerns neutron shielding. We take a simple model of neutrons penetrating a lead wall. It is assumed that each neutron enters the lead wall at a right angle to the wall and travels a unit distance. Then it collides with a lead atom and

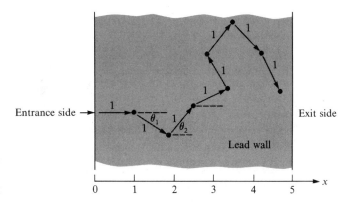

FIGURE 11.9
Neutron shielding
experiment

rebounds in a random direction. Again, it travels a unit distance before colliding with another lead atom. It rebounds in a random direction and so on. Assume that after eight collisions, all the neutron's energy is spent. Assume also that the lead wall is 5 units thick in the x direction and for all practical purposes infinitely thick in the y direction. The question is: *What percentage of neutrons can be expected to emerge from the other side of the lead wall?* (See Figure 11.9.)

Let x be the distance measured from the initial surface where the neutron enters. From trigonometry, we recall that in a right triangle with hypotenuse 1, one side is $\cos \theta$. Also note that $\cos \theta \leq 0$ when $\pi/2 \leq \theta \leq \pi$ (see Figure 11.10). The first collision occurs at a point where $x = 1$. The second occurs at a point where $x = 1 + \cos \theta_1$. The third collision occurs at a point where $x = 1 + \cos \theta_1 + \cos \theta_2$, and so on. If $x \geq 5$, the neutron has exited. If $x < 5$ for all eight collisions, the wall has shielded the area from that particular neutron. For a Monte Carlo simulation, we can use random angles θ_i in the interval $(0, \pi)$ because of symmetry. The simulation program then follows:

program *shielding*
integer parameter $n \leftarrow 5000$, *iprt* $\leftarrow 1000$
real array $(r_{ij})_{n \times 7}$
integer i, j, m
real x, per
$m \leftarrow 0$
call *random* ((r_{ij}))
for $i = 1$ **to** n **do**
  $x \leftarrow 1$
  for j = 1 **to** 7 **do**
    $x \leftarrow x + \cos(\pi r_{ij})$
    if $x \leq 0$ **then** exit loop j
    if $x \geq 5$ **then**
      $m \leftarrow m + 1$
      exit loop j
    end if
  end do

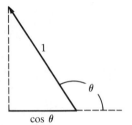

FIGURE 11.10 $\cos \theta$ $\cos \theta$

if mod($i, iprt$) = 0 **then**
    $per \leftarrow 100\, \text{real}(m)/\text{real}(i)$
    output i, per
end if
end do
end program *shielding*

After running this program, we can say that approximately 1.85% of the neutrons can be expected to emerge from the lead wall.

COMPUTER PROBLEMS 11.3

1. A point (a, b) is chosen at random in a rectangle defined by inequalities $|a| \leq 1$ and $|b| \leq 2$. What is the probability that the resulting quadratic equation $ax^2 + bx + 1 = 0$ has *real* roots? Find the answer both analytically and by the Monte Carlo method.

2. Compute the average distance between two points in the circle $x^2 + y^2 = 1$. To solve this, generate N random pairs of points (x_i, y_i) and (v_i, w_i) in the circle and compute

$$N^{-1} \sum_{i=1}^{N} \left[(x_i - v_i)^2 + (y_i - w_i)^2 \right]^{1/2}$$

3. (The French railroad system) Define the distance between two points (x_1, y_1) and (x_2, y_2) in the plane to be $\sqrt{(x_1 - x_2)^2 + (y_1 - y_2)^2}$ if the points are on a straight line through the origin but $\sqrt{x_1^2 + y_1^2} + \sqrt{x_2^2 + y_2^2}$ in all other cases. Draw a picture to illustrate. Compute the average distance between two points randomly selected in a unit circle centered at the origin.

4. Consider a circle of radius 1. A point is chosen at random inside the circle, and a chord that has the chosen point as midpoint is drawn. What is the probability that the chord will have length greater than $\frac{3}{2}$? Solve the problem analytically and by the Monte Carlo method.

5. Two points are selected at random on the circumference of a circle. What is the average distance from the center of the circle to the center of gravity of the two points?

6. Consider the cardioid given by $(x^2 + y^2 + x)^2 = (x^2 + y^2)$. Write a program to find the average distance, *staying within the cardioid*, between two points randomly selected within the figure. Use 1000 points and print intermediate estimates.

7. Find the length of the lemniscate whose equation in polar coordinates is given by $r^2 = \cos 2\theta$. *Hint*: In polar coordinates, $ds^2 = dr^2 + r^2 d\theta^2$.

8. Suppose that a die is loaded so that the six faces are not equally likely to turn up when the die is rolled. The probabilities associated with the six faces are as follows:

Outcome	1	2	3	4	5	6
Probability	0.15	0.2	0.25	0.15	0.1	0.15

Write and run a program to simulate 1500 throws of such a die.

9. Consider a pair of loaded dice as described in the text. By a Monte Carlo simulation, determine the probability of throwing a 12 in 25 throws of the dice.

10. Consider a neutron-shielding problem similar to the one in the text but modified as follows: Imagine the neutron beam impinging on the wall 1 unit above its base. The wall can be very high. Neutrons cannot escape from the top, but they can escape from the bottom as well as from the exit side. Find the percentage of escaping neutrons.

11. Rewrite the routine(s) for the birthday problem using some other scheme for determining whether or not there is a match.

12. Write a program to estimate the probability that three random points on the edges of a square form an obtuse triangle (see the figure). *Hint*: Use the Law of Cosines: $\cos \theta = (b^2 + c^2 - a^2)/2bc$.

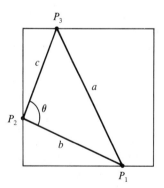

13. A **histogram** is a graphical device for displaying frequencies by means of rectangles whose heights are proportional to frequencies. For example, in throwing two dice 3600 times, the resulting sums $2, 3, \ldots, 12$ should occur with frequencies close to those shown in the histogram below. By means of a Monte Carlo simulation, obtain a histogram for the frequency of digits $0, 1, \ldots, 9$ that appear in 1000 random numbers.

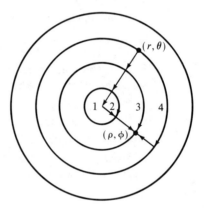

14. Consider a circular city of diameter 20 kilometers (see the figure below). Radiating from the center are 36 straight roads, spaced 10° apart in angle. There are also 20 circular roads spaced 1 kilometer apart. What is the average distance, measured along the roads, between road intersection points in the city?

15. A particle breaks off from a random point on a rotating flywheel. Referring to the figure below, determine the probability of the particle hitting the window. Perform a Monte Carlo simulation to compute the probability in an experimental way.

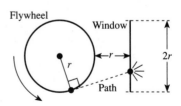

16. Write a program to simulate the following phenomenon: A particle is moving in the xy-plane under the effect of a random force. It starts at $(0,0)$. At the end of each second, it moves 1 unit in a random direction. We want to record in a table its position at the end of each second, taking altogether 1000 seconds.

17. (A random walk) On a windy night, a drunkard begins walking at the origin of a two-dimensional coordinate system. His steps are 1 unit in length and are random in the following way: With probability $\frac{1}{6}$ he takes a step east, with probability $\frac{1}{4}$ he takes a step north, with probability $\frac{1}{4}$ he takes a step south, and with probability $\frac{1}{3}$ he takes a step west. What is the probability that after 50 steps he will be more than 20 units distant from the origin? Write a program to simulate this problem.

18. (Another random walk) Consider the lattice points (points with integer coordinates) in the square $0 \leq x \leq 6, 0 \leq y \leq 6$. A particle starts at the point $(4, 4)$ and moves in the following way: At each step it moves with equal probability to one of the four adjacent lattice points. What is the probability that when the particle first crosses the boundary of the square, it crosses the bottom side? Use Monte Carlo simulation.

19. What is the probability that within 20 generations the Kzovck family name will die out? Use the following data: In the first generation there is only one male Kzovck. In each succeeding generation the probability that a male Kzovck will have exactly one male offspring is $\frac{4}{11}$, the probability that he will have exactly two is $\frac{1}{11}$, and the probability that he will have more than two is 0.

20. Write a program that simulates the random shuffle of a deck of 52 cards.

21. A merry-go-round with a total of 24 horses allows children to jump on at three gates and jump off at only one gate while it continues to turn slowly. If the children get on and off randomly (at most one per gate), how many revolutions go by before someone must wait longer than one revolution to ride? Assume a probability of $\frac{1}{2}$ that a child gets on or off.

22. Run the programs given in this section and determine whether the results are reasonable.

23. In the unit cube $\{(x, y, z) : 0 \leq x \leq 1, 0 \leq y \leq 1, 0 \leq z \leq 1\}$, if two points are randomly chosen, then what is the expected distance between them?

24. The lattice points in the plane are defined as those points whose coordinates are integers. A circle of diameter 1.5 is dropped on the plane in such a way that its center is a uniformly distributed random point in the square $0 \leq x \leq 1$, $0 \leq y \leq 1$. What is the probability that two or more lattice points lie inside the circle? Use the Monte Carlo simulation to compute an approximate answer.

25. Write a program to simulate a traffic flow problem similar to the one in the example that begins this chapter.

26. Modify and rerun the programs in this section so that large arrays are not used.

12

BOUNDARY VALUE PROBLEMS FOR ORDINARY DIFFERENTIAL EQUATIONS

In the design of pivots and bearings, the mechanical engineer encounters the following problem: The cross section of a pivot is determined by a curve $y = y(x)$ that must pass through two fixed points, $(0, 1)$ and $(1, a)$, as in Figure 12.1. Moreover, for optimal performance (principally low friction), the unknown function must minimize the value of a certain integral

$$\int_0^1 [y(y')^2 + b(x)y^2]\, dx$$

in which $b(x)$ is a known function. From this, it is possible to obtain a second-order differential equation (the so-called Euler equation) for y. The differential equation with its initial and terminal values is

$$\begin{cases} -(y')^2 - 2b(x)y + 2yy'' = 0 \\ y(0) = 1 \qquad y(1) = a \end{cases}$$

This is a nonlinear two-point boundary value problem, and methods for solving it numerically are discussed in this chapter.

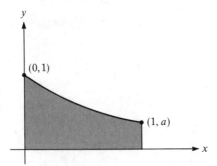

y

$(0, 1)$

$(1, a)$

x

FIGURE 12.1

In previous chapters, we dealt with the initial-value problem for ordinary differential equations, but now we consider another type of numerical problem involving ordinary differential equations. A **boundary value problem** is exemplified by a second-order ordinary differential equation whose solution function is prescribed at the endpoints of the interval of interest. An instance of such a problem is

$$\begin{cases} x'' = -x \\ x(0) = 1 \quad x\left(\dfrac{\pi}{2}\right) = -3 \end{cases}$$

Here we have a differential equation whose general solution involves two arbitrary parameters. To specify a particular solution, two conditions must be given. If this were an initial-value problem, x and x' would be specified at some initial value. In this problem, however, we are given two points of the form $(t, x(t))$ through which the solution curve passes—namely, $(0, 1)$ and $(\pi/2, -3)$. The general solution of the differential equation is $x(t) = c_1 \sin(t) + c_2 \cos(t)$, and the two conditions (known as **boundary values**) enable us to determine that $c_1 = -3$ and $c_2 = 1$.

To verify this explicit solution, we can use the following Maple statements

```
ode := (D@@2)(x)(t) = -x(t);
dsolve(ode, x(t));
```

12.1 Shooting Method

Now suppose that we have a similar problem in which we are unable to determine the general solution as above. We take as our model the problem

$$\begin{cases} x''(t) = f(t, x(t), x'(t)) \\ x(a) = \alpha \quad x(b) = \beta \end{cases} \tag{1}$$

A step-by-step numerical solution of Problem (1) by the methods of Chapter 9 requires two initial conditions, but in Problem (1) only one condition is present at $t = a$. This fact makes a problem like (1) considerably more difficult than an initial-value problem. Several ways to attack it are considered in this chapter. Existence and uniqueness theorems for solutions of two-point boundary value problems can be found in Keller [1976].

One way to proceed in solving Equation (1) is to guess $x'(a)$, then carry out the solution of the resulting initial-value problem as far as b, and pray that the computed solution agrees with β; that is, $x(b) = \beta$. If it does not (which is quite likely), we can go back and change our guess for $x'(a)$. Repeating this procedure until we hit the target β may be a good method if we can learn something from the various trials. There are systematic ways of utilizing this information, and the resulting method is known by the nickname **shooting**.

We observe that the final value $x(b)$ of the solution of our initial-value problem depends on the guess that was made for $x'(a)$. Everything else remains fixed in this problem. Thus, the differential equation $x'' = f(t, x, x')$ and the first initial value, $x(a) = \alpha$, do not change. If we assign a real value z to the missing initial condition,

$$x'(a) = z$$

then the initial-value problem can be solved numerically. The value of x at b is now a function of z, which we denote by $\phi(z)$. In other words, for each choice of z, we obtain a new value for $x(b)$ and ϕ is the name of the function with this behavior. We know very little about $\phi(z)$, but we can compute or evaluate it. It is, however, an *expensive* function to evaluate because each value of $\phi(z)$ is obtained only after solving an initial-value problem.

Shooting Method Algorithm

To summarize, a function $\phi(z)$ is computed as follows: Solve the initial-value problem

$$\begin{cases} x'' = f(t, x(t), x'(t)) \\ x(a) = \alpha \qquad x'(a) = z \end{cases}$$

on the interval $[a, b]$. Let

$$\phi(z) = x(b)$$

Our objective is to adjust z until we find a value for which

$$\phi(z) = \beta$$

One way to do so is to use linear interpolation between $\phi(z_1)$ and $\phi(z_2)$, where z_1 and z_2 are two guesses for the initial condition $x'(a)$. That is, given two values of ϕ, we pretend that ϕ is a linear function and determine an appropriate value of z based on this hypothesis. A sketch of the values of z versus $\phi(z)$ might look like Figure 12.2. The strategy just outlined is the secant method for finding a zero of $\phi(z) - \beta$.

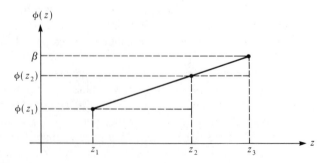

FIGURE 12.2

To obtain an estimating formula for the next value z_3, we compute $\phi(z_1)$ and $\phi(z_2)$ based on values z_1 and z_2, respectively. By considering similar triangles, we have

$$\frac{z_3 - z_2}{\beta - \phi(z_2)} = \frac{z_2 - z_1}{\phi(z_2) - \phi(z_1)}$$

from which

$$z_3 = z_2 + [\beta - \phi(z_2)]\left[\frac{z_2 - z_1}{\phi(z_2) - \phi(z_1)}\right]$$

We can repeat this process and generate the sequence

$$z_{n+1} = z_n + [\beta - \phi(z_n)]\left[\frac{z_n - z_{n-1}}{\phi(z_n) - \phi(z_{n-1})}\right] \qquad (n \geq 2) \qquad \text{(2)}$$

all based on two starting values z_1 and z_2.

This procedure for solving the two-point boundary value problem

$$\begin{cases} x'' = f(t, x, x') \\ x(a) = \alpha \qquad x(b) = \beta \end{cases} \qquad \text{(3)}$$

is then as follows: Solve the initial-value problem

$$\begin{cases} x'' = f(t, x, x') \\ x(a) = \alpha \qquad x'(a) = z \end{cases} \qquad \text{(4)}$$

from $t = a$ to $t = b$, letting the value of the solution at b be denoted by $\phi(z)$. Do this twice with two different values of z, say z_1 and z_2, and compute $\phi(z_1)$ and $\phi(z_2)$. Now calculate a new z, called z_3, by Formula (2). Then compute $\phi(z_3)$ by again solving (4). Obtain z_4 from z_2 and z_3 in the same way, and so on. Monitor

$$\phi(z_{n+1}) - \beta$$

to see whether progress is being made. When it is satisfactorily small, stop. Note that the numerically obtained values $x(t_i)$ for $a \leq t_i \leq b$ must be saved until better ones are obtained (that is, one whose terminal value $x(b)$ is closer to β than the present one) because the objective in solving Problem (3) is to obtain values $x(t)$ for values of t between a and b.

The shooting method may be very time-consuming if each solution of the associated initial-value problem involves a small value for the step size h. Consequently, we use a relatively large value of h until $|\phi(z_{n+1}) - \beta|$ is sufficiently small and then reduce h to obtain the required accuracy.

Modifications and Refinements

Many modifications and refinements are possible. For instance, when $\phi(z_{n+1})$ is near β, one can use higher-order interpolation formulas to estimate successive values of z_i. Suppose, for example, that instead of utilizing two values $\phi(z_1)$ and $\phi(z_2)$ to obtain z_3, we utilize four values:

$$\phi(z_1) \qquad \phi(z_2) \qquad \phi(z_3) \qquad \phi(z_4)$$

to estimate z_5. We could set up a cubic interpolating polynomial p_3 for the data

$$
\begin{array}{c|c|c|c}
z_1 & z_2 & z_3 & z_4 \\
\hline
\phi(z_1) & \phi(z_2) & \phi(z_3) & \phi(z_4)
\end{array}
\tag{5}
$$

and solve

$$p_3(z_5) = \beta$$

for z_5. Since p_3 is a cubic, this would entail some additional work. A better way is to set up a polynomial $\widehat{p}_3$ to interpolate the data

$$
\begin{array}{c|c|c|c}
\phi(z_1) & \phi(z_2) & \phi(z_3) & \phi(z_4) \\
\hline
z_1 & z_2 & z_3 & z_4
\end{array}
\tag{6}
$$

and then use $\widehat{p}_3(\beta)$ as the estimate for z_5. This procedure is known as **inverse interpolation**. (See Section 4.1.)

Further remarks on the shooting method will be made in the next section after the discussion of an alternative procedure.

PROBLEMS 12.1

1. Verify that $x = (2t + 1)e^t$ is the solution to each of the following problems.

$$\begin{cases} x'' = x + 4e^t \\ x(0) = 1 \qquad x\left(\tfrac{1}{2}\right) = 2e^{1/2} \end{cases} \qquad \begin{cases} x'' = x' + x - (2t - 1)e^t \\ x(1) = 3e \qquad x(2) = 5e^2 \end{cases}$$

2. Verify that $x = c_1 e^t + c_2 e^{-t}$ solves the boundary value problem

$$\begin{cases} x'' = x \\ x(0) = 1 \qquad x(1) = 2 \end{cases}$$

if appropriate values of c_1 and c_2 are chosen.

3. Solve these boundary value problems by adjusting the general solution of the differential equation.

a. $x'' = x$ $x(0) = 0$ $x(\pi) = 1$
b. $x'' = t^2$ $x(0) = 1$ $x(1) = -1$

4. a. Determine all pairs (α, β) for which the problem

$$x'' = -x \qquad x(0) = \alpha \qquad x\left(\frac{\pi}{2}\right) = \beta$$

has a solution.

b. Repeat part **a** for $x(0) = \alpha$ and $x(\pi) = \beta$.

5. Verify the following algorithm for the inverse interpolation technique suggested in the text. Here we have set $\phi_i = \phi(z_i)$.

$$u = \frac{z_2 - z_1}{\phi_2 - \phi_1}$$

$$s = \frac{z_3 - z_2}{\phi_3 - \phi_2}$$

$$e = \frac{z_4 - z_3}{\phi_4 - \phi_3}$$

$$v = \frac{s - u}{\phi_3 - \phi_1}$$

$$r = \frac{e - s}{\phi_4 - \phi_2}$$

$$w = \frac{r - v}{\phi_4 - \phi_1}$$

$$z_5 = z_1 + (\beta - \phi_1)\{u + (\beta - \phi_2)[v + w(\beta - \phi_3)]\}$$

6. Let $\phi(z)$ denote $x\left(\frac{\pi}{2}\right)$, where x is the solution of the initial-value problem.

$$\begin{cases} x'' = -x \\ x(0) = 0 \qquad x'(0) = z \end{cases}$$

What is $\phi(z)$?

7. Determine the function ϕ explicitly in the case of this two-point boundary value problem.

$$\begin{cases} x'' = -x \\ x(0) = 1 \qquad x\left(\frac{\pi}{2}\right) = 3 \end{cases}$$

8. (Continuation) Repeat Problem 7 for $x'' = -(x')^2/x$ with $x(1) = 3$ and $x(2) = 5$. Using your result, solve the boundary value problem. *Hint:* The general solution of the differential equation is $x(t) = c_1\sqrt{c_2 + t}$.

9. Determine the function ϕ explicitly in the case of this two-point boundary value problem.

$$\begin{cases} x'' = x \\ x(-1) = e \qquad x'(1) = \dfrac{1}{2}e \end{cases}$$

10. Boundary value problems may involve differential equations of order higher than 2. For example,

$$\begin{cases} x''' = f(t, x, x', x'') \\ x(a) = \alpha \qquad x'(a) = \gamma \qquad x(b) = \beta \end{cases}$$

List the ways this problem can be solved using the shooting method.

11. Solve analytically this three-point boundary value problem.

$$\begin{cases} x''' = -e^t + 4(t + 1)^{-3} \\ x(0) = -1 \qquad x(1) = 3 - e + 2\ln 2 \qquad x(2) = 6 - e^2 + 2\ln 3 \end{cases}$$

12. Solve

$$\begin{cases} x'' = -x \\ x(0) = 2 \qquad x(\pi) = 3 \end{cases}$$

analytically and analyze any difficulties.

13. Show that the following two problems are equivalent in the sense that a solution of one is easily obtained from a solution of the other.

$$\begin{cases} y'' = f(t, y) \\ y(0) = \alpha \qquad y(1) = \beta \end{cases} \qquad \begin{cases} z'' = f(t, z + \alpha - \alpha t + \beta t) \\ z(0) = 0 \qquad z(1) = 0 \end{cases}$$

14. Discuss in general terms the numerical solution of the following two-point boundary value problems. Recommend specific methods for each, being sure to take advantage of any special structure.

a. $$\begin{cases} x'' = \sin t + (e^t\sqrt{t^2 + 1})x + (\cos t)x' \\ x(0) = 0 \qquad x(1) = 5 \end{cases}$$

b. $$\begin{cases} x_1' = x_1^2 + (t - 3)x_1 + \sin t \\ x_2' = x_2^3 + \sqrt{t^2 + 1} + (\cos t)x_1 \\ x_1(0) = 1 \qquad x_2(2) = 3 \end{cases}$$

15. What is $\phi(z)$ in the case of this boundary value problem?

$$\begin{cases} x'' = -x \\ x(0) = 1 \qquad x(\pi) = 3 \end{cases}$$

Explain the implications.

COMPUTER PROBLEMS 12.1

1. The nonlinear two-point boundary value problem

$$\begin{cases} x'' = e^x \\ x(0) = \alpha \qquad x(1) = \beta \end{cases}$$

has the closed-form solution

$$x = \ln c_1 - 2\ln\left\{\cos\left[\left(\frac{1}{2}c_1\right)^{1/2} t + c_2\right]\right\}$$

where c_1 and c_2 are the solutions of

$$\begin{cases} \alpha = \ln c_1 - 2\ln\cos c_2 \\ \beta = \ln c_1 - 2\ln\left\{\cos\left[\left(\frac{1}{2}c_1\right)^{1/2} + c_2\right]\right\} \end{cases}$$

Use the shooting method to solve this problem with $\alpha = \beta = \ln 8\pi^2$. Start with $z_1 = -25/2$ and $z_2 = -23/2$. Determine c_1 and c_2 so that a comparison with the true solution can be made. *Remark*: The corresponding discretization method, as discussed in the next section, involves a system of nonlinear equations with no closed-form solution.

2. Write a program to solve the example that begins this chapter for specific a and $b(x)$, such as $a = \frac{1}{4}$ and $b(x) = x^2$.

12.2 A Discretization Method
Finite Difference Approximations

We turn now to a completely different approach to solving the two-point boundary value problem—one based on a direct discretization of the differential equation. The problem that we want to solve is

$$\begin{cases} x'' = f(t, x, x') \\ x(a) = \alpha \qquad x(b) = \beta \end{cases} \tag{1}$$

Select a set of equally spaced points $t_0, t_1, \ldots, t_n$ on the interval $[a, b]$ by letting

$$t_i = a + ih \qquad \text{with} \qquad h = \frac{b - a}{n} \qquad (0 \leq i \leq n)$$

Next, approximate the derivatives, using the standard central difference Formulas (5) and (20) from Section 4.3:

$$x'(t) \approx \frac{1}{2h}[x(t + h) - x(t - h)] \tag{2}$$

$$x''(t) \approx \frac{1}{h^2}[x(t + h) - 2x(t) + x(t - h)]$$

The approximate value of $x(t_i)$ is denoted by x_i. Hence, the problem becomes

$$
\begin{cases}
x_0 = \alpha \\
\dfrac{1}{h^2}(x_{i-1} - 2x_i + x_{i+1}) = f(t_i, x_i, \dfrac{1}{2h}(x_{i+1} - x_{i-1})) \qquad (1 \leq i \leq n - 1) \quad \textbf{(3)} \\
x_n = \beta
\end{cases}
$$

This is usually a nonlinear system of equations in the $n - 1$ unknowns $x_1, x_2, \ldots, x_{n-1}$ because f generally involves the x_i's in a nonlinear way. The solution of such a system is seldom easy.

The Linear Case

In some cases, System (3) *is* linear. This situation occurs exactly when f in Equation (1) has the form

$$f(t, x, x') = u(t) + v(t)x + w(t)x' \tag{4}$$

In this special case, the principal equation in System (3) looks like this:

$$\frac{1}{h^2}(x_{i-1} - 2x_i + x_{i+1}) = u(t_i) + v(t_i)x_i + w(t_i)\left[\frac{1}{2h}(x_{i+1} - x_{i-1})\right]$$

or, equivalently,

$$-\left(1 + \frac{h}{2}w_i\right)x_{i-1} + (2 + h^2 v_i)x_i - \left(1 - \frac{h}{2}w_i\right)x_{i+1} = -h^2 u_i \tag{5}$$

where $u_i = u(t_i)$, $v_i = v(t_i)$, and $w_i = w(t_i)$. Now let

$$\begin{cases} a_i = -\left(1 + \dfrac{h}{2}w_i\right) \\[2mm] d_i = 2 + h^2 v_i \\[2mm] c_i = -\left(1 - \dfrac{h}{2}w_i\right) \\[2mm] b_i = -h^2 u_i \end{cases} \quad (0 \leq i \leq n)$$

Then the principal equation becomes

$$a_i x_{i-1} + d_i x_i + c_i x_{i+1} = b_i$$

The equations corresponding to $i = 1$ and $i = n - 1$ are different because we know x_0 and x_n. The system, therefore, can be written as

$$\begin{cases} d_1 x_1 \quad + c_1 x_2 \quad = b_1 - a_1 \alpha \\ a_i x_{i-1} + d_i x_i \quad + c_i x_{i+1} = b_i \\ a_{n-1} x_{n-2} + d_{n-1} x_{n-1} \quad = b_{n-1} - c_{n-1}\beta \end{cases} \quad (2 \leq i \leq n - 2) \quad \textbf{(6)}$$

In matrix form, System (6) looks like this:

$$\begin{bmatrix} d_1 & c_1 & & & & \\ a_2 & d_2 & c_2 & & & \\ & a_3 & d_3 & c_3 & & \\ & & \ddots & \ddots & \ddots & \\ & & & a_{n-2} & d_{n-2} & c_{n-2} \\ & & & & a_{n-1} & d_{n-1} \end{bmatrix} \begin{bmatrix} x_1 \\ x_2 \\ x_3 \\ \vdots \\ x_{n-2} \\ x_{n-1} \end{bmatrix} = \begin{bmatrix} b_1 - a_1 \alpha \\ b_2 \\ b_3 \\ \vdots \\ b_{n-2} \\ b_{n-1} - c_{n-1}\beta \end{bmatrix}$$

Since this system is tridiagonal, we can solve it with the special Gaussian procedure *tri* for tridiagonal systems developed in Section 6.3. (That procedure does not include pivoting, however, and may fail in cases where procedure *gauss* would succeed. See Problem **5**.)

Pseudocode and Numerical Example

The ideas just explained are now used to write a program for a specific test case. The problem is of the form (1) with f a linear function as in Equation (4):

$$\begin{cases} x'' = e^t - 3\sin(t) + x' - x \\ x(1) = 1.09737\,491 \qquad x(2) = 8.63749\,661 \end{cases} \quad \textbf{(7)}$$

The solution, known in advance to be $x(t) = e^t - 3\cos(t)$, can be used to check the computer solution. We use the discretization technique described earlier and procedure *tri* for solving the resulting linear system.

First, we decide to use 100 points, including endpoints $a = 1$ and $b = 2$. Thus, $n = 99$, $h = \frac{1}{99}$, and $t_i = 1 + ih$ for $0 \leq i \leq 99$. Then we have $t_0 = 1$, $x_0 = x_0(t_0) = 1.09737\,491$, $t_{99} = 2$, and $x_{99} = x(t_{99}) = 8.63749\,661$. The unknowns in our problem are the remaining values of x_i—namely, $x_1, x_2, \ldots, x_{98}$. We discretize the derivatives by the central difference Formulas (2) and obtain a linear system of type (3). Our principal equation is of the form (5) and is

$$-\left(1 + \frac{h}{2}\right) x_{i-1} + (2 - h^2) x_i - \left(1 - \frac{h}{2}\right) x_{i+1} = -h^2 \left[e^{t_i} - 3\sin(t_i)\right]$$

The boundary values given are correct only to the number of digits shown.

We generalize the pseudocode so that with only a few changes it can accommodate any two-point boundary value problem of type (1) with the right-hand side of form (4). Here $u(x)$, $v(x)$, and $w(x)$ are statement functions.

```
program bvp1
integer parameter n ← 99
real parameter t_a ← 1, t_b ← 2, α ← 1.09737 491, β ← 8.63749 661
real array (a_i)_n, (b_i)_n, (c_i)_n, (d_i)_n, (y_i)_n
integer i
real error, h, t, u, v, w, x
u(x) = e^x − 3 sin(x)
v(x) = −1
w(x) = 1
h ← (t_b − t_a)/n
for i = 1 to n − 1 do
    t ← t_a + ih
    a_i ← −[1 + (h/2)w(t)]
    d_i ← 2 + h²v(t)
    c_i ← −[1 − (h/2)w(t)]
    b_i ← −h²u(t)
end do
b_1 ← b_1 − a_1α
b_{n-1} ← b_{n-1} − c_{n-1}β
for i = 1 to n − 1 do
    a_i ← a_{i+1}
end
call tri(n − 1, (a_i), (d_i), (c_i), (b_i), (y_i))
error ← e^{t_a} − 3 cos(t_a) − α
output t_a, α, error
for i = 1 to n − 1 step 9 do
    t ← t_a + ih
    error ← e^t − 3 cos(t) − y_i
    output t, y_i, error
end do
```

$error \leftarrow e^{t_b} - 3\cos(t_b) - \beta$
output $b, \beta, error$
end program *bvp1*

The computer results are as follows:

t-Value	Solution	Error
1.00000 00	1.09737 49	0.00
1.09090 91	1.59203 02	-8.83×10^{-5}
1.18181 82	2.12274 17	-1.74×10^{-4}
1.27272 73	2.68980 86	-2.56×10^{-4}
1.36363 64	3.29367 04	-3.28×10^{-4}
1.45454 55	3.93494 53	-3.76×10^{-4}
1.54545 45	4.61449 10	-4.06×10^{-4}
1.63636 36	5.33343 17	-4.13×10^{-4}
1.72727 27	6.09319 59	-3.89×10^{-4}
1.81818 18	6.89557 22	-3.16×10^{-4}
1.90909 10	7.74277 78	-1.88×10^{-4}
2.00000 00	8.63749 69	0.00

Shooting Method in Linear Case

We have just seen that this discretization method (also called a **finite-difference method**) is rather simple in the case of the linear two-point boundary value problem:

$$\begin{cases} x'' = u(t) + v(t)x + w(t)x' \\ x(a) = \alpha \qquad x(b) = \beta \end{cases} \tag{8}$$

The shooting method is also especially simple in this case. Recall that the shooting method requires us to solve an initial-value problem:

$$\begin{cases} x'' = u(t) + v(t)x + w(t)x' \\ x(a) = \alpha \qquad x(a) = z \end{cases} \tag{9}$$

and interpret the terminal value $x(b)$ as a function of z. We call that function ϕ and seek a value of z for which $\phi(z) = \beta$. For the linear Problem (9), ϕ is a *linear* function of z, and so Figure 12.2 in Section 12.1 is actually realistic. Consequently, we need only solve Problem (9) with two values of z in order to determine the function precisely. To establish these facts, let us do a little more analysis.

Suppose that we have solved Problem (9) twice with particular values z_1 and z_2. Let the solutions so obtained be denoted by $x_1(t)$ and $x_2(t)$. Then we claim that the function

$$g(t) = \lambda x_1(t) + (1 - \lambda)x_2(t) \tag{10}$$

has properties

$$\begin{cases} g'' = u + vg + wg' \\ g(a) = \alpha \end{cases}$$

which are left to the reader to verify. (The value of λ in this analysis is a constant but is completely arbitrary.)

The function g nearly solves the two-point boundary value Problem (8), and g contains a parameter λ at our disposal. Imposing the condition $g(b) = \beta$, we obtain

$$\lambda x_1(b) + (1 - \lambda)x_2(b) = \beta$$

from which

$$\lambda = \frac{\beta - x_2(b)}{x_1(b) - x_2(b)}$$

Perhaps the simplest way to implement these ideas is to solve two initial-value problems:

$$\begin{cases} x'' = u(t) + v(t)x + w(t)x' \\ x(a) = \alpha \qquad x'(a) = 0 \end{cases}$$

and

$$\begin{cases} y'' = u(t) + v(t)y + w(t)y' \\ y(a) = \alpha \qquad y'(a) = 1 \end{cases}$$

Then the solution to the original two-point boundary value Problem (8) is

$$\lambda x(t) + (1 - \lambda)y(t) \qquad \text{with} \qquad \lambda = \frac{\beta - y(b)}{x(b) - y(b)} \tag{11}$$

In the computer realization of this procedure, we must save the entire solution curves x and y. They are stored in arrays (x_i) and (y_i).

Pseudocode and Numerical Example

As an example of the shooting method, consider the problem of Equation (7). We solve the two problems

$$\begin{cases} x'' = e^t - 3\sin(t) + x' - x \\ x(1) = 1.09737\,491 \\ x'(1) = 0 \end{cases} \qquad \begin{cases} y'' = e^t - 3\sin(t) + y' - y \\ y(1) = 1.09737\,491 \\ y'(1) = 1 \end{cases} \tag{12}$$

by using the fourth-order Runge-Kutta method. To do so, we introduce variables

$$x_0 = t \qquad x_1 = x \qquad x_2 = x'$$

Then the first initial-value problem is

$$
\begin{bmatrix} x_0' \\ x_1' \\ x_2' \end{bmatrix} = \begin{bmatrix} 1 \\ x_2 \\ e^{x_0} - 3\sin(x_0) + x_2 - x_1 \end{bmatrix} \qquad \begin{bmatrix} x_0(1) \\ x_1(1) \\ x_2(1) \end{bmatrix} = \begin{bmatrix} 1 \\ 1.09737\,491 \\ 0 \end{bmatrix}
$$

Now let

$$
y_0 = t \qquad y_1 = y \qquad y_2 = y'
$$

The second initial-value problem that we must solve is similar except that we modify the initial vector:

$$
\begin{bmatrix} y_0' \\ y_1' \\ y_2' \end{bmatrix} = \begin{bmatrix} 1 \\ y_2 \\ e^{y_0} - 3\sin(y_0) + y_2 - y_1 \end{bmatrix} \qquad \begin{bmatrix} y_0(1) \\ y_1(1) \\ y_2(1) \end{bmatrix} = \begin{bmatrix} 1 \\ 1.09737\,491 \\ 1 \end{bmatrix}
$$

It is more efficient to solve these two problems together as a single system. Introducing

$$
x_3 = y \qquad x_4 = y'
$$

into the first system, we have

$$
\begin{bmatrix} x_0' \\ x_1' \\ x_2' \\ x_3' \\ x_4' \end{bmatrix} = \begin{bmatrix} 1 \\ x_2 \\ e^{x_0} - 3\sin(x_0) + x_2 - x_1 \\ x_4 \\ e^{x_0} - 3\sin(x_0) + x_4 - x_3 \end{bmatrix} \qquad \begin{bmatrix} x_0(1) \\ x_1(1) \\ x_2(1) \\ x_3(1) \\ x_4(1) \end{bmatrix} = \begin{bmatrix} 1 \\ 1.09737\,491 \\ 0 \\ 1.09737\,491 \\ 1 \end{bmatrix}
$$

Clearly, the $x_1(t)$ and $x_3(t)$ components of the solution vector at each t satisfy the first and second problems, respectively. Consequently, the solution is

$$
\lambda x_1(t_i) + (1 - \lambda)x_3(t_i) \qquad (1 \le i \le n - 1)
$$

where

$$
\lambda = \frac{8.63749\,661 - x_3(2)}{x_1(2) - x_3(2)}
$$

We use 100 points as before, so that $n = 99$.

program *bvp2*
integer parameter $n \leftarrow 99$, $m \leftarrow 4$
real parameter $a \leftarrow 1$, $b \leftarrow 2$, $\alpha \leftarrow 1.09737\,491$, $\beta \leftarrow 8.63749\,661$
real array $(x_i)_{0:m}, (x1_i)_{0:n}, (x3_i)_{0:n}$
integer i

```
real error, h, p, q, t
x ← (1, α, 0, α, 1)
h ← (b − a)/n
t ← a
for i = 1 to n do
    call rk4sys(m, (x_i), h, 1)
    (x1)_i ← x_1
    (x3)_i ← x_3
    t ← a + ih
end do
p ← [β − (x3)_n]/[(x1)_n − (x3)_n]
q ← 1 − p
for i = 1 to n do
    (x1)_i ← p (x1)_i + q (x3)_i
end do
error ← e^a − 3 cos(a) − α
output a, α, error
for i = 9 to n step 9 do
    t ← a + ih
    error ← e^t − 3 cos(t) − (x1)_i
    output t, (x1)_i, error
end do
end program bvp2

procedure xpsys(m, (x_i), (f_i))
real array (x_i)_{0:m}, (f_i)_{0:m}
f_0 ← 1
f_1 ← x_2
f_2 ← e^{x_0} − 3 sin(x_0) + x_2 − x_1
f_3 ← x_4
f_4 ← e^{x_0} − 3 sin(x_0) + x_4 − x_3
end procedure xpsys
```

The final computer results are as shown.

t-Value	Solution	Error
1.0000000	1.0973749	0.00
1.0909091	1.5919409	9.54×10^{-7}
1.1818182	2.1225657	1.91×10^{-6}
1.2727273	2.6895509	1.43×10^{-6}
1.3636364	3.2933426	2.38×10^{-7}
1.4545455	3.9345679	9.54×10^{-7}
1.5454545	4.6140857	-4.77×10^{-7}
1.6363636	5.3330178	4.77×10^{-7}
1.7272727	6.0928054	1.91×10^{-6}
1.8181818	6.8952556	9.54×10^{-7}
1.9090910	7.7425890	9.54×10^{-7}
2.0000000	8.6374969	0.00

Notice that the errors are smaller than those obtained in the discretization method for the same problem.

To verify these numerical results, we can use Maple as follows to obtain the explicit solution.

```
ode := (D@@2)(x)(t) = exp(t)-3*sin(t)+D(x)(t)-x(t);
dsolve(ode, x(t));
```

In our brief discussion of two-point boundary value problems, we have not touched upon the difficult question of the existence of solutions. Sometimes a boundary value problem has no solution in spite of having smooth coefficients. An example is given in Problem **4b** of Section 12.1. This behavior contrasts sharply with that of initial-value problems. These matters are beyond the scope of this book but are treated, for example, in Keller [1976] and Stoer and Bulirsch [1993].

PROBLEMS 12.2

1. If standard finite-difference approximations to derivatives are used to solve a two-point boundary value problem with $x'' = t + 2x - x'$, what is the typical equation in the resulting linear system of equations?

2. Consider the two-point boundary value problem

$$\begin{cases} x'' = -x \\ x(0) = 0 \qquad x(1) = 1 \end{cases}$$

Set up and solve the tridiagonal system that arises from the finite-difference method when $h = \frac{1}{4}$. Explain any differences from the analytic solution $x(\frac{1}{4}) \approx 0.29401$, $x(\frac{1}{2}) \approx 0.56975$, and $x(\frac{3}{4}) \approx 0.81006$.

3. Verify that Equation (11) gives the solution of boundary value Problem (8).

4. Consider the two-point boundary value problem

$$\begin{cases} x'' = x^2 - t + tx \\ x(0) = 1 \qquad x(1) = 3 \end{cases}$$

Suppose that we have solved two initial-value problems

$$\begin{cases} u'' = u^2 - t + tu \\ u(0) = 1 \qquad u'(0) = 1 \end{cases} \qquad \begin{cases} v'' = v^2 - t + tv \\ v(0) = 1 \qquad v'(0) = 2 \end{cases}$$

numerically and have found as terminal values $u(1) = 2$ and $v(1) = 3.5$. What is a reasonable initial-value problem to try *next* in attempting to solve the original two-point value problem?

5. Consider the tridiagonal System (6). Show that if $v_i > 0$, then some choice of h exists for which the matrix is diagonally dominant.

6. Establish the properties claimed for the function g in Equation (10).

7. Show that for the simple problem

$$\begin{cases} x'' = -x \\ x(a) = \alpha \qquad x(b) = \beta \end{cases}$$

the tridiagonal system to be solved can be written as

$$\begin{cases} (2 - h^2)x_1 \quad - x_2 \quad = \alpha \\ -x_{i-1} + (2 - h^2)x_i \quad - x_{i+1} = 0 \qquad (2 \le i \le n - 2) \\ -x_{n-2} + (2 - h^2)x_{n-1} \quad = \beta \end{cases}$$

8. Write down the system of equations $Ax = b$ that results from using the usual second-order central difference approximation to solve

$$\begin{cases} x'' = (1 + t)x \\ x(0) = 0 \qquad x(1) = 1 \end{cases}$$

9. Let u be a solution of the initial-value problem

$$\begin{cases} u'' = e^t u + t^2 u' \\ u(1) = 0 \qquad u'(1) = 1 \end{cases}$$

How do we solve the following two-point boundary value problem by utilizing u?

$$\begin{cases} x'' = e^t x + t^2 x' \\ x(1) = 0 \qquad x(2) = 7 \end{cases}$$

10. How would you solve the problem

$$\begin{cases} x' = f(t, x) \\ Ax(a) + Bx(b) = C \end{cases}$$

where a, b, A, B, and C are given real numbers? (Assume A and B are not both zero.)

COMPUTER PROBLEMS 12.2

1. Explain the main steps in setting up a program to solve this two-point boundary value problem by the finite-difference method.

$$\begin{cases} x'' = x \sin t + x' \cos t - e^t \\ x(0) = 0 \qquad x(1) = 1 \end{cases}$$

Show any preliminary work that must be done before programming. Exploit the linearity of the differential equation. Program and compare the results when different values of n are used—say, $n = 10$, 100, and 1000.

2. Solve the following two-point boundary value problem numerically. For comparisons, the exact solutions are given.

a. $\begin{cases} x'' = \dfrac{(1-t)x+1}{(1+t)^2} \\ x(0) = 1 \qquad x(1) = 0.5 \end{cases}$ *Solution:* $x = \dfrac{1}{1+t}$

b. $\begin{cases} x'' = \dfrac{1}{3}\left[(2-t)e^{2x} + (1+t)^{-1}\right] \\ x(0) = 0 \qquad x(1) = -\log 2 \end{cases}$ *Solution:* $x = -\log(1+t)$

3. Solve the boundary value problem

$$\begin{cases} x'' = -x + tx' - 2t\cos t + t \\ x(0) = 0 \qquad x(\pi) = \pi \end{cases}$$

by discretization. Compare with the exact solution, which is $x(t) = t + 2\sin t$.

4. Repeat Computer Problem **2** of Section 12.1 using a discretization method.

13 PARTIAL DIFFERENTIAL EQUATIONS

In the theory of elasticity, it is shown that the stress in a cylindrical beam under torsion can be derived from a function $u(x, y)$ that satisfies the Poisson equation

$$\frac{\partial^2 u}{\partial x^2} + \frac{\partial^2 u}{\partial y^2} + 2 = 0$$

In the case of a beam whose cross section is the square defined by $|x| \leq 1$, $|y| \leq 1$, the function u must satisfy Poisson's equation *inside* the square and must be zero at each point on the *perimeter* of the square. By using the methods of this chapter, we can construct a table of approximate values of u.

Many physical phenomena can be modeled mathematically by differential equations. When the function being studied involves two or more independent variables, the differential equation is usually a *partial* differential equation. Since functions of several variables are intrinsically more complicated than those of one variable, partial differential equations can lead to the most challenging of numerical problems. In fact, their numerical solution is one type of scientific calculation in which the resources of the biggest and fastest computing systems easily become taxed. We shall see later why this is so.

Some Partial Differential Equations from Applied Problems

Some important partial differential equations and the physical phenomena that they govern are listed here.

1. The **wave equation** in three spatial variables (x, y, z) and time t is

$$\frac{\partial^2 u}{\partial x^2} + \frac{\partial^2 u}{\partial y^2} + \frac{\partial^2 u}{\partial z^2} - \frac{\partial^2 u}{\partial t^2} = 0$$

The function u represents the displacement at time t of a particle whose position at rest is (x, y, z). With appropriate boundary conditions, this equation governs vibrations of a three-dimensional elastic body.

2. The **heat equation** is

$$\frac{\partial^2 u}{\partial x^2} + \frac{\partial^2 u}{\partial y^2} + \frac{\partial^2 u}{\partial z^2} - \frac{\partial u}{\partial t} = 0$$

The function u represents the temperature at time t in a physical body at the point that has coordinates (x, y, z).

3. Laplace's equation is

$$\frac{\partial^2 u}{\partial x^2} + \frac{\partial^2 u}{\partial y^2} + \frac{\partial^2 u}{\partial z^2} = 0$$

It governs the steady-state distribution of heat in a body or the steady-state distribution of electrical charge in a body. Laplace's equation also governs gravitational, electric, and magnetic potentials and velocity potentials in irrotational flows of incompressible fluids. The form of Laplace's equation given above applies to rectangular coordinates. In cylindrical and spherical coordinates, it takes these respective forms:

$$\frac{\partial^2 u}{\partial r^2} + \frac{1}{r}\frac{\partial u}{\partial r} + \frac{1}{r^2}\frac{\partial^2 u}{\partial \phi^2} + \frac{\partial^2 u}{\partial z^2} = 0$$

$$\frac{1}{r}\frac{\partial^2}{\partial r^2}(ru) + \frac{1}{r^2 \sin\theta}\frac{\partial}{\partial\theta}\left(\sin\theta\frac{\partial u}{\partial\theta}\right) + \frac{1}{r^2 \sin^2\theta}\frac{\partial^2 u}{\partial\phi^2} = 0$$

4. The **biharmonic equation** is

$$\frac{\partial^4 u}{\partial x^4} + 2\frac{\partial^4 u}{\partial x^2\,\partial y^2} + \frac{\partial^4 u}{\partial y^4} = 0$$

It occurs in the study of elastic stress, and from its solution the shearing and normal stresses can be derived for an elastic body.

5. The **Navier-Stokes equations** are

$$\frac{\partial u}{\partial t} + u\frac{\partial u}{\partial x} + v\frac{\partial u}{\partial y} + \frac{\partial p}{\partial x} = \frac{\partial^2 u}{\partial x^2} + \frac{\partial^2 u}{\partial y^2}$$

$$\frac{\partial v}{\partial t} + u\frac{\partial v}{\partial x} + v\frac{\partial v}{\partial y} + \frac{\partial p}{\partial y} = \frac{\partial^2 v}{\partial x^2} + \frac{\partial^2 v}{\partial y^2}$$

Here u and v are components of the velocity vector in a fluid flow. The function p is the pressure, and the fluid is assumed to be incompressible but viscous.

Additional examples from quantum mechanics, electromagnetism, hydrodynamics, elasticity, and so on could also be given, but the five partial differential equations shown already exhibit a great diversity. The Navier-Stokes equation, in particular, illustrates a very complicated problem: a pair of nonlinear, simultaneous partial differential equations.

To specify a unique solution to a partial differential equation, additional conditions must be imposed on the solution function. Typically, these conditions occur in the form of boundary values that are prescribed on all or part of the perimeter of the region in which the solution is sought. The nature of the boundary and the boundary values are usually the determining factors in setting up an appropriate numerical scheme for obtaining the approximate solution.

13.1 Parabolic Problems

Heat Equation Model Problem

In this section, we consider a model problem of modest scope to introduce some of the essential ideas. For technical reasons, the problem is said to be of the **parabolic type**. In it we have the heat equation in one spatial variable accompanied by boundary conditions appropriate to a certain physical phenomenon:

$$
\begin{cases}
\dfrac{\partial^2}{\partial x^2} u(x,t) = \dfrac{\partial}{\partial t} u(x,t) \\[2mm]
\quad u(0,t) = u(1,t) = 0 \\[2mm]
\quad u(x,0) = \sin \pi x
\end{cases}
\tag{1}
$$

These equations govern the temperature $u(x,t)$ in a thin rod of length 1 when the ends are held at temperature 0, under the assumption that the initial temperature in the rod is given by the function $\sin \pi x$ (see Figure 13.1). In the xt-plane, the region in which the solution is sought is described by inequalities $0 \leq x \leq 1$ and $t \geq 0$. On the boundary of this region (shaded in Figure 13.2), the values of u have been prescribed.

Finite-Difference Method

A principal approach to the numerical solution of such a problem is the **finite-difference method**. It proceeds by replacing the derivatives in the equation by finite differences. Two formulas from Section 4.3 are useful in this context:

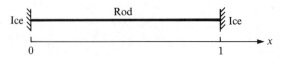

FIGURE 13.1
Heated rod

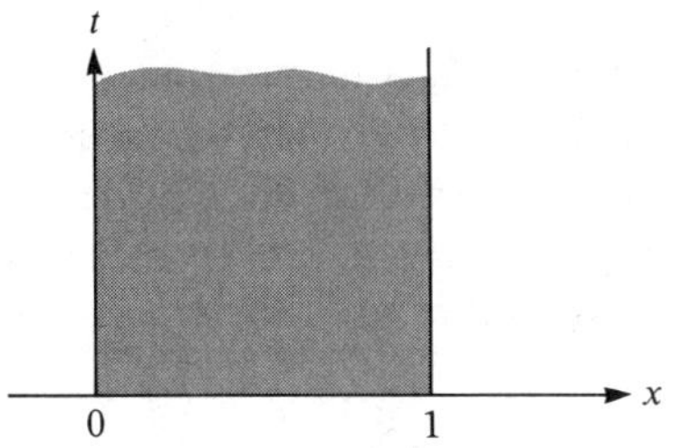

FIGURE 13.2

$$f'(x) \approx \frac{1}{h}[f(x + h) - f(x)]$$

$$f''(x) \approx \frac{1}{h^2}[f(x + h) - 2f(x) + f(x - h)]$$

If the formulas are used in the differential Equation (1), with possibly different step lengths h and k, the result is

$$\frac{1}{h^2}[u(x + h, t) - 2u(x, t) + u(x - h, t)] = \frac{1}{k}[u(x, t + k) - u(x, t)] \qquad (2)$$

This equation is now interpreted as a means of advancing the solution step by step in the t variable. That is, if $u(x, t)$ is known for $0 \leq x \leq 1$ and $0 \leq t \leq t_0$, then Equation (2) allows us to evaluate the solution for $t = t_0 + k$.

Equation (2) can be rewritten in the form

$$u(x, t + k) = \sigma u(x + h, t) + (1 - 2\sigma)u(x, t) + \sigma u(x - h, t) \qquad (3)$$

where

$$\sigma = \frac{k}{h^2}$$

A sketch showing the location of the four points involved in this equation is given in Figure 13.3. Since the solution is known on the boundary of the region, it is possible to compute an approximate solution inside the region by systematically using Equation (3). It is, of course, an *approximate* solution because Equation (2) is only a finite-difference analog of Equation (1).

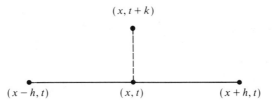

FIGURE 13.3 $(x - h, t)$

To obtain an approximate solution on a computer, we select values for h and k and use Equation (3). An analysis of this procedure, which is outside the scope of this text, shows that for *stability* of the computation, the coefficient $1 - 2\sigma$ in Equation (3) should be nonnegative. (If this condition is not met, errors made at one step will probably be magnified at subsequent steps, ultimately spoiling the solution.) The reader is referred to Forsythe and Wasow [1960] for a discussion of stability.

Pseudocode for Explicit Method

For utmost simplicity, we select $h = 0.1$ and $k = 0.005$. Coefficient σ is now 0.5. This choice makes the coefficient $1 - 2\sigma$ equal to zero. Our pseudocode first prints $u(ih, 0)$ for $0 \leq i \leq 10$ because they are known boundary values. Then it computes and prints $u(ih, k)$ for $0 \leq i \leq 10$ using Equation (3) and boundary values $u(0, t) = u(1, t) = 0$. This procedure is continued until t reaches the value 0.1. The single subscripted arrays (u_i) and (v_i) are used to store the values of the approximate solution at t and $t + k$, respectively. Since the analytic solution of the problem is $u(x, t) = e^{-\pi^2 t} \sin \pi x$ (see Problem **3**), the error can be printed out at each step.

The procedure described is an example of an *explicit method*. The approximate values of $u(x, t + k)$ are calculated explicitly in terms of $u(x, t)$. Not only is this situation atypical, but even in this problem the procedure is rather slow because considerations of stability force us to select

$$k \leq \frac{h^2}{2}$$

Since h must be rather small in order to represent the derivative accurately by the finite-difference formula, the corresponding k must be extremely small. Values $h = 0.1$ and $k = 0.005$ are representative, as are $h = 0.01$ and $k = 0.00005$. With such small values of k, an inordinate amount of computation is necessary to make much progress in the t variable.

```
program parabolic1
integer parameter n ← 10, m ← 20
real parameter h ← 0.1, k ← 0.005
real parameter u0 ← 0, v0 ← 0, un ← 0, vn ← 0
real array (ui)0:n, (vi)0:n
integer i, j
for i = 1 to n − 1 do
    ui ← sin(πih)
end do
output (ui)
for j = 1 to m do
    for i = 1 to n − 1 do
        vi ← (ui−1 + ui+1)/2
    end do
```

```
        output (v_i)
        t ← jk
        for i = 1 to n − 1 do
            u_i ← e^{−π²t} sin(πih) − v_i
        end do
        output (u_i)
        for i = 1 to n − 1 do
            u_i ← v_i
        end do
    end do
end program parabolic1
```

Crank-Nicolson Method

An alternative procedure of the implicit type goes by the name of its inventors, John Crank and Phyllis Nicolson, and is based on a simple variant of Equation (2):

$$\frac{1}{h^2}[u(x+h,t) - 2u(x,t) + u(x-h,t)] = \frac{1}{k}[u(x,t) - u(x,t-k)] \qquad (4)$$

If a numerical solution at grid points $x = ih$, $t = jk$ has been obtained up to a certain level in the t variable, Equation (4) governs the values of u on the next t level. Therefore, Equation (4) should be rewritten as

$$-u(x-h,t) + ru(x,t) - u(x+h,t) = su(x,t-k) \qquad (5)$$

in which

$$r = 2 + s \qquad \text{and} \qquad s = \frac{h^2}{k}$$

The locations of the four points in this equation are shown in Figure 13.4.

On the t level, u is unknown, but on the $(t-k)$ level, u is known. So we can introduce unknowns $u_i = u(ih,t)$ and known quantities $b_i = su(ih, t-k)$ and write Equation (5) in matrix form:

$$\begin{bmatrix} r & -1 & & & & \\ -1 & r & -1 & & & \\ & -1 & r & -1 & & \\ & & \ddots & \ddots & \ddots & \\ & & & -1 & r & -1 \\ & & & & -1 & r \end{bmatrix} \begin{bmatrix} u_1 \\ u_2 \\ u_3 \\ \vdots \\ u_{n-2} \\ u_{n-1} \end{bmatrix} = \begin{bmatrix} b_1 \\ b_2 \\ b_3 \\ \vdots \\ b_{n-2} \\ b_{n-1} \end{bmatrix} \qquad (6)$$

The simplifying assumption that $u(0,t) = u(1,t) = 0$ has been used here. Also, $h = 1/n$. The system of equations is tridiagonal and diagonally dominant because $|r| = 2 + h^2/k > 2$. Hence, it can be solved by the efficient method of Section 6.3.

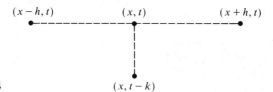

FIGURE 13.4

An elementary argument shows that this method is stable. We shall see that if the initial values $u(x, 0)$ lie in an interval $[\alpha, \beta]$, then values subsequently calculated by using Equation (5) will also lie in $[\alpha, \beta]$, thereby ruling out any unstable growth. Since the solution is built up line by line in a uniform way, we need only verify that the values on the first computed line, $u(x, k)$, lie in $[\alpha, \beta]$. Let j be the index of the largest u_i that occurs on this line $t = k$. Then

$$-u_{j-1} + ru_j - u_{j+1} = b_j$$

Since u_j is the largest of the u's, $u_{j-1} \leqq u_j$ and $u_{j+1} \leqq u_j$. Thus,

$$ru_j = b_j + u_{j-1} + u_{j+1} \leqq b_j + 2u_j$$

Since $r = 2 + s$ and $b_j = su(jh, 0)$, the previous inequality leads at once to

$$u_j \leqq u(jh, 0) \leqq \beta$$

Since u_j is the largest of the u_i, we have

$$u_i \leqq \beta \qquad \text{for all } i$$

Similarly,

$$u_i \geqq \alpha \qquad \text{for all } i$$

thus establishing our assertion.

Pseudocode for Crank-Nicolson Method

A pseudocode to carry out the **Crank-Nicolson method** on the model program is given next. In it, $h = 0.1$, $k = h^2/2$, and the solution is continued until $t = 0.1$. The value of r is 4 and $s = 2$. It is easier to compute and print only the values of u at interior points on each horizontal line. At boundary points, we have $u(0, t) = u(1, t) = 0$. The program calls procedure *tri* from Section 6.3.

program *parabolic2*
integer parameter $n \leftarrow 10$, $m \leftarrow 20$
real parameter $h \leftarrow 0.1$, $k \leftarrow 0.005$
real array $(c_i)_{n-1}, (d_i)_{n-1}, (u_i)_{n-1}, (v_i)_{n-1}$
integer i, j

```
real r, s, t
s ← h²/k
r ← 2 + s
for i = 1 to n − 1 do
    dᵢ ← r
    cᵢ ← −1
    uᵢ ← sin(πih)
end do
output (uᵢ)
for j = 1 to m do
    for i = 1 to n − 1 do
        dᵢ ← r
        vᵢ ← suᵢ
    end for
    call tri(n − 1, (cᵢ), (dᵢ), (cᵢ), (vᵢ), (vᵢ))
    output (vᵢ)
    t ← jk
    for i = 1 to n − 1 do
        uᵢ ← e^(−π²t) sin(πih) − vᵢ
    end do
    output (uᵢ)
    for i = 1 to n − 1 do
        uᵢ ← vᵢ
    end do
end do
end program parabolic2
```

We used the same values for h and k in the pseudocode for two methods (explicit and Crank-Nicolson), so a fair comparison can be made of the outputs. Because the Crank-Nicolson method is stable, a much larger k could have been used.

Alternative Version of Crank-Nicolson Method

Another version of the Crank-Nicolson method is obtained as follows: The central differences at $(x, t − \frac{1}{2}k)$ in Equation (4) produce

$$\frac{1}{h^2}\left[u\left(x + h, t − \frac{k}{2}\right) − 2u\left(x, t − \frac{k}{2}\right) + u\left(x − h, t − \frac{k}{2}\right)\right]$$

$$= \frac{1}{k}[u(x, t) − u(x, t − k)]$$

Since the u values are known only at integer multiples of k, terms like $u(x, t − \frac{1}{2}k)$ are replaced by the average of u values at adjacent grid points; that is,

$$u\left(x, t − \frac{k}{2}\right) \approx \frac{1}{2}[u(x, t) + u(x, t − k)]$$

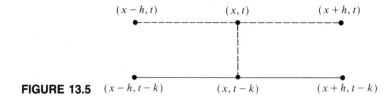

FIGURE 13.5

So we have

$$\frac{1}{2h^2}[u(x+h,t) - 2u(x,t) + u(x-h,t) + u(x+h,t-k)$$

$$-2u(x,t-k) + u(x-h,t-k)] = \frac{1}{k}[u(x,t) - u(x,t-k)]$$

The computational form of this equation is

$$-u(x-h,t) + 2(1+s)u(x,t) - u(x+h,t)$$

$$= u(x-h,t-k) + 2(s-1)u(x,t-k) + u(x+h,t-k) \tag{7}$$

where

$$s = \frac{h^2}{k}$$

The six points in this equation are shown in Figure 13.5. This leads to a tridiagonal system of form (6) with

$$r = 2(1+s)$$

and

$$b_i = u((i-1)h, t-k) + 2(s-1)u(ih, t-k) + u((i+1)h, t-k)$$

PROBLEMS 13.1

1. A second-order linear differential equation with two variables has the form

$$A\frac{\partial^2 u}{\partial x^2} + B\frac{\partial^2 u}{\partial x\,\partial y} + C\frac{\partial^2 u}{\partial y^2} + \cdots = 0$$

Here A, B, and C are functions of x and y, and the terms not written are of lower order. The equation is said to be *elliptic*, *parabolic*, or *hyperbolic* at a point (x, y), depending on whether $B^2 - 4AC$ is negative, zero, or positive, respectively. Classify each of these equations in this manner.

a. $u_{xx} + u_{yy} + u_x + \sin xu_y - u = x^2 + y^2$

b. $u_{xx} - u_{yy} + 2u_x + 2u_y + e^x u = x - y$

c. $u_{xx} = u_y + u - u_x + y$

d. $u_{xy} = u - u_x - u_y$

e. $3u_{xx} + u_{xy} + u_{yy} = e^{xy}$

f. $e^x u_{xx} + \cos yu_{xy} - u_{yy} = 0$

g. $u_{xx} + 2u_{xy} + u_{yy} = 0$

h. $xu_{xx} + yu_{xy} + u_{yy} = 0$

2. Write the two-dimensional form of Laplace's equation in polar coordinates.

3. Show that the function

$$u(x,t) = \sum_{n=1}^{N} c_n e^{-(n\pi)^2 t} \sin n\pi x$$

is a solution of the heat conduction problem $u_{xx} = u_t$ and satisfies the boundary condition

$$u(0,t) = u(1,t) = 0 \qquad u(x,0) = \sum_{n=1}^{N} c_n \sin n\pi x \qquad \text{(for all } N \geq 1)$$

4. Refer to the model problem solved numerically in this section and show that if there is no roundoff, the approximate solution values obtained by using Equation (3) lie in the interval $[0, 1]$. (Assume $1 \geq 2k/h^2$.)

5. Find a solution of Equation (3) that has the form $u(x,t) = a^t \sin \pi x$, where a is a constant.

6. When using Equation (5), how must the linear System (6) be modified for $u(0,t) = c_0$ and $u(1,t) = c_n$ with $c_0 \neq 0$, $c_n \neq 0$? When using Equation (7)?

7. Describe in detail how Problem (1) with boundary conditions $u(0,t) = q(t)$, $u(1,t) = g(t)$, and $u(x,0) = f(x)$ can be solved numerically using System (6). Here q, g, and f are known functions.

8. What finite-difference equation should be a suitable replacement for the equation $\partial^2 u/\partial x^2 = \partial u/\partial t + \partial u/\partial x$ in numerical work?

9. Consider the partial differential equation $\partial u/\partial x + \partial u/\partial t = 0$ with $u = u(x,t)$ in the region $[0, 1] \times [0, \infty]$, subject to the boundary conditions $u(0,t) = 0$ and $u(x,0)$ specified. For fixed t, we discretize only the first term using $(u_{i+1} - u_{i-1})/(2h)$ for $i = 1, 2, \ldots, n - 1$ and $(u_n - u_{n-1})/h$ where $h = 1/n$. Here $u_i = u(x_i, t)$ and $x_i = ih$ with fixed t. In this way, the original problem can be considered a first-order initial-value problem

$$\frac{dy}{dx} + \frac{1}{2h}A y = 0$$

where

$$y = [u_1, u_2, \ldots, u_n]^T \qquad \frac{dy}{dx} = [u'_1, u'_2, \ldots, u'_n]^T \qquad u'_i = \frac{\partial u_i}{\partial t}$$

Determine the $n \times n$ matrix A.

10. Refer to the discussion of the stability of the Crank-Nicolson procedure, and establish the inequality $u_i \geqq \alpha$.

COMPUTER PROBLEMS 13.1

1. Solve the same heat conduction problem as in the text except use $h = 2^{-4}$, $k = 2^{-10}$, and $u(x, 0) = x(1 - x)$. Carry out the solution until $t = 0.0125$.

2. Modify the Crank-Nicolson code in the text so that it uses the alternative scheme (7). Compare the two programs on the same problems with the same spacing.

3. Recode and test the pseudocode in this section using a computer language that supports vector operations.

4. Run the Crank-Nicolson code with different choices of h and k, in particular, letting k be much larger. Try $k = h$, for example.

13.2 Hyperbolic Problems

Wave Equation Model Problem

The wave equation with one space variable

$$\frac{\partial^2 u}{\partial x^2} = \frac{\partial^2 u}{\partial t^2} \tag{1}$$

governs the vibration of a string (transverse vibration in a plane) or the vibration in a rod (longitudinal vibration). It is an example of a second-order linear differential equation of the hyperbolic type. If Equation (1) is used to model the vibrating string, then $u(x, t)$ represents the deflection at time t of a point on the string whose coordinate is x when the string is at rest.

To pose a definite model problem, we suppose the points on the string have coordinates x in the interval $0 \leqq x \leqq 1$ (see Figure 13.6). Let us suppose that at time

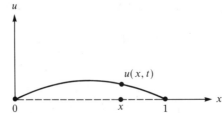

FIGURE 13.6
Vibrating string

$t = 0$, the deflections satisfy equations $u(x, 0) = f(x)$ and $u_t(x, 0) = 0$. Assume also that the ends of the string remain fixed. Then $u(0, t) = u(1, t) = 0$. A fully defined boundary value problem, then, is

$$\begin{cases} u_{xx} - u_{tt} = 0 \\ \quad u(x, 0) = f(x) \\ \quad u_t(x, 0) = 0 \\ \quad u(0, t) = u(1, t) = 0 \end{cases} \tag{2}$$

The region in the xt-plane where a solution is sought is the semi-infinite strip defined by inequalities $0 \leq x \leq 1$ and $t \geq 0$. As in the heat conduction problem of Section 13.1, the values of the unknown function are prescribed on the boundary of the region shown (see Figure 13.7).

Analytic Solution

The model problem in (2) is so simple that it can be immediately solved. Indeed, the solution is

$$u(x, t) = \frac{1}{2}[f(x + t) + f(x - t)] \tag{3}$$

provided that f possesses two derivatives and has been extended to the whole real line by defining

$$f(-x) = -f(x) \qquad \text{and} \qquad f(x + 2) = f(x)$$

To verify that Equation (3) is a solution, we compute derivatives using the chain rule:

$$u_x = \frac{1}{2}[f'(x + t) + f'(x - t)] \qquad u_t = \frac{1}{2}[f'(x + t) - f'(x - t)]$$

$$u_{xx} = \frac{1}{2}[f''(x + t) + f''(x - t)] \qquad u_{tt} = \frac{1}{2}[f''(x + t) + f''(x - t)]$$

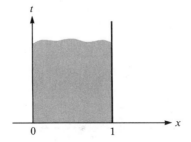

FIGURE 13.7

Obviously,

$$u_{xx} = u_{tt}$$

also

$$u(x, 0) = f(x)$$

Furthermore, we have

$$u_t(x, 0) = \frac{1}{2}[f'(x) - f'(x)] = 0$$

In checking endpoint conditions, we use the formulas by which f was extended:

$$u(0, t) = \frac{1}{2}[f(t) + f(-t)] = 0$$

$$u(1, t) = \frac{1}{2}[f(1 + t) + f(1 - t)]$$

$$= \frac{1}{2}[f(1 + t) - f(t - 1)]$$

$$= \frac{1}{2}[f(1 + t) - f(t - 1 + 2)] = 0$$

The extension of f from its original domain to the entire real line makes it an **odd periodic** function of period 2. *Odd* means that

$$f(x) = -f(-x)$$

and the *periodicity* is expressed by

$$f(x + 2) = f(x)$$

for all x. To compute $u(x, t)$, we need to know f at only two points on the x-axis, $x + t$ and $x - t$, as in Figure 13.8.

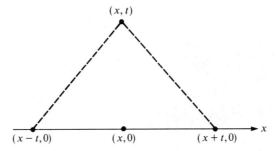

FIGURE 13.8
$(x - t, 0)$ $(x, 0)$ $(x + t, 0)$

Numerical Solution

The model problem is used next to illustrate again the principle of numerical solution. Choosing step sizes h and k for x and t, respectively, and using the familiar approximations for derivatives, we have from Equation (1)

$$\frac{1}{h^2}[u(x+h,t) - 2u(x,t) + u(x-h,t)]$$

$$= \frac{1}{k^2}[u(x,t+k) - 2u(x,t) + u(x,t-k)]$$

which can be rearranged as

$$u(x,t+k) = \rho u(x+h,t) + 2(1-\rho)u(x,t)$$
$$+ \rho u(x-h,t) - u(x,t-k) \qquad \textbf{(4)}$$

Here

$$\rho = \frac{k^2}{h^2}$$

Figure 13.9 shows the point $(x, t+k)$ and the nearby points that enter into Equation (4).

The boundary conditions in Problem (2) can be written

$$\begin{cases} u(x,0) = f(x) \\[2mm] \dfrac{1}{k}[u(x,k) - u(x,0)] = 0 \\[2mm] u(0,t) = u(1,t) = 0 \end{cases} \qquad \textbf{(5)}$$

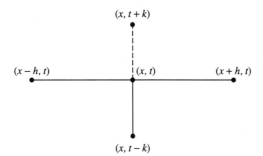

FIGURE 13.9

The problem defined by Equations (4) and (5) can be solved by beginning at the line $t = 0$ where u is known and then progressing one line at a time with $t = k$, $t = 2k, t = 3k, \ldots$. Note that, because of (5), our approximate solution satisfies

$$u(x, k) = u(x, 0) = f(x) \tag{6}$$

The use of the $\mathcal{O}(h)$ approximation for u_t leads to low accuracy in the computed solution to Problem (2). Suppose that there is a row of grid points $(x, -k)$. Letting $t = 0$ in Equation (4), we have

$$u(x, k) = \rho u(x + h, 0) + 2(1 - \rho)u(x, 0) + \rho u(x - h, 0) - u(x, -k)$$

Now the central difference approximation

$$\frac{1}{2k}[u(x, k) - u(x, -k)] = 0$$

for

$$u_t(x, 0) = 0$$

can be used to eliminate the fictitious grid point $(x, -k)$. So instead of Equation (6), we set

$$u(x, k) = \frac{\rho}{2}[f(x + h) + f(x - h)] + (1 - \rho)f(x) \tag{7}$$

because $u(x, 0) = f(x)$. Values of $u(x, nk)$, $n \geq 2$, can now be computed from Equation (4).

Pseudocode

A pseudocode to carry out this numerical process is given next. For simplicity, three one-dimensional arrays (u_i), (v_i), and (w_i) are used: (u_i) represents the solution being computed on the new t line; (v_i) and (w_i) represent solutions on the preceding two t lines.

```
program hyperbolic
integer parameter n ← 10, m ← 20
real parameter h ← 0.1, k ← 0.05
real array (u_i)_{0:n}, (v_i)_{o:f} n, (w_i)_{0:n}
integer i, j
real t, x, ρ
u_0 ← 0; v_0 ← 0; w_0 ← 0; u_n ← 0; v_n ← 0; w_n ← 0
ρ ← (k/h)^2
```

```
for i = 1 to n − 1 do
    x ← ih
    w_i ← f(x)
    v_i ← ½{ρ[f(x − h) + f(x + h)] + 2(1 − ρ)f(x)}
end do
for j = 2 to m do
    for i = 1 to n − 1 do
        u_i ← ρ(v_{i+1} + v_{i−1}) + 2(1 − ρ)v_i − w_i
    end do
    output j, (u_i)
    for i = 1 to n − 1 do
        w_i ← v_i
        v_i ← u_i
        t ← jk
        x ← ih
        u_i ← true_solution(x, t) − v_i
    end do
    output j, (u_i)
end do
end program hyperbolic


real function f(x)
real x
f ← sin(πx)
end function


real function true_solution(x, t)
real t, x
true_solution ← sin(πx) cos(πt)
end function
```

This pseudocode requires accompanying functions to compute values of $f(x)$ and the true solution. It is assumed that the x interval is $[0, 1]$, but when h or n is changed, the interval can be $[0, b]$; that is, $nh = b$. The numerical solution is printed on the t lines that correspond to $1k, 2k, \ldots, mk$.

More advanced treatments show that the ratios

$$\rho = \frac{k^2}{h^2}$$

must not exceed 1 if the solution of the finite-difference equations is to converge to a solution of the differential problem as $k \rightarrow 0$ and $h \rightarrow 0$. Furthermore, if $\rho > 1$, roundoff errors that occur at one stage of the computation would probably be magnified at later stages and thereby ruin the numerical solution.

PROBLEMS 13.2

1. What is the solution of the boundary value problem

$$u_{xx} = u_{tt} \qquad u(x,0) = x(1-x) \qquad u(0,t) = u(1,t) = 0$$

at the point where $x = 3.3$ and $t = 4$?

2. Show that the function $u(x,t) = f(x+at) + g(x-at)$ satisfies the wave equation $a^2 u_{xx} = u_{tt}$.

3. (Continuation) Using the idea in Problem **2**, solve this boundary value problem:

$$u_{xx} = u_{tt} \qquad u(x,0) = F(x) \qquad u_t(x,0) = G(x) \qquad u(0,t) = u(1,t) = 0$$

4. Show that the boundary value problem

$$u_{xx} = u_{tt} \qquad u(x,0) = 2f(x) \qquad u_t(x,0) = 2g(x)$$

has the solution

$$u(x,t) = f(x+t) + f(x-t) + G(x+t) - G(x-t)$$

where G is an antiderivative of g. Assume that $-\infty < x < \infty$ and $t \geq 0$.

5. (Continuation) Solve Problem **4** on a finite x interval, say $0 \leq x \leq 1$, adding boundary condition $u(0,t) = u(1,t) = 0$. In this case, f and g are defined for only $0 \leq x \leq 1$.

COMPUTER PROBLEMS 13.2

1. Given $f(x)$ defined on $[0,1]$, write and test a function for calculating the extended f that obeys the equations $f(-x) = -f(x)$ and $f(x+2) = f(x)$.

2. (Continuation) Write a program to compute the solution of $u(x,t)$ at any given point (x,t) for the boundary value problem of Equation (2).

3. Compare the accuracy of the computed solution, using first Equation (6) and then Equation (7), in the computer program in the text.

4. Use the program in the text to solve boundary value Problem (2) with

$$f(x) = \frac{1}{4}\left(\frac{1}{2} - \left|x - \frac{1}{2}\right|\right) \qquad h = \frac{1}{16} \qquad k = \frac{1}{32}$$

5. Modify the code in the text to solve boundary value Problem (2) when $u_t(x,0) = g(x)$. *Hint*: Equations (5) and (7) will be slightly different (a fact that affects only the initial loop in the program).

6. (Continuation) Use the program that you wrote for Computer Problem **5** to solve the following boundary value problem.

$$
\begin{cases}
u_{xx} = u_{tt} & (0 \le x \le 1,\ t \ge 0) \\
u(x,0) = \sin \pi x \\
u_t(x,0) = \dfrac{1}{4} \sin 2\pi x \\
u(0,t) = u(1,t) = 0
\end{cases}
$$

7. Modify the code in the text to solve the following boundary value problem.

$$
\begin{cases}
u_{xx} = u_{tt} & (-1 \le x \le 1,\ t \ge 0) \\
u(x,0) = |x| - 1 \\
u_t(x,0) = 0 \\
u(-1,t) = u(1,t) = 0
\end{cases}
$$

8. Modify the code in the text to avoid storage of the (v_i) and (u_i) arrays.

9. Simplify the code in the text for the special case $\rho = 1$. Compare the numerical solution at the same grid points for a problem when $\rho = 1$ and $\rho \ne 1$.

13.3 Elliptic Problems

One of the most important partial differential equations in mathematical physics and engineering is **Laplace's equation**, which has the following form in two variables:

$$
\nabla^2 u \equiv \frac{\partial^2 u}{\partial x^2} + \frac{\partial^2 u}{\partial y^2} = 0
$$

Closely related to it is **Poisson's equation**:

$$
\nabla^2 u = g(x,y)
$$

These are examples of **elliptic** equations. (Refer to Problem **1** of Section 13.1 for the classification of equations.) The boundary conditions associated with elliptic equations generally differ from those for parabolic and hyperbolic equations. A model problem is considered here to illustrate the numerical procedures often used.

Helmholtz Equation Model Problem

Suppose that a function $u = u(x,y)$ of two variables is the solution to a certain physical problem. This function is unknown but has some properties that, theoretically, determine it uniquely. We assume that on a given region R in the xy-plane,

$$
\begin{cases}
\nabla^2 u + fu = g \\
u(x,y) \quad \text{known on the boundary of } R
\end{cases} \tag{1}
$$

Here $f = f(x, y)$ and $g = g(x, y)$ are given continuous functions defined in R. The boundary values could be given by a third function

$$u(x, y) = q(x, y)$$

on the perimeter of R. When f is a constant, this partial differential equation is called the **Helmholtz equation.**

Finite-Difference Method

As before, we find an approximate solution of such a problem by the finite-difference method. The first step is to select approximate formulas for the derivatives in our problem. In the present situation, we use the standard formula

$$f''(x) \approx \frac{1}{h^2}[f(x + h) - 2f(x) + f(x - h)] \tag{2}$$

derived in Section 4.3. If it is used on a function of two variables, we obtain the **five-point formula:**

$$\nabla^2 u \approx \frac{1}{h^2}[u(x + h, y) + u(x - h, y) + u(x, y + h) + u(x, y - h) - 4u(x, y)] \tag{3}$$

This formula involves the five points displayed in Figure 13.10.

The local error inherent in the five-point formula is

$$-\frac{h^2}{12}\left[\frac{\partial^4 u}{\partial x^4}(\xi, y) + \frac{\partial^4 u}{\partial y^4}(x, \eta)\right] \tag{4}$$

and, for this reason, Formula (3) is said to provide an approximation of order $\mathcal{O}(h^2)$. In other words, if grids are used with smaller and smaller spacing, $h \to 0$, then the error committed in replacing $\nabla^2 u$ by its finite-difference approximation goes to

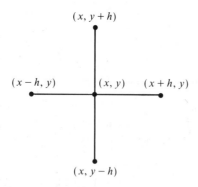

FIGURE 13.10
Five-point stencil

zero as rapidly as does h^2. Equation (3) is called the five-point formula because it involves values of u at (x, y) and at the four nearest grid points.

It should be emphasized that when the differential equation in (1) is replaced by the finite-difference analog, we have changed the problem. Even if the analogous finite-difference problem is solved with complete precision, the solution is that of a problem that only *simulates* the original one. This simulation of one problem by another becomes better and better as h is made to decrease to zero, but the computing cost will inevitably increase.

We should also note that other representations of the derivatives can be used. For example, the **nine-point formula** is

$$
\nabla^2 u \approx \frac{1}{6h^2} [4u(x + h, y) + 4u(x - h, y) + 4u(x, y + h) + 4u(x, y - h)
$$

$$
+ u(x + h, y + h) + u(x - h, y + h) + u(x + h, y - h) \tag{5}
$$

$$
+ u(x - h, y - h) - 20u(x, y)]
$$

This formula is of order $\mathcal{O}(h^2)$. In the special case that u is a **harmonic function** (which means it is a solution of Laplace's equation), the nine-point formula is of order $\mathcal{O}(h^6)$. For additional details, see Forsythe and Wasow [1960, pp. 194–195]. Hence, it is an extremely accurate approximation when using finite-difference methods and solving the Poisson equation $\nabla^2 u = g$, with g a harmonic function. For more general problems, the nine-point Formula (5) has the same order error term as the five-point Formula (3) [namely, $\mathcal{O}(h^2)$] and would not be an improvement over it.

If the mesh spacing is not regular (say, h_1, h_2, h_3, and h_4 are the left, bottom, right, and top spacing, respectively), then it is not difficult to show that at (x, y),

$$
\nabla^2 u \approx \frac{1}{\frac{1}{2}h_1 h_3 (h_1 + h_3)} [h_1 u(x + h_3, y) + h_3 u(x - h_1, y)]
$$

$$
+ \frac{1}{\frac{1}{2}h_2 h_4 (h_2 + h_4)} [h_2 u(x, y + h_4) + h_4 u(x, y - h_2)] \tag{6}
$$

$$
- 2 \left(\frac{1}{h_1 h_3} + \frac{1}{h_2 h_4} \right) u(x, y)
$$

which is only of order h when $h_1 = \alpha_i h$ for $0 < \alpha_i < 1$. This formula is usually used near boundary points, as in Figure 13.11. If the mesh is small, however, the boundary points can be moved over slightly to avoid the use of (6). This perturbation of the region R (in most cases for small h) produces an error no greater than that introduced by using the irregular scheme (6).

Returning to the model Problem (1), we cover the region R by mesh points

$$
x_i = ih \qquad y_j = jh \qquad (i, j \geq 0) \tag{7}
$$

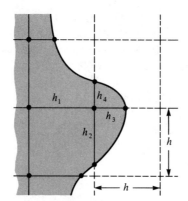

FIGURE 13.11

At this time, it is convenient to introduce an abbreviated notation:

$$u_{ij} = u(x_i, y_i) \qquad f_{ij} = f(x_i, y_i) \qquad g_{ij} = g(x_i, y_j) \qquad \textbf{(8)}$$

With it the five-point formula takes on a simple form at the point (x_i, y_j):

$$(\nabla^2 u)_{ij} \approx \frac{1}{h^2}(u_{i+1,j} + u_{i-1,j} + u_{i,j+1} + u_{i,j-1} - 4u_{ij}) \qquad \textbf{(9)}$$

If this approximation is made in the differential Equation (1), the result is (the reader should verify it)

$$- u_{i+1,j} - u_{i-1,j} - u_{i,j+1} - u_{i,j-1} + \left(4 - h^2 f_{ij}\right) u_{ij} = -h^2 g_{ij} \qquad \textbf{(10)}$$

The coefficients of this equation can be illustrated by a five-point star in which each point corresponds to the coefficient of u in the grid (see Figure 13.12).

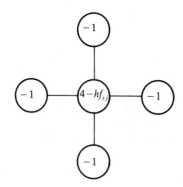

FIGURE 13.12
Five-point star

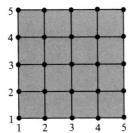

FIGURE 13.13

To be specific, we assume that the region R is a square and the grid has spacing $h = \frac{1}{4}$ (see Figure 13.13). We obtain a single linear equation of the form (10) for each of the nine interior grid points. These nine equations are as follows:

$$-u_{12} - u_{32} - u_{23} - u_{21} + (4 - h^2 f_{22})u_{22} = -h^2 g_{22}$$

$$-u_{22} - u_{42} - u_{33} - u_{31} + (4 - h^2 f_{32})u_{32} = -h^2 g_{32}$$

$$-u_{52} - u_{32} - u_{43} - u_{41} + (4 - h^2 f_{42})u_{42} = -h^2 g_{42}$$

$$-u_{13} - u_{33} - u_{24} - u_{22} + (4 - h^2 f_{23})u_{23} = -h^2 g_{23}$$

$$-u_{43} - u_{23} - u_{34} - u_{32} + (4 - h^2 f_{33})u_{33} = -h^2 g_{33}$$

$$-u_{53} - u_{33} - u_{44} - u_{42} + (4 - h^2 f_{43})u_{43} = -h^2 g_{43}$$

$$-u_{34} - u_{14} - u_{25} - u_{23} + (4 - h^2 f_{24})u_{24} = -h^2 g_{24}$$

$$-u_{44} - u_{24} - u_{35} - u_{33} + (4 - h^2 f_{34})u_{34} = -h^2 g_{34}$$

$$-u_{54} - u_{34} - u_{45} - u_{43} + (4 - h^2 f_{44})u_{44} = -h^2 g_{44}$$

This system of equations could be solved through Gaussian elimination, but let us examine them more closely. There are 45 coefficients. Since u is known at the boundary points, we move these 12 terms to the right-hand side, leaving only 33 nonzero entries out of 81 in our 9×9 system. The standard Gaussian elimination causes a great deal of fill-in in the forward elimination phase—that is, zero entries being replaced by nonzero values. So we seek a method that retains the sparse structure of this system. To illustrate how sparse this system of equations is, we write it in matrix notation:

$$Au = b \tag{11}$$

Suppose that we order the unknowns from left to right and bottom to top:

$$u = [u_{22}, u_{32}, u_{42}, u_{23}, u_{33}, u_{43}, u_{24}, u_{34}, u_{44}]^T \tag{12}$$

This is called the **natural ordering**. Now the coefficient matrix is

$$A = \begin{bmatrix} 4 - h^2 f_{22} & -1 & 0 & -1 & 0 & 0 & 0 & 0 & 0 \\ -1 & 4 - h^2 f_{32} & -1 & 0 & -1 & 0 & 0 & 0 & 0 \\ 0 & -1 & 4 - h^2 f_{42} & 0 & 0 & -1 & 0 & 0 & 0 \\ -1 & 0 & 0 & 4 - h^2 f_{23} & -1 & 0 & -1 & 0 & 0 \\ 0 & -1 & 0 & -1 & 4 - h^2 f_{33} & -1 & 0 & -1 & 0 \\ 0 & 0 & -1 & 0 & -1 & 4 - h^2 f_{43} & 0 & 0 & -1 \\ 0 & 0 & 0 & -1 & 0 & 0 & 4 - h^2 f_{24} & -1 & 0 \\ 0 & 0 & 0 & 0 & -1 & 0 & -1 & 4 - h^2 f_{34} & -1 \\ 0 & 0 & 0 & 0 & 0 & -1 & 0 & -1 & 4 - h^2 f_{44} \end{bmatrix}$$

and the right-hand side is

$$b = \begin{bmatrix} -h^2 g_{22} + u_{12} + u_{21} \\ -h^2 g_{32} + u_{31} \\ -h^2 g_{42} + u_{52} + u_{41} \\ -h^2 g_{23} + u_{13} \\ -h^2 g_{33} \\ -h^2 g_{43} + u_{53} \\ -h^2 g_{24} + u_{14} + u_{25} \\ -h^2 g_{34} + u_{35} \\ -h^2 g_{44} + u_{54} + u_{45} \end{bmatrix}$$

Notice that if $f(x, y) < 0$, then A is a diagonally dominant matrix.

Gauss-Seidel Iterative Method

Since the equations are similar in form, iterative methods are often used to solve such sparse systems. Solving for the diagonal unknown, we have from Equation (10) the **Gauss-Seidel method** or **iteration** given by

$$u_{ij}^{(n+1)} = \frac{1}{4 - h^2 f_{ij}} \left(u_{i+1,j}^{(n)} + u_{i-1,j}^{(n+1)} + u_{i,j+1}^{(n)} + u_{i,j-1}^{(n+1)} - h^2 g_{ij} \right)$$

If we have approximate values of the unkowns at each grid point, this equation can be used to generate new values. We call $u^{(n)}$ the current values of the unknowns at iteration n and $u^{(n+1)}$ the value in the next iteration. Moreover, the new values are used in this equation as soon as they become available.

The pseudocode for this method on a rectangle is as follows:

procedure *seidel*$(a_x, a_y, n_x, n_y, h, itmax, (u_{ij}))$
real array $(u_{ij})_{0:n_x, 0:n_y}$
integer $i, j, k, n_x, n_y, itmax$
real a_x, a_y, x, y
for $k = 1$ **to** *itmax* **do**
    for $j = 1$ **to** $n_y - 1$ **do**
        $y \leftarrow a_y + jh$

```
    for i = 1 to n_x − 1 do
        x ← a_x + ih
        v ← u_{i+1,j} + u_{i−1,j} + u_{i,j+1} + u_{i,j−1}
        u_{ij} ← (v − h²g(x, y))/(4 − h²f(x, y))
    end do
    end do
end do
end procedure seidel
```

In using this procedure, one must decide on the number of iterative steps to be computed, *itmax*. The coordinates of the lower left-hand corner of the rectangle, (a_x, a_y), and the step size h are specified. The number of x grid points is n_x and the number of y grid points is n_y.

Numerical Example and Pseudocode

Let us illustrate this procedure on the boundary value problem

$$\begin{cases} \nabla^2 u − 25u = 0 & \text{inside } R \text{ (unit square)} \\ u = 1 & \text{on the boundary of } R \end{cases} \tag{13}$$

This problem has the known solution $u = \frac{1}{2}[\cosh(5x) + \cosh(5y)]/\cosh(5)$. A driver pseudocode for the Gauss-Seidel procedure, starting with $u = 1$ and taking 20 iterations, is given next. Notice that only 81 words of storage are needed for the array in solving the 49×49 linear system iteratively. Here $h = \frac{1}{8}$.

```
program elliptic
integer parameter n_x ← 8, n_y ← 8, itmax ← 20
real parameter a_x ← 0, b_x ← 1, a_y ← 0, b_y ← 1
real array (u_{ij})_{0:n_x,0:n_y}
integer i, j
real h, x, y
h ← (b_x − a_x)/n_x
for j = 0 to n_y do
    y ← a_y + jh
    u_{0j} ← bndy(a_x, y)
    u_{n_x,j} ← bndy(b_x, y)
end do
for i = 0 to n_x do
    x ← a_x + ih
    u_{i0} ← bndy(x, a_y)
    u_{i,n_y} ← bndy(x, b_y)
end do
for j = 1 to n_y − 1 do
    y ← a_y + jh
```

```
      for i = 1 to n_x − 1
         x ← a_x + ih
         u_ij ← ustart(x, y)
      end do
   end do
   output (u_ij)
   call seidel(a_x, a_y, n_x, n_y, h, itmax, (u_ij))
   output itmax, (u_ij)
   for j = 0 to n_y do
      y ← a_y + jh
      for i = 0 to n_x do
         x ← a_x + ih
         u_ij ← |true_solution(x, y) − u_ij|
      end do
   end do
   output itmax, (u_ij)
   end program elliptic
```

The accompanying functions for this model problem are given next.

```
real function f(x, y)
real x, y
f ← −25
end function
```

```
real function g(x, y)
real x, y
g ← 0
end function
```

```
real function bndy(x, y)
real x, y
bndy ← true_solution(x, y)
end function
```

```
real function ustart(x, y)
real x, y
ustart ← 1
end function
```

```
real function true_solution(x, y)
real x, y
true_solution ← ½[cosh(5x) + cosh(5y)]/ cosh(5)
end function
```

After 20 iterations, the values at the 49 interior grid points are:

0.017083	0.022274	0.032368	0.051407	0.087076	0.15370	0.27777
0.022274	0.027825	0.038123	0.057262	0.092958	0.15950	0.28324
0.032368	0.038123	0.048539	0.067737	0.10344	0.16991	0.29343
0.051407	0.057262	0.067737	0.086982	0.12272	0.18917	0.31257
0.087076	0.092958	0.10344	0.12272	0.15850	0.22498	0.34833
0.15370	0.15950	0.16991	0.18917	0.22498	0.29154	0.41496
0.27777	0.28324	0.29343	0.31257	0.34833	0.41496	0.53870

which differ from the known solution of Problem (13) by

0.89×10^{-3}	0.15×10^{-2}	0.18×10^{-2}	0.20×10^{-2}	0.22×10^{-2}	0.23×10^{-2}	0.20×10^{-2}
0.15×10^{-2}	0.24×10^{-2}	0.29×10^{-2}	0.32×10^{-2}	0.34×10^{-2}	0.35×10^{-2}	0.29×10^{-2}
0.18×10^{-2}	0.29×10^{-2}	0.36×10^{-2}	0.39×10^{-2}	0.41×10^{-2}	0.41×10^{-2}	0.33×10^{-2}
0.20×10^{-2}	0.32×10^{-2}	0.39×10^{-2}	0.43×10^{-2}	0.46×10^{-2}	0.45×10^{-2}	0.36×10^{-2}
0.22×10^{-2}	0.34×10^{-2}	0.41×10^{-2}	0.46×10^{-2}	0.49×10^{-2}	0.48×10^{-2}	0.39×10^{-2}
0.23×10^{-2}	0.35×10^{-2}	0.41×10^{-2}	0.45×10^{-2}	0.48×10^{-2}	0.49×10^{-2}	0.40×10^{-2}
0.20×10^{-2}	0.29×10^{-2}	0.33×10^{-2}	0.36×10^{-2}	0.39×10^{-2}	0.40×10^{-2}	0.34×10^{-2}

These differences are quite large. (Why?)

This example is a good illustration of the fact that the numerical problem being solved is the system of linear Equations (11), which is a discrete approximation to the continuous boundary value Problem (13). When comparing the true solution of (13) with the computed solution of the system, remember the discretization error involved in making the approximation. This error is $\mathcal{O}(h^2)$. With h as large as $h = \frac{1}{8}$, most of the errors in the computed solution are due to the discretization error! To obtain a better agreement between the discrete and continuous problems, select a much smaller mesh size. Of course, the resulting linear system will have a coefficient matrix that is extremely large and quite sparse. Iterative methods are ideal for solving such systems that arise from partial differential equations. For additional information, see Hageman and Young [1981].

PROBLEMS 13.3

1. Establish the formula for the error in the
   a. five-point formula, Equation (3),
   b. nine-point formula, Equation (5).

2. Establish the irregular five-point Formula (6) and its error term.

3. Write the matrices that occur in Equation (11) when the unknowns are ordered according to the vector $\boldsymbol{u} = [u_{22}, u_{42}, u_{33}, u_{24}, u_{44}, u_{32}, u_{23}, u_{43}, u_{34}]^T$. This is known as **red-black** or **checkerboard ordering**.

4. a. Verify Equation (10).
   b. Verify that the solution of Equation (13) is as given in the text.

5. Consider the problem of solving the partial differential equation

$$20u_{xx} - 30u_{yy} + \frac{5}{x+y}u_x + \frac{1}{y}u_y = 69$$

in a region R with u prescribed on the boundary. Derive a five-point finite-difference equation of order $\mathcal{O}(h^2)$ that corresponds to this equation at some interior point (x_i, y_j).

6. Solve this boundary value problem to estimate $u(\frac{1}{2}, \frac{1}{2})$ and $u(0, \frac{1}{2})$:

$$\begin{cases} \nabla^2 u = 0 & (x, y) \in R \\ u = x & (x, y) \in \partial R \end{cases}$$

The region R with boundary ∂R is shown in the figure (the arc is circular). Use $h = \frac{1}{2}$. *Note*: This problem (any many others in this text) can be stated in physical terms also. For example, in this problem we are finding the steady-state temperature in a beam of cross section R if the surface of the beam is held at temperature $u(x, y) = x$.

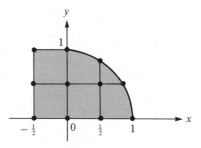

7. Consider the boundary value problem

$$\begin{cases} \nabla^2 u = 9(x^2 + y^2) & (x, y) \in R \\ u = x - y & (x, y) \in \partial R_1 \end{cases}$$

for the region in the unit square with $h = \frac{1}{3}$ in the figure below. Here ∂R is the boundary of R, $\partial R_2 = \{(x, y) \in \partial R : \frac{2}{3} \le x < 1, \ \frac{2}{3} \le y < 1\}$, and

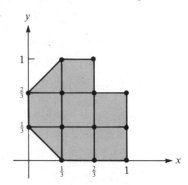

$\partial R_1 = \partial R - \partial R_2$. At the mesh points, determine the system of linear equations that yields an approximate value for $u(x, y)$. Write the system in the form $\mathbf{Au = b}$.

8. Determine the linear system to be solved if the nine-point Formula (5) is used as the approximation in the problem of Equation (1). Notice the pattern in the coefficient matrix with both the five- and nine-point formulas when unknowns in each row are grouped together. (Draw dotted lines through A to form 3×3 submatrices.)

9. In Equation (11), show that A is diagonally dominant when $f(x, y) \leqq 0$.

10. What is the linear system if the nine-point formula

$$\nabla^2 u \approx \frac{1}{12h^2}\big[16u(x + h, y) + 16u(x - h, y) + 16u(x, y + h) + 16u(x, y - h)$$

$$- u(x + 2h, y) - u(x - 2h, y) - u(x, y + 2h) - u(x, y - 2h)$$

$$- 64u(x, y)\big]$$

is used? What are the advantages and disadvantages of using it? *Hint:* It has accuracy $\mathcal{O}(h^4)$.

COMPUTER PROBLEMS 13.3

1. Print the system of linear equations for solving Equation (13) with $h = \frac{1}{4}$ and $\frac{1}{8}$. Solve these systems using procedures *gauss* and *solve* of Chapter 6.

2. Try the Gauss-Seidel routine on the problem

$$\begin{cases} \nabla^2 u = 2e^{x+y} & (x, y) \in R \\ \quad u = e^{x+y} & (x, y) \in \partial R \end{cases}$$

R is the rectangle shown in the figure. Starting values and mesh sizes are in the following table. Compare your numerical solutions with the exact solutions after *itmax* iterations.

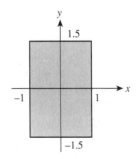

Starting Values	h	itmax
$u = xy$	0.1	15
$u = 0$	0.2	20
$u = (1 + x)(1 + y)$	0.25	40
$u = \left(1 + x + \dfrac{x^2}{2}\right)\left(1 + y + \dfrac{y^2}{2}\right)$	0.05	100
$u = 1 + xy$	0.25	200

3. Modify the Gauss-Seidel procedure to handle the red-black ordering. Redo Computer Problem **2** with this ordering. Does the ordering make any difference? (See Problem **3**.)

4. Rewrite the Gauss-Seidel pseudocode so that it can handle any ordering; that is, introduce an ordering array (ℓ_i). Try several different orderings—natural, red-black, spiral, and diagonal.

5. Consider the heat transfer problem on the irregular region shown in the figure below. The mathematical statement of this problem is as follows:

$$\begin{cases} \dfrac{\partial^2 u}{\partial x^2} + \dfrac{\partial^2 u}{\partial y^2} = 0 & \text{inside} \\[2mm] \dfrac{\partial u}{\partial x} = 0 & \text{sides} \\[2mm] u = 0 & \text{top} \\[2mm] u = 100 & \text{bottom} \end{cases}$$

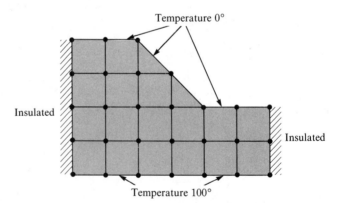

Temperature 0°

Insulated

Insulated

Temperature 100°

Here the partial derivative $\partial u / \partial x$ can be approximated by a divided-difference formula. Establish that the insulated boundaries act like mirrors so that we can assume the temperature is the same as at an adjacent interior grid point. Determine the associated linear system and solve for the temperature u_i with $1 \leq i \leq 10$.

6. Modify procedure *seidel* so that is uses the nine-point Formula (5). Re-solve model Problem (13) and compare results.

7. Solve the example that begins this chapter with $h = \frac{1}{9}$.

8. Solve the boundary value problem

$$\begin{cases} \nabla^2 u + 2u = g & \text{inside } R \\ \qquad\quad u = 0 & \text{on boundary of } R \end{cases}$$

where $g(x, y) = (xy + 1)(xy - x - y) + x^2 + y^2$ and R is the unit square. This problem has the known solution $u = \frac{1}{2}xy(x - 1)(y - 1)$. Use the Gauss-Seidel procedure *seidel* starting with $u = xy$ and take 30 iterations.

9. (Continuation) Using the modified procedure *seidel* of Computer Problem **6** in which the nine-point Formula (5) is used, re-solve this problem. Compare results and explain the difference.

14 MINIMIZATION OF MULTIVARIATE FUNCTIONS

An engineering design problem leads to a function

$$F(x, y) = (\cos x + e^y)^2 + 3(xy)^4$$

in which x and y are parameters to be selected and $F(x, y)$ is a function related to the cost of manufacturing and is to be minimized. Methods for locating optimal points (x, y) in such problems are developed in this chapter.

An important application of calculus is the problem of finding the local minima of a function. Problems of maximization are covered by the theory of minimization because the maxima of F occur at points where $-F$ has its minima. In calculus, the principal technique for minimization is to differentiate the function whose minimum is sought, set the derivative equal to zero, and locate the points that satisfy the resulting equation.

This technique can be used on functions of one or several variables. For example, if a minimum value of $F(x_1, x_2, x_3)$ is sought, we look for the points where all three partial derivatives are simultaneously zero:

$$\frac{\partial F}{\partial x_1} = \frac{\partial F}{\partial x_2} = \frac{\partial F}{\partial x_3} = 0$$

This procedure cannot be readily accepted as a *general-purpose* numerical method because it requires differentiation followed by the solution of one or more equations in one or more variables. This task may be as difficult to accomplish as a direct frontal attack on the original problem.

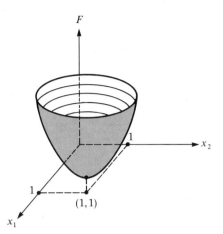

FIGURE 14.1
Elliptic paraboloid

Unconstrained and Constrained Minimization Problems

The minimization problem has two forms: the unconstrained and the constrained. In an unconstrained minimization problem, a function F is defined from the n-dimensional space $\mathbb{R}^n$ into the real line $\mathbb{R}$, and a point $z \in \mathbb{R}^n$ is sought with the property that

$$F(z) \leqq F(x) \qquad \text{for all } x \in \mathbb{R}^n$$

It is convenient to write points in $\mathbb{R}^n$ simply as x, y, z, and so on. If it is necessary to display the components of a point, we write $x = [x_1, x_2, \ldots, x_n]^T$. In a constrained minimization problem, a subset K in $\mathbb{R}^n$ is prescribed and a point $z \in K$ is sought for which

$$F(z) \leqq F(x) \qquad \text{for all } x \in K$$

Such problems are more difficult because of the need to keep the points within the set K. Sometimes the set K is defined in a complicated way.

Consider the elliptic paraboloid $F(x_1, x_2) = x_1^2 + x_2^2 - 2x_1 - 2x_2 + 4$, which is sketched in Figure 14.1. Clearly, the unconstrained minimum occurs at $(1, 1)$ because $F(x_1, x_2) = (x_1 - 1)^2 + (x_2 - 1)^2 + 2$. However, if $K = \{(x_1, x_2) : x_1 \leqq 0, x_2 \leqq 0\}$, the constrained minimum is 4 at $(0, 0)$.

14.1 One-Variable Case

The special case in which a function F is defined on $\mathbb{R}$ is considered first because the more general problem with n variables is often solved by a sequence of one-variable problems.

Suppose that $F : \mathbb{R} \to \mathbb{R}$ and that we seek a point $z \in \mathbb{R}$ with the property that $F(z) \leq F(x)$ for all $x \in \mathbb{R}$. Note that if no assumptions are made about F, this problem is insoluble in its general form. For instance, the function

$$f(x) = \frac{1}{1 + x^2}$$

has no minimum point. Even for relatively well-behaved functions, such as

$$F(x) = x^2 + \sin(53x)$$

numerical methods may encounter some difficulties because of the large number of purely local minima. (See Computer Problem **6**.) Recall that a point z is a **local minimum** point of a function F if there is some neighborhood N of z in which all points satisfy $F(z) \leq F(x)$.

Unimodal *F*

One reasonable assumption is that on some interval $[a, b]$ given to us in advance, F has only a single local minimum. This property is often expressed by saying that F is **unimodal** on $[a, b]$. (*Caution*: In statistics, *unimodal* refers to a single local maximum.) Some unimodal functions are sketched in Figure 14.2.

An important property of a continuous unimodal function, which might be surmised from Figure 14.2, is that it is strictly decreasing up to the minimum point and strictly increasing thereafter. To be convinced of this, let x^* be the minimum point of F on $[a, b]$ and suppose, for instance, that F is not strictly decreasing on the interval $[a, x^*]$. Then points x_1 and x_2 that satisfy $a \leq x_1 < x_2 \leq x^*$ and $F(x_1) \leq F(x_2)$ must exist. Now let x^{**} be a minimum point of F on the interval $[a, x_2]$. (Recall that a continuous function on a closed finite interval attains its minimum value.) We can assume that $x^{**} \neq x_2$ because if x^{**} were initially chosen

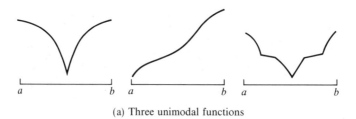

(a) Three unimodal functions

FIGURE 14.2
Examples of unimodal and non-unimodal functions

(b) Three functions that are not unimodal

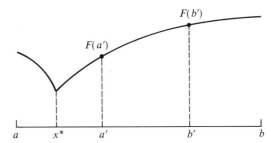

FIGURE 14.3

as x_2, it could be replaced by x_1 inasmuch as $F(x_1) \leq F(x_2)$. But now we see that x^{**} is a local minimum point of F in the interval $[a, b]$ because it is a minimum point of F on $[a, x_2]$ but is not x^* itself. The presence of two local minimum points, of course, contradicts the unimodality of F.

The Fibonacci Search Algorithm

Now we pose a problem concerning the search for a minimum point x^* of a continuous unimodal function F on a given interval $[a, b]$. *How accurately can the true minimum point x^* be computed with only n evaluations of F?* With no evaluations of F, the best that can be said is that $x^* \in [a, b]$; taking the midpoint $\widehat{x} = \frac{1}{2}(b + a)$ as the best estimate gives an error of $|x^* - \widehat{x}| \leq \frac{1}{2}(b - a)$. One evaluation by itself does not improve this situation, and so the best estimate and the error remain the same as in the previous case. Consequently, we need at least two function evaluations to obtain a better estimate.

Suppose that F is evaluated at a' and b' with the results shown in Figure 14.3. If $F(a') < F(b')$, then because F is increasing to the right of x^*, we can be sure that $x^* \in [a, b']$. On the other hand, similar reasoning for the case $F(a') \geq F(b')$ shows that $x^* \in [a', b]$. To make both intervals of uncertainty as small as possible, we move b' to the left and a' to the right. Thus, F should be evaluated at two nearby points on either side of the midpoint, as shown in Figure 14.4. Suppose that

$$a' = \frac{1}{2}(a + b) - 2\delta \quad \text{and} \quad b' = \frac{1}{2}(a + b) + 2\delta$$

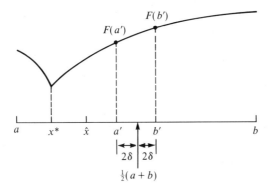

FIGURE 14.4

Taking the midpoint of the appropriate subinterval $[a, b']$ or $[a', b]$ as the best estimate $\hat{x}$ of x^*, we find that the error does not exceed $\frac{1}{4}(b - a) + \delta$. This the reader can easily verify.

For $n = 3$, two evaluations are first made at the $\frac{1}{3}$ and $\frac{2}{3}$ points of the initial interval $[a, b]$; that is,

$$a' = a + \frac{1}{3}(b - a) \quad \text{and} \quad b' = a + \frac{2}{3}(b - a)$$

From the two values $F(a')$ and $F(b')$, it can be determined whether $x^* \in [a, b']$ or $x^* \in [a', b]$. Next, relabel the new interval of uncertainty as $[a, b]$ by setting $a = a'$ if $F(a') \geq F(b')$ or $b = b'$ if $F(a') < F(b')$. We know the value of F at the midpoint of this interval, which is either a' or b', and we can use this information. The third evaluation is made at a point 2δ. The former case is shown in Figure 14.5. Now if $F(a') \geq F(b')$, we set $a = a'$; otherwise, we set $b = b'$. Finally, $\hat{x} = \frac{1}{2}(a + b)$ is the best estimate of x^*. The error does not exceed $\frac{1}{6}(b - a) + \delta$ for the original interval $[a, b]$.

By continuing the search pattern outlined, we find an estimate $\hat{x}$ of x^* with only n evaluations of F and with an error not exceeding

$$\frac{1}{2}\left(\frac{b - a}{\lambda_n}\right) \tag{1}$$

where λ_n is the $(n + 1)$st member of the **Fibonacci sequence**:

$$\begin{cases} \lambda_0 = \lambda_1 = 1 \\ \lambda_k = \lambda_{k-1} + \lambda_{k-2} \qquad (k \geq 2) \end{cases} \tag{2}$$

For example, elements λ_0 through λ_7 are 1, 1, 2, 3, 5, 8, 13, and 21.

In the **Fibonacci search algorithm**, we initially determine the number of steps N for a desired accuracy $\epsilon > \delta$ by selecting N to be the subscript of the smallest Fibonacci number greater than $\frac{1}{2}(b - a)/\epsilon$. We define a sequence of intervals,

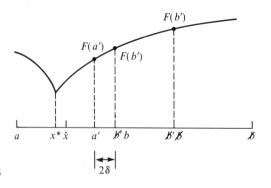

FIGURE 14.5

starting with the given interval $[a, b]$ of length $\ell = b - a$, and, for $k = N, N - 1, \ldots, 3$, use these formulas for updating:

$$\Delta = \left(\frac{\lambda_{k-2}}{\lambda_k}\right)(b - a) \tag{3}$$

$$a' = a + \Delta \qquad b' = b - \Delta$$

$$\begin{cases} a = a' & \text{if } F(a') \geq F(b') \\ b = b' & \text{if } F(a') < F(b') \end{cases}$$

At the step $k = 2$, we set

$$a' = \frac{1}{2}(a + b) - 2\delta \qquad b' = \frac{1}{2}(a + b) + 2\delta$$

$$\begin{cases} a = a' & \text{if } F(a') \geq F(b') \\ b = b' & \text{if } F(a') < F(b') \end{cases}$$

and we have the final interval $[a, b]$, from which we compute $\hat{x} = \frac{1}{2}(a + b)$. This algorithm requires only one function evaluation per step after the initial step.

To verify the algorithm, consider the situation shown in Figure 14.6. Since $\lambda_k = \lambda_{k-1} + \lambda_{k-2}$, we have

$$\ell' = \ell - \Delta = \ell - \left(\frac{\lambda_{k-2}}{\lambda_k}\right)\ell = \left(\frac{\lambda_{k-1}}{\lambda_k}\right)\ell \tag{4}$$

and the length of the interval of uncertainty has been reduced by the factor $(\lambda_{k-1}/\lambda_k)$. The next step yields

$$\Delta' = \left(\frac{\lambda_{k-3}}{\lambda_{k-1}}\right)\ell' \tag{5}$$

and Δ' is actually the distance between a' and b'. Therefore, one of the preceding points at which the function was evaluated is at one end or the other of ℓ; that is,

$$b' - a' = \ell = 2\Delta = \left(\frac{\lambda_k - 2\lambda_{k-2}}{\lambda_k}\right)\ell$$

$$= \left(\frac{\lambda_{k-1} - \lambda_{k-2}}{\lambda_k}\right)\ell = \left(\frac{\lambda_{k-3}}{\lambda_k}\right)\ell$$

$$= \left(\frac{\lambda_{k-3}}{\lambda_{k-1}}\right)\ell' = \Delta'$$

by Equations (2), (4), and (5).

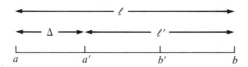

FIGURE 14.6

It is clear by Equation (4) that after $N - 1$ function evaluations, the next-to-last interval has length $(1/\lambda_N)$ times the length of the initial interval $[a, b]$. So the final interval is $(b - a)(1/\lambda_N)$ wide and the maximum error (1) is established. The final step is similar to that outlined, and F is evaluated at a point 2δ away from the midpoint of the next-to-last interval. Finally, set $\hat{x} = \frac{1}{2}(b + a)$ from the last interval $[a, b]$.

One disadvantage of the Fibonacci search is that the algorithm is rather complicated. Also, the desired precision must be given in advance, and the number of steps to be computed for this precision determined before beginning the computation. Thus, the initial evaluation points for the function F depend on N, the number of steps.

The Golden Section Search Algorithm

A similar algorithm that is free of these drawbacks is described next. It has been termed the **golden section search** because it depends on a ratio r known to the early Greeks as the golden section ratio:

$$r = \frac{1}{2}\left(\sqrt{5} - 1\right) \approx 0.61803\ 39887$$

This number satisfies the equation $r^2 = 1 - r$. In each step of this iterative algorithm, an interval $[a, b]$ is available from the previous work. It is an interval known to contain the minimum point x^*, and our objective is to replace it by a smaller interval that is also known to contain x^*. In each step, two values of F are needed:

$$\begin{cases} x = a + r(b - a) & u = F(x) \\ y = a + r^2(b - a) & v = F(y) \end{cases} \tag{6}$$

There are two cases to consider: Either $u > v$ or $u \leq v$. Let us take the first. Figure 14.7 depicts this situation. Since F is assumed unimodal, the minimum of F must be in the interval $[a, x]$. This interval is the input interval at the beginning of the next step. Observe now that within the interval $[a, x]$, one evaluation of F is already available—namely, at y. Also note that

$$a + r(x - a) = y$$

because $x - a = r(b - a)$. In the next step, therefore, y will play the role of x, and we shall need the value of F at the point $a + r^2(x - a)$. So what must be done in this step is to carry out the following replacements *in order*:

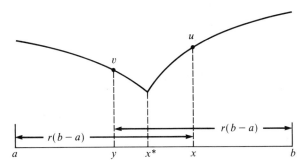

FIGURE 14.7

$$b \leftarrow x$$

$$x \leftarrow y$$

$$u \leftarrow v$$

$$y \leftarrow a + r^2(b - a)$$

$$v \leftarrow F(y)$$

The other case is similar. If $u \leq v$, the picture might be as in Figure 14.8. In this case, the minimum point must lie in $[y, b]$. Within this interval, one value of F is available—namely, at x. Observe that

$$y + r^2(b - y) = x$$

(See Problem **8**.) Thus, x should now be given the role of y, and the value of F is to be computed at $y + r(b - y)$. The following ordered replacements accomplish this:

$$a \leftarrow y$$

$$y \leftarrow x$$

$$v \leftarrow u$$

$$x \leftarrow a + r(b - a)$$

$$u \leftarrow F(x)$$

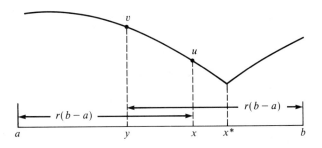

FIGURE 14.8

Problem **9** hints at a shortcoming of this procedure—it is quite slow. Slowness in this context refers to the large number of function evaluations needed to achieve reasonable precision. It can be surmised that this slowness is attributable to the extreme generality of the algorithm. No advantage has been taken of any smoothness that the function F may possess.

Taylor Series Algorithm

Suppose that F is represented by a Taylor series in the vicinity of the point x^*. Then

$$F(x) = F(x^*) + (x - x^*)F'(x^*) + \frac{1}{2}(x - x^*)^2 F''(x^*) + \cdots$$

Since x^* is a minimum point of F, we have $F'(x^*) = 0$. Thus,

$$F(x) \approx F(x^*) + \frac{1}{2}(x - x^*)^2 F''(x^*)$$

This tells us that, in the neighborhood of x^*, $F(x)$ is approximated by a quadratic function whose minimum is also at x^*. Since we do not know x^* and do not want to involve derivatives in our algorithms, a natural stratagem is to interpolate F by a quadratic polynomial. Any three values $(x_i, F(x_i))$, $i = 1, 2, 3$, can be used for this purpose. The minimum point of the resulting quadratic function may be a better approximation to x^* than is x_1, x_2, or x_3. Writing an algorithm that carries out this idea iteratively is not trivial, and many unpleasant cases must be handled. What should be done if the quadratic interpolant has a maximum instead of a minimum, for example? There is also the possibility that $F''(x^*) = 0$, in which case higher-order terms of the Taylor series determine the nature of F near x^*.

Here is the outline of an algorithm for this procedure. At the beginning we have a function F whose minimum is sought. Two starting points x and y are given, as well as two control numbers δ and ϵ. Computing begins by evaluating the two numbers

$$\begin{cases} u = F(x) \\ v = F(y) \end{cases}$$

Now let

$$z = \begin{cases} 2x - y & \text{if } u < v \\ 2y - x & \text{if } u \geq v \end{cases}$$

In either case, the number

$$w = F(z)$$

is to be computed.

At this stage we have three points x, y, and z, together with corresponding function values u, v, and w. In the main iteration step of the algorithm, one of these points and its accompanying function value are replaced by a new point and new function value. The process is repeated until a success or failure is reached.

In the main calculation, a quadratic polynomial q is determined to interpolate F at the three current points x, y, and z. The formulas are discussed below. Next, the point t where $q'(t) = 0$ is determined. Under ideal circumstances, t is a *minimum point* of q and an *approximate minimum* point of F. So one of the x, y, or z should be replaced by t. We are interested in examining $q''(t)$ to determine the shape of the curve q near t.

For the complete description of this algorithm, the formulas for t and $q''(t)$ must be given. They are obtained as follows:

$$
\left\{
\begin{aligned}
a &= \frac{v - u}{y - x} \\[6pt]
b &= \frac{w - v}{z - y} \\[6pt]
c &= \frac{b - a}{z - x} \\[6pt]
t &= \frac{1}{2}\left[x + y - \frac{a}{c} \right] \\[6pt]
q''(t) &= 2c
\end{aligned}
\right.
\tag{7}
$$

Their derivation is outlined in Problem **11**.

The **solution case** occurs if

$$
q''(t) > 0 \quad \text{and} \quad \max\{|t - x|, |t - y|, |t - z|\} < \epsilon
$$

The condition $q''(t) > 0$ indicates, of course, that q' is *increasing* in the vicinity of t so that t is indeed a minimum point of q. The second condition indicates that this estimate, t, of the minimum point of F is within distance ϵ of each of the three points x, y, and z. In this case, t is accepted as a solution.

The **usual case** occurs if

$$
q''(t) > 0 \quad \text{and} \quad \delta \geq \max\{|t - x|, |t - y|, |t - z|\} \geq \epsilon
$$

These inequalities indicate that t is a minimum point of q but not near enough to the three initial points to be accepted as a solution. Also, t is not farther than δ units from each of x, y, and z and can thus be accepted as a reasonable new point. The old point that has the greatest function value is now replaced by t and its function value by $F(t)$.

The **first bad case** occurs if

$$
q''(t) > 0 \quad \text{and} \quad \max\{|t - x|, |t - y|, |t - z|\} > \delta
$$

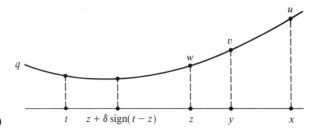

FIGURE 14.9

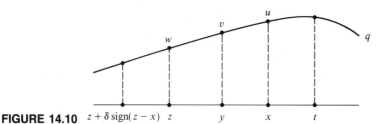

FIGURE 14.10

Here t is a minimum point of q but is so remote that there is some danger in using it as a new point. We identify one of the original three points that is farthest from t, for example, x, and also we identify the point closest to t, say z. Then we replace x by $z + \delta \operatorname{sign}(t - z)$ and u by $F(x)$. Figure 14.9 shows this case. The curve is the graph of q.

The **second bad case** occurs if

$$q''(t) < 0$$

thus indicating that t is a maximum point of q. In this case, identify the greatest and the least among u, v, and w. Suppose, for example, that $u \geq v \geq w$. Then replace x by $z + \delta \operatorname{sign}(z - x)$. An example is shown in Figure 14.10.

PROBLEMS 14.1

1. For the function $F(x_1, x_2, x_3) = x_1^2 + 3x_2^2 + 2x_3^2 - 4x_1 - 6x_2 + 8x_3$, find the unconstrained minimum point. Then find the constrained minimum over the set K defined by inequalities $x_1 \leq 0$, $x_2 \leq 0$, and $x_3 \leq 0$. Next, solve the same problem when K is defined by $x_1 \leq 2$, $x_2 \leq 0$, and $x_3 \leq -2$.

2. For the function $F(x, y) = 13x^2 + 13y^2 - 10xy - 18x - 18y$, find the unconstrained minimum. *Hint:* Try substituting $x = u + v$ and $y = u - v$.

3. If F is unimodal and continuous on the interval $[a, b]$, how many local maxima may F have on $[a, b]$?

4. For the Fibonacci search algorithm, write expressions for $\widehat{x}$ in the cases $n = 2, 3$.

5. Carry out four steps of the Fibonacci search algorithm using $\epsilon = \frac{1}{4}$ to determine the following:

    a. Minimum of $F(x) = x^2 - 6x + 2$ on $[0, 10]$

    b. Minimum of $F(x) = 2x^3 - 9x^2 + 12x + 2$ on $[0, 3]$

    c. Maximum of $F(x) = 2x^3 - 9x^2 + 12x$ on $[0, 2]$

6. Let F be a continuous unimodal function defined on the interval $[a, b]$. Suppose that the values of F are known at n points—namely, $a = t_1 < t_2 < \cdots < t_n = b$. How accurately can one estimate the minimum point x^* from only the values of t_i and $F(t_i)$?

7. The equation satisfied by Fibonacci numbers, namely, $\lambda_n - \lambda_{n-1} - \lambda_{n-2} = 0$, is an example of a linear difference equation with constant coefficients. Solve it by postulating that $\lambda_n = \alpha^n$ and finding that $\alpha_1 = \frac{1}{2}\left(1 + \sqrt{5}\right)$ or $\alpha_2 = \frac{1}{2}\left(1 - \sqrt{5}\right)$ will serve for α. Initial conditions $\lambda_0 = \lambda_1 = 1$ can be met by a solution of the form $\lambda_n = c_1 \alpha_1^n + c_2 \alpha_2^n$. Find c_1 and c_2. Establish that

$$\lim_{n \to \infty} \left(\frac{\lambda_n}{\lambda_{n-1}}\right) = \alpha_1 = \frac{1}{2}\left(1 + \sqrt{5}\right)$$

Show that this agrees with Equations (8) and (11) of Section 3.3.

8. Verify that $y + r^2(b - y) = x$ in the golden section algorithm. *Hint:* Use $r^2 + r = 1$.

9. If F is unimodal on an interval of length ℓ, how many evaluations are necessary in the golden section algorithm to estimate the minimum point with an error of at most 10^{-k}?

10. (Continuation) In Problem **9**, how large must n be if $\ell = 1$ and $k = 10$?

11. Using the divided-difference algorithm on the table

x	y	z
u	v	w

show that the quadratic interpolant in Newton form is

$$q(t) = u + a(t - x) + c(t - x)(t - y)$$

with a, b, and c given by Equation (7). Verify, then, the formulas for t and $q''(t)$ given in (7).

12. If routines can be written easily for F, F', and F'', how can Newton's method be used to locate the minimum point of F? Write down the formula that defines the iterative process. Does it involve F?

13. If routines are available for F and F', how can the secant method be used to minimize F?

14. The reciprocal of the golden section ratio, $s = \frac{1}{2}\left(1 + \sqrt{5}\right)$, has many mystical properties; for example:

a. $s = 1 + \cfrac{1}{1 + \cfrac{1}{1 + \cfrac{1}{1 + \cfrac{1}{1 + \cdots}}}}$

b. $s = \sqrt{1 + \sqrt{1 + \sqrt{1 + \sqrt{1 + \cdots}}}}$

c. $s^n = s^{n-1} + s^{n-2}$

d. $s = s^{-1} + s^{-2} + s^{-3} + \cdots$

Establish these properties.

COMPUTER PROBLEMS 14.1

1. Write a routine to carry out the golden section algorithm for a given function and interval. The search should continue until a preassigned error bound is reached but not beyond 100 steps in any case.

2. (Continuation) Test the routine of Computer Problem **1** on these examples.

   a. $F(x) = \sin x$ on $[0, \pi/2]$

   b. $F(x) = (\arctan x)^2$ on $[-1, 1]$

   c. $F(x) = |\ln x|$ on $[\frac{1}{2}, 4]$

   d. $F(x) = |x|$ on $[-1, 1]$

3. Code and test the following algorithm for approximating the minima of a function F of one variable over an interval $[a, b]$: The algorithm defines a sequence of quadruples $a < a' < b' < b$ by initially setting $a' = \frac{2}{3}a + \frac{1}{3}b$ and $b' = \frac{1}{3}a + \frac{2}{3}b$ and repeatedly updating by $a = a'$, $a' = b'$, and $b' = \frac{1}{2}(b + b')$ if $F(a') > F(b')$; $b = b'$, $a' = \frac{1}{2}(a + a')$, and $b' = a$ if $F(a') < F(b')$; $a = a'$, $b = b'$, $a' = \frac{2}{3}a + \frac{1}{3}b$, and $b' = \frac{1}{3}a + \frac{2}{3}b$ if $F(a') = F(b')$. *Note:* The construction ensures that $a < a' < b' < b$ and the minimum of F always occurs between a and b. Furthermore, only one new function value need be computed at each stage of the calculation after the first unless the case $F(a') = F(b')$ is obtained. The values of a, a', b', and b tend to the same limit, which is a minimum point of F. Notice the similarity to the method of bisection of Section 3.1.

4. Write and test a routine for the Fibonacci search algorithm. Verify that a partial algorithm for the Fibonacci search is as follows: Initially, set

$$\Delta = \left(\frac{\lambda_{N-2}}{\lambda_N}\right)(b - a)$$

$$a' = a + \Delta$$

$$b' = b - \Delta$$

$$u = F(a')$$

$$v = F(b')$$

Then loop on k from $N - 1$ downward to 3, updating as follows:

If $u \geqq v$:

$$
\begin{cases}
a \leftarrow a' \\
a' \leftarrow b' \\
u \leftarrow v \\
\Delta \leftarrow \left(\dfrac{\lambda_{k-2}}{\lambda_k}\right)(b - a) \\
b' \leftarrow b - \Delta \\
v \leftarrow F(b')
\end{cases}
$$

If $v > u$:

$$
\begin{cases}
b \leftarrow b' \\
b' \leftarrow a' \\
v \leftarrow u \\
\Delta \leftarrow \left(\dfrac{\lambda_{k-2}}{\lambda_k}\right)(b - a) \\
a' \leftarrow a + \Delta \\
u \leftarrow F(a')
\end{cases}
$$

Add steps for $k = 2$.

5. (Berman algorithm) Suppose that F is unimodal on $[a, b]$. Then if x_1 and x_2 are any two points such that $a \leqq x_1 < x_2 \leqq b$, we have

$$F(x_1) > F(x_2) \text{ implies } x^* \in (x_1, b]$$
$$F(x_1) = F(x_2) \text{ implies } x^* \in [x_1, x_2]$$
$$F(x_1) < F(x_2) \text{ implies } x^* \in [a, x_2)$$

So by evaluating F at x_1 and x_2 and comparing function values, we are able to reduce the size of the interval known to contain x^*. The simplest approach is to start at the midpoint $x_0 = \frac{1}{2}(a + b)$ and if F is, say, decreasing for $x > x_0$, we test F at $x_0 + ih$, $i = 1, 2, \ldots, q$, with $h = (b - a)/2q$, until we find a point x_1 from which F begins to increase again (or until we reach b). Then we repeat this procedure starting at x_1 and using a smaller step length h/q. Here q is the maximal number of evaluations at each step, say 4. Write a subroutine to perform the Berman algorithm and test it for evaluating the approximate minimization of one-dimensional functions. *Note*: The total number of evaluations of F needed for executing this algorithm up to some iterative step k depends on the location of x^*. If, for example, $x^* = b$, then clearly we need q evaluations at each iteration and hence kq evaluations. This number will decrease the closer x^* is to x_0, and it can be shown that with $q = 4$, the *expected* number of evaluations is three per step. It is interesting to compare the efficiency of the Berman algorithm ($q = 4$) with that of the Fibonacci search algorithm. The expected number of evaluations per step is three, and the uncertainty interval decreases by a factor $4^{-1/3} \approx 0.63$ per evaluation. In comparison, the Fibonacci search algorithm has a reduction factor of $2/\left(1 + \sqrt{5}\right) \approx 0.62$. Of course, the factor 0.63 in the Berman algorithm represents only an average and can be considerably lower but also as high as $4^{-1/4} \approx 0.87$.

6. Select a routine from your program library for finding the minimum point of a function of one variable. Experiment with the function $F(x) = x^2 + \sin(53x)$ to determine whether this routine encounters any difficulties. Use starting values both near and far from the minimum point. (See the figure.)

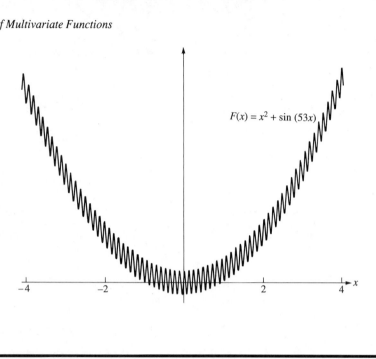

$$F(x) = x^2 + \sin(53x)$$

14.2 Multivariate Case

Now we consider a real-valued function of n real variables $F : \mathbb{R}^n \to \mathbb{R}$. As before, a point x^* is sought such that

$$F(x^*) \leq F(x) \quad \text{for all } x \in \mathbb{R}^n$$

Some of the theory of multivariate functions must be developed in order to understand the rather sophisticated minimization algorithms in current use.

Taylor Series for *F*: Gradient Vector and Hessian Matrix

If the function F possesses partial derivatives of certain low orders (which is usually assumed in the development of these algorithms), then at any given point x a **gradient vector** $G(x)$ is defined with components

$$G_i(x) = \frac{\partial F(x)}{\partial x_i} \qquad (1 \leq i \leq n) \tag{1}$$

and a **Hessian matrix** $H(x)$ is defined with components

$$H_{ij}(x) = \frac{\partial^2 F(x)}{\partial x_i \, \partial x_j} \qquad (1 \leq i, j \leq n) \tag{2}$$

We interpret $G(x)$ as an n-component vector and $H(x)$ as an $n \times n$ matrix, both depending on x.

Using the gradient and Hessian, we can write the first few terms of the Taylor series for F as

$$F(x + h) = F(x) + \sum_{i=1}^{n} G_i(x)h_i + \frac{1}{2} \sum_{i=1}^{n} \sum_{j=1}^{n} h_i H_{ij}(x)h_j + \cdots \tag{3}$$

Equation (3) can also be written in an elegant vector-matrix form:

$$F(x + h) = F(x) + G(x)^T h + \frac{1}{2} h^T H(x)h + \cdots \tag{4}$$

Here x is the fixed point of expansion in $\mathbb{R}^n$ and h is the variable in $\mathbb{R}^n$ with components $h_1, h_2, \ldots, h_n$. The three dots indicate higher-order terms in h that are not needed in this discussion.

A result in calculus states that the *order* in which partial derivatives are taken is immaterial if all partial derivatives that occur are continuous. In the special case of the Hessian matrix, if the second partial derivatives of F are all continuous, then H is a **symmetric matrix**; that is, $H = H^T$ because

$$H_{ij}(x) = \frac{\partial^2 F}{\partial x_i \, \partial x_j} = \frac{\partial^2 F}{\partial x_j \, \partial x_i} = H_{ji}$$

EXAMPLE 1   To illustrate Formula (4), let us compute the first three terms in the Taylor series for the function

$$F(x_1, x_2) = \cos(\pi x_1) + \sin(\pi x_2) + e^{x_1 x_2}$$

taking $(1, 1)$ as the point of expansion.

Solution   Partial derivatives are

$$\frac{\partial F}{\partial x_1} = -\pi \sin(\pi x_1) + x_2 e^{x_1 x_2} \qquad \frac{\partial F}{\partial x_2} = \pi \cos(\pi x_2) + x_1 e^{x_1 x_2}$$

$$\frac{\partial^2 F}{\partial x_1^2} = -\pi^2 \cos(\pi x_1) + x_2^2 e^{x_1 x_2} \qquad \frac{\partial^2 F}{\partial x_2 \, \partial x_1} = (x_1 x_2 + 1)e^{x_1 x_2}$$

$$\frac{\partial^2 F}{\partial x_1 \, \partial x_2} = (x_1 x_2 + 1)e^{x_1 x_2} \qquad \frac{\partial^2 F}{\partial x_2^2} = -\pi^2 \sin(\pi x_2) + x_1^2 e^{x_1 x_2}$$

Note the equality of cross derivatives; that is, $\partial^2 F/\partial x_1 \, \partial x_2 = \partial^2 F/\partial x_2 \, \partial x_1$. At the particular point $x = [1, 1]^T$, we have

$$F(x) = -1 + e \qquad G(x) = \begin{bmatrix} e \\ -\pi + e \end{bmatrix} \qquad H(x) = \begin{bmatrix} \pi^2 + e & 2e \\ 2e & e \end{bmatrix}$$

So by Equation (4),

$$F(1 + h_1, 1 + h_2) = -1 + e + [e, -\pi + e]\begin{bmatrix} h_1 \\ h_2 \end{bmatrix}$$

$$+ \frac{1}{2}[h_1, h_2]\begin{bmatrix} \pi^2 + e & 2e \\ 2e & e \end{bmatrix}\begin{bmatrix} h_1 \\ h_2 \end{bmatrix} + \cdots$$

or, equivalently, by Equation (3),

$$F(1 + h_1, 1 + h_2) = -1 + e + eh_1 + (-\pi + e)h_2$$

$$+ \frac{1}{2}\left[\left(\pi^2 + e\right)h_1^2 + (2e)h_1h_2 + (2e)h_2h_1 + eh_2^2\right] + \cdots \quad \square$$

We can verify these calculations using the following Maple statements.

```
with(linalg):
F := cos(Pi*x) + sin(Pi*y) + exp(x*y);
G := grad(F, [x,y]);
H := hessian(F, [x,y]);
x := 1; y :=1;
map(eval, F);
map(eval, G);
map(eval, H);
```

Alternative Form of Taylor Series

Another form of the Taylor series is useful. First, let z be the point of expansion and then let $h = x - z$. Now from Equation (4),

$$F(x) = F(z) + G(z)^T(x - z) + \frac{1}{2}(x - z)^T H(z)(x - z) + \cdots \qquad (5)$$

We illustrate with two special types of functions.

First, the **linear function** has the form

$$F(x) = c + \sum_{i=1}^{n} b_i x_i = c + b^T x$$

for appropriate coefficients $c, b_1, b_2, \ldots, b_n$. Clearly, the gradient and Hessian are $G_i(z) = b_i$ and $H_{ij}(z) = 0$, so Equation (5) yields

$$F(x) = F(z) + \sum_{i=1}^{n} b_i(x_i - z_i) = F(z) + b^T(x - z)$$

Second, consider a general **quadratic function**. For simplicity, we take only two variables. The form of the function is

$$F(x_1, x_2) = c + (b_1 x_1 + b_2 x_2) + \frac{1}{2} \left(a_{11} x_1^2 + 2 a_{12} x_1 x_2 + a_{22} x_2^2 \right) \tag{6}$$

which can be interpreted as the Taylor series for F when the point of expansion is $(0, 0)$. To verify this assertion, the partial derivatives must be computed and evaluated at $(0, 0)$:

$$\frac{\partial F}{\partial x_1} = b_1 + a_{11} x_1 + a_{12} x_2 \qquad \frac{\partial F}{\partial x_2} = b_2 + a_{22} x_2 + a_{12} x_1$$

$$\frac{\partial^2 F}{\partial x_1^2} = a_{11} \qquad \frac{\partial^2 F}{\partial x_1 \partial x_2} = a_{12}$$

$$\frac{\partial^2 F}{\partial x_2 \partial x_1} = a_{12} \qquad \frac{\partial^2 F}{\partial x_2^2} = a_{22}$$

Letting $z = [0, 0]^T$, we obtain from Equation (5)

$$F(x) = c + [b_1, b_2] \begin{bmatrix} x_1 \\ x_2 \end{bmatrix} + \frac{1}{2} [x_1, x_2] \begin{bmatrix} a_{11} & a_{12} \\ a_{12} & a_{22} \end{bmatrix} \begin{bmatrix} x_1 \\ x_2 \end{bmatrix}$$

This is the matrix form of the original quadratic function of two variables. It can also be written as

$$F(x) = c + b^T x + \frac{1}{2} x^T A x \tag{7}$$

where c is a scalar, b a vector, and A a matrix. Equation (7) holds for a general quadratic function of n variables, with b an n-component vector and A an $n \times n$ matrix.

Returning to Equation (3), we now write out the complicated double sum in complete detail to assist in understanding it:

$$x^T H x = \sum_{i=1}^{n} \sum_{j=1}^{n} x_i H_{ij} x_j = \begin{cases} \sum_{j=1}^{n} x_1 H_{1j} x_j \\ + \sum_{j=1}^{n} x_2 H_{2j} x_j \\ + \cdots \\ + \cdots \\ + \sum_{j=1}^{n} x_n H_{nj} x_j \end{cases}$$

$$
= \left\{
\begin{array}{l}
x_1 H_{11} x_1 + x_1 H_{12} x_2 + \cdots + x_1 H_{1n} x_n \\
+ \, x_2 H_{21} x_1 + x_2 H_{22} x_2 + \cdots + x_2 H_{2n} x_n \\
+ \cdots \qquad\qquad\qquad\quad + \cdots \\
+ \cdots \qquad\qquad\qquad\quad + \cdots \\
+ \, x_n H_{n1} x_1 + x_n H_{n2} x_2 + \cdots + x_n H_{nn} x_n
\end{array}
\right\}
$$

Thus, $x^T H x$ can be interpreted as the sum of all n^2 terms in a square array of which the (i, j) element is $x_i H_{ij} x_j$.

Steepest Descent Procedure

A crucial property of the gradient vector $G(x)$ is that it points in the direction of the most rapid increase in the function F—the direction of steepest ascent. Conversely, $-G(x)$ points in the direction of the steepest descent. This fact is so important that it is worth a few words of justification. Suppose that h is a unit vector, $\sum_{i=1}^n h_i^2 = 1$. Now the rate of change of F (at x) in the direction h is defined naturally by

$$
\frac{d}{dt} F(x + th) \bigg|_{t=0}
$$

This rate of change can be evaluated by using Equation (4). From that equation, it follows that

$$
F(x + th) = F(x) + t G(x)^T h + \frac{1}{2} t^2 h^T H(x) h + \cdots \tag{8}
$$

Differentiation with respect to t leads to

$$
\frac{d}{dt} F(x + th) = G(x)^T h + t h^T H(x) h + \cdots \tag{9}
$$

By letting $t = 0$ here, we see that the rate of change of F in the direction h is nothing else than

$$
G(x)^T h
$$

Now we ask: *For what unit vector h is the rate of change a maximum?* The simplest path to the answer is to invoke the powerful **Cauchy-Schwarz inequality**:

$$
\sum_{i=1}^n u_i v_i \leqq \left(\sum_{i=1}^n u_i^2 \right)^{1/2} \left(\sum_{i=1}^n v_i^2 \right)^{1/2} \tag{10}
$$

with equality holding only if one of the vectors u or v is a nonnegative multiple of the other. Applying this to

$$G(x)^T h = \sum_{i=1}^{n} G_i(x) h_i$$

and remembering that $\sum_{i=1}^{n} h_i^2 = 1$, we conclude that the maximum occurs when h is a positive multiple of $G(x)$—that is, when h points in the direction of G.

Based on the foregoing discussion, a minimization procedure called **steepest descent** can be described. At any given point x, the gradient vector $G(x)$ is calculated. Then a one-dimensional minimization problem is solved by determining the value t^* for which the function

$$\phi(t) = F(x + tG(x))$$

is a minimum. Then we replace x by $x + t^*G(x)$ and begin anew.

The method of steepest descent is not usually competitive with other methods but it has the advantage of simplicity. One way of speeding it up is described in Computer Problem **2**.

Contour Diagrams

In understanding how these methods work on functions of two variables, it is often helpful to draw contour diagrams. A **contour** of a function F is a set of the form

$$\{x : F(x) = c\}$$

where c is a given constant. For example, the contours of function

$$F(x) = 25x_1^2 + x_2^2$$

are ellipses, as shown in Figure 14.11. Contours are also called **level sets** by some authors. At any point on a contour, the gradient of F is perpendicular to the curve. So, in general, the path of steepest descent may look like Figure 14.12.

More Advanced Algorithms

To explain more advanced algorithms, we consider a general real-valued function F of n variables. Suppose that we have obtained the first three terms in the Taylor series of F in the vicinity of a point z. *How can they be used to guess the minimum point of F?* Obviously, we could ignore all terms beyond the quadratic terms and find the minimum of the resulting quadratic function:

$$F(x + z) = F(z) + G(z)^T x + \frac{1}{2}x^T H(z)x + \cdots \tag{11}$$

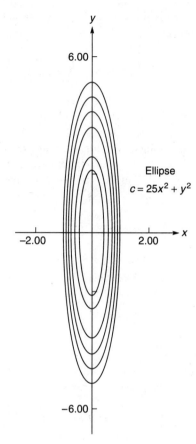

FIGURE 14.11

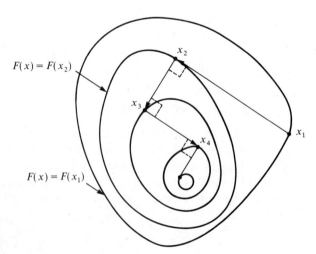

FIGURE 14.12
Path of steepest descent

Here z is fixed and x is the variable. To find the minimum of this quadratic function of x, we must compute the first partial derivatives and set them equal to zero. Denoting this quadratic function by Q and simplifying the notation slightly, we have

$$Q(x) = F(z) + \sum_{i=1}^{n} G_i x_i + \frac{1}{2} \sum_{i=1}^{n} \sum_{j=1}^{n} x_i H_{ij} x_j \tag{12}$$

from which it follows that

$$\frac{\partial Q}{\partial x_k} = G_k + \sum_{j=1}^{n} H_{kj} x_j \qquad (1 \leqq k \leqq n) \tag{13}$$

(See Problem **13**.) The point x that is sought is thus a solution of the system of n equations

$$\sum_{j=1}^{n} H_{kj} x_j = -G_k \qquad (1 \leqq k \leqq n)$$

or, equivalently,

$$H(z)x = -G(z) \tag{14}$$

The preceding analysis suggests the following iterative procedure for locating a minimum point of a function F: Start with a point z that is a current estimate of the minimum point. Compute the gradient and Hessian of F at the point z. They can be denoted by G and H, respectively. Of course, G is an n-component vector of numbers and H is an $n \times n$ matrix of numbers. Then solve the matrix equation

$$Hx = -G$$

obtaining an n-component vector x. Replace z by $z + x$ and return to the beginning of the procedure.

Minimum, Maximum, and Saddle Points

There are many reasons for expecting trouble from the iterative procedure just outlined. One especially noisome aspect is that we can expect to find a point only where the first partial derivatives of F vanish; it need not be minimum point. It is what we call a **stationary point**. Such points can be classified into three types: *minimum point, maximum point*, and *saddle point*. They can be illustrated by simple quadratic surfaces familiar from analytic geometry:

1. Minimum of $F(x, y) = x^2 + y^2$ at $(0, 0)$ (See Figure 14.13a.)
2. Maximum of $F(x, y) = 1 - x^2 - y^2$ at $(0, 0)$ (See Figure 14.13b.)
3. Saddle point of $F(x, y) = x^2 - y^2$ at $(0, 0)$ (See Figure 14.13c.)

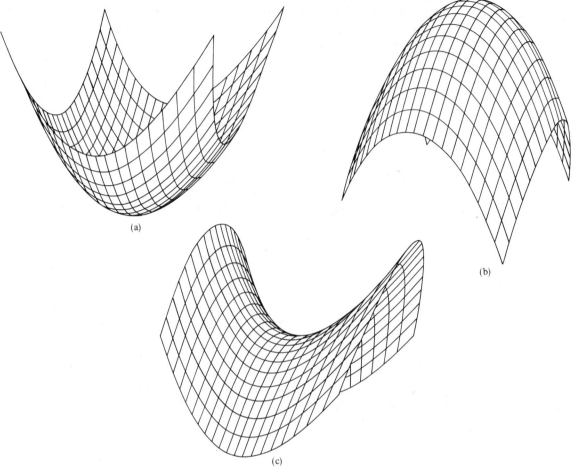

FIGURE 14.13
Simple quadric surfaces

Positive Definite Matrix

If z is a stationary point of F, then

$$G(z) = 0$$

Moreover, a criterion ensuring that Q, as defined in Equation (12), has a minimum point is this:

THEOREM

If the matrix H has the property that $x^T H x > 0$ for every nonzero vector x, then the quadratic function Q has a minimum point.

(See Problem **15**.) A matrix that has this property is said to be **positive definite**. Notice that this theorem involves only second-degree terms in the quadratic function Q.

As examples of quadratic functions that do not have minima, consider the following:

1. $-x_1^2 - x_2^2 + 13x_1 + 6x_2 + 12$

2. $x_1^2 - x_2^2 + 3x_1 + 5x_2 + 7$

3. $x_1^2 - 2x_1x_2 + x_1 + 2x_2 + 3$

4. $2x_1 + 4x_2 + 6$

In the first two examples, let $x_1 = 0$ and $x_2 \to \infty$. In the third, let $x_1 = x_2 \to \infty$. In the last, let $x_1 = 0$ and $x_2 \to -\infty$. In each case, the function values approach $-\infty$, and no global minimum can exist.

The algorithms currently recommended for minimization are of a type called **quasi-Newton**. The principal example is an algorithm introduced in 1959 by Davidon called the **variable metric algorithm**. Subsequently, important modifications and improvements were made by others, such as R. Fletcher, M. J. D. Powell, C. G. Broyden, P. E. Gill, and W. Murray. For a practical guide to optimization, see Gill, Murray, and Wright [1981]. These algorithms proceed iteratively, assuming in each step that a local quadratic approximation is known for the function F whose minimum is sought. The minimum of this quadratic function either provides the new point directly or is used to determine a line along which a one-dimensional search can be carried out. In implementation of the algorithm, the gradient can be either provided in the form of a procedure or computed numerically by finite differences. The Hessian H is not computed, but an estimate of its LU factorization is kept up to date as the process continues.

PROBLEMS 14.2

1. Determine whether these functions have minimum values in $\mathbb{R}^2$.

    a. $x_1^2 - x_1x_2 + x_2^2 + 3x_1 + 6x_2 - 4$

    b. $x_1^2 - 3x_1x_2 + x_2^2 + 7x_1 + 3x_2 + 5$

    c. $2x_1^2 - 3x_1x_2 + x_2^2 + 4x_1 - x_2 + 6$

    d. $ax_1^2 - 2bx_1x_2 + cx_2^2 + dx_1 + ex_2 + f$

    Hint: Use the method of completing the square.

2. Locate the minimum point of $3x^2 - 2xy + y^2 + 3x - 46 + 7$ by finding the gradient and Hessian and solving the appropriate linear equations.

3. Using $(0,0)$ as the point of expansion, write the first three terms of the Taylor series for $F(x, y) = e^x \cos y - y \ln(x + 1)$.

4. Using $(1, 1)$ as the point of expansion, write the first three terms of the Taylor series for $F(x, y) = 2x^2 - 4xy + 7y^2 - 3x + 5y$.

5. The Taylor series expansion about zero can be written as

$$F(\mathbf{x}) = F(0) + G(0)^T \mathbf{x} + \frac{1}{2}\mathbf{x}^T H(0)\mathbf{x} + \cdots$$

Show that the Taylor series about z can be written in a similar form by using matrix-vector notation; that is,

$$F(x) = F(z) + G(z)^T X + \frac{1}{2} X^T \mathcal{H}(z) X + \cdots$$

where

$$X = \begin{bmatrix} x \\ z \end{bmatrix} \qquad G(z) = \begin{bmatrix} G(z) \\ -G(z) \end{bmatrix} \qquad \mathcal{H}(z) = \begin{bmatrix} H(z) & -H(z) \\ -H(z) & H(z) \end{bmatrix}$$

6. Show that the gradient of $F(x, y)$ is perpendicular to the contour. *Hint*: Interpret the equation $F(x, y) = c$ as defining y as a function of x. Then by the chain rule,

$$\frac{\partial F}{\partial x} + \frac{\partial F}{\partial y} \frac{dy}{dx} = 0$$

From it obtain the slope of the tangent to the contour.

7. Consider the function

$$F(x_1, x_2, x_3) = 3e^{x_1 x_2} - x_3 \cos x_1 + x_2 \ln x_3$$

   a. Determine the gradient vector and Hessian matrix.

   b. Derive the first three terms of the Taylor series expansion about $(0, 1, 1)$.

   c. What linear system should be solved for a reasonable guess as to the minimum point for F? What is the value of F at this point?

8. It is asserted that the Hessian of an unknown function F at a certain point is

$$\begin{bmatrix} 3 & 2 \\ 1 & 4 \end{bmatrix}$$

What conclusion can be drawn about F?

9. What are the gradients of the following functions at the points indicated?

   a. $F(x, y) = x^2 y - 2x + y$ at $(1, 0)$

   b. $F(x, y, z) = xy + yz^2 + x^2 z$ at $(1, 2, 1)$

10. Consider $F(x, y, z) = y^2 z^2 (1 + \sin^2 x) + (y + 1)^2 (z + 3)^2$. We want to find the minimum of the function. The program to be used requires the gradient of the function. What formulas must we program for the gradient?

11. Let F be a function of two variables whose gradient at $(0, 0)$ is $[-5, 1]^T$ and whose Hessian is

$$\begin{bmatrix} 6 & -1 \\ -1 & 2 \end{bmatrix}$$

Make a reasonable guess as to the minimum point of F. Explain.

12. Write the function $F(x_1, x_2) = 3x_1^2 + 6x_1x_2 - 2x_2^2 + 5x_1 + 3x_2 + 7$ in the form of Equation (7) with appropriate A, b, and c. Show in matrix form the linear equations that must be solved in order to find a point where the first partial derivatives of F vanish. Finally, solve these equations to locate this point numerically.

13. Verify Equation (13). In differentiating the double sum in Equation (12), first write all terms that contain x_k. Then differentiate and use the symmetry of the matrix H.

14. Consider the quadratic function Q in Equation (12). Show that if H is positive definite, then the stationary point is a minimum point.

15. (General quadratic equation) Generalize Equation (6) to n variables. Show that a general quadratic function $Q(x)$ of n variables can be written in the matrix-vector form of Equation (7), where A is an $n \times n$ symmetric matrix, b a vector of length n, and c a scalar. Establish that the gradient and Hessian are

$$G(x) = Ax + b \quad \text{and} \quad H(x) = A$$

respectively.

16. Let A be an $n \times n$ symmetric matrix and define an upper triangular matrix $U = (u_{ij})$ by putting

$$u_{ij} = \begin{cases} a_{ij} & \text{if } i = j \\ 2a_{ij} & \text{if } i < j \\ 0 & \text{if } i > j \end{cases}$$

Show that $x^T U x = x^T A x$ for all vectors x.

17. Show that the general quadratic function $Q(x)$ of n variables can be written

$$Q(x) = c + b^T x + \frac{1}{2} x^T U x$$

where U is an upper triangular matrix. Can this simplify the work of finding the stationary point of Q?

18. Show that the gradient and Hessian satisfy the equation

$$H(z)(x - z) = G(x) - G(z)$$

for a general quadratic function of n variables.

19. Using Taylor series, show that a general quadratic function of n variables can be written in block form

$$Q(x) = \frac{1}{2} X^T \mathcal{A} X + \mathcal{B}^T X + c$$

where

$$X = \begin{bmatrix} x \\ z \end{bmatrix} \qquad \mathcal{A} = \begin{bmatrix} A & -A \\ -A & A \end{bmatrix} \qquad \mathcal{B} = \begin{bmatrix} b \\ -b \end{bmatrix}$$

Here z is the point of expansion.

20. (Least squares problem) Consider the function

$$F(x) = (b - Ax)^T(b - Ax) + \alpha x^T x$$

where A is a real $m \times n$ matrix, b a real column vector of order m, and α a positive real number. We want the minimum point of F for given A, b, and α. Show that

$$F(x + h) - F(x) = (Ah)^T(Ah) + \alpha h^T h \geq 0$$

for h a vector of order n, provided that

$$(A^T A + \alpha I)x = A^T b$$

This means that any solution of this linear system minimizes $F(x)$; hence, this is the normal equation.

COMPUTER PROBLEMS 14.2

1. Select a routine from your program library for minimizing a function of many variables without the need to program derivatives. Test it on one or more of the following well-known functions. The ordering of our variables are (x, y, z, w).

   a. Rosenbrock's: $100(y - x^2)^2 + (1 - x)^2$. Start at $(-1.2, 1.0)$.

   b. Powell's: $(x + 10y)^2 + 5(z - w)^2 + (y - 2z)^4 + 10(x - w)^4$. Start at $(3, -1, 0, 1)$.

   c. Fletcher and Powell's: $100(z - 10\phi)^2 + \left(\sqrt{x^2 + y^2} - 1\right)^2 + z^2$ in which ϕ is an angle determined from (x, y) by $\cos 2\pi\phi = x/\sqrt{x^2 + y^2}$ and $\sin 2\pi\phi = y\sqrt{x^2 + y^2}$, where $-\pi/2 < 2\pi\phi \leq 3\pi/2$. Start at $(1, 1, 1)$.

   d. Powell's: $x^2 + 2y^2 + 3z^2 + 4w^2 + (x + y + z + w)^4$. Start at $(1, -1, 1, 1)$.

   e. Wood's: $100(x^2 - y)^2 + (1 - x)^2 + 90(z^2 - w)^2 + (1 - z)^2 + 10(y - 1)^2 + (w - 1)^2 + 19.8(y - 1)(w - 1)$. Start at $(-3, -1, -3, -1)$.

2. (Accelerated steepest descent) This version of steepest descent is superior to the basic one. A sequence of points $x_1, x_2, \ldots$ is generated as follows: Point x_1 is specified as the starting point. Then x_2 is obtained by one step of steepest descent from x_1. In the general step, if $x_1, x_2, \ldots, x_m$ have been obtained, we find a point z by steepest descent from x_m. Then x_{m+1} is taken as the minimum point on the line $x_{m-1} + t(z - z_{m-1})$. Program and test this algorithm on one of the examples in Computer Problem **1**.

3. Find a routine in your program library and solve the minimization problem that begins this chapter.

4. We want to find the minimum of $F(x, y, z) = z^2 \cos x + x^2 y^2 + x^2 e^z$ using a computer program that requires procedures for the gradient of F together with F. Write the procedures needed. Find the minimum using a preprogrammed code that uses the gradient.

5. Assume that

   procedure $xmin(f, (grad_i), n, (xi), (g_{ij}))$

   is available to compute the minimum value of a function of two variables. Suppose that this routine requires not only the function but also its gradient. If we are going to use this routine with the function $F(x, y) = e^x \cos^2(xy)$, what procedure will be needed? Write the appropriate code. Find the minimum using a preprogrammed code that uses the gradient.

15 LINEAR PROGRAMMING

In the study of how the U.S. economy is affected by changes in the supply and cost of energy, it has been found appropriate to use a linear programming model. This is a large system of linear inequalities that govern the variables in the model, together with a linear function of these variables to be maximized. Typically the variables are the activity levels of various processes in the economy, such as the number of barrels of oil pumped per day or the number of men's shirts produced per day. A model that contains reasonable detail could easily involve thousands of variables and thousands of linear inequalities. Such problems are discussed in this chapter, and some guidance is offered on how to use existing software.

15.1 Standard Forms and Duality

First Primal Form

Linear programming is a branch of mathematics that deals with finding extreme values of linear functions when the variables are constrained by linear inequalities. Any problem of this type can be put into a standard form known as "first primal form" (see the box on the top of the next page) by simple manipulations (to be discussed later).

In matrix notation, the linear programming problem in first primal form looks like this:

$$\text{maximize} : c^T x$$

$$\text{constraints} : \begin{cases} Ax \leqq b \\ x \geqq 0 \end{cases} \tag{1}$$

FIRST PRIMAL FORM

Given data c_j, a_{ij}, b_i $(1 \leqq j \leqq n, 1 \leqq i \leqq m)$, we wish to determine the x_j's $(1 \leqq j \leqq n)$ that maximize the linear function

$$\sum_{j=1}^{n} c_j x_j$$

subject to the constraints

$$\begin{cases} \sum_{j=1}^{n} a_{ij} x_j \leqq b_i & (1 \leqq i \leqq m) \\ x_j \geqq 0 & (1 \leqq j \leqq n) \end{cases}$$

Here c and x are n-component vectors, b is an m-component vector, and A an $m \times n$ matrix. A **vector inequality** $u \leqq v$ means that u and v are vectors with the same number of components and that *all* the individual components satisfy the inequality $u_i \leqq v_i$. The linear function $c^T x$ is called the **objective function**.

In a linear programming problem, the set of all vectors that satisfy the constraints is called the **feasible set**, and its elements are the **feasible points**. So in the preceding notation, the feasible set is

$$K = \{x \in \mathbb{R}^n : x \geqq 0 \quad \text{and} \quad Ax \leqq b\}$$

A more precise (and concise) statement of the linear programming problem, then, is: Determine $x^* \in K$ such that $c^T x^* \geqq c^T x$ for all $x \in K$.

Numerical Example

To get an idea of the type of practical problem that can be solved by linear programming, consider a simple example of optimization. Suppose that a certain factory uses two raw materials to produce two products. Suppose also that the following are true:

1. Each unit of the first product requires 5 units of the first raw material and 3 of the second.

2. Each unit of the second product requires 3 units of the first raw material and 6 of the second.

3. On hand are 15 units of the first raw material and 18 of the second.

4. The profits on sales of the products are 2 per unit and 3 per unit, respectively.

How should the raw materials be used to realize a maximum profit? In order to answer this question, variables x_1 and x_2 are introduced to represent the number

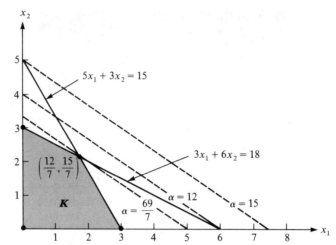

FIGURE 15.1
Graphical
solution method

of units of the two products to be manufactured. In terms of these variables, the profit is

$$2x_1 + 3x_2 \tag{2}$$

The process uses up $5x_1 + 3x_2$ units of the first raw material and $3x_1 + 6x_2$ units of the second. The limitations in fact 3 above are expressed by these inequalities:

$$\begin{cases} 5x_1 + 3x_2 \leq 15 \\ 3x_1 + 6x_2 \leq 18 \end{cases} \tag{3}$$

Of course, $x_1 \geq 0$ and $x_2 \geq 0$. Thus, the solution to the problem is a vector $x \geq 0$ that maximizes the objective function (2) while satisfying the constraints (3). More precisely, among all vectors x in the set

$$K = \{x : x \geq 0, 5x_1 + 3x_2 \leq 15, 3x_1 + 6x_2 \leq 18\}$$

we want the one that makes $2x_1 + 3x_2$ as large as possible.

Because the number of variables in this example is only two, the problem can be solved graphically. To locate the solution, we begin by graphing the set K. This is the shaded region in Figure 15.1. Then we draw some of the lines $2x_1 + 3x_2 = \alpha$, where α is given various values. These lines are dashed in the figure and labeled with the values of α. Finally, we select one of these lines with a maximum α that intersects K. That intersection is the solution point and a vertex of K. It is obtained numerically by solving simultaneously the equations $5x_1 + 3x_2 = 15$ and $3x_1 + 6x_2 = 18$. Thus, $x = \left[\frac{12}{7}, \frac{15}{7} \right]^T$, and the corresponding profit from Equation (2) is $2\left(\frac{12}{7}\right) + 3\left(\frac{15}{7}\right) = \frac{69}{7}$.

We can use Maple as follows to solve this problem.

```
with(simplex):
obj := 2*x + 3*y;
consts := {5*x + 3*y <= 15,
           3*x + 6*y <= 18};
maximize(obj, consts union {x >= 0, y >= 0});
```

We obtain $x = \frac{12}{7}$ and $y = \frac{15}{7}$.

Note in this example that the units used, whether dollars, pesos, pounds, or kilograms, do not matter for the mathematical method as long as they are used consistently. Notice also that x_1 and x_2 are permitted to be arbitrary real numbers. The problem would be quite different if only integer values were acceptable as a solution. This situation would occur if the products being produced consisted of indivisible units, such as a manufactured article. If the integer constraint is imposed, only points with integer coordinates inside K are acceptable. So $(0, 3)$ is the best of them. Observe particularly that we *cannot* simply round off the solution $(1.71, 2.14)$ to the nearest integers to solve the problem with integer constraints. The point $(2, 2)$ lies just outside K. However, if the company could alter the constraints slightly by increasing the amount of the first raw material to 16, the integer solution $(2, 2)$ would be allowable. Special programs for *integer* linear programming are available but are outside the scope of this book.

Observe how the solution would be altered if our profit or objective function were $2x_1 + x_2$. In this case, the dashed lines in the figure would have a different slope (namely, -2) and a different vertex of the shaded region would occur as the solution—namely, $(3, 0)$. A characteristic feature of linear programming problems is that the solutions (if any exist) can always be found among the vertices.

Transforming Problems into First Primal Form

A linear programming problem not already in the first primal form can be put into that form by some standard techniques:

1. If the original problem calls for the minimization of the linear function $c^T x$, this is the same as maximizing $(-c)^T x$.

2. If the original problem contains a constraint like $a^T x \geqq \beta$, it can be replaced by the constraint $(-a)^T x \leqq -\beta$.

3. If the objective function contains a constant, this fact has no effect on the solution. For example, the maximum of $c^T x + \lambda$ occurs for the same x as the maximum of $c^T x$.

4. If the original problem contains equality constraints, each can be replaced by two inequality constraints. Thus, the equation $a^T x = \beta$ is equivalent to $a^T x \leqq \beta$ and $a^T x \geqq \beta$.

5. If the original problem does not require a variable (say x_i) to be nonnegative, we can replace x_i by the difference of two nonnegative variables, say $x_i = u_i - v_i$, where $u_i \geqq 0$ and $v_i \geqq 0$.

Here is an example that illustrates all five techniques. Consider the linear programming problem

$$\text{minimize} : 2x_1 + 3x_2 - x_3 + 4$$

$$\text{constraints} : \begin{cases} x_1 - x_2 + 4x_3 \geqq 2 \\ x_1 + x_2 + x_3 = 15 \\ x_2 \geqq 0 \geqq x_3 \end{cases}$$

It is equivalent to the following problem in first primal form:

$$\text{maximize} : -2u + 2v - 3z - w$$

$$\text{constraints} : \begin{cases} -u + v + z + 4w \leqq -2 \\ u - v + z - w \leqq 15 \\ -u + v - z + w \leqq -15 \\ u \geqq 0 \quad v \geqq 0 \quad z \geqq 0 \quad w \geqq 0 \end{cases}$$

Dual Problem

Corresponding to a given linear programming problem in first primal form is another problem called its **dual**. It is obtained from the original primal problem

$$\text{maximize} : c^T x$$

(P)
$$\text{constraints} : \begin{cases} Ax \leqq b \\ x \geqq 0 \end{cases}$$

by defining the dual to be the problem

$$\text{minimize} : b^T y$$

(D)
$$\text{constraints} : \begin{cases} A^T y \geqq c \\ y \geqq 0 \end{cases}$$

For example, the dual of this problem

$$\text{maximize} : 2x_1 + 3x_2$$

$$\text{constraints} : \begin{cases} 4x_1 + 5x_2 \leqq 6 \\ 7x_1 + 8x_2 \leqq 9 \\ 10x_1 + 11x_2 \leqq 12 \\ x_1 \geqq 0 \quad x_2 \geqq 0 \end{cases}$$

is

$$\text{minimize} : 6y_1 + 9y_2 + 12y_3$$

$$\text{constraints} : \begin{cases} 4y_1 + 7y_2 + 10y_3 \geqq 2 \\ 5y_1 + 8y_2 + 11y_3 \geqq 3 \\ y_1 \geqq 0 \quad y_2 \geqq 0 \quad y_3 \geqq 0 \end{cases}$$

Note that, in general, the dual problem has different dimensions from those of the original problem. Thus, the number of *inequalities* in the original problem becomes the number of *variables* in the dual problem.

An elementary relationship between the original primal problem and its dual is as follows:

THEOREM

If x satisfies the constraints of the primal problem and y satisfies the constraints of its dual, then $c^T x \leqq b^T y$. Consequently, if $c^T x = b^T y$, then x and y are solutions of the primal problem and the dual problem, respectively.

Proof   By the assumptions made, $x \geqq 0$, $Ax \leqq b$, $y \geqq 0$, and $A^T y \geqq c$. Consequently,

$$c^T x \leqq (A^T y)^T x = y^T A x \leqq y^T b = b^T y \qquad \blacksquare$$

This relationship can be used to estimate the number $\lambda = \max\{c^T x : x \geqq 0$ and $Ax \leqq b\}$. (This number is often termed the **value** of the linear programming problem.) To estimate λ, take any x and y that satisfy $x \geqq 0$, $y \geqq 0$, $Ax \leqq b$, and $A^T y \geqq c$. Then $c^T x \leqq \lambda \leqq b^T y$.

The importance of the dual problem stems from the fact that the extreme values in the primal and dual problems are the same. Formally stated,

---

DUALITY THEOREM

If the original problem has a solution x^*, then the dual problem has a solution y^*; furthermore, $c^T x^* = b^T y^*$.

This result is nicely illustrated by the numerical example from the beginning of this section. The dual to that problem is

$$\text{minimize} : 15y_1 + 18y_2$$

$$\text{constraints} : \begin{cases} 5y_1 + 3y_2 \geqq 2 \\ 3y_1 + 6y_2 \geqq 3 \\ y_1 \geqq 0 \quad y_2 \geqq 0 \end{cases}$$

The graph of this problem is given in Figure 15.2. Moving the line $15y_1 + 18y_2 = \alpha$, we see that the vertex $\left(\frac{1}{7}, \frac{3}{7}\right)$ is the minimum point. The values of the objective functions are indeed identical because $15\left(\frac{1}{7}\right) + 18\left(\frac{3}{7}\right) = \frac{69}{7}$. Moreover, the solution $x = \left[\frac{12}{7}, \frac{15}{7}\right]^T$ and $y = \left[\frac{1}{7}, \frac{3}{7}\right]^T$ can be related, but we shall not cover this.

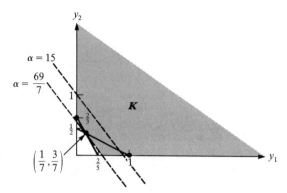

FIGURE 15.2
Graphical method
of the dual problem

We can use Maple as follows to solve this problem.

```
with(simplex):
obj := 15*x + 18*y;
consts := {5*x + 3*y >= 2, 3*x + 6*y >= 3};
minimize(obj, consts, NONNEGATIVE);
```

We obtain $x = \frac{1}{7}$ and $y = \frac{3}{7}$.

Second Primal Form

Returning to the general problem in the first primal form, we introduce additional
nonnegative variables $x_{n+1}, x_{n+2}, \dots, x_{n+m}$, known as **slack variables**, so that some
of the inequalities can be written as equalities. Using this device, we can put the
original problem into the following standard form:

SECOND PRIMAL FORM

Maximize the linear function

$$\sum_{j=1}^{n} c_j x_j$$

subject to the constraints

$$\begin{cases} \displaystyle\sum_{j=1}^{n} a_{ij}x_j + x_{n+i} = b_i & (1 \leqq i \leqq m) \\ x_j \geqq 0 & (1 \leqq j \leqq m + n) \end{cases}$$

Using matrix notation, we have

$$\text{maximize} : \boldsymbol{c}^T \boldsymbol{x}$$

$$\text{constraints} : \begin{cases} \boldsymbol{Ax} = \boldsymbol{b} \\ \boldsymbol{x} \geqq 0 \end{cases}$$

Here it is assumed that the $m \times n$ matrix $\boldsymbol{A}$ contains an $m \times m$ identity matrix in its last m columns and that the last m entries of $\boldsymbol{c}$ are 0. Also, note that when a problem in first primal form is changed to second primal form, we increase the number of variables and thus alter the quantities n, $\boldsymbol{x}$, $\boldsymbol{c}$, and $\boldsymbol{A}$. That is, a problem in the first primal form with n variables would contain $n + m$ variables in the second form.

To illustrate the transformation of a problem from first to second primal form, consider the example introduced at the beginning of this section:

$$\text{maximize} : 2x_1 + 3x_2$$

$$\text{constraints} : \begin{cases} 5x_1 + 3x_2 \leqq 15 \\ 3x_1 + 6x_2 \leqq 18 \\ x_1 \geqq 0 \quad x_2 \geqq 0 \end{cases}$$

Two slack variables x_3 and x_4 are introduced to take up the slack in two of the inequalities. The new problem in second primal form is then

$$\text{maximize} : 2x_1 + 3x_2 + 0x_3 + 0x_4$$

$$\text{constraints} : \begin{cases} 5x_1 + 3x_2 + x_3 = 15 \\ 3x_1 + 6x_2 + x_4 = 18 \\ x_1 \geqq 0 \quad x_2 \geqq 0 \quad x_3 \geqq 0 \quad x_4 \geqq 0 \end{cases}$$

Problems involving absolute values of the variables or absolute values of linear expressions can often be turned into linear programming problems. To illustrate, consider the problem of minimizing $|x - y|$ subject to linear constraints on x and y. We can introduce a new variable $z \geqq 0$ and then impose constraints $x - y \leqq z$, $-x + y \leqq z$. Then we seek to minimize the linear form $0x + 0y + 1z$.

PROBLEMS 15.1

1. Put the following problem into first primal form:

$$\text{minimize} : |x_1 + 2x_2 - x_3|$$

$$\text{constraints} : \begin{cases} x_1 + 3x_2 - x_3 \leqq 8 \\ 2x_1 - 4x_2 - x_3 \geqq 1 \\ |4x_1 + 5x_2 + 6x_3| \leqq 12 \\ x_1 \geqq 0 \quad x_2 \geqq 0 \quad x_3 \geqq 0 \end{cases}$$

Hint: $|A| \leqq B$ can be written as $-B \leqq A \leqq B$.

2. A program is available for solving linear programming problems in first primal form. Put the following problem into that form.

$$\text{minimize} : 5x_1 + 6x_2 - 2x_3 + 8$$

$$\text{constraints} : \begin{cases} 2x_1 - 3x_2 & \geq 5 \\ x_1 + x_2 & \leq 15 \\ 2x_1 - x_2 + x_3 \leq 25 \\ x_1 + x_2 - x_3 \geq 1 \\ x_1 \geq 0 \quad x_2 \geq 0 \quad x_3 \geq 0 \end{cases}$$

3. Consider the following linear programming problems:

   a. maximize : $2x_1 + 3x_2$

   $$\text{constraints} : \begin{cases} x_1 + 2x_2 \geq -6 \\ -x_1 + 3x_2 \leq 3 \\ |2x_1 - 5x_2| \leq 5 \\ x_1 \geq 0 \quad x_2 \geq 0 \end{cases}$$

   b. minimize : $7x_1 + x_2 - x_3 + 4$

   $$\text{constraints} : \begin{cases} x_1 - x_2 + x_3 \geq 2 \\ x_1 + x_2 + x_3 \leq 10 \\ -2x_1 - x_2 \quad \leq -4 \\ x_1 \geq 0 \quad x_2 \geq 0 \end{cases}$$

   Rewrite each problem in first primal form and give the dual problem.

4. Sketch the feasible region for the following constraints:

   $$\begin{cases} x - y \leq 2 \\ x + y \leq 3 \\ 2x + y \geq 3 \\ x \geq 0 \quad y \geq 0 \end{cases}$$

   a. By substituting the vertices into the objective function

   $$z(x, y) = x + 2y$$

   determine the minimum value of this function on the feasible region.

   b. Let

   $$z(x, y) = \left(x - \frac{1}{2}\right)^2 + \left(y - \frac{1}{2}\right)^2$$

   Show that the minimum value of z over the feasible region does not occur at a vertex.

5. Put the following linear programming problems into first primal form. What is the dual of each?

   a.   minimize : $2x + y - 3z + 1$

   $$\text{constraints} : \begin{cases} x - y \geq 3 \\ |x - z| \leq 2 \\ x \geq 0 \quad y \geq 0 \end{cases}$$

   b.   minimize : $3x - 2y + 5z + 3$

   $$\text{constraints} : \begin{cases} x + y + z \geq 4 \\ x - y - z = 2 \\ x \geq 0 \quad y \geq 0 \quad z \geq 0 \end{cases}$$

   c.   maximize : $3x + 2y$

   $$\text{constraints} : \begin{cases} 6x + 5y \leq 17 \\ 2x + 11y \leq 23 \\ x \leq 0 \end{cases}$$

6. Consider the following linear programming problem:

   $$\text{maximize} : 2x_1 + 2x_2 - 6x_3 - x_4$$

   $$\text{constraints} : \begin{cases} 3x_1 \qquad\qquad\qquad + x_4 = 25 \\ x_1 + x_2 + x_3 + x_4 = 20 \\ 4x_1 \qquad + 6x_3 \qquad \geq 5 \\ 2x_1 \qquad + 3x_3 + 2x_4 \geq 0 \\ x_1 \geq 0 \quad x_2 \geq 0 \quad x_3 \geq 0 \quad x_4 \geq 0 \end{cases}$$

   a. Reformulate this problem in second primal form.

   b. Formulate the dual problem.

7. Solve the following linear programming problem graphically:

   $$\text{maximize} : 3x_1 + 5x_2$$

   $$\text{constraints} : \begin{cases} x_1 \qquad\quad \leq 4 \\ \qquad x_2 \leq 6 \\ 3x_1 + 2x_2 \leq 18 \\ x_1 \geq 0 \quad x_2 \geq 0 \end{cases}$$

8. (Continuation) Solve the dual problem of Problem 7.

9. Show that the dual problem may be written as:

   $$\text{maximize} : b^T y$$

   $$\text{constraints} : \begin{cases} y^T A \geq c^T \\ y \geq 0 \end{cases}$$

10. Describe how $\max\{|x - y - 3|, |2x + y + 4|, |x + 2y - 7|\}$ can be minimized by using a linear programming code.

11. Show how this problem can be solved by linear programming:

$$\text{minimize} : |x - y|$$

$$\text{constraints} : \begin{cases} x \leq 3y \\ x \geq y \\ y \leq x - 2 \end{cases}$$

12. Consider the linear programming problem

$$\text{minimize} : x_1 + x_4 + 25$$

$$\text{constraints} : \begin{cases} 2x_1 + 2x_2 + x_3 < 7 \\ 2x_1 - 3x_2 + x_4 = 4 \\ x_2 - x_4 > 1 \\ 3x_2 - 8x_3 + x_4 = 5 \\ x_1, x_2, x_3, x_4 \geq 0 \end{cases}$$

Write in matrix-vector form the dual problem and the second primal problem.

13. Solve each of the linear programming problems by the graphical method. Determine x to

$$\text{maximize} : c^T x$$

$$\text{constraints} : \begin{cases} Ax \leq b \\ x \geq 0 \end{cases}$$

Here nonunique and unbounded "solutions" may be obtained.

a. $c = [2, -4]^T$ $\quad A = \begin{bmatrix} -3 & -5 \\ 4 & 9 \end{bmatrix}$ $\quad b = [-15, 36]$

b. $c = \left[2, \dfrac{1}{2}\right]^T$ $\quad A = \begin{bmatrix} 6 & 5 \\ 4 & 1 \end{bmatrix}$ $\quad b = [30, 12]^T$

c. $c = [3, 2]^T$ $\quad A = \begin{bmatrix} -3 & 2 \\ -4 & 9 \end{bmatrix}$ $\quad b = [6, 36]^T$

d. $c = [2, -3]^T$ $\quad A = \begin{bmatrix} -1 & 1 \\ 0 & 1 \end{bmatrix}$ $\quad b = [0, 5]^T$

e. $c = [-4, 11]^T$ $\quad A = \begin{bmatrix} -3 & 4 \\ -4 & 11 \end{bmatrix}$ $\quad b = [12, 44]^T$

f. $c = [-3, 4]^T$ $\quad A = \begin{bmatrix} 2 & 3 \\ -4 & -5 \end{bmatrix}$ $\quad b = [6, -20]^T$

g. $c = [2, 1]^T$ $\quad A = \begin{bmatrix} 1 & 1 \\ 1 & 2 \end{bmatrix}$ $\quad b = [0, -2]^T$

h. $c = [3, 1]^T$ $\quad A = \begin{bmatrix} 2 & 4 \\ 5 & 3 \end{bmatrix}$ $\quad b = [21, 18]^T$

14. Solve the following linear programming problem by hand, using a graph for help:

$$\text{maximize} : 4x + 4y + z$$

$$\text{constraints} : \begin{cases} 3x + 2y + z = 12 \\ 7x + 7y + 2z \leq 144 \\ 7x + 5y + 2z \leq 80 \\ 11x + 7y + 3z \leq 132 \\ x \geq 0 \quad y \geq 0 \end{cases}$$

Hint: Use the equation to eliminate z from all other expressions. Solve the resulting two-dimensional problem.

15. Put this linear programming problem into second primal form. You may want to make changes of variables. If so, include a *dictionary* relating new and old variables.

$$\text{minimize} : \varepsilon_1 + \varepsilon_2 + \varepsilon_3$$

$$\text{constraints} : \begin{cases} |3x + 4y + 6| \leq \varepsilon_1 \\ |2x - 8y - 4| \leq \varepsilon_2 \\ |-x - 3y + 5| \leq \varepsilon_3 \\ \varepsilon_1 > 0 \quad \varepsilon_2 > 0 \quad \varepsilon_3 > 0 \quad x > 0 \quad y > 0 \end{cases}$$

Solve the resulting problem.

16. Consider the linear programming problem

$$\text{maximize} : c_1 x_1 + c_2 x_2$$

$$\text{constraints} : \begin{cases} a_1 x_1 + a_2 x_2 \leq b \\ x_1 \geq 0 \quad x_2 \geq 0 \end{cases}$$

In the special case when all data are positive, show that the dual problem has the same extreme value as the original problem.

17. Suppose that a linear programming problem in first primal form has the property that $c^T x$ is not bounded on the feasible set. What conclusion can be drawn about the dual problem?

COMPUTER PROBLEMS 15.1

1. A western shop wishes to purchase 300 felt and 200 straw cowboy hats. Bids have been received from three wholesalers. Texas Hatters has agreed to supply not more than 200 hats, Lone Star Hatters not more than 250, and Lariat Ranch Wear not more than 150. The owner of the shop has estimated that his profit per hat sold from Texas Hatters would be $3/felt and $4/straw, from Lone Star Hatters $3.80/felt and $3.50/straw, and from Lariat Ranch Wear $4/felt and $3.60/straw. Set up a linear programming problem to maximize the owner's profits. Solve by using a program from your software library.

2. The ABC Drug Company makes two types of liquid painkiller that have brand names Relieve (R) and Ease (E) and contain a different mixture of three basic drugs, A, B, and C, produced by the company. Each bottle of R requires $\frac{7}{9}$ unit of drug A, $\frac{1}{2}$ unit of drug B, and $\frac{3}{4}$ unit of drug C. Each bottle of E requires $\frac{4}{9}$ unit of drug A, $\frac{5}{2}$ unit of drug B, and $\frac{1}{4}$ unit of drug C. The company is able to produce each day only 5 units of drug A, 7 units of drug B, and 9 units of C. Moreover, Federal Drug Administration regulations stipulate that the number of bottles of R manufactured cannot exceed twice the number of bottles of E. The profit margin for each bottle of E and R is $7 and $3, respectively. Set up the linear programming problem in first primal form in order to determine the number of bottles of the two painkillers that the company should produce each day so as to maximize their profits. Solve by using available software.

3. Suppose that the university student government wishes to charter planes to transport at least 750 students to the bowl game. Two airlines, A1 and A2, agree to supply aircraft for the trip. A1 has five aircraft available carrying 75 passengers each, and A2 has three aircraft available carrying 250 passengers each. The cost per aircraft is $900 and $3250 for the trip from A1 and A2, respectively. The student government wants to charter at most six aircraft. How many of each type should be chartered to minimize the cost of the airlift? How much should the student government charge in order to make 50¢ profit per student? Solve by the graphical method and verify by using a routine from your program library.

4. (Continuation) Rework Computer Problem **3** in the following two possibly different ways:

   a. The number of students going on the airlift is maximized.

   b. The cost per student is minimized.

5. (A diet problem) A university dining hall wishes to provide at least 5 units of vitamin C and 3 units of vitamin E per serving. Three foods are available containing these vitamins. Food f_1 contains 2.5 and 1.25 units per ounce of C and E, respectively, whereas food f_2 contains just the opposite amounts. The third food f_3 contains an equal amount of each vitamin at 1 unit per ounce. Food f_1 costs 25¢ per ounce, food f_2 costs 56¢ per ounce, and food f_3 costs 10¢ per ounce. The dietitian wishes to provide the meal at a minimum cost per serving that satisfies the minimum vitamin requirements. Set up this linear programming problem in second primal form. Solve with the aid of a code from your computer program library.

15.2 Simplex Method

The principal algorithm used in solving linear programming problems is the **simplex method**. Here enough of the background of this method is described so that the reader can use available computer programs that incorporate it.

Consider a linear programming problem in second primal form:

$$\text{maximize} : c^T x$$

$$\text{constraints} : \begin{cases} Ax = b \\ x \geqq 0 \end{cases}$$

It is assumed that c and x are n-component vectors, b an m-component vector, and A an $m \times n$ matrix. Also, it is assumed that $b \geqq 0$ and that A contains an $m \times m$ identity matrix in its last m columns. As before, we define the set of feasible points as

$$K = \{x \in \mathbb{R}^n : Ax = b, x \geqq 0\}$$

The points of K are exactly the points that are competing to maximize $c^T x$.

Vertices in K and Linearly Independent Columns of A

The set K is a polyhedral set in $\mathbb{R}^n$, and the algorithm to be described proceeds from vertex to vertex in K, always increasing the value of $c^T x$ as it goes from one to another. Let us give a precise definition of *vertex*. A point x in K is called a **vertex** if it is impossible to express it as $x = \frac{1}{2}(u + v)$, with both u and v in K and $u \neq v$. In other words, x is not the midpoint of any line segment whose endpoints lie in K.

We denote by $a^{(1)}, a^{(2)}, \ldots, a^{(n)}$ the column vectors constituting the matrix A. The following theorem relates the columns of A to the vertices of K:

THEOREM

Let $x \in K$ and define $J(x) = \{i : x_i > 0\}$. Then the following are equivalent:

1. x is a vertex of K.
2. The set $\{a^{(i)} : i \in J(x)\}$ is linearly independent.

Proof   If statement **1** is false, then we can write $x = \frac{1}{2}(u + v)$, with $u \in K$, $v \in K$, and $u \neq v$. For every index i not in the set $J(x)$, we have $x_i = 0$, $u_i \geqq 0$, $v_i \geqq 0$, and $x_i = \frac{1}{2}(u_i + v_i)$. This forces u_i and v_i to be zero. Thus, all the nonzero components of u and v correspond to indices i in $J(x)$. Since u and v belong to K,

$$b = Au = \sum_{i=1}^n u_i a^{(i)} = \sum_{i \in J(x)} u_i a^{(i)}$$

and

$$b = Av = \sum_{i=1}^n v_i a^{(i)} = \sum_{i \in J(x)} v_i a^{(i)}$$

Hence,

$$\sum_{i \in J(x)} (u_i - v_i) \, a^{(i)} = 0$$

showing the linear dependence of the set $\{a^{(i)} : i \in J(x)\}$. Thus, statement **2** is false. Consequently, statement **2** implies statement **1**.

For the converse, assume that statement **2** is false. From the linear dependence of column vectors $a^{(i)}$ for $i \in J(x)$, we have

$$\sum_{i \in J(x)} y_i \, a^{(i)} = 0 \quad \text{with} \quad \sum_{i \in J(x)} |y_i| \neq 0$$

for appropriate coefficients y_i. For each $i \notin J(x)$, let $y_i = 0$. Form the vector y with components y_i for $i = 1, 2, \ldots, n$. Then, for any λ, we see that because $x \in K$,

$$A(x + \lambda y) = \sum_{i=1}^{n} (x_i \pm \lambda y_i) \, a^{(i)} = \sum_{i=1}^{n} x_i \, a^{(i)} \pm \lambda \sum_{i \in J(x)} y_i \, a^{(i)} = Ax = b$$

Now select the real number λ positive but so small that $x + \lambda y \geq 0$ and $x - \lambda y \geq 0$. [To see that it is possible, consider separately the components for $i \in J(x)$ and $i \notin J(x)$.] The resulting vectors, $u = x + \lambda y$ and $v = x - \lambda y$ and $v = x - \lambda y$, belong to K. They differ, and obviously $x = \frac{1}{2}(u + v)$. Thus, x is not a vertex of K; that is, statement **1** is false. So statement **1** implies statement **2**.   ∎

Given a linear programming problem, there are three possibilities:

1. There are no feasible points; that is, the set K is empty.
2. K is not empty and $c^T x$ is not bounded on K.
3. K is not empty and $c^T x$ is bounded on K.

It is true (but not obvious) that in case **3** there is a point x in K such that $c^T x \geq c^T y$ for all y in K. We have assumed that our problem is in the second primal form so that possibility 1 cannot occur. Indeed, A contains an $m \times m$ identity matrix and so has the form

$$A = \begin{bmatrix} a_{11} & a_{12} & \cdots & a_{1k} & 1 & 0 & \cdots & 0 \\ a_{21} & a_{22} & \cdots & a_{2k} & 0 & 1 & \cdots & 0 \\ \vdots & \vdots & \ddots & \vdots & \vdots & \vdots & \ddots & \vdots \\ a_{m1} & a_{m2} & \cdots & a_{mk} & 0 & 0 & \cdots & 1 \end{bmatrix}$$

where $k = n - m$. Consequently, we can *construct* a feasible point x easily by setting $x_1 = x_2 = \cdots = x_k = 0$ and $x_{k+1} = b_1$, $x_{k+2} = b_2$, and so on. It is then clear that $Ax = b$. The inequality $x \geq 0$ follows from our initial assumption that $b \geq 0$.

Simplex Method

Here is a brief outline of the simplex method for solving linear programming problems. It involves a sequence of exchanges so that the trial solution proceeds systematically from one vertex to another in K. This procedure is stopped when the value of $c^T x$ is no longer increased as a result of the exchange.

Simplex Algorithm

Select a small positive value for ε. In each step, we have a set of m indices $\{k_1, k_2, \ldots, k_m\}$.

1. Put columns $a^{(k_1)}, a^{(k_2)}, \ldots, a^{(k_m)}$ into B and solve $Bx = b$.

2. If $x_i > 0$ for $1 \leqq i \leqq m$, continue. Otherwise exit because the algorithm has failed.

3. Set $e = [c_{k_1}, c_{k_2}, \ldots, c_{k_m}]^T$ and solve $B^T y = e$.

4. Choose any s in $\{1, 2, \ldots, n\}$ but not in $\{k_1, k_2, \ldots, k_m\}$ for which $c_s - y^T a^{(s)}$ is greatest.

5. If $c_s - y^T a^{(s)} < \varepsilon$, exit because x is the solution.

6. Solve $Bz = a^{(s)}$.

7. If $z_i \leqq \varepsilon$ for $1 \leqq i \leqq m$, then exit because the objective function is unbounded on K.

8. Among the ratios x_i/z_i that have $z_i > 0$ for $1 \leqq i \leqq m$, let x_r/z_r be the smallest. In case of a tie, let r be the first occurrence.

9. Replace k_r by s and go to step **1**.

A few remarks on this algorithm are in order. In the beginning, select the indices $k_1, k_2, \ldots, k_m$ so that $a^{(k_1)}, a^{(k_2)}, \ldots, a^{(k_m)}$ form an $m \times m$ identity matrix. At step **5**, where we say x *is a solution*, we mean that the vector $v = (v_i)$ given by $v_{k_i} = x_i$ for $1 \leqq i \leqq n$ and $v_i = 0$ for $i \notin \{k_1, k_2, \ldots, k_m\}$ is the solution. A convenient choice for the tolerance ε that occurs in steps **5** and **7** might be 10^{-6}.

In any reasonable implementation of the simplex method, advantage must be taken of the fact that succeeding occurrences of step **1** are very similar. In fact, only one column of B changes at a time. Similar remarks hold for steps **3** and **6**.

We do not recommend that the reader attempt to program the simplex algorithm. Efficient codes, refined over many years of experience, are usually available in software libraries. Many of them can provide solutions to a given problem *and* to its dual with very little additional computing. Sometimes this feature can be exploited to decrease the execution time of a problem. To see why, consider a linear programming problem in first primal form:

$$\text{maximize} : c^T x$$

(P)

$$\text{constraints} : \begin{cases} Ax \leqq b \\ x \geqq 0 \end{cases}$$

As usual, we assume that x is an n vector and that A is an $m \times n$ matrix. When the simplex algorithm is applied to this problem, it performs an iterative process on an $m \times m$ matrix denoted by B in the preceding description. If the number of inequality constraints m is very large relative to n, then the dual problem may be easier to solve, since the B matrices for it will be of dimension $n \times n$. Indeed, the dual problem is

$$\text{minimize} : b^T y$$

(D)

$$\text{constraints} : \begin{cases} A^T y \geqq c \\ y \geqq 0 \end{cases}$$

and the number of inequality constraints here is n. An example of this technique appears in the next section.

PROBLEMS 15.2

1. Show that the linear programming problem

$$\text{maximize} : c^T x$$

$$\text{constraints} : Ax \leqq b$$

can be put into first primal form by increasing the number of variables by just one. *Hint*: Replace x_j by $y_j - y_0$.

2. Show that the set K can have only a finite number of vertices.

3. Suppose that u and v are solution points for a linear programming problem and that $x = \frac{1}{2}(u + v)$. Show that x is also a solution.

4. Using the simplex method as described, solve the numerical example in the text.

5. Using standard manipulations, put the dual problem (D) into first and second primal forms.

6. Show how a code for solving a linear programming problem in first primal form can be used to solve a system of n linear equations in n variables.

7. Using standard techniques, put the dual problem (D) into first primal form (P); then take the dual of it. What is the result?

COMPUTER PROBLEMS 15.2

Select a linear programming code from your computing center library and use it to solve these problems.

1. $\quad \text{minimize} : 8x_1 + 6x_2 + 6x_3 + 9x_4$

$$\text{constraints} : \begin{cases} x_1 + 2x_2 \quad\quad\quad + x_4 \geqq 2 \\ 3x_1 + \quad x_2 \quad\quad\quad + x_4 \geqq 4 \\ \quad\quad\quad\quad\quad\quad x_3 + x_4 \geqq 1 \\ x_1 \quad\quad\quad + x_3 \quad\quad \geqq 1 \\ x_1 \geqq 0, \quad x_2 \geqq 0, \quad x_3 \geqq 0, \quad x_4 \geqq 0 \end{cases}$$

2.   minimize : $10x_1 - 5x_2 - 4x_3 + 7x_4 + x_5$

$$\text{constraints :} \begin{cases} 4x_1 - 3x_2 - x_3 + 4x_4 + x_5 = 1 \\ -x_1 + 2x_2 + 2x_3 + x_4 + 3x_5 = 4 \\ x_1 \geq 0 \quad x_2 \geq 0 \quad x_3 \geq 0 \quad x_4 \geq 0 \quad x_5 \geq 0 \end{cases}$$

3.   maximize : $2x_1 + 4x_2 + 3x_3$

$$\text{constraints :} \begin{cases} 4x_1 + 2x_2 + 3x_3 \leq 15 \\ 3x_1 + 2x_2 + x_3 \leq 7 \\ x_1 + x_2 + 2x_3 \leq 6 \\ x_1 \geq 0 \quad x_2 \geq 0 \quad x_3 \geq 0 \end{cases}$$

15.3   Approximate Solution of Inconsistent Linear Systems

Linear programming can be used for the approximate solution of systems of linear equations that are inconsistent. An $m \times n$ system of equations

$$\sum_{j=1}^{n} a_{ij}x_j = b_i \qquad (1 \leq i \leq m)$$

is said to be **inconsistent** if there is no vector $x = [x_1, x_2, \ldots, x_n]^T$ that simultaneously satisfies all m equations in the system. For instance, the system

$$\begin{cases} 2x_1 + 3x_2 = 4 \\ x_1 - x_2 = 2 \\ x_1 + 2x_2 = 7 \end{cases} \qquad \textbf{(1)}$$

is inconsistent, as can be seen by attempting to carry out the Gaussian elimination process.

ℓ_1 Problem

Since no vector x can solve an inconsistent system of equations, the **residuals**

$$r_i = \sum_{j=1}^{n} a_{ij}x_j - b_i \qquad (1 \leq i \leq m)$$

cannot be made to vanish simultaneously. Hence, $\sum_{i=1}^{m} |r_i| > 0$. Now it is natural to ask for an x vector that renders the expression $\sum_{i=1}^{m} |r_i|$ as small as possible. This problem is called the ℓ_1 problem for this system of equations. Other criteria, leading to different approximate solutions, might be to minimize $\sum_{i=1}^{m} r_i^2$ or $\max_{1 \leq i \leq m} |r_i|$. Chapter 10 discusses in detail the problem of minimizing $\sum_{i=1}^{m} r_i^2$.

The minimization of $\sum_{i=1}^{n} |r_i|$ by appropriate choice of the x vector is a problem for which special algorithms have been designed (see Barrodale and Roberts [1974]). However, if one of these special programs is not available or if the problem is of small scope, linear programming can be used.

A simple, direct restatement of the problem is

$$\text{minimize} : \sum_{i=1}^{m} \varepsilon_i$$

$$\text{constraints} : \begin{cases} \sum_{j=1}^{n} a_{ij}x_j - b_i \leq \varepsilon_i & (1 \leq i \leq m) \\ -\sum_{j=1}^{n} a_{ij}x_j + b_i \leq \varepsilon_i & (1 \leq i \leq m) \end{cases} \tag{2}$$

If a linear programming code is at hand in which the variables are not required to be nonnegative, then it can be used on Problem (2). If the variables must be nonnegative, the following technique can be applied. Introduce a variable y_{n+1} and write $x_j = y_j - y_{n+1}$. Then define $a_{i,n+1} = -\sum_{j=1}^{n} a_{ij}$. This step creates an additional column in the matrix A. Now consider the linear programming problem

$$\text{maximize} : -\sum_{i=1}^{m} \varepsilon_i$$

$$\text{constraints} : \begin{cases} \sum_{j=1}^{n+1} a_{ij}y_j - \varepsilon_i \leq b_i & (1 \leq i \leq m) \\ -\sum_{j=1}^{n+1} a_{ij}y_j - \varepsilon_i \leq -b_i & (1 \leq i \leq m) \\ y \geq 0 \quad \varepsilon \geq 0 \end{cases} \tag{3}$$

which is in first primal form with $m + n + 1$ variables and $2m$ inequality constraints.

It is not hard to verify that Problem (3) is equivalent to Problem (2). The main point is that

$$\sum_{j=1}^{n+1} a_{ij}y_j = \sum_{j=1}^{n} a_{ij}(x_j + y_{n+1}) + a_{i,n+1}y_{n+1}$$

$$= \sum_{j=1}^{n} a_{ij}x_j + y_{n+1}\sum_{j=1}^{n} a_{ij} + y_{n+1}\left(-\sum_{j=1}^{n} a_{ij}\right)$$

$$= \sum_{j=1}^{n} a_{ij}x_j$$

Another technique can be used to replace the $2m$ inequality constraints in (3) by a set of m equality constraints. We write

$$\varepsilon_i = |r_i| = u_i + v_i$$

where $u_i = r_i$ and $v_i = 0$ if $r_i \geq 0$, but $v_i = -r_i$ and $u_i = 0$ if $r_i < 0$. The resulting linear programming problem is

$$\text{maximize} : -\sum_{i=1}^{m} u_i - \sum_{i=1}^{m} v_i$$

$$\text{constraints} : \begin{cases} \displaystyle\sum_{j=1}^{n+1} a_{ij}y_j - u_i + v_i = b_i & (1 \leq i \leq m) \\[2mm] u \geq 0 \quad v \geq 0 \quad y \geq 0 \end{cases}$$

Using the preceding formulas, we have

$$\begin{aligned} r_i &= \sum_{j=1}^{n} a_{ij}x_j - b_i = \sum_{j=1}^{n} a_{ij}(y_j - y_{n+1}) - b_i \\ &= \sum_{j=1}^{n} a_{ij}y_j - y_{n+1}\sum_{j=1}^{n} a_{ij} - b_i \\ &= \sum_{j=1}^{n+1} a_{ij}y_j - b_i = u_i - v_i \end{aligned}$$

From it we conclude that $r_i + v_i = u_i \geq 0$. Now v_i and u_i should be as small as possible, consistent with this restriction, because we are attempting to minimize $\sum_{i=1}^{m}(u_i + v_i)$. So if $r_i \geq 0$, we take $v_i \geq 0$ and $u_i = r_i$, whereas if $r_i < 0$, we take $v_i = -r_i$ and $u_i = 0$. In either case, $|r_i| = u_i + v_i$. Thus, minimizing $\sum_{i=1}^{m}(u_i + v_i)$ is the same as minimizing $\sum_{i=1}^{m} |r_i|$.

The example of the inconsistent linear system given by (1) could be solved in the ℓ_1 sense by solving the linear programming problem

$$\text{minimize} : u_1 + v_1 + u_2 + v_2 + u_3 + v_3$$

$$\text{constraints} : \begin{cases} 2y_1 + 3y_2 - 5y_3 - u_1 + v_1 = 4 \\ y_1 - y_2 \qquad\quad - u_2 + v_2 = 2 \\ y_1 + 2y_2 - 3y_3 - u_3 + v_3 = 7 \\ y_1, y_2, y_3 \geq 0 \quad u_1, u_2, u_3 \geq 0 \quad v_1, v_2, v_3 \geq 0 \end{cases}$$

The solution is

$$\begin{array}{ccc} u_1 = 0 & u_2 = 0 & u_3 = 0 \\ v_1 = 0 & v_2 = 0 & v_3 = 5 \\ y_1 = 2 & y_2 = 0 & y_3 = 0 \end{array}$$

From it we recover the ℓ_1 solution of System (1) in the form

$$
\begin{array}{ll}
x_1 = y_1 - y_3 = 2 & r_1 = u_1 - v_1 = 0 \\
x_2 = y_2 - y_3 = 0 & r_2 = u_2 - v_2 = 0 \\
& r_3 = u_3 - v_3 = -5
\end{array}
$$

We can use Maple as follows to solve this problem.

```
with(simplex):
obj := u1 + v1 + u2 + v2 + u3 + v3;
consts := {2*y1 + 3*y2 - 5*y3 - u1 + v1 = 4,
           y1 - y2 - u2 + v2 = 2,
           y1 + 2*y2 - 3*y3 - u3 + v3 = 7};
minimize(obj, consts, NONNEGATIVE);
```

We obtain $u_1 = v_1 = u_2 = v_2 = u_3 = y_2 = y_3 = 0$, $v_3 = 5$, and $y_1 = 2$.

ℓ_∞ Problem

Consider again a system of m linear equations in n unknowns

$$
\sum_{j=1}^{n} a_{ij}x_j = b_i \qquad (1 \leq i \leq m)
$$

If the system is inconsistent, we know that the residuals $r_i = \sum_{j=1}^{n} a_{ij}x_j - b_i$ cannot all be zero for any x vector. So the quantity $\varepsilon = \max_{1 \leq i \leq m} |r_i|$ is positive. The problem of making ε a minimum is called the ℓ_∞ problem for the system of equations. An equivalent linear programming problem is

minimize : ε

$$
\text{constraints} :
\begin{cases}
\displaystyle\sum_{j=1}^{n} a_{ij}x_j - \varepsilon \leq b_i & (1 \leq i \leq m) \\[3mm]
\displaystyle-\sum_{j=1}^{n} a_{ij}x_j - \varepsilon \leq -b_i & (1 \leq i \leq m)
\end{cases}
$$

If a linear programming code is available in which the variables need not be greater than or equal to zero, then it can be used to solve the ℓ_∞ problem as formulated above. If the variables must be nonnegative, we first introduce a variable y_{n+1} so large that the quantities $y_j = x_j + y_{n+1}$ are positive. Next, we solve the linear programming problem

minimize : ε

$$
\text{constraints} :
\begin{cases}
\displaystyle\sum_{j=1}^{n+1} a_{ij}y_j - \varepsilon \leq b_i & (1 \leq i \leq m) \\[3mm]
\displaystyle-\sum_{j=1}^{n+1} a_{ij}y_j - \varepsilon \leq -b_i & (1 \leq i \leq m) \\[3mm]
\varepsilon \geq 0 \quad y_j \geq 0 & (1 \leq j \leq n+1)
\end{cases}
\tag{4}
$$

Here we have again defined $a_{i,n+1} = -\sum_{j=1}^{n} a_{ij}$.

For our example (1), the solution that minimizes the quantity

$$\max\{|2x_1 + 3x_2 - 4|, |x_1 - x_2 - 2|, |x_1 + 2x_2 - 7|\}$$

is obtained from the linear programming problem

$$\text{minimize} : \varepsilon$$

$$\text{constraints} : \begin{cases} 2y_1 + 3y_2 - 5y_3 - \varepsilon \leqq 4 \\ y_1 - y_2 \qquad\quad - \varepsilon \leqq 2 \\ y_1 + 2y_2 - 3y_3 - \varepsilon \leqq 7 \\ -2y_1 - 3y_2 + 5y_3 - \varepsilon \leqq -4 \\ -y_1 + y_2 \qquad\quad - \varepsilon \leqq -2 \\ -y_1 - 2y_2 + 3y_3 - \varepsilon \leqq -7 \\ y_1, y_2, y_3 \geqq 0 \quad \varepsilon \geqq 0 \end{cases} \tag{5}$$

The solution is

$$y_1 = \frac{8}{9} \quad y_2 = \frac{5}{3} \quad y_3 = 0 \quad \varepsilon = \frac{25}{9}$$

From it the ℓ_∞ solution of (1) is recovered as follows:

$$x_1 = y_1 - y_3 = \frac{8}{9} \qquad x_2 = y_2 - y_3 - \frac{15}{9}$$

We can use Maple as follows to solve this problem.

```
with(simplex):
obj := e;
consts := { 2*y1 + 3*y2 - 5*y3 - e <=  4,
            y1 -    y2          - e <=  2,
            y1 + 2*y2 - 3*y3 - e <=  7,
          -2*y1 - 3*y2 + 5*y3 - e <= -4,
           -y1 +    y2          - e <= -2,
           -y1 - 2*y2 + 3*y3 - e <= -7};
minimize(obj, consts union {y1>=0, y2>=0,
                            y3>=0, e>=0});
```

We obtain $y_1 = \frac{8}{9}$, $y_2 = \frac{5}{3}$, $y_3 = 0$, and $\varepsilon = \frac{25}{9}$.

In problems like (4), m is often much larger than n. Thus, in accordance with remarks made in Section 15.2, it may be preferable to solve the dual of Problem (4) because it would have $2m$ variables but only $n + 2$ inequality constraints. To illustrate, the dual of Problem (5) is

$$\text{maximize} : 4u_1 + 2u_2 + 7u_3 - 4u_4 - 2u_5 - 7u_6$$

$$\text{constraints} : \begin{cases} 2u_1 + u_2 + u_3 - 2u_4 - u_5 - u_6 \geqq 0 \\ 3u_1 - u_2 + 2u_3 - 3u_4 + u_5 - 2u_6 \geqq 0 \\ -5u_1 \qquad\quad - 3u_3 + 5u_4 \qquad + 3u_6 \geqq 0 \\ -u_1 - u_2 - u_3 - u_4 - u_5 - u_6 \geqq -1 \\ u_i \geqq 0 \quad (1 \leqq i \leqq 6) \end{cases}$$

The three types of approximate solution that have been discussed (for an overdetermined system of linear equations) are useful in different situations. Broadly speaking, an ℓ_∞ solution is preferred when the data are known to be accurate. An ℓ_2 solution is preferred when the data are contaminated with errors that are believed to conform to the normal probability distribution. The ℓ_1 solution is often used when data are suspected of containing *wild* points—points that result from gross errors, such as the incorrect placement of a decimal point. Additional information can be found in Rice and White [1964]. (The ℓ_2 problem is discussed in Chapter 10 also.)

PROBLEMS 15.3

1. Consider the inconsistent linear system

$$\begin{cases} 5x_1 + 2x_2 && = 6 \\ x_1 + x_2 + x_3 = 2 \\ 7x_2 - 5x_3 = 11 \\ 6x_1 && + 9x_3 = 9 \end{cases}$$

Write the following with nonnegative variables.

a. The equivalent linear programming problem for solving the system in the ℓ_1 sense.

b. The equivalent linear programming problem for solving the system in the ℓ_∞ sense.

2. (Continuation) Repeat Problem **1** for the system

$$\begin{cases} 3x + y = 7 \\ x - y = 11 \\ x + 6y = 13 \\ -x + 3y = -12 \end{cases}$$

3. We want to find a polynomial p of degree n that approximates a function f as well as possible *from below*; that is, we want $0 \le f - p \le \varepsilon$ for minimum ε. Show how p could be obtained with reasonable precision by solving a linear programming problem.

4. In order to solve the ℓ_1 problem for the system of equations

$$\begin{cases} x - y = 4 \\ 2x - 3y = 7 \\ x + y = 2 \end{cases}$$

we can solve a linear programming problem. What is it?

COMPUTER PROBLEMS 15.3

1. Obtain numerical answers for parts **a** and **b** of Problem **1** in the preceding set.

2. (Continuation) Repeat for Problem **2**.

3. Find a polynomial of degree 4 that represents the function e^x in the following sense: Select 20 equally spaced points x_i in interval $[0, 1]$ and require the polynomial to minimize the expression $\max_{1 \leq i \leq 20} |e^{x_i} - p(x_i)|$. *Hint:* This is the same as solving 20 equations in five variables in the ℓ_∞ sense. The ith equation is $A + Bx_i + Cx_i^2 + Dx_i^3 + Ex_i^4 = e^{x_i}$ and the unknowns are A, B, C, D, and E.

Appendix A

LINEAR ALGEBRA CONCEPTS AND NOTATION

The two concepts from linear algebra that we are most concerned with are *vectors* and *matrices* because of their usefulness in compressing complicated expressions into a compact notation. The vectors and matrices in this text are most often *real*, since they consist of real numbers. These concepts easily generalize to *complex* vectors and matrices.

Vectors

A **vector** $x \in \mathbb{R}^n$ can be thought of as a one-dimensional array of numbers and is written as

$$
x = \begin{bmatrix} x_1 \\ x_2 \\ \vdots \\ x_n \end{bmatrix}
$$

where x_i is called the ith **element**, **entry**, or **component**. An alternative notation useful in pseudocodes is $x = (x_i)_n$. Sometimes x is said to be a **column vector** to distinguish it from a **row vector** y written as

$$
y = [y_1, y_2, \ldots, y_n]
$$

For example, here are some vectors

$$
\begin{bmatrix} \frac{1}{5} \\ 3 \\ -\frac{5}{6} \\ \frac{2}{7} \end{bmatrix}
\qquad
[\pi, \quad e, \quad 5, -4]
\qquad
\begin{bmatrix} \frac{1}{2} \\ \frac{1}{3} \end{bmatrix}
$$

To save space, a column vector x can be written as a row vector such as

$$
x = [x_1, x_2, \ldots, x_n]^T \qquad \text{or} \qquad x^T = [x_1, x_2, \ldots, x_n]
$$

by adding a T (for *transpose*) to indicate that we are interchanging or transposing a row or column vector. As an example, we have

$$[1 \quad 2 \quad 3 \quad 4]^T = \begin{bmatrix} 1 \\ 2 \\ 3 \\ 4 \end{bmatrix}$$

Many operations involving vectors are component-by-component operations. For vectors x and y

$$x = \begin{bmatrix} x_1 \\ x_2 \\ \vdots \\ x_n \end{bmatrix} \qquad y = \begin{bmatrix} y_1 \\ y_2 \\ \vdots \\ y_n \end{bmatrix}$$

the following definitions apply.

Equality

$$x = y \text{ if and only if } x_i = y_i \text{ for all } i \ (1 \leq i \leq n)$$

Inequality

$$x < y \text{ if and only if } x_i < y_i \text{ for all } i \ (1 \leq i \leq n)$$

Addition/Subtraction

$$x \pm y = \begin{bmatrix} x_1 \pm y_1 \\ x_2 \pm y_2 \\ \vdots \\ x_n \pm y_n \end{bmatrix}$$

Scalar Product

$$\alpha x = \begin{bmatrix} \alpha x_1 \\ \alpha x_2 \\ \vdots \\ \alpha x_n \end{bmatrix} \qquad \text{for } \alpha \text{ a constant or scalar}$$

Here is an example:

$$\begin{bmatrix} 2 \\ 4 \\ 6 \\ 8 \end{bmatrix} = 2 \begin{bmatrix} 0 \\ 2 \\ 0 \\ 4 \end{bmatrix} + \begin{bmatrix} 2 \\ 0 \\ 6 \\ 0 \end{bmatrix}$$

For m vectors $x^{(1)}, x^{(2)}, \ldots, x^{(m)}$ and m scalars $\alpha_1, \alpha_2, \ldots, \alpha_m$, we define a **linear combination** as

$$\sum_{i=1}^{m} \alpha_i x^{(i)} = \alpha_1 x^{(1)} + \alpha_2 x^{(2)} + \cdots + \alpha_m x^{(m)} = \begin{bmatrix} \sum_{i=1}^{m} \alpha_i x_1^{(i)} \\ \sum_{i=1}^{m} \alpha_i x_2^{(i)} \\ \vdots \\ \sum_{i=1}^{m} \alpha_i x_n^{(i)} \end{bmatrix}$$

Special vectors are the standard **unit vectors**

$$e^{(1)} = \begin{bmatrix} 1 \\ 0 \\ 0 \\ \vdots \\ 0 \end{bmatrix} \qquad e^{(2)} = \begin{bmatrix} 0 \\ 1 \\ 0 \\ \vdots \\ 0 \end{bmatrix} \qquad \cdots \qquad e^{(n)} = \begin{bmatrix} 0 \\ 0 \\ 0 \\ \vdots \\ 1 \end{bmatrix}$$

Clearly,

$$\sum_{i=1}^{n} \alpha_i e^{(i)} = \begin{bmatrix} \alpha_1 \\ \alpha_2 \\ \vdots \\ \alpha_n \end{bmatrix}$$

Hence, any vector x can be written as a linear combination of the standard unit vectors

$$x = x_1 e^{(1)} + x_2 e^{(2)} + \cdots + x_n e^{(n)} = \sum_{i=1}^{n} x_i e^{(i)}$$

As an example, notice that

$$\begin{bmatrix} 1 \\ 2 \\ 3 \\ 4 \end{bmatrix} = \begin{bmatrix} 1 \\ 0 \\ 0 \\ 0 \end{bmatrix} + 2 \begin{bmatrix} 0 \\ 1 \\ 0 \\ 0 \end{bmatrix} + 3 \begin{bmatrix} 0 \\ 0 \\ 1 \\ 0 \end{bmatrix} + 4 \begin{bmatrix} 0 \\ 0 \\ 0 \\ 1 \end{bmatrix}$$

The **dot product** or **inner product** of vectors x and y is the number

$$x^T y = [x_1, x_2, \ldots, x_n] \begin{bmatrix} y_1 \\ y_2 \\ \vdots \\ y_n \end{bmatrix} = \sum_{i=1}^{n} x_i y_i$$

As an example, we see that

$$[1, 1, 1, 1] \begin{bmatrix} 1 \\ 1 \\ 1 \\ 1 \end{bmatrix} = 4$$

Matrices

A **matrix** is a two-dimensional array of numbers that can be written as

$$A = \begin{bmatrix} a_{11} & a_{12} & \cdots & a_{1m} \\ a_{21} & a_{22} & \cdots & a_{2m} \\ \vdots & \vdots & \ddots & \vdots \\ a_{n1} & a_{n2} & \cdots & a_{nm} \end{bmatrix}$$

where a_{ij} called the **element** or **entry** in the ith row and jth column. An alternative notation is $A = (a_{ij})_{n \times m}$. A column vector is also an $n \times 1$ matrix and a row vector is also a $1 \times m$ matrix. For example, here are three matrices:

$$\begin{bmatrix} \frac{1}{5} & \frac{2}{7} & -1 \\ 3 & 2 & \frac{1}{8} \\ -\frac{5}{6} & \frac{2}{5} & 3 \end{bmatrix} \qquad \begin{bmatrix} 1 & 6 & \frac{9}{8} & -5 \end{bmatrix} \qquad \begin{bmatrix} \frac{11}{2} & \frac{4}{9} \\ \frac{2}{3} & -\frac{7}{8} \\ \pi & e \\ 1/\pi & 1/e \end{bmatrix}$$

The entries in A can be grouped into column vectors:

$$A = \begin{bmatrix} \begin{bmatrix} a_{11} \\ a_{21} \\ \vdots \\ a_{n1} \end{bmatrix} & \begin{bmatrix} a_{12} \\ a_{22} \\ \vdots \\ a_{n2} \end{bmatrix} & \cdots & \begin{bmatrix} a_{1m} \\ a_{2m} \\ \vdots \\ a_{nm} \end{bmatrix} \end{bmatrix} = \begin{bmatrix} a^{(1)} & a^{(2)} & \cdots & a^{(m)} \end{bmatrix}$$

where $a^{(j)}$ is the jth column vector. Also, A can be grouped into row vectors:

$$A = \begin{bmatrix} [a_{11} & a_{12} & \cdots & a_{1m}] \\ [a_{21} & a_{22} & \cdots & a_{2m}] \\ & & \vdots \\ [a_{n1} & a_{n2} & \cdots & a_{nm}] \end{bmatrix} = \begin{bmatrix} A^{(1)} \\ A^{(2)} \\ \vdots \\ A^{(n)} \end{bmatrix}$$

where $A^{(i)}$ is the ith row vector. Notice that

$$\begin{bmatrix} 1 & 5 & 9 & 13 \\ 2 & 6 & 10 & 14 \\ 3 & 7 & 11 & 15 \\ 4 & 8 & 12 & 16 \end{bmatrix} = \begin{bmatrix} \begin{bmatrix} 1 \\ 2 \\ 3 \\ 4 \end{bmatrix} & \begin{bmatrix} 5 \\ 6 \\ 7 \\ 8 \end{bmatrix} & \begin{bmatrix} 9 \\ 10 \\ 11 \\ 12 \end{bmatrix} & \begin{bmatrix} 13 \\ 14 \\ 15 \\ 16 \end{bmatrix} \end{bmatrix} = \begin{bmatrix} [1 & 5 & 9 & 13] \\ [2 & 6 & 10 & 14] \\ [3 & 7 & 11 & 15] \\ [4 & 8 & 12 & 16] \end{bmatrix}$$

An $n \times n$ matrix of special importance is the **identity** matrix, denoted by I, composed of all 0's except that the main diagonal consists of 1's:

$$I = \begin{bmatrix} 1 & 0 & \cdots & 0 \\ 0 & 1 & \cdots & 0 \\ \vdots & \vdots & \ddots & \vdots \\ 0 & 0 & \cdots & 1 \end{bmatrix} = \begin{bmatrix} e^{(1)} & e^{(2)} & \cdots & e^{(n)} \end{bmatrix}$$

A matrix of this same general form with entries d_i on the main diagonal is called a **diagonal** matrix and is written as

$$D = \begin{bmatrix} d_1 & & & \\ & d_2 & & \\ & & \ddots & \\ & & & d_n \end{bmatrix} = \mathrm{diag}(d_1, d_2, \ldots, d_n)$$

where the blank space indicates 0 entries. A **tridiagonal** matrix is a square matrix of the form

$$T = \begin{bmatrix} d_1 & c_1 & & & & \\ a_1 & d_2 & c_2 & & & \\ & a_2 & d_3 & c_3 & & \\ & & \ddots & \ddots & \ddots & \\ & & & a_{n-2} & d_{n-1} & c_{n-1} \\ & & & & a_{n-1} & d_n \end{bmatrix}$$

where the diagonal entries $\{a_i\}$, $\{d_i\}$, and $\{c_i\}$ are called the **subdiagonal, main diagonal**, and **superdiagonal**, respectively.

For the general $n \times n$ matrix $A = (a_{ij})$, A is a diagonal matrix if $a_{ij} = 0$ whenever $i \neq j$, and A is a tridiagonal matrix if $a_{ij} = 0$ whenever $|i - j| \geq 2$. The matrix A is a **lower triangular matrix** whenever $a_{ij} = 0$ for all $i < j$ and is an **upper triangular matrix** whenever $a_{ij} = 0$ for all $i > j$. Examples of identity, diagonal, tridiagonal, lower triangular, and upper triangular matrices, respectively, are:

$$\begin{bmatrix} 1 & 0 & 0 & 0 \\ 0 & 1 & 0 & 0 \\ 0 & 0 & 1 & 0 \\ 0 & 0 & 0 & 1 \end{bmatrix} \quad \begin{bmatrix} 3 & 0 & 0 & 0 \\ 0 & 5 & 0 & 0 \\ 0 & 0 & 7 & 0 \\ 0 & 0 & 0 & 9 \end{bmatrix} \quad \begin{bmatrix} 5 & 3 & 0 & 0 & 0 \\ 2 & 5 & 3 & 0 & 0 \\ 0 & 2 & 9 & 2 & 0 \\ 0 & 0 & 3 & 7 & 2 \\ 0 & 0 & 0 & 3 & 7 \end{bmatrix}$$

$$\begin{bmatrix} 6 & 0 & 0 & 0 \\ 3 & 6 & 0 & 0 \\ 4 & -2 & 7 & 0 \\ 5 & -3 & 9 & 21 \end{bmatrix} \quad \begin{bmatrix} 1 & -1 & 2 & 1 \\ 0 & 5 & -5 & 1 \\ 0 & 0 & 9 & -3 \\ 0 & 0 & 0 & 2 \end{bmatrix}$$

As with vectors, many operations involving matrices correspond to component operations. For matrices A and B,

$$A = \begin{bmatrix} a_{11} & a_{12} & \cdots & a_{1m} \\ a_{21} & a_{22} & \cdots & a_{2m} \\ \vdots & \vdots & \ddots & \vdots \\ a_{n1} & a_{n2} & \cdots & a_{nm} \end{bmatrix} \qquad B = \begin{bmatrix} b_{11} & b_{12} & \cdots & b_{1m} \\ b_{21} & b_{22} & \cdots & b_{2m} \\ \vdots & \vdots & \ddots & \vdots \\ b_{n1} & b_{n2} & \cdots & b_{nm} \end{bmatrix}$$

The following definitions apply:

Equality

$A = B$ if and only if $a_{ij} = b_{ij}$ for all i ($1 \leq i \leq n$) and all j ($1 \leq j \leq m$)

Inequality

$A < B$ if and only if $a_{ij} < b_{ij}$ for all i ($1 \leq i \leq n$) and all j ($1 \leq j \leq m$)

Addition/Subtraction

$$A \pm B = \begin{bmatrix} a_{11} \pm b_{11} & a_{12} \pm b_{12} & \cdots & a_{1m} \pm b_{1m} \\ a_{21} \pm b_{21} & a_{22} \pm b_{22} & \cdots & a_{2m} \pm b_{2m} \\ \vdots & \vdots & \ddots & \vdots \\ a_{n1} \pm b_{n1} & a_{n2} \pm b_{n2} & \cdots & a_{nm} \pm b_{nm} \end{bmatrix}$$

Scalar Product

$$\alpha A = \begin{bmatrix} \alpha a_{11} & \alpha a_{12} & \cdots & \alpha a_{1m} \\ \alpha a_{21} & \alpha a_{22} & \cdots & \alpha a_{2m} \\ \vdots & \vdots & \ddots & \vdots \\ \alpha a_{n1} & \alpha a_{n2} & \cdots & \alpha a_{nm} \end{bmatrix} \qquad \text{for } \alpha \text{ a constant}$$

As an example, we have

$$\begin{bmatrix} \frac{1}{5} & \frac{7}{5} & -1 \\ -3 & 2 & -8 \\ \frac{6}{5} & \frac{2}{5} & -3 \end{bmatrix} = \frac{1}{5} \begin{bmatrix} 1 & 7 & 0 \\ 0 & 10 & 0 \\ 6 & 2 & 0 \end{bmatrix} - \begin{bmatrix} 0 & 0 & 1 \\ 3 & 0 & 8 \\ 0 & 0 & 3 \end{bmatrix}$$

Matrix-Vector Product

The product of an $n \times m$ matrix A and an $m \times 1$ vector b is of special interest. Considering the matrix A in terms of its columns, we have

$$Ab = [a^{(1)} \quad a^{(2)} \quad \cdots \quad a^{(m)}] \begin{bmatrix} b_1 \\ b_2 \\ \vdots \\ b_m \end{bmatrix}$$

$$= b_1 a^{(1)} + b_2 a^{(2)} + \cdots + b_m a^{(m)}$$

$$= \sum_{i=1}^{m} b_i a^{(i)}$$

Thus, Ab is a vector and can be thought of as a linear combination of the columns of A with coefficients the entries of b. Considering matrix A in terms of its rows, we have

$$Ab = \begin{bmatrix} A^{(1)} \\ A^{(2)} \\ \vdots \\ A^{(n)} \end{bmatrix} b = \begin{bmatrix} A^{(1)}b \\ A^{(2)}b \\ \vdots \\ A^{(n)}b \end{bmatrix}$$

Thus, the jth element of Ab can be viewed as the dot product of the jth row of A and the vector b.

Matrix Product

The product of the matrix $A = (a_{ij})_{n \times m}$ and the matrix $B = (b_{ij})_{m \times r}$ is the matrix $C = (c_{ij})_{n \times r}$ such that

$$AB = C$$

where

$$c_{ij} = a_{i1}b_{1j} + a_{i2}b_{2j} + \cdots + a_{im}b_{mj} = \sum_{k=1}^{m} a_{ik}b_{kj} \qquad (1 \le i \le 1,\ \le j \le r)$$

The element c_{ij} is the dot product of the ith row vector of A

$$A^{(i)} = [a_{i1}, a_{i2}, \ldots, a_{im}]$$

and the jth column vector of B

$$b^{(j)} = \begin{bmatrix} b_{1j} \\ b_{2j} \\ \vdots \\ b_{mj} \end{bmatrix}$$

that is

$$c_{ij} = A^{(i)}b^{(j)}$$

Similarly, the matrix product AB can be thought of in two different ways. We can write either

$$AB = A\begin{bmatrix} b^{(1)} & b^{(2)} & \cdots & b^{(r)} \end{bmatrix} \tag{1}$$

$$= \begin{bmatrix} Ab^{(1)} & Ab^{(2)} & \cdots & Ab^{(r)} \end{bmatrix}$$

$$= C$$

or

$$AB = \begin{bmatrix} A^{(1)} \\ A^{(2)} \\ \vdots \\ A^{(n)} \end{bmatrix} B = \begin{bmatrix} A^{(1)}B \\ A^{(2)}B \\ \vdots \\ A^{(n)}B \end{bmatrix} = C \tag{2}$$

Equation (1) implies that the jth column of $C = AB$ is

$$c^{(j)} = Ab^{(j)}$$

That is, each column of C is the result of *post*multiplying A by the jth column of B. In other words, each column of C can be obtained by taking inner products of a column of B with all rows of A:

$$c^{(j)} = Ab^{(j)} = \begin{bmatrix} \leftarrow \\ \leftarrow \\ \vdots \\ \leftarrow \end{bmatrix}\begin{bmatrix} b_{1j} \\ b_{2j} \\ \vdots \\ b_{mj} \end{bmatrix} = \begin{bmatrix} c_{1j} \\ c_{2j} \\ \vdots \\ c_{nj} \end{bmatrix}$$

The long left-arrow means an inner product is formed across the elements in the row—that is, $c_{ij} = \sum_{k=1}^{n} a_{ik}b_{kj}$. Equation (2) implies that the ith row of the result C of multiplying A times B is

$$C^{(i)} = A^{(i)}B$$

That is, each row of C is the result of *pre*multiplying B by the ith row of A. In other words, each row of C can be obtained by taking inner products of a row of A with all columns of B:

$$C^{(i)} = A^{(i)}B = [a_{i1} \quad a_{i2} \quad \cdots \quad a_{im}]\begin{bmatrix} \uparrow & \uparrow & \cdots & \uparrow \end{bmatrix}$$

$$= [c_{i1} \quad c_{i2} \quad \cdots \quad c_{ir}]$$

The long up-arrow means an inner product is formed from the elements in the column—that is, $c_{ij} = \sum_{k=1}^{n} a_{ik}b_{kj}$.

As an example, we can determine the matrix product columnwise as

$$\begin{bmatrix} 3 & 1 & 7 \\ 2 & 4 & -5 \\ 1 & -3 & 2 \end{bmatrix}\begin{bmatrix} -1 & -3 & 2 \\ 1 & 1 & 1 \\ -3 & -2 & 1 \end{bmatrix} = \begin{bmatrix} c^{(1)} & c^{(2)} & c^{(3)} \end{bmatrix}$$

where

$$c^{(1)} = \begin{bmatrix} 3 & 1 & 7 \\ 2 & 4 & -5 \\ 1 & -3 & 2 \end{bmatrix} \begin{bmatrix} -1 \\ 1 \\ -3 \end{bmatrix} = \begin{bmatrix} -23 \\ 17 \\ -10 \end{bmatrix}$$

$$c^{(2)} = \begin{bmatrix} 3 & 1 & 7 \\ 2 & 4 & -5 \\ 1 & -3 & 2 \end{bmatrix} \begin{bmatrix} -3 \\ 1 \\ -2 \end{bmatrix} = \begin{bmatrix} -22 \\ 8 \\ -10 \end{bmatrix}$$

$$c^{(3)} = \begin{bmatrix} 3 & 1 & 7 \\ 2 & 4 & -5 \\ 1 & -3 & 2 \end{bmatrix} \begin{bmatrix} 2 \\ 1 \\ 1 \end{bmatrix} = \begin{bmatrix} 14 \\ 3 \\ 1 \end{bmatrix}$$

or we can determine it rowwise as

$$\begin{bmatrix} 3 & 1 & 7 \\ 2 & 4 & -5 \\ 1 & -3 & 2 \end{bmatrix} \begin{bmatrix} -1 & -3 & 2 \\ 1 & 1 & 1 \\ -3 & -2 & 1 \end{bmatrix} = \begin{bmatrix} C^{(1)} \\ C^{(2)} \\ C^{(3)} \end{bmatrix}$$

where

$$C^{(1)} = \begin{bmatrix} 3 & 1 & 7 \end{bmatrix} \begin{bmatrix} -1 & -3 & 2 \\ 1 & 1 & 1 \\ -3 & -2 & 1 \end{bmatrix} = \begin{bmatrix} -23 & -22 & 14 \end{bmatrix}$$

$$C^{(2)} = \begin{bmatrix} 2 & 4 & -5 \end{bmatrix} \begin{bmatrix} -1 & -3 & 2 \\ 1 & 1 & 1 \\ -3 & -2 & 1 \end{bmatrix} = \begin{bmatrix} 17 & 8 & 3 \end{bmatrix}$$

$$C^{(3)} = \begin{bmatrix} 1 & -3 & 2 \end{bmatrix} \begin{bmatrix} -1 & -3 & 2 \\ 1 & 1 & 1 \\ -3 & -2 & 1 \end{bmatrix} = \begin{bmatrix} -10 & -10 & 1 \end{bmatrix}$$

Other Concepts

The **transpose** of the $n \times m$ matrix A, denoted A^T, is obtained by interchanging the rows and columns of $A = (a_{ij})_{n \times n}$:

$$A^T = \begin{bmatrix} A^{(1)} \\ A^{(2)} \\ \vdots \\ A^{(n)} \end{bmatrix}^T = \begin{bmatrix} A^{(1)^T} & A^{(2)^T} & \cdots & A^{(n)^T} \end{bmatrix}$$

or

$$A^T = \begin{bmatrix} a^{(1)} & a^{(2)} & \cdots & a^{(m)} \end{bmatrix}^T = \begin{bmatrix} a^{(1)^T} \\ a^{(2)^T} \\ \vdots \\ a^{(m)^T} \end{bmatrix}$$

Hence, A^T is the $m \times n$ matrix:

$$A^T = \begin{bmatrix} a_{11} & a_{21} & \cdots & a_{n1} \\ a_{12} & a_{22} & \cdots & a_{n2} \\ \vdots & \vdots & & \vdots \\ a_{1m} & a_{2m} & \cdots & a_{nm} \end{bmatrix} = (a_{ji})_{m \times n}$$

As an example, we have

$$\begin{bmatrix} 2 & 4 & 9 \\ 5 & 7 & 3 \\ 10 & 6 & 2 \end{bmatrix}^T = \begin{bmatrix} 2 & 5 & 10 \\ 4 & 7 & 6 \\ 9 & 3 & 2 \end{bmatrix}$$

An $n \times n$ matrix A is **symmetric** if $a_{ij} = a_{ji}$ for all i ($1 \leq i \leq n$) and all j ($1 \leq j \leq n$). In other words, A is symmetric if $A = A^T$.

Some useful properties for matrices of compatible sizes are as follows:

1. $AB \neq BA$ (in general)
2. $AI = IA = A$
3. $A0 = 0A = 0$
4. $(A^T)^T = A$
5. $(A + B)^T = A^T + B^T$
6. $(AB)^T = B^T A^T$

If A and B are square matrices that satisfy $AB = BA = I$, then B is said to be the **inverse** of A, which is denoted A^{-1}.

To illustrate property **1**, form the following products and observe that matrix multiplication is not commutative:

$$\begin{bmatrix} 3 & 1 & 7 \\ 2 & 4 & -5 \\ 1 & -3 & 2 \end{bmatrix} \begin{bmatrix} -1 & -3 & 2 \\ 1 & 1 & 1 \\ -3 & -2 & 1 \end{bmatrix}$$

$$\begin{bmatrix} -1 & -3 & 2 \\ 1 & 1 & 1 \\ -3 & -2 & 1 \end{bmatrix} \begin{bmatrix} 3 & 1 & 7 \\ 2 & 4 & -5 \\ 1 & -3 & 2 \end{bmatrix}$$

Also, verify that $AA^{-1} = A^{-1}A = I$ for

$$A = \begin{bmatrix} 1 & 1 & 1 \\ -1 & 3 & 2 \\ 2 & 1 & 1 \end{bmatrix}$$

and

$$A^{-1} = \begin{bmatrix} 1 & 0 & 1 \\ -5 & 1 & 3 \\ 7 & -1 & 4 \end{bmatrix}$$

As our final set of examples, we have the product of a matrix times a vector and of two matrices:

$$\begin{bmatrix} 3 & 2 & -1 \\ 5 & 3 & 2 \\ -1 & 1 & -3 \end{bmatrix}\begin{bmatrix} x_1 \\ x_2 \\ x_3 \end{bmatrix} = \begin{bmatrix} 3x_1 + 2x_2 - x_3 \\ 5x_1 + 3x_2 + 2x_3 \\ -x_1 + x_2 - 3x_3 \end{bmatrix}$$

$$\begin{bmatrix} 1 & 0 & 0 \\ -\frac{5}{3} & 1 & 0 \\ -8 & 5 & 1 \end{bmatrix}\begin{bmatrix} 3 & 2 & -1 \\ 5 & 3 & 2 \\ -1 & 1 & -3 \end{bmatrix} = \begin{bmatrix} 3 & 2 & -1 \\ 0 & -\frac{1}{3} & \frac{11}{3} \\ 0 & 0 & 15 \end{bmatrix}$$

The reader should verify them and note how they relate to solving the following problem using naive Gaussian elimination (see Section 6.1):

$$\begin{cases} 3x_1 + 2x_2 - x_3 = 7 \\ 5x_1 + 3x_2 + 2x_3 = 4 \\ -x_1 + x_2 - 3x_3 = -1 \end{cases}$$

As well, compute the products shown and relate them to this problem:

$$\begin{bmatrix} 1 & 0 & 0 \\ -\frac{5}{8} & 1 & 0 \\ -8 & 5 & 1 \end{bmatrix}\begin{bmatrix} 7 \\ 4 \\ -1 \end{bmatrix}$$

$$\begin{bmatrix} \frac{1}{3} & 2 & -\frac{7}{15} \\ 0 & -3 & \frac{11}{15} \\ 0 & 0 & \frac{1}{15} \end{bmatrix}\begin{bmatrix} 3 & 2 & -1 \\ 0 & -\frac{1}{3} & \frac{11}{3} \\ 0 & 0 & 15 \end{bmatrix}$$

$$\begin{bmatrix} \frac{1}{3} & 2 & -\frac{7}{15} \\ 0 & -3 & \frac{11}{15} \\ 0 & 0 & \frac{1}{15} \end{bmatrix}\begin{bmatrix} 7 \\ -\frac{23}{3} \\ -37 \end{bmatrix}$$

ANSWERS FOR SELECTED PROBLEMS

PROBLEMS 1.1

1. two other ways: $pi \leftarrow 2.0 \arcsin(1.0)$
$pi \leftarrow 2.0 \arccos(0.0)$

3a. $sum \leftarrow 0$
for $i = 1$ **to** n **do**
  for $j = 1$ **to** n **do**
    $sum \leftarrow sum + a_{ij}$
  end do
end do

4. $n(n+1)/2$

5. n multiplications and n additions/subtractions

12. $z = 1 + \sum_{i=2}^{n} \prod_{j=2}^{i} b_j$

COMPUTER PROBLEMS 1.1

3. $\exp(1.0) \approx 2.71828\,18284\,6$
8. computation deviates from theory when $a_1 = 10^{-12}, 10^{-8}, 10^{-4}, 10^{20}$, for example
9. x may underflow and be set to zero
12a. the computation m/n may result in truncation so that $x \neq y$ **14.** 40 different spellings

PROBLEMS 1.2

1. $\cosh x = \sum_{k=0}^{\infty} \frac{x^{2k}}{(2k)!}$; $\cosh 0.7 \approx 1.25517$ **2a.** $e^{\cos x} = e\left(1 - \frac{x^2}{2} + \cdots\right)$

2b. $\sin(\cos x) = (\sin 1) - (\cos 1)\left(\frac{x^2}{2}\right) + \cdots$

3. $\cos\left(\frac{\pi}{3} + h\right) = \frac{1}{2}\sum_{k=0}^{\infty}(-1)^k \frac{h^{2k}}{(2k)!} + \frac{\sqrt{3}}{2}\sum_{k=1}^{\infty}(-1)^k \frac{h^{2k-1}}{(2k-1)!}$; $\cos(60.001°) \approx 0.49998\,488$ **4.** $m = 2$

5.   at least 18 terms   **6.**   yes; by using this formula, we avoid the series for e^{-x} and use the one for e^{x}

8. $\ln(1 - x) = -\displaystyle\sum_{k=1}^{\infty} \frac{x^{k}}{k}; \quad \ln\left(\frac{1 + x}{1 - x}\right) = 2\sum_{k=1}^{\infty} \frac{x^{2k-1}}{(2k - 1)}$

9. $x = \dfrac{1}{3}, \quad \ln 2 = 0.69313$ (four terms);   at least ten terms   **12b.** $\displaystyle\lim_{x \to 0} \frac{\arctan x}{x} = 1$

12c. $\displaystyle\lim_{x \to \pi} \frac{\cos x + 1}{\sin x} = 0$

13.   at least 38 terms   **14.** $\text{erf}(x) = \dfrac{2}{\sqrt{\pi}}\left[x - \dfrac{x^{3}}{3} + \dfrac{x^{5}}{5(2!)} - \dfrac{x^{7}}{7(3!)} + \cdots\right]; \quad \text{erf}(1) \approx 0.8382$

15. 10^{10} **16.** 10^{5} **17.** $\sin(45.0005°) \approx 0.70711\,295$

19. $\sin x + \cos x = 1 + x - \dfrac{x^{2}}{2} - \dfrac{x^{3}}{6} + \cdots; \quad (\sin x)(\cos x) = x - \dfrac{2}{3}x^{3} + \dfrac{2}{15}x^{5} - \dfrac{4}{315}x^{7} + \cdots;$

$\sin(0.001) + \cos(0.001) \approx 1.00099\,94998\,3; \quad \sin(0.0006)\cos(0.0006) \approx 0.00059\,99998\,57$

21.   at least seven terms   **22.**   at least 100 terms

24. $-\dfrac{5}{8}h^{4}$ **26a.**   first derivative $+\infty$ at 0   **26b.**   first derivative not continuous

26e.   function $-\infty$ at 0   **28.** $\dfrac{1}{8}\left(x - \dfrac{17}{4}\right)$ **29.** $s \leftarrow 0$
$\qquad$ **for** $i = 2$ **to** n **do**
$\qquad\qquad s \leftarrow s + \log(i)$
$\qquad\qquad$ **output** i, s
$\qquad$ **end do**

COMPUTER PROBLEMS   1.2

4.

	$c = 1$	$c = 10^{8}$
x_1	0	-1
x_2	-10^{8}	-10^{8}

9. g converges faster (in five iterations)   **11.** $\lambda_{50} = 1\,25862\,69025$

12. $\alpha_{50} = 1\,73937\,96001$

PROBLEMS   2.1

1. $441.68164\,0625$ (decimal)   **3a.** $(27.45075\,341\ldots)_{8}$ **3b.** $(113.16662\,13\ldots)_{8}$
3c. $(71.24426\,416\ldots)_{8}$ **4.** 613.40625 (decimal)   **5c.** $(101\,111)_{2}$ **5e.** $(110\,011)_{2}$
5g. $(33.72664)_{8}$ **6.** $(0.3146\,3146\ldots)_{8}$ **10.** $(479)_{10} = (111\,011\,111)_{2}$
12. $e \approx (2.718)_{10} = (010.101\,101\,111\,100\,111\ldots)_{2}$

PROBLEMS   2.2

3.   1   **5.** $1.00005; \quad 1.0$ **6.** $|x| < 5 \times 10^{-5}$ **7.** β^{1-n} **24.** $\approx 3 \times 2^{-24}$

25. $\approx 2^{-22}$ **36.** $\dfrac{1}{2} \times 10^{-12}$ rounding; 10^{-12} chopping   **37.**   9%

41. $x = \dfrac{6032}{9990}; \quad x = \dfrac{6032}{10010}$ **42.** 6×10^{-5}

PROBLEMS 2.3

1a. near $\pi/2$, sine curve is relatively flat   **6.** $y = \dfrac{\cos^2 x}{1 + \sin x}$ **8.** $|x| < \sqrt{6\varepsilon}$, ε machine precision

9. $x_1 \approx 10^5$, $x_2 \approx 10^{-5}$ **13.** $f(x) = -\dfrac{1}{2}x^3 - \dfrac{1}{2}x^4$; $f(0.0125) \approx -9.888 \times 10^{-7}$

15. $f(x) = \dfrac{1}{\sqrt{1 + x^2} + 1} + 3 - 1.7x^2$; $f(0) = 3.5$ **17.** $f(x) = \dfrac{1}{\sqrt{x^2 + 1} + x}$

18. $f(x) = \begin{cases} \ln(x + \sqrt{x^2 + 1}) & x > 0 \\ 0 & x = 0 \\ -\ln(-x + \sqrt{x^2 + 1}) & x < 0 \end{cases}$ **20.** $z = \dfrac{x^4}{\sqrt{x^4 + 4} + 2}$

23. $f(x) \approx 1 - x + \dfrac{x^2}{3} - \dfrac{x^3}{6}$; $f(0.008) \approx 0.992020915$

COMPUTER PROBLEMS 2.3

7. $|x| < 10^{-15}$ **8.** $p_{50} = 2.85987$ **9.** no solution;   0;   0;   any solution;

− 1.	0.
− 0.10208 42383	− 4.89791 57617
4.00000 00001	4.0009 99999
− 0.10208 42383	− 4.89791 57617
1.0000 00000	1.00000 0000E34
1.99683 77223	2.00316 22777

14.

x	exponent(x)	n
0	1.0	1
1	2.71828 18285	10
−1	0.36787 94412	10
0.5	1.64872 12707	8
−0.123	0.88426 36626	5
−25.5	$8.42346\,37545 \times 10^{-12}$	25
3.14159	23.14063 12270	17
−1776	0	25

PROBLEMS 3.1

3. $\left\{ 0, \pm\dfrac{\pi}{2}, \pm\pi, \pm\dfrac{3\pi}{2}, \pm 2\pi, \ldots \right\}$ **6.** $x = 0$ **8.** 0.61906;   1.51213   **13.** 20 steps

14. $\left\{ -\left(\dfrac{\pi}{4} + \delta\right), 0, \dfrac{\pi}{4} + \varepsilon, \dfrac{3\pi}{4} + \varepsilon, \dfrac{5\pi}{4} + \varepsilon, \ldots \right\}$, where $\delta \approx 0.2$ and ε starts at approximately 0.4 and decreases

COMPUTER PROBLEMS 3.1

4. $1, 2, 3, 3 - 2i, 3 + 2i, 5 + 5i, 5 - 5i, 16$ **9.** 2.365

PROBLEMS 3.2

3. $x_{n+1} = \dfrac{1}{2}[x_n + 1/(Rx_n)]$ **4.** 0.79;   1.6   **5.** $x_{n+1} = x_n - \dfrac{f(x_n)f'(x_n)}{[f'(x_n)]^2 - f(x_n)f''(x_n)}$

7. $x_{n+1} = x_n - \dfrac{f'(x_n)}{f''(x_n)} + \dfrac{\sqrt{[f'(x_n)]^2 - 2f(x_n)f''(x_n)}}{f''(x_n)}$ **9.** $y = \dfrac{\sqrt{2}}{2}x + \dfrac{\sqrt{2}}{2}\left(1 - \dfrac{\pi}{4}\right)$

23. $e_{n+1} = e_n^2 \left[\dfrac{\dfrac{f^{(m+1)}(\eta_n)}{m!} - \dfrac{f^{(m+1)}(\xi_n)}{(m+1)(m-1)!}}{\dfrac{f^{(m)}(r)}{(m-1)!} + \dfrac{e_n f^{(m+1)}(\eta_n)}{m!}} \right]$ **24.** $e_{n+1} = \dfrac{1}{2} e_n^2 \dfrac{f''}{g}$ **25.** diverges

27. $|x_0| < \sqrt{3}$ **29.** Newton's method cycles if $x_0 \neq 0$.   **30.** $x \leftarrow R$
for $n = 1$ to 4 do
$\qquad x \leftarrow (2x + Rx^2)/3$
end do

35. $x_{n+1} = [(m-1)x_n^m + R]/(mx_n^{m-1});$ $x_{n+1} = x_n[(m+1)R - x_n^m]/(mR)$ **36.** $|g'(r)| < 1$ if $0 < \omega < 2$

COMPUTER PROBLEMS   3.2

4. 0.32796 77853 31818 36223 77546   **5.** 2.09455 14815 42326 59148 23865 40579
8. 1.83928 67552   **9.** 0.47033 169   **10a.** 1.89549 42670 340   **10b.** 1.99266 68631 307
10c. 2.58280 14730 552   **10d.** 0.51097 34293 8857   **17.** 3.13108;   3.15145   (two nearby roots)

PROBLEMS   3.3

4. $e_{n+1} = \left[1 - \left(\dfrac{x_n - x_0}{f(x_n) - f(x_0)} \right) f'(\xi_n) \right] e_n$ **10a.** linear   **10d.** quadratic

12. show $|\xi - x_{n+1}| \leq c|\xi - x_n|$ **13.** $\sqrt{2}$ **14.** $x = 4.510187$

COMPUTER PROBLEMS   3.3

1a. 1.53209   **1b.** 1.23618   **2.** 1.36880 81078 21373   **4.** 20.80485 4
6. $-0.45896;$ 3.73308

PROBLEMS   4.1

2. $p_4(x) = -1 + (x-1)\left(\dfrac{2}{3} + (x-2)\left(\dfrac{1}{8} + (x - 2.5)\left(\dfrac{3}{4} + (x-3)\dfrac{11}{6} \right) \right) \right)$

6. $q(x) = x^4 - x^3 + x^2 - x + 1 - \dfrac{31}{120}(x+2)(x+1)(x)(x-1)(x-2)$ **7a.** $x^3 - 3x^2 + 2x + 1$

8. $p(x) = x - 2.5$ **12.** $2 + x(-1 + (x-1)(1 - (x-3)x))$

13. $p_4(x) = -1 + 2(x+2) - (x+2)(x+1) + (x+2)(x+1)x;$ $p_2(x) = 1 + 2(x+1)x$

19. 1.5727; no advantage   **21.** $p(x) = -\dfrac{3}{5}x^3 - \dfrac{2}{5}x^2 + 1$ **22.** 0.38099; 0.077848

25. $p_3(x) = 7 - 2x + x^3$ **27.** $\ell_3(x) = -(x-4)(x^2 - 1)/8$

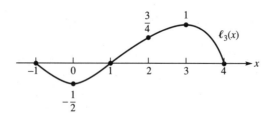

36. 0.85527; 0.87006

COMPUTER PROBLEMS   4.1

2. $p(x) = 2 + 46(x-1) + 89(x-1)(x-2) + 6(x-1)(x-2)(x-3) + 4(x-1)(x-2)(x-3)(x-4)$

PROBLEMS 4.2

4. 1.25×10^{-5} **5.** errors: 7.7×10^{-6}, 6.1×10^{-6} **6.** 498   **7.** 4.3×10^{-14}

8. 2.6×10^{-6} **11.** $n \geq 8$ **14.** $f[x_0, x_1, x_2, x_3, x_4] = 0$ **15.** yes

PROBLEMS 4.3

1. $-hf''(\xi)$ **9.** error term $= -\dfrac{h^2}{6}\left[f'''(\xi_1) + \dfrac{1}{2}f^{(4)}(\xi_2) \right]$ for some $\xi_i \in (x-h, x+h)$

10a. $-\dfrac{2}{3}h^2 f'''(\xi)$ **11a.** $-\dfrac{h^2}{4}f^{(5)}(\xi)$ **12.** $p'\left(\dfrac{x_0 + x_1}{2} \right) = \dfrac{f(x_1) - f(x_0)}{x_1 - x_0}$

14. $\alpha = 1$, error term $= -\dfrac{h^2}{6}f'''(\xi)$; $\alpha \neq 1$, error term $= -(\alpha - 1)\dfrac{h}{2}f''(\xi)$

20. $L \approx \left\{ \left[\phi\left(\dfrac{h}{2}\right) \right]^2 - \phi(h)\phi\left(\dfrac{h}{3}\right) \right\} \Big/ \left\{ 2\phi\left(\dfrac{h}{2}\right) - \phi(h) - \phi\left(\dfrac{h}{3}\right) \right\}$

COMPUTER PROBLEMS 4.3

1. $0.20211\,58503$

PROBLEMS 5.1

5. $L(f; P) \leq M(f; P) \leq U(f; P)$ **7.** $0.00010\,00025\,0006$ **10.** $n \geq 113477$

11. $U - L = \dfrac{1}{n-1}[f(1) - f(0)]$

COMPUTER PROBLEMS 5.1

1. $0.94598\,385$; $0.94723\,395$ **3.** 3.28968

PROBLEMS 5.2

6.

h	2	1	1/2	1/4
L	0	0	1/2	3/4
U	2	2	3/2	5/4
T	2	1	1	1

9. $f(x) = x^n$ $(n > 3)$ on $[0, 1]$, with partition $\{0, 1\}$

12. $n \geq 1156$ **15.** $-(b - a)hf'(\xi)/2$ **16.** $n \geq 16\,07439$;   too small

18. $T = \dfrac{1}{n^3}\left[\dfrac{1}{6}(n-1)(2n-1)n \right] + \dfrac{1}{2n}$ **29a.** $\dfrac{1}{24}h^3 f''(\xi)$ **29b.** $\dfrac{1}{24}\displaystyle\sum_{i=1}^{n} h_i^3 f''(\xi_i)$

29c. $\dfrac{b-a}{24}h^2 f''(\xi)$

COMPUTER PROBLEMS 5.2

2a. 2   **2b.** 1.71828   **2c.** 0.43882

PROBLEMS 5.3

7. not well   **10.** $1 + 2^{m-1}$ **13.** $R(2, 2) = \dfrac{2h}{45}[7f(a) + 32f(a + h) + 12f(a + 2h) + 32f(a + 3h) + 7f(b)]$

20. show $\displaystyle\int_a^b f(x)\,dx - R(n-1, 0) \approx c4^{-n}$ **21.** let $m = 1$ and let $n \to \infty$ in Formula (2)

25. $Z = \dfrac{4096}{2835} f\left(\dfrac{h}{8}\right) - \dfrac{1344}{2835} f\left(\dfrac{h}{4}\right) + \dfrac{84}{2835} f\left(\dfrac{h}{2}\right) - \dfrac{1}{2835} f(h)$ **27.** $x_{n+1} + n^3(x_{n+1} - x_n)/(3n^2 + 3n + 1)$

COMPUTER PROBLEMS   5.3

3. $R(7,7) = 0.49996\,9819$ **5.** $R(7,7) = 0.76519\,7687$ **7.** $R(5,0) = 1.81379\,9364$

8. $\dfrac{2}{9} = 0.22222\ldots$ **9.** $0.62135\,732$

PROBLEMS   5.4

2. $\displaystyle\int_a^b f(x)\,dx = \dfrac{16}{15} S_{2(n-1)} - \dfrac{1}{15} S_{n-1}$ **3.** $-\dfrac{3}{80} h^5 f^{(4)}(\xi)$ **5.** $\dfrac{\pi}{4}$ **7a.** 7.1667

7b. 7.0833 **7c.** 7.0777

PROBLEMS   5.5

5. $\alpha = \gamma = \dfrac{4}{3}, \quad \beta = -\dfrac{2}{3}$ **6.** $A = (b - a), \quad B = \dfrac{1}{2}(b - a)^2$ **7.** $\dfrac{5h}{12} f(a) + \dfrac{2h}{3} f(a + h) - \dfrac{h}{12} f(a + 2h)$

9. $\alpha = \sqrt{\dfrac{3}{5}}, \quad a = c = \dfrac{5}{9}, \quad b = \dfrac{8}{9}$ **10.** $w_1 = w_2 = \dfrac{h}{2}, \quad w_3 = w_4 = -\dfrac{h^3}{24}$

11. $A = 2h, \quad B = 0; \quad C = \dfrac{h^3}{3}$

COMPUTER PROBLEMS   5.5

1. 1.4183

PROBLEMS   6.1

1a. $x_1 = 58, \quad x_2 \approx -19.37, \quad x_3 \approx -26.84$ **1b.** $x \approx 2.096, \quad y \approx -1.288, \quad z \approx -0.05769$
1c. $x_1 \approx -7.233, \quad x_2 \approx 1.133, \quad x_3 \approx 2.433, \quad x_4 = 4.5$ **4a.** no solution

4b. infinite number of solutions   **5.** $\begin{cases} y = 341.7 \\ x = -692.8 \end{cases} \quad \begin{cases} y = 356.39081 \\ x = -722.6526 \end{cases}$

COMPUTER PROBLEMS   6.1

6. $z = [2i, i, i, i]^T, \lambda = 1 + 5i; \quad z = [1, 2, 1, 1]^T, \lambda = 2 + 6i; \quad z = [-i, -i, 0, -i]^T, \lambda = -3 - 7i;$
$z = [1, 1, 1, 0]^T, \lambda = -4 - 8i$

7a. $(3.75, 90°); \quad (3.27, -65.7°); \quad (0.775, 172.9°)$ **7b.** $(2.5, -90°); \quad (2.08, 56.3°); \quad (1.55, -60.2°)$

PROBLEMS   6.2

2. $\begin{bmatrix} 1/2 & 5/2 & -4 & -1 \\ 1/4 & -1/2 & -5/19 & -62/19 \\ 3/4 & 9/10 & 38/5 & 9/10 \\ 4 & 1 & 0 & 4 \end{bmatrix}$ **3.** $x = [1/3, 3, 1/3]^T$ **7.** $\begin{bmatrix} 1/4 & 5/2 & 7/4 & 1/2 \\ 4 & 2 & 1 & 2 \\ 1/2 & 0 & 5/9 & 17/9 \\ 1/4 & 3/5 & 27/10 & 1/5 \end{bmatrix}$

16. $n(n + 1)$ **17.** $\left[\dfrac{29}{10}(n^2 - 1) + \dfrac{7}{30} n(n - 1)(2n - 1)\right] 10^{-6}$ seconds

18.

n	10	10^2	10^3	10^4
Time	$\frac{1}{3} \times 10^{-3}$ sec	$\frac{1}{3}$ sec	5.56 min	3.86 days
Cost	$0.005¢$	$5¢$	$46.30	$46,296.30

22a. $x_1 = \dfrac{5}{9}, \quad x_2 = \dfrac{2}{9}, \quad x_3 = \dfrac{1}{9} \times 10^{-9}$

22b. $x_1 = 1, \quad x_2 = -\dfrac{2}{9}, \quad x_3 = \dfrac{1}{9} \times 10^{-9}$

COMPUTER PROBLEMS   6.2

2. $[3.4606, 1.5610, -2.9342, -0.4301]^T$ **3.** $[6.7831, 3.5914, -6.4451, -1.5179]^T$
4. $2 \leqq n \leqq 10, x_i \approx 1$ for all i;   for large n, many $x_i \neq 1$ **5.** $b_i = n^2 + 2(i-1)$
6. $x_2 = 1, \quad x_i = 0, \quad$ for $i \neq 2$

PROBLEMS   6.3

1a. $5n - 4$ **2.** $n + 2nk - k(k+1)$ **6.** $D^{-1}AD =$ tridiagonal $[\pm\sqrt{a_{i-1}c_{i-1}},\ d_i,\ \pm\sqrt{a_ic_i}]$

COMPUTER PROBLEMS   6.3

3. $\begin{cases} d_i \leftarrow d_i - 1/d_{i-1} \\ b_i \leftarrow b_i - b_{i-1}/d_{i-1} \quad (2 \leqq i \leqq n) \end{cases} \begin{cases} d_n \leftarrow b_n \\ d_i \leftarrow (b_i - d_{i+1})/d_i \quad (n-1 \geqq j \geqq 1) \end{cases}$

4. $\begin{cases} d_1 = 1 \\ d_i = 1 - (4d_{i-1})^{-1} \quad (2 \leqq i \leqq 100) \end{cases}$ **11a.** $\begin{cases} x_1 \leftarrow b_1/a_{11} \\ x_i \leftarrow \left(b_i - \displaystyle\sum_{j=1}^{n-1} a_{ij}x_j \right) \Big/ a_{ii} \quad (2 \leqq i \leqq n) \end{cases}$

12. $\begin{cases} c_i \leftarrow c_i/d_i \\ b_i \leftarrow b_i/d_i \\ d_{i+1} \leftarrow d_{i+1} - a_{i+1}c_i \\ b_{i+1} \leftarrow b_{i+1} - a_{i+1}b_i \quad (1 \leqq i \leqq n-1) \end{cases}$ $\begin{cases} b_n \leftarrow b_n/d_n \\ b_i \leftarrow b_i - c_ib_{i+1} \quad (1 = n-1, \ldots, 1) \end{cases}$

PROBLEMS   6.4

3a. $M = \begin{bmatrix} 1 & 0 & 0 & 0 & 0 \\ 0 & 1 & 0 & 0 & 0 \\ 0 & -2 & 1 & 0 & 0 \\ 0 & 0 & -2 & 1 & 0 \\ -4 & 0 & 0 & 0 & 1 \end{bmatrix}$ $U = \begin{bmatrix} 25 & 0 & 0 & 0 & 1 \\ 0 & 27 & 4 & 3 & 2 \\ 0 & 0 & 50 & -6 & -4 \\ 0 & 0 & 0 & 0 & 0 \\ 0 & 0 & 0 & 0 & 20 \end{bmatrix}$ **4b.** $A = \begin{bmatrix} 3 & 2 & 1 \\ 2 & 2 & 1 \\ 1 & 1 & 1 \end{bmatrix}$

5a. $M = \begin{bmatrix} 1 & 0 & 0 & 0 \\ 0 & 1 & 0 & 0 \\ 0 & -x/b & 1 & 0 \\ -w/a & (xy)/(bc) & -y/c & 1 \end{bmatrix}$ $U = \begin{bmatrix} a & 0 & 0 & z \\ 0 & b & 0 & 0 \\ 0 & 0 & c & 0 \\ 0 & 0 & 0 & d - (wz)/a \end{bmatrix}$

5b. $L' = \begin{bmatrix} a & 0 & 0 & 0 \\ 0 & b & 0 & 0 \\ 0 & x & c & 0 \\ 0 & 0 & y & d - (wz)/a \end{bmatrix}$ $U' = \begin{bmatrix} 1 & 0 & 0 & z/a \\ 0 & 1 & 0 & 0 \\ 0 & 0 & 1 & 0 \\ 0 & 0 & 0 & 1 \end{bmatrix}$

6a. $LU = \begin{bmatrix} 1 & 0 & 0 & 0 \\ -1/4 & 1 & 0 & 0 \\ -1/4 & -1/15 & 1 & 0 \\ 0 & -4/15 & -2/7 & 1 \end{bmatrix} \begin{bmatrix} 4 & -1 & -1 & 0 \\ 0 & 15/4 & -1/4 & -1 \\ 0 & 0 & 56/15 & -16/15 \\ 0 & 0 & 0 & 24/7 \end{bmatrix}$ **7.** 192

9. $U = \begin{bmatrix} 1 & 0 & 0 & 1 \\ 0 & 1 & 0 & -2 \\ 0 & 0 & 1 & 4 \\ 0 & 0 & 0 & -8 \end{bmatrix} \quad L = \begin{bmatrix} 1 & 0 & 0 & 0 \\ 1 & 1 & 0 & 0 \\ -1 & 1 & 1 & 0 \\ 1 & -1 & 1 & 1 \end{bmatrix}$

10a. $LDU' = \begin{bmatrix} 1 & 0 & 0 \\ 1 & 1 & 0 \\ 3 & -1 & 1 \end{bmatrix} \begin{bmatrix} 2 & 0 & 0 \\ 0 & -2 & 0 \\ 0 & 0 & 3 \end{bmatrix} \begin{bmatrix} 1 & -1/2 & 1 \\ 0 & 1 & -1/2 \\ 0 & 0 & 1 \end{bmatrix}$ **10b.** $x = [-1, 2, 1]^T$

11a. $L = \begin{bmatrix} 1 & 0 & 0 \\ 2 & 1 & 0 \\ -1 & 3 & 1 \end{bmatrix} \quad D = \begin{bmatrix} -2 & 0 & 0 \\ 0 & 1 & 0 \\ 0 & 0 & -1 \end{bmatrix} \quad U' = \begin{bmatrix} 1 & -1/2 & 1 \\ 0 & 1 & 1 \\ 0 & 0 & 1 \end{bmatrix}$ **11b.** $x = [-1, 1, 1]^T$

13. $A^{-1} = \dfrac{1}{15} \begin{bmatrix} 11 & -5 & -7 \\ -13 & 10 & 11 \\ -8 & 5 & 1 \end{bmatrix}$

COMPUTER PROBLEMS 6.4

3.   *Case 4*: $p_5(A) = \begin{bmatrix} 536 & -668 & 458 & -186 \\ -668 & 994 & -854 & 458 \\ 458 & -854 & 994 & -668 \\ -186 & 458 & -668 & 536 \end{bmatrix}$

PROBLEMS 7.1

1. yes   **6.** in Problem 5, the bracketed expression is $f'(\xi_1) - f'(\xi_2)$ and in magnitude does not exceed $2Q$

9. $n - 1 \geq 10^6 (\pi/\sqrt{8})$ **17.** $\begin{cases} Q_0(x) = -(x+1)^2 + 2, & Q_1(x) = -2x + 1, & Q_2(x) = 8(x - \frac{1}{2})^2 - 2(x - \frac{1}{2}) \\ Q_3(x) = -5(x-1)^2 + 6(x-1) + 1, & Q_4(x) = 12(x-2)^2 - 4(x-2) + 2 \end{cases}$

PROBLEMS 7.2

4. $a = -4,$ $b = -6,$ $c = -3,$ $d = -1,$ $e = -3$ **6.** no   **12.** $a = 3,$ $b = 3,$ $c = 1$
13. no   **15.** $a = -1,$ $b = 3,$ $c = -2,$ $d = 2$ **17.** $n + 3$ **19.** f is not a cubic spline
22. $p_3(x) = x - 0.0175x^2 + 0.1927x^3;$ no   **26.** S is linear

32. $S_0(x) = \left(-\dfrac{5}{7}\right)(x - 1)^3 + \left(\dfrac{12}{7}\right)(x - 1)$

$S_1(x) = \left(\dfrac{6}{7}\right)(x - 2)^3 - \left(\dfrac{5}{7}\right)(3 - x)^3 - \left(\dfrac{6}{7}\right)(x - 2) + \left(\dfrac{12}{7}\right)(3 - x)$

$S_2(x) = \left(-\dfrac{5}{7}\right)(x - 3)^3 + \left(\dfrac{6}{7}\right)(4 - x)^3 + \left(\dfrac{12}{7}\right)(x - 3) - \left(\dfrac{6}{7}\right)(4 - x)$

$S_3(x) = \left(-\dfrac{5}{7}\right)(5 - x)^3 + \left(\dfrac{12}{7}\right)(5 - x)$

PROBLEMS 7.3

2. $f_n(x) = \cos(n \arccos x)$

3. $B_i^2(x) = \begin{cases} \dfrac{(x - t_i)^2}{(t_{i+2} - t_i)(t_{i+1} - t_i)}, & \text{on } [t_i, t_{i+1}] \\[3mm] \dfrac{(x - t_i)(t_{i+2} - x)}{(t_{i+2} - t_i)(t_{i+2} - t_{i+1})} + \dfrac{(t_{i+3} - x)(x - t_{i+1})}{(t_{i+3} - t_{i+1})(t_{i+2} - t_{i+1})}, & \text{on } [t_{i+1}, t_{i+2}] \\[3mm] \dfrac{(t_{i+3} - x)^2}{(t_{i+3} - t_{i+1})(t_{i+3} - t_{i+2})}, & \text{on } [t_{i+2}, t_{i+3}] \\[3mm] 0, & \text{elsewhere} \end{cases}$

5. $\displaystyle\sum_{i=-\infty}^{\infty} f(t_i) B_i^0(x)$ **14.** $n - k \leq i \leq m - 1$ **15.** use induction on k and $B_{i+i}^{k+i}(x) = 0$ on $[t_i, t_{i+1}]$

16. no   **17.** no   **19.** $\displaystyle\sum_{i=-\infty}^{\infty} t_{i+1} B_i^1(x)$

PROBLEMS 7.4

5. no   **7.** let $C_i^2 = t_{i+1}t_{i+2}$, then $C_i^1 = xt_{i+1}$, and $C_i^0 = x^2$ **9.** $B_i^k(t_j) = 0$ iff $t_j \geq t_{i+k+1}$ or $t_j \leq t_i$

10. $x = (t_{i+3}t_{i+2} - t_i t_{i+1})/(t_{i+3} + t_{i+2} - t_{i+1} - t_i)$

COMPUTER PROBLEMS 7.4

1. 47040

PROBLEMS 8.1

1a. $x = \dfrac{1}{4}t^4 + \dfrac{7}{3}t^3 - \dfrac{2}{3}t^{3/2} + c$ **1b.** $x = ce^t$ **1e.** $x = c_1 e^t + c_2 e^{-t}$ or $x = c_1 \cosh t + c_2 \sinh t$

2a. $x = \dfrac{1}{3}t^3 + \dfrac{3}{4}t^{4/3} + 7$ **3c.** $x = \displaystyle\sum_{n=0}^{\infty} (-1)^n \dfrac{t^{2n+1}}{(2n+1)(2n+1)!} + c$ **3d.** $x = e^{-t^2/2}\left[\int t^2 e^{t^2/2}\, dt + c\right]$

5. let $p(t) = a_0 + a_1 t + a_2 t^2 + \cdots$ and determine a_i

8. $a_0 = a_1 = a_2 = 0$, $\quad a_3 = \dfrac{1}{3}$, $\quad a_k + k a_{k+2} = 0$ $\quad (k \geq 2)$

9. $t = 10$, error $= 2.2 \times 10^4 \varepsilon$; $\quad t = 20$, error $= 4.8 \times 10^8 \varepsilon$ **10.** $x^{(4)} = 18xx'x'' + 6(x')^3 + 3x^2 x'''$

COMPUTER PROBLEMS 8.1

1. $x(2.77) = 385.79118$ **2b.** $x(1.75) = 0.63299\,9983$ **2c.** $x(5) = -0.20873\,51554$
3. $x(10) = 22026.47$ **4a.** error at $t = 1$ is 1.8×10^{-10} **5.** $x(0) = 0.03245\,34427$
7. $x(1) = 1.64872\,12691$ **9.** $x(0) = 1.67984\,09205 \times 10^{-3}$ **10.** $x(0) = -3.75940\,73450$

PROBLEMS 8.2

2c. $f(t, x) = +\sqrt{x/(1 - t^2)}$ **3.** $x(-0.2) = 1.92$ **8.** solve $\dfrac{df}{dx} = e^{-x^2}$, $\quad f(0) = 0$

10. $h^3\left(\dfrac{1}{6} - \dfrac{\alpha}{4}\right)D^2 f + \dfrac{h^3}{6}f_x Df$ where $D = \dfrac{\partial}{\partial t} + f\dfrac{\partial}{\partial x}$ and $D^2 = \dfrac{\partial^2}{\partial t^2} + 2f\dfrac{\partial^2}{\partial x\, \partial t} + f^2\dfrac{\partial^2}{\partial x^2}$

11. $h = 1/1024$

14b. $x^{(4)} = D^3 f + f_x D^2 f + 3Df_x Df + f_x^2 Df$ where $D^3 = \dfrac{\partial^3}{\partial t^3} + 3f\dfrac{\partial^3}{\partial x\, \partial t^2} + 3f^2\dfrac{\partial^3}{\partial t\, \partial x^2} + f^3\dfrac{\partial^3}{\partial x^2}$

17. Taylor series of $f(x, y) = g(x) + h(y)$ about (a, b) is equal to the Taylor series of $g(x)$ about a plus that of $h(y)$ about b

21. $A = 1$, $B = h - k$, $C = (h - k)^2$

22. $f(x + h, y + k) \approx \left(1 + 2xh + k + (1 + 2x^2)h^2 + 2hkx + \frac{1}{2}k^2\right)f$; $f(0.001, 0.998) \approx 2.71285\,34$

COMPUTER PROBLEMS   8.2

2. $x(1) = 1.5708$ **3b.** $n = 7$; $x(2) = 0.82356\,78972$ (RK), $0.82356\,78970$ (T)
3c. $n = 7$; $x(2) = -0.49999\,99998$ (RK), $-0.50000\,00012$ (T)   **4.** $x(1) = 0.60653 = x(3)$
5. $x(3) = 1.5$ **6.** $x(0) = 1.0 = x(1.6)$ **8.** $x(1) = 3.95249$ **9.** $x(10) = 1.344 \times 10^{43}$

PROBLEMS   8.3

2. $\frac{\partial}{\partial s}x(9, s) = e^{252} \approx 10^{109}$ **3a.**   all t **3c.**   positive t **3e.**   no t

6.   divergent for all t **7.** $a = \frac{24}{13}$, $b = -\frac{11}{13}$, $c = \frac{2}{13}$, $d = \frac{10}{13}$, $e = -\frac{2}{39}h^2$

8. $a = 1, b = c = \frac{h}{2}$;   error term is $\mathcal{O}(h^3)$

COMPUTER PROBLEMS   8.3

2. $x\left(\frac{1}{2}\right) = 2.25$ **3.** $x\left(-\frac{1}{2}\right) = -4.5$

8. $y(e) = -6.38905\,60989$ where $y(x) = [1 - \ln v(x)]v(x)$ **9.** $0.21938\,39244$

10. $0.99530\,87432$ **15.** $\text{Si}(1) = 0.94608\,30703$

PROBLEMS   9.1

1. $X' = \begin{bmatrix} 1 \\ x_1^2 + \log x_2 + x_0^2 \\ e^{x_2} - \cos x_1 + \sin(x_0 x_1) - (x_1 x_2)^7 \end{bmatrix}$, $X(0) = [0, 1, 3]^T$

COMPUTER PROBLEMS   9.1

1. $x(1) = 2.46869\,39399$, $y(1) = 1.28735\,52872$ **2.** $x(0.38) = 1.90723 \times 10^{12}$, $y(0.38) = -8.28807 \times 10^4$

4. $x(-1) = 3.36788$, $y(-1) = 2.36788$ **5.** $x_1\left(\frac{\pi}{2}\right) = x_4\left(\frac{\pi}{2}\right) = 0$, $x_2\left(\frac{\pi}{2}\right) = 1$, $x_3\left(\frac{\pi}{2}\right) = -1$

7. $x(6) = 4.39411$, $y(6) = 3.10378$

PROBLEMS   9.2

1. $X' = \begin{bmatrix} 1 \\ x_2 \\ x_3 \\ x_4 \\ x_4^2 + \cos(x_2 x_3) - \sin(x_0 x_1) + \log(x_1/x_0) \end{bmatrix}$, $X(0) = [0, 1, 3, 4, 5]^T$

2. $X' = \begin{bmatrix} 1 \\ x_2 \\ x_3 \\ 2x_2 + \log x_3 + \cos x_1 \end{bmatrix}$, $X(0) = [0, 1, -3, 5]^T$ **8.** $X' = \begin{bmatrix} 1 \\ x_2 \\ x_2 - x_1 \end{bmatrix}$, $X(0) = [0, 0, 1]^T$

PROBLEMS 10.1

1. $y(x) = 1$ **3.** $a = (1 + 2e)/(1 + 2e^2)$, $b = 1$ **5.** $a = 2.1$, $b = 0.9$

7. $c = \sum_{k=0}^{m} y_k \log x_k \Big/ \sum_{k=0}^{m} (\log x_k)^2$ **13.** $y = (6x - 5)/10$

16. $a \approx 2.5929$, $b \approx -0.32583$, $c \approx 0.022738$ **18.** $a = 1$, $b = \dfrac{1}{3}$ **19.** $y(x) = (10x^2 + 20)/35$

PROBLEMS 10.2

2. $\begin{cases} w_{n+2} = w_{n+1} = 0 \\ w_k = c_k + 3xw_{k+1} + 2w_{k+2} \\ f(x) = w_0 - (1 + 2x)w_1 \end{cases}$ $(k = n, n - 1, \ldots, 0)$ **8.** $\begin{cases} g_0(x) = 1 \\ g_1(x) = (x + 1)/2 \\ g_j(x) = (x + 1)g_{j-1}(x) - g_{j-2}(x) \end{cases}$ $(j \geqq 2)$

10. $n + 2$ multiplications, $2n + 1$ additions/subtractions if $2x$ is computed as $x + x$

12. n multiplications, $2n$ additions/subtractions

COMPUTER PROBLEMS 10.2

7. $a_{ij} = \begin{cases} 0 & i \neq j \\ 1 & i = j > 1 \\ 2 & i = j = 1 \end{cases}$

PROBLEMS 10.3

2. coefficient matrix for the normal equations has elements $a_{ij} = \dfrac{1}{i + j - 1}$ by (5)   **3.** $c = 0$

4. $y = b^x$ **6.** $c = \ln 2$ **8.** $x = -1$, $y = \dfrac{20}{13}$ **9a.** $c = \dfrac{24}{\pi^3}$ **9b.** $c = 3$

14. no   **16.** $\begin{bmatrix} \pi & 0 & 2 \\ 0 & \pi/2 & 0 \\ 2 & 0 & \pi/2 \end{bmatrix} \begin{bmatrix} a \\ b \\ c \end{bmatrix} = \begin{bmatrix} (1/2)(e^{2\pi} - 1) \\ -(2/5)(e^{2\pi} + 1) \\ (1/5)(e^{2\pi} + 1) \end{bmatrix}$

COMPUTER PROBLEMS 10.3

1. $a = 2$, $b = 3$

COMPUTER PROBLEMS 11.1

8. 31.6%   **11.** sequence not periodic

13.

0	1	2	3	4	5	6	7	8	9
97	93	97	107	90	115	88	101	113	99

15. 5.6%   **16.** 200

COMPUTER PROBLEMS 11.2

2. 1.71828   **4.** $\dfrac{4}{3}\pi$ **5.** 40.9   **7.** 0.518   **9.** 1.11

10. 2.00034 6869   **14.** 0.635   **17b.** 8.3

COMPUTER PROBLEMS   11.3

1. $\dfrac{2}{3}$ **2.** 0.898   **4.** $\dfrac{7}{16}$ **6.** 1.05   **7.** 5.24   **9.** 0.996

12. 0.6394   **14.** 11.6 kilometers   **15.** 0.14758   **17.** 0.009
21. 24.2 revolutions   **23.** 0.6617

PROBLEMS   12.1

2. $c_1 = (1 - 2e)/(1 - e^2)$, $c_2 = (2e - e^2)/(1 - e^2)$ **3a.** $x(t) = (e^{\pi+t} - e^{\pi-t})/(e^{2\pi} - 1)$
3b. $x(t) = (t^4 - 25t + 12)/12$ **4a.** $x(t) = \beta \sin t + \alpha \cos t$ for all (α, β)
4b. $x(t) = c_1 \sin t + \alpha \cos t$ for all $\alpha + \beta = 0$ with c_1 arbitrary   **7.** $\phi(z) = z$ **8.** $\phi(z) = \sqrt{9 + 6z}$
9. $\phi(z) = (e^5 + e + ze^4 - z)/(2e^2)$ **11.** $x(t) = -e^t + 2\ln(t + 1) + 3t$

PROBLEMS   12.2

1. $-\left(1 - \dfrac{h}{2}\right)x_{i-1} + 2(1 + h^2)x_1 - \left(1 - \dfrac{h}{2}\right)x_{i+1} = -h^2 t$ **2.** $x_1 \approx 0.29427$, $x_2 \approx 0.57016$, $x_3 \approx 0.81040$

4. $x'(0) = \dfrac{5}{3}$ **8.** $-x_{i-1} + [2 + (1 + t_i)h^2]x_i - x_{i+1} = 0$

9. $x(t) = u(t) + (1 - \lambda)v(t)$ where $\lambda = [7 - v(2)]/[u(7) - v(7)]$

PROBLEMS   13.1

1a. elliptic   **1c.** parabolic   **1f.** hyperbolic   **2.** $\dfrac{1}{r}\dfrac{\partial}{\partial r}\left(r\dfrac{\partial u}{\partial r}\right) + \dfrac{1}{r^2}\dfrac{\partial^2 u}{\partial \theta^2} = 0$

5. $a = [1 + 2kh^{-2}(\cos \pi h - 1)]^{1/k}$
8. $u(x, t + k) = \dfrac{k}{h^2}(1 - h)u(x + h, t) + \dfrac{k}{h^2}\left(\dfrac{h^2}{k} + h - 2\right)u(x, t) + \dfrac{k}{h^2}u(x - h, t)$

9. $A = \begin{bmatrix} 0 & 1 & & & \\ -1 & 0 & 1 & & \\ & \ddots & \ddots & \ddots & \\ & & -1 & 0 & 1 \\ & & & -2 & 2 \end{bmatrix}$

PROBLEMS   13.2

1. -0.21 **3.** $u(x, t) = \dfrac{1}{2}[F(x + t) - F(-x + t)] + \dfrac{1}{2}[\overline{G}(x + t) - \overline{G}(-x + t)]$ where $\overline{G}$ is the antiderivative of G

COMPUTER PROBLEMS   13.2

1.   **real function** $fbar(x)$
    real x, y
    $y \leftarrow$ integer$(-(1 + x)/2)$
    $y \leftarrow x + 2y$
    if $y < 0$ **then**
        $fbar \leftarrow -f(-y)$
    else
        $fbar \leftarrow f(y)$
    end if
    end function $fbar$

PROBLEMS   13.3

5. $\left(20 + \dfrac{2.5h}{x_i + y_j}\right)u_{i+1,j} + \left(20 - \dfrac{2.5h}{x_i + y_j}\right)u_{i-1,j} + \left(-30 + \dfrac{0.5h}{y_j}\right)u_{i,j+1} + \left(-30 + \dfrac{0.5h}{y_j}\right)u_{i,j-1} + 20u_{ij} = 69h^2$

6. $u\left(0, \dfrac{1}{2}\right) \approx -8.932 \times 10^{-3}$; $u\left(\dfrac{1}{2}, \dfrac{1}{2}\right) \approx 4.643 \times 10^{-1}$ **7.** $A = \begin{bmatrix} -4 & 1 & 1 & 0 \\ 1 & -4 & 0 & 1 \\ 1 & 0 & -4 & 1 \\ 0 & 1 & 1 & -4 \end{bmatrix}$

COMPUTER PROBLEMS   13.3

5. 18.41° 13.75°
41.47° 36.60° 24.41°
69.41° 66.77° 61.05° 53.01° 51.00°

PROBLEMS   14.1

1. $F(2, 1, -2) = -15$; $F(0, 0, -2) = -8$; $F(2, 0, -2) = -12$ **2.** $F\left(\dfrac{9}{8}, \dfrac{9}{8}\right) = -20.25$

4. Case $n = 2$: $\begin{cases} \widehat{x} = (3a + b)/4 + \delta & \text{if } a \leqq x^* \leqq b' \\ \widehat{x} = (a + 3b)/4 - \delta & \text{if } a' \leqq x^* \leqq b \end{cases}$ **5a.** exact solution $F(3) = -7$

7. $c_1 = \alpha_1/\sqrt{5}$, $c_2 = -\alpha_2/\sqrt{5}$ **9.** $n \geqq 1 + (k + \log \ell - \log 2)/|\log r|$ **10.** $n \geqq 48$

PROBLEMS   14.2

1a. yes   **1b.** no   **2.** $\left(\dfrac{1}{4}, \dfrac{9}{4}\right)$ **3.** $F(x, y) = 1 + x - xy + \dfrac{1}{2}x^2 - \dfrac{1}{2}y^2 + \cdots$

7b. $F(x) = \dfrac{3}{2} - \dfrac{1}{2}x_2 + 3x_1x_2 + x_2x_3 + 2x_1^2 - \dfrac{1}{2}x_3^2 + \cdots$ **10.** $G = \begin{bmatrix} 2y^2z^2 \sin x \cos x \\ 2yz^2(1 + \sin^2 x) + 2(y + 1)(z + 3)^2 \\ 2y^2z(1 + \sin^2 x) + 2(y + 1)^2(z + 3) \end{bmatrix}$

12. $\left(-\dfrac{19}{30}, -\dfrac{1}{5}\right)$

PROBLEMS   15.1

4a. minimum value 1.5 at $(1.5, 0)$ **7.** maximum of 36 at $(2, 6)$ **8.** minimum of 36 at $(0, 3, 1)$

11. minimum 2 for $(x, x - 2)$ where $x \geqq 3$ **13a.** maximum of 18 at $(9, 0)$

13c. unbounded solution   **13f.** no solution   **13h.** maximum of $\dfrac{21}{4}$ at $\left(0, \dfrac{21}{4}\right)$

14. maximum of 100 at $(24, 32, -124)$

COMPUTER PROBLEMS   15.1

1.

	Felt	Straw
Texas Hatters	0	200
Lone Star Hatters	150	0
Lariat Ranch Wear	150	0

3. $13.50

5. cost 50¢ for 1.6 ounces of food f_1, 1 ounce of food f_3, and none of food f_2

PROBLEMS 15.2

1. maximize: $\sum_{j=0}^{n} c_j y_j$ 　　　　Here $c_0 = -\sum_{j=1}^{n} c_j$ and $a_{i0} = -\sum_{j=1}^{n} a_{ij}$. 　　**2.** at most 2^n

　　constraints: $\begin{cases} \sum_{j=0}^{n} a_{ij} y_j \le b_i \\ y_i \ge 0 \quad (0 \le i \le n) \end{cases}$

5. first primal form :

　　maximize: $-b^T y$

　　constraints: $\begin{cases} -A^T y \le -c \\ y \ge 0 \end{cases}$

COMPUTER PROBLEMS 15.2

2. $x = [0, 0, 5/3, 2/3, 0]^T$ 　　**3.** $x = [0, 8/3, 5/3]^T$

PROBLEMS 15.3

1a. maximize: $-\sum_{i=1}^{4} (u_i + v_i)$

　　constraints: $\begin{cases} 5y_1 + 2y_2 \quad\quad - 7y_4 - u_1 + v_1 = 6 \\ y_1 + y_2 + y_3 - 3y_4 - u_2 + v_2 = 2 \\ \quad\quad 7y_2 - 5y_3 - 2y_4 - u_3 + v_3 = 11 \\ 6y_1 \quad\quad + 9y_3 - 15y_4 - u_4 + v_4 = 9 \\ u \ge 0 \quad v \ge 0 \quad y \ge 0 \end{cases}$

1b. minimize: ε

　　constraints: $\begin{cases} 5y_1 + 2y_2 \quad\quad - 7y_4 - \varepsilon \le 6 \\ y_1 + y_2 + y_3 - 3y_4 - \varepsilon \le 2 \\ \quad\quad 7y_2 - 5y_3 - 2y_4 - \varepsilon \le 11 \\ 6y_1 \quad\quad + 9y_3 - 15y_4 - \varepsilon \le 9 \\ -5y_1 - 2y_2 \quad\quad + 7y_4 - \varepsilon \le -6 \\ -y_1 - y_2 - y_3 + 3y_4 - \varepsilon \le -2 \\ \quad\quad - 7y_2 + 5y_3 + 2y_4 - \varepsilon \le -11 \\ -6y_1 \quad\quad - 9y_3 + 15y_4 - \varepsilon \le -9 \\ \varepsilon \ge 0 \quad y_j \ge 0 \quad (1 \le i \le 4) \end{cases}$

COMPUTER PROBLEMS 15.3

1a. $x_1 = 0.353, \quad x_2 = 2.118, \quad x_3 = 9.765$ 　　**1b.** $x_1 = 0.671, \quad x_2 = 1.768, \quad x_3 = 0.453$

3. $p(x) = 1.0001 + 0.9978x + 0.51307x^2 + 0.13592x^3 + 0.071344x^4$

BIBLIOGRAPHY

Abell, M. L., and J. P. Braselton. 1993. *The Mathematica Handbook*. New York: Academic Press.

Abramowitz, M., and I. A. Stegun (eds.). 1964. *Handbook of Mathematical Functions with Formulas, Graphs, and Mathematical Tables*. National Bureau of Standards. New York: Dover, 1965 (reprint).

Acton, F. S. 1959. *Analysis of Straight-Line Data*. New York: Wiley. New York: Dover, 1966 (reprint).

Ames, W. F. 1992. *Numerical Methods for Partial Differential Equations*, 3rd ed. New York: Academic Press.

Armstrong, R. D., and J. Godfrey. 1979. "Two linear programming algorithms for the linear discrete ℓ_1 norm problem." *Mathematics of Computation* **33**, 289–300.

Atkinson, K. 1993. *Elementary Numerical Analysis*. New York: Wiley.

Barrodale, I., and C. Phillips. 1975. "Solution of an overdetermined system of linear equations in the Chebyshev norm." *Association for Computing Machinery Transactions on Mathematical Software* **1**, 264–270.

Barrodale, I., and F. D. K. Roberts. 1974. "Solution of an overdetermined system of equations in the ℓ_1 norm." *Communications of the Association for Computing Machinery* **17**, 319–320.

Barrodale, I., F. D. K. Roberts, and B. L. Ehle. 1971. *Elementary Computer Applications*. New York: Wiley.

Bartels, R. H. 1971. "A stabilization of the simplex method." *Numerische Mathematik* **16**, 414–434.

Beale, E. M. L. 1988. *Introduction to Optimization*. New York: Wiley.

Bloomfield, P., and W. Steiger. 1983. *Least Absolute Deviations, Theory, Applications, and Algorithms*, Boston: Birkhäuser.

de Boor, C. 1971. "CADRE: An algorithm for numerical quadrature." In *Mathematical Software*, edited by J. R. Rice, 417–449. New York: Academic Press.

de Boor, C. 1978. *A Practical Guide to Splines*. New York: Springer-Verlag.

Borwein, J. M., and P. B. Borwein. 1984. "The arithmetic-geometric mean and fast computation of elementary functions," *Society of Industrial and Applied Mathematics Reviews* **26**, 351–366.

Branham, R. 1990. *Scientific Data Analysis: An Introduction to Overdetermined Systems*. New York: Springer-Verlag.

Brent, R. P. 1976. "Fast multiple precision evaluation of elementary functions." *Journal of the Association for Computing Machinery* **23**, 242–251.

Buchanan, J. L., and P. R. Turner. 1992. *Numerical Methods and Analysis*. New York: McGraw-Hill.

Burden, R. L., and J. D. Faires. 1993. *Numerical Analysis*, 5th ed. Boston: PWS Publishers.

Butcher, J. C. 1987. *The Numerical Analysis of Ordinary Differential Equations: Runge-Kutta and General Linear Methods*, New York: Wiley.

Chaitlin, G. J. 1975. "Randomness and mathematical proof." *Scientific American*, May, 47–52.

Collatz, L. 1966. *The Numerical Treatment of Differential Equations*, 3rd ed. Berlin: Springer-Verlag.

Conte, S. D., and C. de Boor. 1980. *Elementary Numerical Analysis*, 3rd ed. New York: McGraw-Hill.

Cooper, L., and D. Steinberg. 1974. *Methods and Applications of Linear Programming*. Philadelphia: Saunders.

Dahlquist, G., and A. Björck. 1974. *Numerical Methods*. Englewood Cliffs, New Jersey: Prentice-Hall.

Davis, P. J., and P. Rabinowitz. 1984. *Methods of Numerical Integration*, 2nd ed. New York: Academic Press.

Dennis, J. E., and R. Schnable. 1983. *Quasi-Newton Methods for Nonlinear Problems*. Englewood Cliffs, New Jersey: Prentice-Hall.

Devitt, J. S. 1993. *Calculus with Maple V*. Pacific Grove, California: Brooks/Cole.

Dongarra, J. J., I. S. Duff, D. C. Sorenson, and H. van der Vorst. 1990. *Solving Linear Systems on Vector and Shared Memory Computers*. Philadelphia: Society for Industrial and Applied Mathematics.

Dorn, W. S., and D. D. McCracken. 1972. *Numerical Methods with FORTRAN IV Case Studies*. New York: Wiley.

Ellis, W., Jr., E. W. Johnson, E. Lodi, and D. Schwalbe. 1992. *Maple V Flight Manual: Tutorials for Calculus, Linear Algebra, and Differential Equations*. Pacific Grove, California: Brooks/Cole.

Ellis, W., Jr., and E. Lodi. 1991. *A Tutorial Introduction to Derive*. Pacific Grove, California: Brooks/Cole.

Ellis, W., Jr., and E. Lodi. 1991. *A Tutorial Introduction to Mathematica*. Pacific Grove, California: Brooks/Cole.

Evans, G. W., G. F. Wallace, and G. L. Sutherland. 1967. *Simulation Using Digital Computers*. Englewood Cliffs, New Jersey: Prentice-Hall.

Faires, J. D., and R. L. Burden. 1993. *Numerical Methods*. Boston: PWS Publishers.

Fehlberg, E. 1969. "Klassische Runge-Kutta formeln fünfter und siebenter ordnung mit schrittweitenkontrolle." *Computing* **4**, 93–106.

Flehinger, B. J. 1966. "On the probability that a random integer has initial digit A." *American Mathematical Monthly* **73**, 1056–1061.

Fletcher, R. 1976. *Practical Methods of Optimization*. New York: Wiley.

Forsythe, G. E. 1957. "Generation and use of orthogonal polynomials for data-fitting with a digital computer." *Society for Industrial and Applied Mathematics Journal* **5**, 74–88.

Forsythe, G. E., M. A. Malcolm, and C. B. Moler. 1977. *Computer Methods for Mathematical Computations*. Englewood Cliffs, New Jersey: Prentice-Hall.

Forsythe, G. E., and C. B. Moler. 1967. *Computer Solution of Linear Algebraic Systems*. Englewood Cliffs, New Jersey: Prentice-Hall.

Forsythe, G. E., and W. R. Wasow. 1960. *Finite Difference Methods for Partial Differential Equations*. New York: Wiley.

Fröberg, C.-E. 1969. *Introduction to Numerical Analysis*. Reading, Massachusetts: Addison-Wesley.

Gallivan, K. A., M. Heath, E. Ng, B. Peyton, R. Plemmons, J. Ortega, C. Romine, A. Sameh, and R. Voigt. 1990. *Parallel Algorithms for Matrix Computations*. Philadelphia: Society of Industrial and Applied Mathematics.

Gear, C. W. 1971. *Numerical Initial Value Problems in Ordinary Differential Equations*. Englewood Cliffs, New Jersey: Prentice-Hall.

Gerald, C. F. 1978. *Applied Numerical Analysis*. Reading, Massachusetts: Addison-Wesley.

Gill, P. E., W. Murray, and M. H. Wright. 1981. *Practical Optimization*. New York: Academic Press.

Gleick, J. 1992. *Genius: The Life and Science of Richard Feynman*. New York: Pantheon.

Goldberg, D. 1991. "What every computer scientists should know about floating-point arithmetic," *ACM Computing Surveys* **23**, 5–48.

Golub, G. H., and J. M. Ortega. 1993. *Scientific Computing*. New York: Academic Press.

Golub, G. H., and C. F. van Loan. 1983. *Matrix Computations*. Baltimore: Johns Hopkins University Press.

Good, I. J. 1972. "What is the most amazing approximate integer in the universe?" *Pi Mu Epsilon Journal* **5**, 314–315.

Hageman, L. A., and D. M. Young. 1981. *Applied Iterative Methods*. New York: Academic Press.

Hämmerlin, G., and K.-H. Hoffmann. 1991. *Numerical Mathematics*. New York: Springer-Verlag.

Hammersley, J. M., and D. C. Handscomb. 1964. *Monte Carlo Methods*. London: Methuen.

Hansen, T., G. L. Mullen, and H. Niederreiter. 1993. "Good parameters for a class of node sets in quasi-Monte Carlo integration." *Mathematics of Computation* **61**, 225–234.

Henrici, P. 1962. *Discrete Variable Methods in Ordinary Differential Equations*. New York: Wiley.

Hildebrand, F. B. 1974. *Introduction to Numerical Analysis*. New York: McGraw-Hill.

Hodges, A. 1983. *Alan Turing: The Enigma*. New York: Simon & Schuster.

Householder, A. S. 1970. *The Numerical Treatment of a Single Nonlinear Equation*. New York: McGraw-Hill.

Hull, T. E., and A. R. Dobell. 1962. "Random number generators." *Society for Industrial and Applied Mathematics Journal on Applied Mathematics* **4**, 230–254.

Hull, T. E., W. H. Enright, B. M. Fellen, and A. E. Sedgwick. 1972. "Comparing numerical methods for ordinary differential equations." *Society for Industrial and Applied Mathematics Journal on Numerical Analysis* **9**, 603–637.

Isaacson, E., and H. B. Keller. 1966. *Analysis of Numerical Methods*. New York: Wiley.

Jennings, A. 1977. *Matrix Computation for Engineers and Scientists*. New York: Wiley.

Keller, H. B. 1968. *Numerical Methods for Two-Point Boundary-Value Problems*. Toronto: Blaisdell.

Keller, H. B. 1976. *Numerical Solution of Two-Point Boundary Value Problems*. Philadelphia: Society of Industrial and Applied Mathematics.

Kernighan, B. W., and P. J. Plauger. 1974. *The Elements of Programming Style*. New York: McGraw-Hill.

Kincaid, D., and W. Cheney. 1990. *Numerical Analysis: Mathematics of Scientific Computing*. Pacific Grove, California: Brooks/Cole.

Kincaid, D. R., and D. M. Young. 1979. "Survey of iterative methods." In *Encyclopedia of Computer Science and Technology*, edited by J. Belzer, A. G. Holzman, and A. Kent. New York: Dekker.

Kinderman, A. J., and J. F. Monahan. 1977. "Computer generation of random variables using the ratio of uniform deviates." *Association of Computing Machinery Transactions on Mathematical Software* **3**, 257–260.

Lambert, J. D. 1973. *Computational Methods in Ordinary Differential Equations*. New York: Wiley.

Lapidus, L., and J. H. Seinfeld. 1971. *Numerical Solution of Ordinary Differential Equations*. New York: Academic Press.

Lawson, C. L., and R. J. Hanson. 1974. *Solving Least-Squares Problems*. Englewood Cliffs, New Jersey: Prentice-Hall.

Leva, J. L. 1992. "A fast normal random number generator." *Association of Computing Machinery Transactions on Mathematical Software* **18**, 449–455.

Lootsam, F. A. (ed.). 1972. *Numerical Methods for Nonlinear Optimization*. New York: Academic Press.

MacLeod, M. A. 1973. "Improved computation of cubic natural splines with equi-spaced knots." *Mathematics of Computation* **27**, 107–109.

Maron, M. J. 1991. *Numerical Analysis: A Practical Approach*. Boston: PWS Publishers.

Marsaglia, G. 1968. "Random numbers fall mainly in the planes." *Proceedings of the National Academy of Sciences* **61**, 25–28.

Nerinckx, D., and A. Haegemans. 1976. "A comparison of nonlinear equation solvers." *Journal of Computational and Applied Mathematics* **2**, 145–148.

Nering, E. D., and A. W. Tucker. 1992. *Linear Programs and Related Problems*. New York: Acdemic Press.

Niederreiter, H. 1978. "Quasi-Monte Carlo methods." *Bulletin of the American Mathematical Society* **84**, 957–1041.

Niederreiter, H. 1992. *Random Number Generation and Quasi-Monte Carlo Methods*. Philadelphia: Society of Industrial and Applied Mathematics.

Nievergelt, J., J. G. Farrar, and E. M. Reingold. 1974. *Computer Approaches to Mathematical Problems*. Englewood Cliffs, New Jersey: Prentice-Hall.

Noble, B., and J. W. Daniel. 1977. *Applied Linear Algebra*. Englewood Cliffs, New Jersey: Prentice-Hall.

Orchard-Hays, W. 1968. *Advanced Linear Programming Computing Techniques*. New York: McGraw-Hill.

Ortega, J. M., and W. C. Rheinboldt. 1970. *Iterative Solution of Nonlinear Equations in Several Variables*. New York: Academic Press.

Ostrowski, A. M. 1966. *Solution of Equations and Systems of Equations*, 2nd ed. New York: Academic Press.

Phillips, G. M., and P. J. Taylor. 1973. *Theory and Applications of Numerical Analysis*. New York: Academic Press.

Rabinowitz, P. 1968. "Applications of linear programming to numerical analysis." *Society for Industrial and Applied Mathematics Review* **10**, 121–159.

Rabinowitz, P. 1970. *Numerical Methods for Nonlinear Algebraic Equations*. London: Gordon & Breach.

Raimi, R. A. 1969. "On the distribution of first significant figures." *American Mathematical Monthly* **76**, 342–347.

Ralston, A. 1965. *A First Course in Numerical Analysis*. New York: McGraw-Hill.

Ralston, A., and C. L. Meek (eds.). 1976. *Encyclopedia of Computer Science*. New York: Petrocelli/Charter.

Rice, J. R. 1971. "SQUARS: An algorithm for least squares approximation." In *Mathematical Software*, edited by J. R. Rice. New York: Academic Press.

Rice, J. R. 1983. *Numerical Methods, Software, and Analysis*. New York: McGraw-Hill.

Rice, J. R., and J. S. White. 1964. "Norms for smoothing and estimation." *Society for Industrial and Applied Mathematics Review* **6**, 243–256.

Rivlin, T. J. 1990. *The Chebyshev Polynomials*, 2nd ed. New York: Wiley.

Salamin, E. 1976. "Computation of π using arithmetic-geometric mean." *Mathematics of Computation* **30**, 565–570.

Scheid, F. 1968. *Theory and Problems of Numerical Analysis*. New York: McGraw-Hill.

Schoenberg, I. J. 1946. "Contributions to the problem of approximation of equidistant data by analytic functions." *Quarterly of Applied Mathematics* **4**, 45–99, 112–141.

Schoenberg, I. J. 1967. "On spline functions." In *Inequalities*, edited by O. Shisha, 255–291. New York: Academic Press.

Schrage, L. 1979. "A more portable Fortran random number generator." *Association for Computing Machinery Transactions on Mathematical Software* **5**, 132–138.

Schrijver, A. 1986. *Theory of Linear and Integer Programming*. Somerset, New Jersey: Wiley.

Schultz, M. H. 1973. *Spline Analysis*. Englewood Cliffs, New Jersey: Prentice-Hall.

Shampine, L. F., and M. K. Gordon. 1975. *Computer Solution of Ordinary Differential Equations*. San Francisco: W. H. Freeman.

Skeel, R. D., and J. B. Keiper. 1992. *Elementary Numerical Computing with Mathematica*. New York: McGraw-Hill.

Smith, G. D. 1965. *Solution of Partial Differential Equations*. London: Oxford University Press.

Späth, H. 1992. *Mathematical Algorithms for Linear Regression*. New York: Academic Press.

Stetter, H. J. 1973. *Analysis of Discretization Methods for Ordinary Differential Equations*. Berlin: Springer-Verlag.

Stewart, G. W. 1973. *Introduction to Matrix Computations*. New York: Academic Press.

Stoer, J., and R. Bulirsch. 1993. *Introduction to Numerical Analysis*, 2nd ed. New York: Springer-Verlag.

Street, R. L. 1973. *The Analysis and Solution of Partial Differential Equations*. Pacific Grove, California: Brooks/Cole.

Stroud, A. H. 1974. *Numerical Quadrature and Solution of Ordinary Differential Equations*. New York: Springer-Verlag.

Subbotin, Y. N. 1967. "On piecewise-polynomial approximation." *Mat. Zametcki* **1**, 63–70. (Translation: 1967. *Math. Notes* **1**, 41–46.)

DeTemple, D. W. 1993. "A Quicker Convergence to Euler's Constant," *American Mathematical Monthly* **100**, 468–470.

Törn, A. and A. Zilinskas. 1989. *Global Optimization*. New York: Springer.

Traub, J. F. 1964. *Iterative Methods for the Solution of Equations*. Englewood Cliffs, New Jersey: Prentice-Hall.

Turner, P. R. 1982. "The distribution of leading significant digits." *Journal of the Institute of Mathematics and Its Applications* **2**, 407–412.

Watkins, D. S. 1991. *Fundamentals of Matrix Computation*. New York: Wiley.

Whittaker, E., and G. Robinson. 1944. *The Calculus of Observation*, 4th ed. London: Blackie. New York: Dover, 1967 (reprint).

Wilkinson, J. H. 1963. *Rounding Errors in Algebraic Processes*. Englewood Cliffs, New Jersey: Prentice-Hall.

Young, D. M., and R. T. Gregory. 1972. *A Survey of Numerical Mathematics*, Vols. 1–2. Reading, Massachusetts: Addison-Wesley. New York: Dover 1988 (reprint).

INDEX

A page number followed by a number in parentheses prefixed by 'Pb or CPb' refers to a problem or a computer problem on the given page. For example, 136 (CPb 4.2.7) refers to page 136, Computer Problem 7, Section 4.2.

Formulas and Definitions from Integral Calculus

$$\int x^\alpha \, dx = x^{\alpha+1}/(\alpha + 1) + C \qquad (\alpha \neq 1)$$

$$\int e^x \, dx = e^x + C$$

$$\int x^{-1} \, dx = \ln|x| + C$$

$$\int \sin x \, dx = -\cos x + C$$

$$\int \cos x \, dx = \sin x + C$$

$$\int \sec^2 x \, dx = \tan x + C$$

$$\int \sec x \tan x \, dx = \sec x + C$$

$$\int \frac{dx}{\sqrt{a^2 - x^2}} = \arcsin \frac{x}{a} + C \qquad (a \neq 1)$$

$$\int \frac{dx}{a^2 + x^2} = \frac{1}{a} \arctan \frac{x}{a} + C \qquad (a \neq 1)$$

$$\int \sinh x \, dx = \cosh x + C$$

$$\int \cosh x \, dx = \sinh x + C$$

$$\int \tan x \, dx = \ln|\sec x| + C$$

$$\int \sec x \, dx = \ln|\sec x + \tan x| + C$$

$$\int (x^2 + a^2)^{-1/2} \, dx = \ln|\sqrt{x^2 + a^2} + x| + C$$

$$\int (x^2 \pm a^2)^{1/2} \, dx = \frac{x}{2}\sqrt{x^2 \pm a^2} \pm \frac{a^2}{2} \ln|x + \sqrt{x^2 \pm a^2}| + C$$

$$\int \ln x \, dx = x \ln|x| - x + C$$

$$\int \sin^2 x \, dx = \frac{x}{2} - \frac{\sin 2x}{4} + C$$

$$\int \cos^2 x \, dx = \frac{x}{2} + \frac{\sin 2x}{4} + C$$

$$\int \frac{dx}{x(ax + b)} = \frac{1}{b} \ln\left|\frac{x}{ax + b}\right| + C$$

$$\int u \, dv = uv - \int v \, du$$

$$\int F'(g(x))g'(x) \, dx = F(g(x)) + C$$

Fundamental Theorem of Calculus

$$\frac{d}{dx} \int_a^x f(t) \, dt = f(x)$$